Principles of
Soundscape Ecology

PRINCIPLES OF SOUNDSCAPE ECOLOGY

Discovering Our Sonic World

BRYAN C. PIJANOWSKI

The University of Chicago Press

Chicago and London

The University of Chicago Press, Chicago 60637
The University of Chicago Press, Ltd., London
© 2024 by Bryan C. Pijanowski
Published 2024
Printed in the United States of America

33 32 31 30 29 28 27 26 25 24   1 2 3 4 5

ISBN-13: 978-0-226-82427-7 (cloth)
ISBN-13: 978-0-226-82429-1 (paper)
ISBN-13: 978-0-226-82428-4 (e-book)
DOI: https://doi.org/10.7208/chicago/9780226824284.001.0001

Library of Congress Cataloging-in-Publication Data

Names: Pijanowski, Bryan C., author.
Title: Principles of soundscape ecology : discovering our sonic world / Bryan C.
    Pijanowski.
Description: Chicago : The University of Chicago Press, 2024. | Includes
    bibliographical references.
Identifiers: LCCN 2023035268 | ISBN 9780226824277 (cloth) |
    ISBN 9780226824291 (paperback) | ISBN 9780226824284 (ebook)
Subjects: LCSH: Sound—Physiological effect. | Sound—Psychological aspects. |
    Acoustic phenomena in nature. | Hearing. | BISAC: SCIENCE / Life Sciences /
    Ecology | SCIENCE / Acoustics & Sound
Classification: LCC QC225.7 .P55 2024 | DDC 591.59/4—dc23/eng/20230802
LC record available at https://lccn.loc.gov/2023035268

♾ This paper meets the requirements of ANSI/NISO Z39.48-1992 (Permanence of
Paper).

To Dawn, Alex, and Benjamin

Contents

Preface

Those who dwell, as scientists or laymen, among the beauties and
mysteries of the Earth are never alone or weary of life.
—Rachel Carson

When you begin to listen deeply to the world around you, your life changes.
For those of us who have discovered this marvelous sonic world as a space
to explore, the infinite possibilities of what sound can do to uncover the
mysteries of this fascinating planet become daunting, but also exhilarat-
ing. As an academic, one cannot help but notice that the topic of sound can
be found in just about every university department: biology, anthropology,
music, engineering, physics, psychology, atmospheric sciences, English and
other language studies, media studies, education, medicine, geography . . .
the list goes on. There are a myriad of ways to connect the disciplines using
sound. At the same time, these multi- and transdisciplinary perspectives
on sound in nature and society make one quickly realize that sound can
facilitate exciting new ways of discovering our world. To those of you new to
sound, welcome to the marvelous sonic dimension of whatever you study;
I hope that you find this work to be a useful resource to learn and explore.

This book presents the principles of soundscape ecology as organized
around four important components necessary to fully describe the field:
fundamentals, concepts, methods, and applications. I was motivated to
write the book because my incoming graduate students needed a comprehensive guide that provided broad, but detailed, content about this rapidly
evolving field. But I think it can also serve as an excellent reference for those
advancing the field of soundscape ecology and provide the basis for practi-

tioners who are interested in understanding how this field can be applied. Soundscape ecology has led to new discoveries in nearly all ecosystems and has addressed important societal challenges such as assessing biodiversity trends, reconnecting people with nature, understanding what is needed to create more healthy sonic spaces for living, and discovering how music connects people and nature in various cultures around the world.

Over ten years have passed since I led a series of papers with colleagues that laid out the fundamental concepts of soundscape ecology. Since that time, thousands of papers have been published by researchers around the world that have advanced concepts, tools, and applications of soundscape ecology in terrestrial, aquatic, and urban systems. My own work has evolved considerably, too, as I became more curious about soundscapes not solely to study ecological concepts, but also to understand how sound is perceived by people and animals and how soundscapes can inspire people to care about nature. Forays into the interplay of science and art have led to exciting insights shared with scholars in other disciplines. As a result, I have worked in nearly all of Earth's major terrestrial biomes, where I have met amazing scientists and humanists who are interested in making scientific discoveries and creating compositions that can change the world. The soundscapes of these places, especially the wild spaces, have provided me with deep intellectual and emotional experiences few have ever had, and with a realization that Earth's acoustic heritage is complex and rich with information about how the planet is changing.

Creating a textbook that spans such diverse disciplines in a way that is cohesive and scaffolds knowledge is not easy. Indeed, it is difficult to be an expert in all these fields. I have attempted to study and synthesize knowledge from leaders in many of these fields, and I have often found inspiration in their work that has motivated me to become a better soundscape ecologist, scientist, and transdisciplinary scholar. I have often reached out to friends and colleagues for advice and mini-"tutorials" on particular topics. In some cases, their mentorship has helped me to understand and appreciate the great scholarship that occurs in all fields that support the diverse dimensions of soundscape ecology. Those very helpful people include Stuart Gage, Mark Lomolino, Jennifer Post, Maria Fernandez-Gimenez, Robin Reid, Laura Zanotti, Chantsaa Jamsranjav, Byamba Ichinkhorloo, Tim Keitt, Meredith Cobb, Michelle Dennis, Christopher Raymond, Dave Ebert, Brad Lisle, Dan Shepardson, Barbara Flagg, Tim Archer, Jennifer Miksis-Olds, Christian Butzke, Jinha Jung, Jingjing Liang, Alex Pijanowski, Al Rebar, Alex Piel, Zhao Zhao, Christopher Anderson, Andrea Raya Rey,

Alejandro Valenzuela, Swapan Kumar Sarker, Bernie Krause, Ulmar Grafe, Marc Caffee, Jonathan Beever, Almo Farina, David Conlin, Ian Agranat, Jeff Titon, Aaron Allen, Pat Smoker, and Jarrod Doucette. I have also been fortunate to have had some exceptionally talented students who have challenged me over the years and have made me a better scientist. These students include Amandine Gasc, Kristen Bellisario, Sarah Dumyahn, Kostas Alexandridis, Ben Gottesman, Dante Francomano, Matt Harris, Maryam Ghadiri, David Savage, Burak Pekin, Taylor Broadhead, Brian Napoletano, Luis Villanueva-Rivera, Sam Lima, Ruth Bowers-Sword, Santiago Ruiz Guzman, and Francisco Rivas Fuenzalida. Funding from the National Science Foundation, US National Park Service, US Fisheries and Wildlife Services, NASA, the USDA Hatch/McIntire-Stennis Program, and, at Purdue University, the Office of the Executive Vice President for Research and Partnerships, the College of Agriculture's Dean's Office, the Office of Graduate Studies, the Center for the Environment and Institute for a Sustainable Future, and the Department of Forestry and Natural Resources made it possible for me to do "wildly" transdisciplinary work with a global reach.

I am especially indebted to those at the University of Chicago Press who first conceived of a book like this that would support the growth of a burgeoning new field that spans ecology, social sciences, data sciences, and the humanities. Special thanks go to Joseph Calamia, Nicholas Lilly, and Tamara Ghattas at the Press. Two anonymous reviewers offered exceptional comments on a draft version of the book. Norma Sims Roche provided impeccable copyedits and suggestions for improvement throughout the book and greatly improved the readability of the text. The illustrations were beautifully produced by Barbara Schoeberl, whose simple, clean style fits the book exceedingly well. All errors, omissions, and misstatements are, however, ultimately owned by me.

Bryan C. Pijanowski

West Lafayette, Indiana

January 15, 2023

FUNDAMENTALS

Part I covers three areas: the history of the field of soundscape ecology and the term "soundscape," the physics of sound, and descriptions of all types of sound sources. Chapter 1 reviews the history of the concept of soundscapes and soundscape ecology. Chapter 2, "The Physics of Sound," introduces the reader to fundamental terms used throughout the book and should be studied carefully if this is your first time studying acoustics. Chapter 3, "Sources of Sound in the Soundscape," provides a comprehensive survey of sounds from three types of sources, biological, geophysical, and anthropogenic, with a focus on biological sounds.

Introduction

1

OVERVIEW. What is a soundscape and what is soundscape ecology? Answers to these questions are drawn from a variety of scholars in diverse disciplines. This chapter summarizes the historical development and use of the term "soundscape." Interestingly, the term has been used independently by scholars in the fields of geography, music, and ecology. The chapter then compares several other areas of scholarship that are closely related to soundscape ecology: bioacoustics, acoustic ecology, and ecoacoustics. Finally, it discusses the aims of soundscape ecology, as it is a field that has focused on addressing some of the greatest environmental and societal challenges of the twenty-first century.

KEYWORDS: bioacoustics, ecoacoustics, sensory landscape, soundscape, soundscape ecology

---

1.1. Sounds Are a Universal Indicator of Change

On Earth, sounds—produced by animals, humans, and the atmosphere—are found almost everywhere. There are even sounds produced by inanimate objects and the movement of Earth itself: by earthquakes, glaciers, and sand dunes, for example. These sounds, from diverse sources, reflect ecological and social processes that are occurring at a given place and time. That sounds are created by processes—those that are biological, geophysical, and anthropological—means that sounds reflect change and are thus a universal characteristic of dynamic systems. Many scholars have recently recognized this and are now using sound as a measure of change in ecological, social, and social-ecological systems.

This book aims to present an overview of the principles of soundscape ecology, which is the science of measuring all a place's sounds and linking them to ongoing processes occurring there to understand the status and the dynamic nature of the place. It also considers how organisms, including people, perceive sounds. This book is designed to synthesize the work done in this area since the field was first proposed ten years ago (Pijanowski et al. 2011a) and to be useful to a broad range of students, researchers, and natural resource managers. The goals of this book are to summarize the fundamental concepts needed to conduct work in soundscape ecology, present what we know about how organisms and people perceive sound, provide a summary of the ways to measure and analyze soundscapes, and explore the applications that have been the major focus of soundscape ecology. To accomplish this, the book has been organized around four major parts: fundamentals, concepts, methods, and applications.

1.2. What Is a Soundscape?

There have been numerous definitions of the term "soundscape" in the literature. The suffix *scape* refers to an area, scene, space, or view, so *soundscape* is a term describing the sounds occurring over an area. The term is meant to be analogous to "landscape," "seascape," and "cityscape," all of which focus on the visual aspects of an area. "Soundscape," of course, focuses on an area's sonic aspects.

The concept of a soundscape has been advanced by scholars in many disciplines over the years, but the definitions put forth have many elements in common. Granö (1929; see also Porteus and Mastin 1985; Granö 1997), a Finnish geographer, is credited with the first use of an analogous term, *sensory landscape*, that was part of his broader description of a field he called *pure geography* or *landscape science*. His study focused on characterizing the diverse attributes of natural and human-dominated landscapes as they are perceived through the human senses (Granö 1997, 11). Of the five scientifically identified senses (vision, hearing, touch, smell, and taste), he focused on the first three in considerable detail. To demonstrate how sensory modalities could be used to study landscapes, he mapped the auditory spaces of the Finnish island of Valosaari (Granö 1997, 127) (Fig. 1.1A). This mapping, which included sounds from people, birds, and cowbells, noted the quality and frequency of the sounds and whether they came from nearby or far away. He also described sounds in terms of when they occurred and how often, making sounds an attribute of a place, all perceived in the context of a landscape.

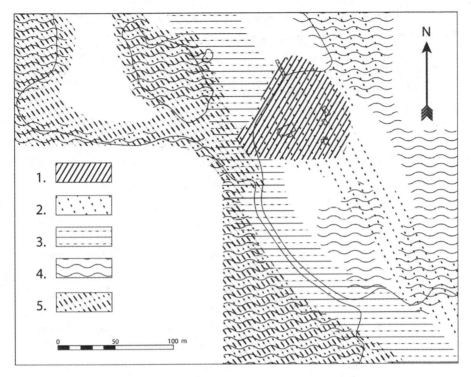

1.
2.
3.
4.
5.

0   50   100 m

N

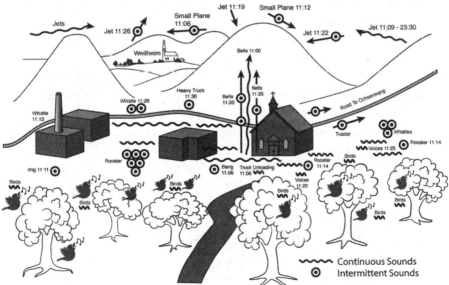

Jets

Jet 11:26

Small Plane 11:06

Jet 11:19

Small Plane 11:12

Jet 11:22

Jet 11:09 - 23:30

Weilheim

Bells 11:00

Whistle 11:26

Heavy Truck 11:30

Bells 11:30

Bells 11:35

Road To Ochsenwang

Whistle 11:13

Tractor

Whistles

Voices 11:25

Rooster 11:14

ring 11:11

Rooster

Bang 11:06

Truck Unloading 11:06

Rooster 11:14

Birds

Voices 11:20

Birds

Birds

Birds

Birds

Birds

Birds

Birds

~~~ Continuous Sounds
⊙ Intermittent Sounds

Figure 1.1 Early illustrations of the concept of sound in landscapes. (A) Each of the five patterns represents a different sound. (Modified from Granö 1997.) (B) Numerals show the times when sounds occurred. (Modified from Schafer 1969.)

The first use of the term "soundscape" in English dates to the late 1960s, when it was used by Southworth, who studied how visually impaired people navigated cities using sound. Southworth (1967, 2) defined a *soundscape* as "the quality and type of [all] sounds and their arrangements in space and time." The focus of his work was to develop an analytical framework for (1) the types of sounds that occur in a place, (2) how sounds are perceived by people, especially when contrasted with visual input, and (3) how sound-scapes and their perception can inform the design of cities. Southworth (1967) developed a taxonomy of sound sources, which was a catalog of everything that makes sound; form qualities, which included intensity, quality, temporal dynamics, and location and orientation of sounds; and sound synthesis, which referred to how sound affects the informativeness, uniqueness, and level of delight experienced by a person. His early work culminated in principles of city design in which both visual and acoustic elements are important aspects of creating purposeful city spaces.

A few years later, Canadian musician and composer R. Murray Schafer (1969, 1977) also used the term "soundscape," and he is most often credited for popularizing it in his seminal book *The Soundscape: Our Sonic Environment and the Tuning of the World* (1977, 1994). At one level, Schafer (1994, 7) defined *soundscape* simply as the "acoustic environment." To Schafer, a soundscape is the actual sonic environment of a place or a musical composition that creates an acoustic experience of an environment for a person. Schafer was motivated in his work to create public awareness of the impact that humans have on the soundscape, but also to understand how people connect to the environment through sound and the important role that music plays in this connection (see the work of close Schafer colleagues Truax [2019] and Westerkamp [2002]). He argued that music is "a search for the harmonizing influence of sounds in the world about us" (1994, 6). His concern was that society is changing the sounds of the natural system, replacing them with sounds (e.g., road noise) that have no value to the listener, thereby creating a passive listening experience for people and, as Schafer described, a "universal deafness" of society (1977, 3). Schafer's hand-drawn soundscape of Bissinger, Germany, has become a classic illustration of a soundscape (Fig. 1.1B). His sketch illustrates sound sources and the characteristics of the associated sounds (e.g., duration and movement) accompanied by a visual depiction of the landscape. Schafer's and his colleagues' work is presented in Chapters 6 and 14.

In the mid-1980s, a natural sounds recording expert, Bernie Krause, extended Schafer's work into Earth's "wild spaces." Referring to them as *wild soundscapes*, Krause emphasized that their biological sounds were "shaped" by evolution and ecology. Paying homage to Schafer's use of musical metaphors, he used terms such as "orchestration" of animal sounds. Krause (1987, 1993) emphasized the role that soundscapes can play in the assessment of landscape condition. As habitats change even slightly, these changes can be reflected in the soundscape. He also referenced the work of many conservation biologists, such as Aldo Leopold, who described how humans affect environmental quality. Although Krause is not an ecologist by formal training, his work has had tremendous impact on the direction of ecological research in this field.

By the late 1990s, research on urban soundscapes (see Kang et al. 2016; Botteldooren et al. 2011) developed from the work of Southworth (1967) in urban planning, Schafer's emphasis on human experiences of the sonic environment (see Adams et al. 2006; Botteldooren et al. 2006), studies on noise management in cities (Brown 2010), and work in the social sciences (e.g., Dubois et al. 2006). The work on urban soundscapes (Kang 2006; Kang et al. 2016) has focused on three major areas: (1) classification and measurement of sound sources common to the urban environment, (2) methods to standardize how humans perceive sounds in urban environments (Dubois et al. 2006; Brown et al. 2011; Brown 2011), and (3) ways that sensors can be used to monitor sounds, particularly noise levels, and then map them for the purposes of determining possible interventions (Park et al. 2014, 2017; Salamon et al. 2014). The motivation for studying urban soundscapes has been to improve the quality of the sounds that people experience in the built environment. Chapter 13 summarizes the current work on urban soundscapes.

More recently, Pijanowski et al. (2011a, 2011b) provided a revised definition of the term "soundscape," using it in a way that allows ecologists to study the ecosystem dynamics of landscapes. They defined a soundscape as "the collection of all biological, geophysical, and anthropogenic sounds that emanate from a landscape and which vary across space and time, reflecting important ecosystem processes and human activities." This definition attempts to align sound-source categories with the biotic, abiotic, and social ecosystem components traditionally identified in ecosystem ecology (Odum and Odum 1953; Odum and Barrett 1971). The terms "biophony," "geophony," and "anthrophony" (Pijanowski et al. 2011a) were introduced as part of this detailed soundscape definition, which recognized that the sounds within each sound-source category have an acoustic structure that represents the dynamics and patterns associated with a part of the sonic

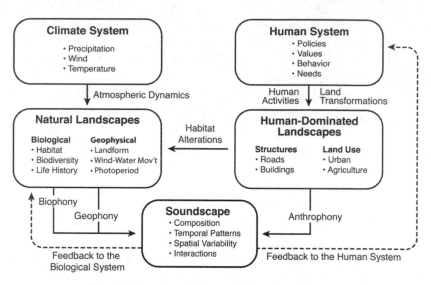

Figure 1.2 Components of terrestrial soundscape systems.

environment. A review of the fundamentals of the physics and science of sound sources is presented in Chapters 2 and 3.

Following Pijanowski et al. (2011a), Figure 1.2 represents the five key system components that create a terrestrial soundscape at any given place and time. The climate system, with its precipitation, wind, and temperature regimes, creates the patterns of biological habitats and geophysical landforms found in natural landscapes. The human system is driven by policies, values, behaviors, and needs (Smith and Pijanowski 2014) that often transform the land for urban use or food production. Natural landscapes are home to geophysical dynamics, which create geophysical sounds, and to biological components—namely, habitats where sound propagates and organisms that produce sounds. These three types of sound sources—biological, geophysical, and anthropogenic—integrate to form a soundscape, with its unique composition, temporal patterns, spatial variation, and sonic interactions that are often important to biological and human systems alike. In this system, there are two major sources of feedback. The first is feedback into the human system through both sound perception and sound exposure. Sounds, in turn, influence policies, values (e.g., property values), behaviors, and needs. The second is feedback into the natural system, where sounds from other animals and from the geophysical environment become important sources of information for animals. Chapter 11 focuses on the soundscape ecology research that has been conducted in terrestrial systems.

Since the concept of soundscapes was applied to terrestrial ecosystems by Pijanowski et al. (2011a, 2011b), there has been considerable success in applying it to aquatic systems. Ocean systems, in particular, have been extensively studied (Pieretti et al. 2017; MacWilliam and Hawkins 2013; Staaterman et al. 2014; Kaplan and Mooney 2016; Miksis-Olds et al. 2018), as have freshwater ponds and lakes (Rountree et al. 2019; Gottesman et al. 2020b; Linke et al. 2020) and, in a limited way, river systems (Kacem et al. 2020). An aquatic soundscape (Fig. 1.3) has components similar to but, not surprisingly, sound sources different from those of a terrestrial soundscape. The Earth system drives the dynamics of aquatic soundscapes, particularly via photoperiods, lunar cycles, climate dynamics, and especially in the marine system, ocean currents. Natural aquatic systems can be divided into two kinds of components: geophysical and biological. The geophysical environment is composed of temperature, salinity, turbidity, depth, and flow, all of which influence the structure of an aquatic habitat, its plant and animal communities, and their life-history patterns. Geophysical sounds of aquatic systems include rain, waves, and in some cases, earthquakes (Montgomery and Radford 2017). Humans affect aquatic systems in a variety of ways, but primarily either through sound-producing objects, like ships, oil and gas platforms, dams, and wind turbines, or through activities, like harvesting food (e.g., seaweed), fishing, introducing invasive species, and dredging. These effects alter the natural aquatic system by changing its physical structure or its species composition. Sounds from the three sound-source categories, biological, geophysical, and anthropogenic, combine to create an aquatic soundscape with a specific composition that changes across time and space and has its own important sonic interactions. Chapter 12 focuses on the soundscape ecology research that has been conducted in aquatic systems.

Also at the core of Pijanowski's (2011b) ecological definition of soundscapes is the recognition that there are known interactions between the three major sound-source categories (Fig. 1.4). For example, sounds from the human environment are known to affect animal communication. In particular, urban noise has been shown to force birds to shift their singing into quieter periods—namely, nighttime—or to sing at higher frequencies so they can be heard above the sounds created by an urban environment (Warren et al. 2006). Ambient sounds created by the geophysical environment, such as the sound of wind or breaking waves, also create an environment where animals need to alter their communication so that masking does not occur. Chapters 11 and 12 summarize the work that has been done to quan-

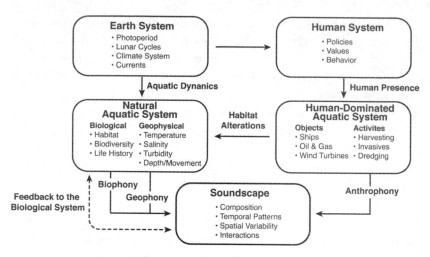

Figure 1.3 Components of aquatic soundscape systems.

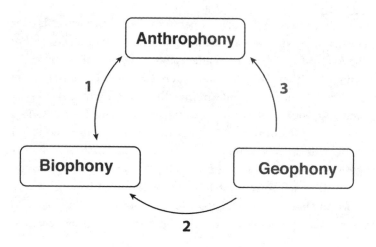

Figure 1.4 Interactions of major sound sources. (Modified from Pijanowski et al. 2011a.)

tify the effect of noise on animal communication and behavior in terrestrial and aquatic systems, respectively.

Scholars in cultural anthropology, human geography, and ethnomusicology have begun studying the impact that sounds have on the perception of place by people, and in particular, those of indigenous communities. The early work of cultural anthropologist Feld (1986), ethnomusicologist Seeger (1986), and geographer Tuan (1977) examined how humans perceive their surroundings through sound and through their total sensorium (i.e., all senses). More recently, music scholars (Titon 2009; Allen and Dawe 2016;

Guyette and Post 2015) have studied the role of natural soundscapes in music and, more broadly, sonic practices (sounds that are produced by humans and are reflective of many aspects of culture). Thus, soundscapes become a sociocultural feature of communities reflected by place.

Around 2016, to create a standard working definition of the term "soundscape," a group of acousticians (see Schulte-Fortkamp 2018) organized a working group (WG 54, Assessment of Soundscape Quality) to create a formal definition for the term. Proposed as International Organization for Standardization (ISO) standard 12931-1:2014, it defined a soundscape as "an acoustic environment as perceived or experienced and/or understood by a person or people, in context." The incentive for developing such a standard definition (Brown et al. 2009; Brown 2011) was the need to quantify sounds, particularly sound levels, in order to create sonic spaces for people that are not stressful or annoying or where the ambient sounds can be enjoyed. ISO 12913-2, which is Part 2 of the standard and which is still under development, is intended to describe the data and metric requirements that would allow assessments of soundscapes. In general, it has been argued (Brown et al. 2016) that moving toward a standard definition is necessary for environmental noise management. Others (MacDonald and Brown 1995) have suggested that using a more neutral term than "noise" for sounds found in a metropolitan environment would remove the bias of labeling any sound as "noise." Ultimately, this definition of *soundscape* focuses on all sounds perceived by people.

1.3. What Is Soundscape Ecology?

Soundscape ecology (Pijanowski et al. 2011a, 2011b; Pijanowski and Farina 2011) was proposed over ten years ago as an emerging field of science that provides ecologists with new concepts, theories, technologies, and methodologies to advance our understanding of the ecological dynamics occurring within landscapes. Of particular interest was the effect that human activities have on these dynamics, as examined through the lens of sound. With the development of passive acoustic recorders (PARs) and numerous measures of the diversity and composition of sound, and with the advancement of computing infrastructure that helps researchers store and analyze large quantities of data, soundscape ecology has become a promising field (Servick 2014). Soundscape ecology has the potential to address topics such as biodiversity decline, the impact of climate change on ecosystems, how changes in land use affect ecosystem dynamics, and the role of sound in providing humans with information about how the environment is changing. Over the past decade, this new area of scientific research (Pijanowski et al. 2011a, 2011b; Pijanowski and Farina 2011) has witnessed an explosion of new tools and applications. This book will bring together the existing work across a variety of fields, including ecology, social sciences, humanities, natural resources and urban planning, human health, and engineering, to lay out the breadth and depth of this highly interdisciplinary field.

1.4. Intellectual Roots of Soundscape Ecology

Soundscape ecology has been presented (Pijanowski et al. 2011a) as a science that brings together concepts, methodologies, terms, and tools from an array of other disciplines (Fig. 1.5), including physics; the sciences that study systems that produce sound (e.g., animal communication, ocean wave mechanics); evolutionary biology (e.g., sensory ecology); an assortment of spatial sciences (e.g., landscape ecology and optical remote sensing); a variety of disciplines and interdisciplinary areas of scholarship in the social sciences and humanities, such as ethnomusicology, sense of place studies, and psychoacoustics; and, because soundscape ecologists produce massive amounts of data, areas of data science such as data mining, machine learning, and environmental engineering (e.g., of sensors and sensor networks).

Although Figure 1.5 is meant as a heuristic illustration of the breadth of disciplines that soundscape ecology draws from, some of the most important contributions have come from interdisciplinary efforts. One exemplary effort is the work of Feld (1986), who examined the role that sound plays in forming sentiment (i.e., emotional dispositions) and sense of place for an indigenous group in New Guinea that forms strong cultural relationships among people, with animals, and with the afterlife. His work combined methods and knowledge from anthropology and ethnomusicology.

The properties of sound have been studied as a major branch of *physics* for centuries. Sound is complex, but it shares many properties with light. Sound, like light, behaves as a wave that moves through space, with frequency and amplitude (see Section 2.1), and its propagation and attenuation have been studied in air, water, and solids. An understanding of how sound behaves in space led ultimately to the development of a variety of sound recording devices, including microphones (Section 2.3).

In nearly all areas of the biological, geophysical, and social sciences, sound has become an area of study. A few of these diverse *sound source sciences* are explored in considerable depth in the rest of the book. The study

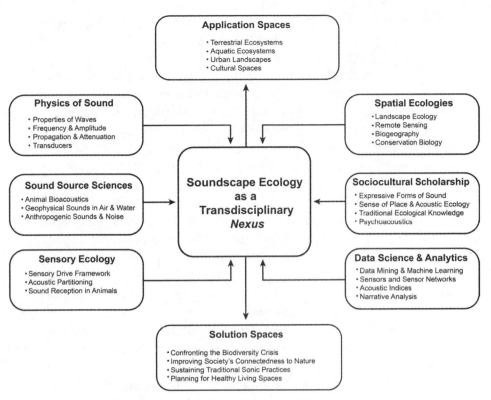

Application Spaces

- Terrestrial Ecosystems
- Aquatic Ecosystems
- Urban Landscapes
- Cultural Spaces

Physics of Sound

- Properties of Waves
- Frequency & Amplitude
- Propagation & Attenuation
- Transducers

Spatial Ecologies

- Landscape Ecology
- Remote Sensing
- Biogeography
- Conservation Biology

Soundscape Ecology as a Transdisciplinary *Nexus*

Sound Source Sciences

- Animal Bioacoustics
- Geophysical Sounds in Air & Water
- Anthropogenic Sounds & Noise

Sociocultural Scholarship

- Expressive Forms of Sound
- Sense of Place & Acoustic Ecology
- Traditional Ecological Knowledge
- Psychoacoustics

Sensory Ecology

- Sensory Drive Framework
- Acoustic Partitioning
- Sound Reception in Animals

Data Science & Analytics

- Data Mining & Machine Learning
- Sensors and Sensor Networks
- Acoustic Indices
- Narrative Analysis

Solution Spaces

- Confronting the Biodiversity Crisis
- Improving Society's Connectedness to Nature
- Sustaining Traditional Sonic Practices
- Planning for Healthy Living Spaces

Figure 1.5 Intellectual roots of soundscape ecology positioning the field as a transdisciplinary nexus of discovery that addresses four major application spaces and four solution spaces.

of animal communication, for example, examines the use of visual, acoustic, olfactory, and tactile signals, how these signals are produced by the sender, and how they are perceived by the receiver (Bradbury and Vehrencamp 1998). The field of bioacoustics, a subdiscipline of animal communication, is defined as the "study of sound in non-human animals" by the Animal Bioacoustics Technical Committee of the Acoustical Society of America (Popper and Dooling 2002). It examines "how sound is produced, propagated, and then perceived by an animal." This subdiscipline includes the study of acoustic communication, sound production mechanisms, auditory anatomy and function, sonar, acoustic tracking, and the effects of human-made and environmental noise on animals (Popper and Dooling 2002). The relationship between ocean wave movement and the sounds that it produces has been extensively studied by geophysical scientists. Chapter 3 in this book attempts to identity as many sound sources as possible and the underlying processes that create those sounds.

Sound production and perception in animals has been a major interest of evolutionary biologists who have focused on comparative studies across species and groups of species. Broadly called *sensory ecology* (Endler 1992), this area of scholarship provides an array of theoretical considerations that focus on the evolutionary origins of animal communication—in particular, the preferences of signal receivers and signal transmitters, their traits, and how these characteristics co-evolved given biological and geophysical constraints. These concepts are presented in Chapter 4.

Research in the area of *spatial ecology* has had, arguably, the greatest influence on the development of soundscape ecology. As originally proposed (Pijanowski et al. 2011a, 2011b), soundscape ecology builds on the rich set of concepts present in landscape ecology (see Forman 1983; Risser et al. 1984; Urban et al. 1987; Turner et al. 1990, 2001; Keitt et al. 1997) and biogeography (see Whittaker et al. 2005; Lomolino et al. 2017). In landscape ecology, a landscape is defined as an "area that is spatially heterogeneous in at least one factor of interest" (Turner 1989); it can be either a terrestrial or an aquatic system, or a combination of the two. Biogeography, a discipline based on the early work of Humboldt (1856), Darwin (1859), and Wallace (1880) (see Lomolino 2016), attempts to describe principles of the geographic template of species distributions, biotic-abiotic interactions, and interspecific interactions. Landscape ecology and biogeography build on the fundamental concepts of community ecology (Mittelbach and McGill 2019), a branch of ecology that focuses on understanding Earth's biodiversity over space and time (Urban et al. 2002), and population ecology. Some studies in soundscape ecology have attempted to determine population sizes using species-specific call rates. These concepts are presented in Chapter 5.

Applications of acoustics in natural resource management have historically focused on the impact of noise on wildlife health as well as on human enjoyment of natural areas (Buxton et al. 2017). This field also includes land use planning, in which the focus is on the types and levels of noise in urban settings (Kang et al. 2016). In marine systems, noise from ships, oil and gas platforms, and wind turbines has been a considerable focus (Popper and Hastings 2009; Montgomery and Radford 2017). It has also been proposed (Dumyahn and Pijanowski 2011a, 2011b) that natural soundscapes are resources in and of themselves and should be managed and protected like other natural resources. The principles of conservation biology are also important considerations in soundscape ecological research. These additional ecological concepts are presented in Chapter 5.

The *cultural and social sciences*, broadly referred to as the social sciences, have also made significant contributions to our understanding of sound-scapes, especially regarding how people perceive individual sounds and col-lections of sounds in a single place. The field of acoustic ecology, which grew out of Schafer's work from the 1970s, presented the overarching concept that humans are principally responsible for the sounds that are created; in a way, people are the composers of all sounds, since they modify landscapes and add sounds through various sonic activities. A few noteworthy exam-ples of soundscape compositions—deliberate artistic creations that blend natural and electronic sounds in ways that reflect the expressive focus in-herent in the humanities—have emerged (Westerkamp 2002).

An area of anthropology called *traditional ecological knowledge* (TEK) has focused on indigenous communities and the knowledge about the envi-ronment that they possess and pass along to younger generations (Berkes 2017). TEK can include how certain sounds indicate a changing environ-ment as well as local songs and other expressive forms of sound produc-tion that embed knowledge about the community and its relationship to the environment.

The field of environmental psychology has provided a context for how people (and, in a few cases, animals) perceive soundscapes. The highly mul-tidisciplinary field of psychoacoustics (Sections 2.4 and 6.7) studies how people perceive sound, from the anatomical structures necessary for human hearing to how the brain processes acoustic stimuli (Howard and Angus 2017). In addition, environmental psychoacoustics studies how people per-ceive natural and urban soundscapes by developing rubrics or scales (Axels-son et al. 2010; Raimbault and Dubois 2005; Davies et al. 2013; Hong and Jeon 2015). The key sociocultural concepts, including perception, acoustic ecology, ethnomusicology, and traditional ecology knowledge, are summa-rized in Chapter 6.

Data science and analytics and associated areas of research in environ-mental engineering are also important contributors of both concepts and tools to soundscape ecology. Some areas, such as digital signal processing and machine learning, have advanced considerably over the past four de-cades (Rabiner and Gold 1975; Mitra and Kaiser 1993; Tan and Jiang 2018). A variety of machine-learning approaches have been employed that allow computers to recognize acoustic patterns in data. Visualization of data has advanced considerably in the past ten years. Studies have also integrated data from multiple-sensor networks (Alías and Alsina-Pagès 2019) across acoustic, visual, and meteorological sensors. ARBIMON (Sieve Analytics 2018), Ecosounds (Rowe et al. 2018), REAL (Kasten et al. 2012), and the

National Science Foundation–funded Pumilio at Purdue (Villanueva-Rivera and Pijanowski 2012) are all acoustic management systems that can store and retrieve a massive amount of acoustic data using standard metadata and file formats. The resulting explosion of technology adoption, seen in the use of Seewave (Sueur et al. 2008) and the SoundEcology package, has led to further tool automation and community resourcing through places such as GitHub. Data science concepts are presented in Chapter 7 and soundscape analytics in Chapter 10.

It is argued here, most specifically in Chapter 8, that soundscape ecology is a transdisciplinary nexus of discovery that borrows from the contributions made across all of these disciplinary and interdisciplinary fields of study. The term *nexus*, derived from the Latin *nectare*, which means "to connect" (Liu et al. 2018), is used to emphasize that the field of soundscape ecology leverages approaches that link multiple distinct bodies of knowledge, across the biological, geophysical, and social sciences and scholarship in the humanities, in order to address critical societal problems. To date, the field has made contributions to our understanding of the patterns and processes that occur in terrestrial (see Chapter 11), aquatic (Chapter 12), and urban systems (Chapter 13), as well as the expressive spaces of human culture (Chapter 14).

1.5. Parallel Developments

Soundscape ecology is distinctly different from other acoustically oriented areas of study that exist today: bioacoustics (Popper and Dooling 2002), ecoacoustics (Sueur and Farina 2015; Farina and Gage 2017), and acoustic ecology (Truax 2019). These areas of study, all of which have contributed to the development of soundscape ecology, are summarized in Figure 1.6. On each *x*-axis are the three major types of sound sources (biophony, geophony, and anthrophony). The *y*-axis shows whether the field looks at sound production, propagation, or reception, with the latter split into three kinds of receivers (microphones, people, and animals). We can use this rubric to compare each of these closely related fields of study with soundscape ecology.

Bioacoustics has a long history (see Popper and Dooling 2002) of examining a variety of dimensions of acoustic communication in animals. For over a century, *bioacoustics* (Erbe and Dent 2017; Ganchev 2017) has focused on how animals produce sound, why they produce sound, and how they hear. It has also examined acoustic forms as a function of taxonomy, of techniques used to identify species by sound, and of interspecies interactions as they relate to sound. The field has evolved from a principally descriptive science to one that focuses on (1) the interplay of communication and behavior;

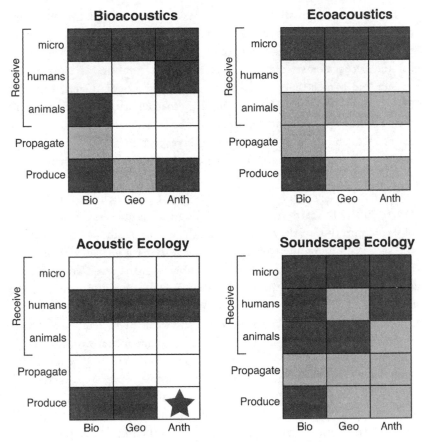

Figure 1.6 Comparison of soundscape ecology with three
closely related fields developed in parallel.

(2) sound production and hearing; (3) the learning of sounds, songs, or calls; (4) the effects of noise on animal behavior and hearing; (5) comparative assessments of related species and groups of animals; (6) intraspecific acoustic assessments to determine the structures, patterns, and variations of repertoires; (7) field experimentation that helps elucidate patterns of sound production, propagation, and perception; (8) the use of advanced audio recording technologies; and (9) acoustic modeling of sound propagation in different environments, such as forests and oceans. In the past fifteen years, research on the impacts of noise on animal communication (Warren et al. 2006; Kight and Swaddle 2011; Popper and Hawkins 2019) has been a primary focus as regulators, natural resource managers, and industry have recognized the deleterious effects human-produced noise can have on individual animals and populations.

Ecoacoustics originated from the work of Sueur et al. (2008), who determined that passive acoustic recording could be used to assess biodiversity patterns across taxa. As another area of ecological research that uses sound to study ecosystems, *ecoacoustics* focuses on natural and anthropogenic sounds and relates them to the status and dynamics of the environment (Farina and Gage 2017). More specifically, ecoacoustics uses acoustic monitoring techniques to provide information about changes in animal diversity, abundance, and behavioral patterns with respect to changes in the environment (Sueur and Farina 2015) and develops ways of quantifying these environmental sounds.

Acoustic ecology emerged from the work of Schafer (1994) and Truax (1978, 2001). The term "soundscape ecology" is used by the acoustic ecology community too; they have defined it as "the study of the effects of the acoustic environment on the physical responses or behavior of those living in it" and have used it interchangeably with the term "acoustic ecology" (Truax and Barrett 2011, 127). The name "acoustic ecology" and the use of the term "ecology" in "soundscape ecology" were not intended to be scientific or ecological. Rather, "ecology" was meant to reference natural sounds. Some of the work in acoustic ecology has focused on the use of ecological sounds to create compositions intended to inspire people to listen to nature, in which the "environmental context is preserved, enhanced, and exploited by the composer" so that the composition forms "a new place of listening" (Westerkamp 1999; see also Duhauptus and Solomos 2014). Much of the work in acoustic ecology was coordinated as a long-term, collaborative project at Simon Fraser University, called the World Soundscape Project, and was broadly organized by the World Forum for Acoustic Ecology (www.wfae.net).

To place soundscape ecology in the context of these other parallel areas of study, we can examine four components of the sound model: production, propagation, reception, and perception, as illustrated in Figure 1.6. To more clearly describe how soundscape ecology differs from bioacoustics, ecoacoustics, and acoustic ecology, the relative roles of each of the sound sources and components of sound are identified in matrix form. Bioacoustics has traditionally focused on the production and reception of sound by animals, with a minor focus on sound propagation (Morton 1975). As pointed out by Popper and Dooling (2002), bioacoustic work has recently shifted to a more applied focus, examining the role that noise plays in masking animal communication, changing animal behavior, or decreasing animal health in terrestrial and aquatic systems. On the other hand, ecoacoustics focuses on sound production by the animals at a place and attempts to understand its relationship to geophysical and anthropogenic sound sources (Sueur and Farina 2015). How animals use acoustic information is also a focus of research in ecoacoustics (Sueur and Farina 2015; Farina and Gage 2017). The propagation of animal sounds is of minor emphasis in this field. Ecoacoustics has primarily focused on the use of passive acoustic recorders, and the data produced are then used to create acoustic metrics that quantify biological sounds in intensity and complexity, sort geophysical sounds into broad types, and sort anthropogenic sounds by source. Acoustic ecology has traditionally focused on all the sounds present at a place and how they are perceived by humans (Truax 2019). Finally, soundscape ecology attempts to understand all sound sources, how they are perceived by animals and people alike, and how sound propagates in the environment (see Pijanowski et al. 2011a, 2011b; Dumyahn and Pijanowski 2011a, 2011b; Pijanowski and Farina 2011). Thus, one of the major differences between ecoacoustics and soundscape ecology is the emphasis on perception by both animals and humans; incorporation of the social element in soundscape ecology comes from its stronger ties to the humanities and social sciences. In addition, soundscape ecology has a breadth of scholarship similar to that of landscape ecology. For example, landscape ecology has developed sub-specialties, including urban landscape ecology (Pickett et al. 2008; Forman 2014), landscape aesthetics (Nassauer 1995a 1995b; Gobster et al. 2007), and landscape sustainability (Antrop 2006; Opdam et al. 2006).

1.6. How Soundscape Ecology May Address Global Environmental and Societal Grand Challenges

Soundscape ecology has the potential to address some of the biggest modern environmental and societal challenges of our time. In addition to the major themes of soundscape ecology presented above, this book will explore several of these solution spaces (see Fig. 1.5). Although not presented separately as sections or chapters of the book, they represent repeating cross-cutting themes that we will encounter again in Part II, Concepts (Chapters 4–8) and Part IV, Applications (Chapters 11–14).

1.6.1. Confronting the Biodiversity Crisis

There are nine planetary boundaries that humans must not cross if we are to ensure a sustainable use of natural resources on Earth (Rockström et al. 2009; Steffan et al. 2015). Of these, the biodiversity threshold is considered the most critical. Yet biodiversity is being rapidly lost. A recent global assessment by the Intergovernmental Science-Policy Platform on Biodiversity and Ecosystem Services (IPBES) (Díaz et al. 2019) describes the dire state of biodiversity trends globally. Approximately 25% of the known animal and plant species, which translates into over a million species, are threatened with extinction. A study of the trends among birds in North America (Rosenberg et al. 2019) showed that populations of most bird species, including both rare and common species, had declined by as much as 30% in the past fifty years. Long-term studies of insect populations (see Hallmann et al. 2017) have suggested that there has been a precipitous decline in insects due to a variety of reasons (Sánchez-Bayo and Wyckhuys 2019). More robust data, methods, and analyses are needed to ensure a better understanding of these trends and their causes (see Thomas et al. 2019) because the data used for many of these studies are incomplete or, in some cases, biased (Christie et al. 2021; Wagner 2019). Terrestrial mammal populations have also been decreasing globally due to the increase in urbanization and agricultural intensification (Pekin and Pijanowski 2012). Disease, habitat loss, and climate change are affecting amphibian populations globally (Haulahan et al. 2000; Blaustein and Kiesecker 2002; Sodhi et al. 2008), and some work suggests that anthropogenic noise is a serious hindrance to breeding for many species (Sun and Narins 2005). A decline in freshwater systems has also been quantified recently, with assessments showing that megafaunal species declined by as much as 88% between 1970 and 2012 (He et al. 2019).

The challenge of monitoring many groups of animals, especially insects and amphibians, is their natural tendency to vary greatly over space and time (Stuart et al. 2004; Pollock et al. 2002). Currently, we do not know how biodiversity varies over local, regional, or global scales as habitats change and populations are overharvested (Velland et al. 2013). Humans rely on biodiversity: as much as 75% of our crops are pollinated by animals; many species are needed to process wastes and disperse seeds to maintain a variety of ecosystem functions; and many species are natural biochemical factories (Cardinale et al. 2012), possibly producing, for example, drugs that will cure cancer. Yet many questions remain unanswered. How are nocturnal animals affected by environmental change? What are the seasonal shifts in animal population patterns, and how do they relate to habitat and climate change? Using satellite remote sensing and other geospatial technologies, we can investigate how patch sizes inside habitats, relative amounts of edge and core habitat, and the spatial configuration of habitats affect animal populations. In short, soundscape ecology is well poised to address the most serious environmental challenge of the twenty-first century.

1.6.2. Improving Society's Connectedness with Nature

In 2007, the United Nations estimated that more than half of the global human population lived in urban settlements (UN 2007), a pattern that reflects centuries of people moving from rural areas to cities. In developed nations, such as the United States, nearly three-quarters of the human population lives in cities. Environmental psychologists (e.g., Martyn and Brymer 2016) and many sociologists (e.g., Simmel 1903, as translated in Simmel 2012) recognize that urban life creates a physical as well as a psychological separation of humans from nature. Potential consequences of this human-nature separation include a decline in concern about the natural world, the exposure of large numbers of people to the serious physical and mental health impacts of noise in urban areas (King and Davis 2003), and a decrease in access to the therapeutic benefits of being in natural areas (see Frumkin et al. 2017). According to some, our children are developing a new condition: "nature deficit disorder" (Louv 2008). Cited as a common problem across much of modern society, this mental condition derives from a lack of exposure to the sensory stimuli provided by natural spaces. The field of soundscape ecology is uniquely poised to bring together concepts, tools, techniques, and methods to answer many related questions. For instance, what role does natural sound, both biological and geophysical, play in connecting society to nature? How might natural soundscapes change our affective and cognitive understanding of the role that nature plays in sustaining our quality of life? How can conservation efforts locally and globally benefit from a greater acoustic connection to the natural world, especially in remote and unusual places?

1.6.3. Sustaining Sonic Practices as a Form of Traditional Ecological Knowledge

Traditional ecological knowledge (TEK) (see Berkes 1993; Berkes et al. 2000) encompasses the "knowledge and insights acquired through extensive observation of an area or species. This may include knowledge passed down in an oral tradition or shared amongst users of a resource" (Huntington 2000). In many cases, knowledge holders are from indigenous groups. Ecomusicologists (see Allen and Dawe 2016; Guyette and Post 2016), scholars of ethnomusicology who study the roles of natural sounds in the culture of a community, have shown that TEK takes a variety of forms, from written language to body gestures to stories, as well as aural traditions, which are broadly called "sonic practices." Sonic practices can include a broad array of culturally important sonic products or vocalizations produced during the traditional activities of daily life. Sonic products can be songs, utterances, poems, and other forms of spoken knowledge; these can be produced in formal (i.e., as performances) or informal settings (e.g., with friends or alone in nature). They also include multispecies communication (e.g., herders communicating in a two-way interactive form with their livestock) and mimicry (e.g., a person creating the sounds of wind or animals). Sounds produced during our normal activities (e.g., the sound of a goat's milk streaming into a bucket) are also included in the sonic practices that reflect culture. Sonic practices also include the knowledge of local acoustic cues that signal how the environment is changing. Sonic practices pass knowledge through generations about certain sounds that are signals of ecosystem change. Many examples exist in cultures around the world: Mongolian herders listen to the sounds of wind, cracking ice, and thunder as well as those of wildlife and livestock. The songs of these herders from the Asian steppes include information about how birds signify bodily transitions as cultural ways of life (Post and Pijanowski 2018). It is now recognized that as time passes, the traditional wisdom of local knowledge holders is lost. With this loss, the long-term knowledge of how to manage landscapes sustainably is threatened. More research is needed to address several fundamental questions: What are the various forms of sonic practices that exist within and across all indigenous groups? What are the ecological similarities and differences that lead to specific forms of sonic practices? Which sonic practices are critical? What are the threats to these sonic practices? What interventions are needed to preserve and sustain them? What is the role of sonic practices in the conservation of local species or of places? What are the consequences of lost sonic practices?

1.6.4. Planning for Healthy Living Spaces

A recent report by the World Health Organization (2021) shows that most cities experience sounds that exceed healthy levels. Levels of sound above 55 decibels are known to cause hypertension in people, disrupt sleep patterns, and decrease the concentration of learners. In many cases, long-term exposure to high levels of sound, especially sounds from cars, trucks, and construction machinery, can cause hearing loss. A study of the health effects of long-term noise exposure on people living in Europe (WHO 2011) found that noise exposure had resulted in the loss of a total 1 million healthy years of life. Studies (Kaplan 1995; Franco et al. 2017) examining individuals exposed to natural sounds showed they were less stressed and had a more positive attitude toward life than their urban counterparts. It has also been shown that noise can increase physiological stress in many mammals and birds. Soundscape ecology can assist in land use planning in urban and other human-dominated landscapes by answering critical questions: How does sound travel in and around cities? Can the built environment be designed to preserve natural sounds, even in human-dominated places? Learning how to intentionally design the soundscapes of cites could improve quality of life for humans and animals alike.

Summary

Sounds are ever present on Earth. The collection of every sound from the three major sound-source categories (biological, geophysical, and anthropogenic) creates a soundscape unique to a place and a specific instant in time. The field of soundscape ecology offers the opportunity to understand the dynamics of landscapes by using sound as a measure of change. Soundscape ecology brings together a variety of concepts, tools, and approaches from diverse fields such as ecology, natural resource management, engineering, the humanities, and the social sciences. As interest in soundscapes, and acoustics in general, has increased over the past few decades, several parallel areas of study have emerged and evolved. However, soundscape ecology, as originally proposed, is the only field that incorporates all sound sources, every major component of sound propagation, and how sound is perceived by organisms.

Discussion Questions

1. How have soundscapes reflected the acoustic environment of landscapes in various fields? In particular, how are the patterns and processes that are common to landscapes reflected in the sounds that occur there?
2. If sounds in a soundscape are classified by their sources, how might you classify the sounds coming from the wind rustling the leaves in a tree?
3. Discuss how the soundscape of a place might change over time. Consider how it might vary over a day, from morning to the middle of the night, and then from season to season. If this is a group discussion, describe how soundscapes vary among the different places you have lived.
4. The study of sound is a part of many disciplines on a university campus. List departments and courses that focus on some aspect of sound, acoustics, or music.

5. Describe how you perceive the following sounds in terms of your psychological attitude toward them: a train whistle, a jackhammer, a cricket, the song of a bird.

Further Reading

Pijanowski, Bryan C., Almo Farina, Stuart H. Gage, Sarah L. Dumyahn, and Bernie L. Krause. "What Is Soundscape Ecology? An Introduction and Overview of an Emerging New Science." In "Soundscape Ecology," ed. Bryan C. Pijanowski and Almo Farina, special issue, *Landscape Ecology* 26, no. 9 (2011): 1213–32.

Pijanowski, Bryan C., Luis J. Villanueva-Rivera, Sarah L. Dumyahn, Almo Farina, Bernie L. Krause, Brian M. Napoletano, Stuart H. Gage, and Nadia Pieretti. "Soundscape Ecology: The Science of Sound in the Landscape." *BioScience* 61, no. 3 (2011): 203–16.

Popper, Arthur N., and Robert J. Dooling. "History of Animal Bioacoustics." *Journal of the Acoustical Society of America* 112, no. 5 (2002): 2368.

Schafer, R. Murray. *The Soundscape: Our Sonic Environment and the Tuning of the World*. Destiny Books, 1994.

Servick, Kelly. "Eavesdropping on Ecosystems." *Science* 343, no. 6173 (2014): 834–37.

The Physics of Sound

2

OVERVIEW. This chapter summarizes the fundamental concepts of sound and explains how sound is captured and recorded as digital files for eventual analysis. A review of the most important terms in acoustics is presented, along with a description of how microphones work. A brief overview of the field of psychoacoustics is provided, with examples of how human responses to sound intensity and frequency are measured. This overview of the physics of sound covers those concepts used often by soundscape ecologists in their analysis of audio files for their research.

KEYWORDS: dB, frequency, mechanical wave, microphone, rarefaction, sound pressure level, spectrogram, vibration, wavelength

---

2.1. Sound as a Wave

2.1.1. Key Terms

To understand how sound exists on Earth, one must first understand how it moves through matter (solids, liquids, or gases). The clearest example considers how sound moves through fluids (gases or liquids) as a wave. At equilibrium, fluids on Earth have a fixed "undisturbed" pressure. Undisturbed air at sea level, for example, has a well-established mean ambient air pressure. A disruption of that resting, or equilibrium, pressure, such as that created by hands clapping, produces an acoustic event called a *mechanical wave* or *pressure wave*, which moves as a front through space.

We define the phenomenon of *sound* as the movement of a mechanical wave through an elastic medium, which can be a gas (e.g., air), a liquid (e.g., water), or a solid (e.g., ice). Three forms of mechanical waves exist: *longitudinal*, *transverse*, and *ripple* waves. A longitudinal wave that moves through the air, for example, can be sensed by the human ear, by other sensory cells of other animals, or by a microphone. Sound can also be defined as an experience: an excitation of the pressure-sensing (e.g., hearing) system in organisms that results in the perception of a pressure wave (Everest and Pohlmann 2022). Thus, we can define sound either (1) as a physical construct or (2) as a sensory experience. Let us consider sound as a physical construct first.

2.1.2. Longitudinal Waves

As a mechanical wave moves through an elastic medium (Fig. 2.1A), a high-density zone of molecules is formed that has a pressure higher than the resting, or ambient, pressure of that medium (called a *compression*). The formation of that high-pressure zone creates another zone that has a pressure lower than the ambient pressure (called a *rarefaction*). Figure 2.1A shows how a sound is produced by a tuning fork. With the tuning fork at rest, all air molecules move through space with essentially random trajectories, which creates the ambient pressure. Striking the tuning fork generates alternating pulses of compression and rarefaction that travel through the air; here, the mechanical wave progresses from left to right away from the sound source. The vibrating movement of the tuning fork is periodic, creating multiple zones of compression and rarefaction. This movement of a mechanical wave through air is the most common form of sound propagation and is referred to as a *longitudinal wave* or a *compression wave*. Vibrations created by the syrinx of a bird or the larynx of a human create longitudinal waves in air that we hear as birdcalls or as human speech. The sounds of thunder, running water, a train whistle, and wind are longitudinal waves as well.

2.1.3. Transverse Waves

Mechanical waves can also move along a solid object, such as a string, causing the object to behave as a *transverse wave* that vibrates "up and down" around a plane (Fig. 2.1B). The height of this transverse wave is proportional to the increase in pressure that is forcing the wave upward, which also generates a lower-pressure area at the bottom of the wave. The physical object that is behaving as a transverse wave produces a longitudinal wave

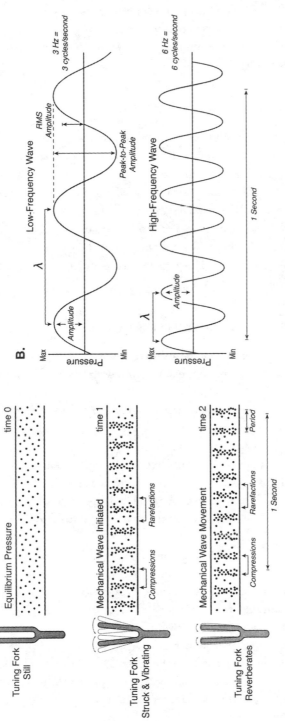

Figure 2.1 Conceptualization of sound as a wave: (A) A longitudinal wave, showing equilibrium pressure (*top*) and then with compression and rarefaction areas (*bottom*). (B) A transverse wave, illustrating wavelength, amplitude, peak-to-peak amplitude, and RMS amplitude as an arrow to the light dotted line.

that moves through air, which is the sound we hear. An example in nature is the sound created by insects that vibrate silk or grass blades as a form of what is called "substrate communication." Transverse waves are also called *shear waves* by physicists.

2.1.4. Ripple Waves

Dropping a stone in a calm pond creates another kind of mechanical wave. *Ripple waves* radiate out from a disturbance created in an elastic medium by an object, such as a stone thrown into a pond. A ripple wave causes molecules in the disturbed medium to move outward in a circular pattern (on a horizontal plane). Ripple waves in one elastic medium (e.g., water) can displace another elastic medium next to it (e.g., air), especially if the movement includes a collision of the wave with a solid (e.g., a beach at the edge of the ocean). Several groups of insects produce water ripples as another form of substrate communication.

2.1.5. Other Phenomena Related to Mechanical Waves

Soundscape ecologists should be aware of several other phenomena related to the movement of sound in the environment as mechanical waves:

> *Vibrations in solids.* Mechanical waves can be transmitted through solids in a variety of ways. In general, physicists refer to waves that travel through solids as *vibrations.* Vibrations moving through solids generally propagate as longitudinal and transverse waves simultaneously. Mechanical waves of this type include the sounds created by ice breaking off glaciers (called *calving*), by movement of sand in dunes, by the digging motions of burrowing animals, and by the activities of wood-boring beetles in trees.
>
> *Shock waves.* When an object creating a mechanical wave is traveling faster through an elastic medium than the mechanical wave itself can travel through that medium, a *shock wave* occurs. Shock waves commonly arise in two contexts, the first of which is a thunderstorm. A bolt of lightning creates a stream of electrons, which causes cold and hot air in motion to hold electric charges and then discharge. The column of air heated by the discharge, moving faster than the speed of sound, creates the sound of thunder. The second common context is a jet airplane flying faster than the speed of sound, which creates a sonic boom. Jets are often rated for

their top flying speed in terms of Mach, which is the speed of sound in air; a jet that moves at twice the speed of sound is traveling at Mach 2.0.

2.1.6. Properties of All Mechanical Waves

Recall that a mechanical wave travels through space as alternating pulses, or cycles, of compression and rarefaction. The time between these pulses is the wave's *period*, denoted as τ (tau) and measured in seconds (s). The distance between these pulses in space is the *wavelength*, denoted as λ (lambda) and measured in meters (m). The number of pulses passing a point per unit time is the wave's *frequency*, usually expressed as cycles per second, or hertz (Hz; named after the German physicist Heinrich Hertz). Fast-moving mechanical waves have high frequencies. Humans can hear sounds in the range of 20–20,000 Hz. Sounds with frequencies above this range are called *ultrasonic* (*ultra* = "above"), and those with frequencies below this range are called *infrasonic* (*infra* = "below" or "farther on"). To reduce the number of zeros in Hz values, frequency is often expressed in kilohertz (kHz; 1 kHz = 1,000 Hz).

In physics, *pressure* is defined as the amount of force applied per unit area. At sea level, the pressure of air at rest at the reference temperature of 15°C is typically 101,325 pascals (Pa). The pascal (named after Blaise Pascal, a French scientist and philosopher) is the SI unit of pressure. This modern unit of pressure is the force of one newton (named after physicist Isaac Newton) per square meter. It should be noted that many older textbooks in bioacoustics and physics use older, non-SI units of pressure, including atmospheres (atm), pounds per square inch (psi), and millimeters of mercury (mmHg).

The measure of a sound's intensity is its *amplitude* or *sound pressure level* (SPL)—in other words, the greater the sound pressure, the louder the sound. Amplitude measures the height (i.e., pressure) of a sound wave relative to the ambient, or background, pressure within the elastic medium through which the wave is passing (illustrated by the horizontal line between the maximum and the minimum pressure in Fig. 2.1B). This version of amplitude or SPL is called *peak pressure* or *0-to-peak pressure*. One can also express SPL as the difference between the maximum and minimum height of a wave, typically expressed in Pa; this measure is referred to as *peak-to-peak amplitude* (see Fig. 2.1B).

Expressing sound pressure in Pa is often not practical because the human ear is extremely sensitive to an enormous range of sound pressure levels; for example, the difference in sound pressure level between the faintest sound within the human hearing range and a sound that would be painfully loud is as much as ten trillion to one (Everest and Pohlmann 2022). Instead, the most common way of expressing sound pressure is to compare the amplitude of the sound of interest with that of the faintest sound that a human can hear. This measure, called *relative amplitude*, is calculated as 10 times the logarithm of the ratio of these two values. Relative amplitudes are often expressed in *decibels* (dB; after Alexander Graham Bell; *deci* = "one-tenth"). Because the ratio is expressed as root powers (i.e., the two values are squared) of the unit, the "2" exponent can be brought out in front of the logarithm and thus multiplied by 10 to get the value 20. The standard equation for calculating the dB is thus

$$L = 20 * \log_{10} \frac{I}{I_0},$$

where I is the intensity (i.e., amplitude) of the sound being measured, I_0 is the intensity of the reference pressure (the faintest sound a human can hear), and L is loudness (in Pa). The faintest sound a human can hear is 20 µPa re 1 m; this value is thus set to 0 dB ($L = 20 * \log_{10} [20/20]$) at a reference frequency of 4 kHz.

To understand the scaling of dB, consider that a sound that is 10 times greater in intensity than the faintest sound humans can hear is 10 dB. One that is 100 times greater in intensity than the faintest sound is 20 dB, and one that is 1,000 times greater is 30 dB. Normal human conversation is approximately 60 dB, and a jet engine at takeoff is typically 120 dB, which is near the upper level of human hearing; anything above this causes pain and permanent hearing loss (Fig. 2.2). The theoretical maximum dB value in air is 194 dB, at which point the rarefaction zone would be a complete vacuum.

Because there needs to be a reference pressure for the measured peak-to-peak values reported, a standard reference pressure of 1 micropascal (µPa; one-millionth of a pascal) at 1 meter from the sound source is used by convention. Thus, a peak-to-peak amplitude of 80 dB would be written out as in this example,

80 dB re 1 µPa at 1 m,

which states that the peak-to-peak amplitude is 80 dB above the ambient reference pressure of 1 µPa when measured at 1 m from the source. The term "re" is short for "reference," and the measures to the right of it list the reference conditions.

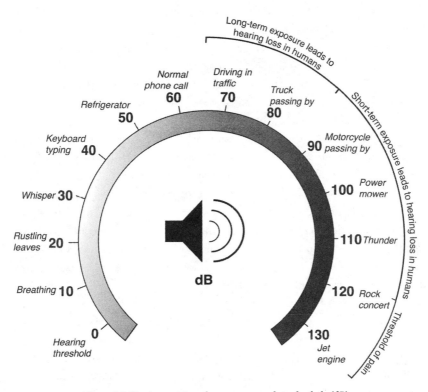

Figure 2.2 The intensities of common sounds in decibels (dB).

Another way to quantify amplitude is to express it as a weighted average of sound pressure level over time. Typically, this measure is calculated in four steps. First, sound pressure level, in μPa, is measured along the time axis, usually at fixed intervals (e.g., 1 s). Second, each of these values is then squared. Third, the average of all the squared pressure values is calculated. The fourth step is to take the square root of the average of all the squared pressure values. This final measure is called the *root mean square* (RMS) *amplitude* (see Fig. 2.1B) and can be expressed in μPa or dB. It thus expresses average loudness over time. RMS amplitude is used frequently in studies of noise because the form of such waves is extremely complex.

Relative amplitudes of sounds recorded by microphones are sometimes measured as the ratio of the amplitude of the recorded sound to the maximum amplitude that the microphone can sense. Some passive acoustic recorders thus store this information as a negative dB value (i.e., the ratio of the recorded sound's amplitude to the maximum amplitude for the microphone gives a value less than 1.0, which is negative on a logarithmic scale). This measure of dB is called *full scale dB* (dBFS).

When mechanical waves pass through fluids, two separate forms of movement occur: the movement of compression zones across space and the bulk movement of molecules along the direction of the longitudinal wave. In technical terms, the longitudinal (compression) wave is measured as pressure, which is a scalar (it has magnitude, but not direction), and the bulk molecular movement is measured as particle velocity, which is a vector (it has magnitude and direction). At the source of the sound, the two movements are out of phase with each other; that is, they move at different speeds. More specifically, at the sound source, the particle velocity is greater than the velocity of the compression wave. Eventually, the compression wave and the bulk movement of air molecules become synchronous; as a physicist would say, they are in phase with each other. The area the waves pass through prior to this phase matching is called the *near field*, and the area they pass through afterward is called the *far field*. The near field is the area within one wavelength of the sound source, and the far field is the area two or more wavelengths away (a transition zone exists between one and two wavelengths' distance). Sensing of near-field sounds is very important to some animals, especially insects and fish (Popper and Fey 1973). Many groups of insects have developed the capacity to detect movement of air or water molecules rather than compression waves per se. This means that they can detect sounds produced within a very short distance but not ambient sounds that are manifested as compression waves; in other words, they do not sense sounds that are produced at a distance.

Audio recordings that store information as data that can be converted to measures of frequency, amplitude (typically as dB), and time are called *full-spectrum* recordings. Some audio recordings store only temporal information for the dominant frequency; these are called *zero-crossing recordings*. Full-spectrum recordings are usually used in soundscape ecology research in order to capture soundscapes' rich information. Some ultrasonic sensors that monitor bats use zero-crossing recording technology to capture their simple, narrowband signals in the ultrasonic range.

Several alternative ways to express dB that adjust sound pressure levels according to the sensitivity of the human ear across all frequencies, called *frequency-weighted dB filters*, have been developed. An A-weighted dB filter, designated as dB(A) or dBA, adjusts for the fact that the human ear is most sensitive to midrange frequencies (1,000–8,000 Hz) by decreasing sound pressure levels across low frequencies (generally those less than 1,000 Hz), reducing their amplitude values. A-weighted dB measures are common in environmental noise studies, especially those that report workplace noise levels to the US Occupational Safety and Health Administration (OSHA). Frequency-weighted dB sound pressure levels are useful for studies that concern how humans normally perceive soundscapes in natural and urban settings.

The graphical representation of amplitude over time is called a *waveform*. Waveforms, which illustrate change in sound pressure levels (plotted on the y-axis) over time (plotted on the x-axis), are great diagnostic tools for soundscape ecologists. For a *pure tone*, which is a sound of a single frequency produced over time, a waveform illustrates the shape of the wave. The shape is typically a sine curve, but can also be a sawtooth, a square, or another shape. A waveform that has a repeating shape, such as a sine curve, is referred to as a *periodic waveform*. Some animals, such as dolphins, emit pure tonal sounds that are sometimes called "whistles" (Caldwell and Caldwell 1977).

When a soundscape is complex, we are often interested in the amplitudes of mechanical waves at different frequencies. It is therefore useful to be able to measure them across an *octave*: an interval between frequencies where one frequency is twice the other. We can calculate dB over an octave by defining the middle frequency, f_0, and the lower and upper frequency bounds. The interval between lower and upper frequencies defines the *octave band*. Octave bands are most often calculated using a reference middle frequency of 1,000 Hz. We can then step up or down octaves by either doubling (going up) or halving (going down) from the middle frequency. Table 2.1A lists common octave middle frequencies and the lower and upper bounds that define each octave band. We can also split octaves into *one-third octave bands* to calculate SPL across narrower frequency ranges (Table 2.1B).

The speed of a compression wave is a function of the density and temperature of the transmission medium. In general, the *speed of sound* is greatest in solids, followed by liquids and then gases. The speed of sound in dry air at 0°C is 331.3 meters per second (m/s); in pure water at 0°C it is 1,402.3 m/s. Thus, sound travels 4.2 times faster in water than in air. Salt water is denser than pure water, and the speed of sound in salt water at 0°C is 1,449.4 m/s, about 3% faster than in fresh water.

Sound pressure levels decline as a mechanical wave moves away from its source. Energy dissipates into the transmission medium in two ways: by attenuation and by scattering. *Scattering* is the specular (i.e., reflected) change in a mechanical wave as portions of the wave begin moving in directions other than the principal direction. The amount of scattering is dependent on the characteristics of surfaces (e.g., water, a leaf, a rock) that a sound

Table 2.1 Frequency bands with middle frequencies and lower and upper bounds

A. Octave bands

Octave middle frequency (Hz)	Lower bound	Upper bound
15.6 (preferred 16.0*)	11.0	22.1
31.3 (preferred 31.5)	22.1	44.2
62.5 (preferred 63)	44.2	88.4
125	88.4	176.8
250	176.8	353.6
500	353.6	707.1
1,000	707.1	1,414.2
2,000	1,412.2	2,828.4
4,000	2,828.4	5,656.9
8,000	5,656.9	11,313.7
16,000	11,313.7	22,627.4

B. One-third octave bands (for a subset of the octave bands listed in part A)

One-third octave or octave (bold) middle frequency (Hz)	Lower bound	Upper bound
125	114.4	140.3
157.5 (preferred 160)	140.3	176.8
198.4 (preferred 200)	176.8	222.7
250	222.7	280.6
315	280.6	353.6
396.9 (preferred 400)	353.6	445.4
500	445.4	561.2
630	561.2	707.1
793.7 (preferred 800)	707.1	890.9
1,000	**890.8**	**1,122.5**

*Preferred values are simply rounded to nearest half value (ending in .0 or .5).

wave hits as well as the frequency of the wave. *Attenuation* is the reduction in intensity that occurs as the wave travels through space. Attenuation is typically expressed as the ratio between the wave's intensities at two points in space, given as dB loss per unit length (e.g., dB/m). In general, sound intensity decreases by 6 with each doubling of distance from the source.

Soundscape ecologists are interested in how animal acoustic communication signals attenuate and scatter through habitats such as tropical rainforests and kelp forests, and how frequencies and amplitudes are influenced by habitat composition. Because sound can travel more easily in water than in air, its attenuation is less in water. In other words, given the same frequency and initial SPL, sounds can travel much farther in water than in air. (It is also true that sounds can travel farther horizontally in water than in air because the surface and bottom of the water body prevent the mechanical wave from propagating in all directions; the wave thus has less three-dimensional space in which to travel.)

Because the sounds in a soundscape span a large range of frequencies, another measure that soundscape ecologists use is *power spectral density* (PSD). PSD is the sum of all sound energy (in dB) within a particular range of frequencies over a defined time period. Often, PSD is calculated for several frequency bands, each of which is characteristic of the calls of an animal. If a soundscape ecologist desires to examine PSD across multiple recordings (e.g., across 10 frequency bands for thousands of recordings), then the maximum PSD, or PSD_{max}, for each band is calculated across all of the recordings, and a relative PSD, or rPSD, is calculated for each recording (i.e., PSD_i/PSD_{max}, where i is the individual PSD recording value); rPSD values range from 0.0 to 1.0.

Because sound can be represented as a three-dimensional measure of time, frequency, and amplitude, many researchers who study sound describe the shape of sound across these three dimensions as spectral shape (Middlebrooks 1997; Zahorian and Jagharghi 1993; Hartmann 1998; Bartsch and Wakefield 2004), spectral characteristics (Bianco et al. 2019, 3591; Buscombe et al. 2014), spectral structure (Mysore 2010), or spectral form (Buscombe et al. 2014). In music, another term, spectral envelope (Schwarz 1998; Warren et al. 2005), is often used to refer to the generalized shape of a complex sound's power spectrum. All of these terms refer to time-varying estimates of frequency and amplitude, but for the rest of this book, the term *spectral shape* will be used.

Bioacousticians and soundscape ecologists examine several kinds of frequencies when they analyze sound production in animals. One is the *base* or *fundamental frequency*, which is the center of the lowest intense frequency present in a complex sound that has several spectral components. The fundamental frequency is often designated as F_0. The *dominant frequency* is the frequency of maximum amplitude across all sounds produced by an organism or object. *Formant frequencies*, which are often cited in studies of primate vocalization and human speech, are multiple frequencies produced as air vibrates within the vocal cords and/or passes through the nasal passages and mouth. Most sounds in the environment do not occur as pure frequencies, so terms such as *broadband* and *narrowband* are used to describe the relative widths of the frequency ranges being examined. To describe a specific range of frequencies with an upper and lower frequency value, the term *bandwidth* is used. Bandwidths are reported in Hz or kHz; for example, the bandwidth between 500 Hz and 3 kHz is 2.5 kHz.

Because sound can be formed as a simple transverse wave, waves at regular frequency intervals produce a phenomenon called *harmonics*. Harmonics are created when sound is produced at frequencies that are integer multiples of the fundamental frequency and the locations of peaks along the mechanical waves are thus aligned, which enhances the intensity of all the waves. For example, harmonics occur when a fundamental frequency of, say, 100 Hz is produced with a second *harmonic frequency* of 200 Hz, a third harmonic frequency of 300 Hz, and so on. The harmonic frequencies are sometimes called *overtones*. As we will see in Chapter 3, harmonics are common in sounds produced by many animals, particularly insects and mammals.

Mechanical waves can oscillate across frequencies or amplitudes. This oscillation is called *modulation*. *Frequency modulation* (also called *spectral modulation*) is common in sounds produced by birds and mammals; it occurs when a sound is produced at a relatively constant intensity, but its frequency varies. *Amplitude modulation* occurs when a sound varies in amplitude but not in frequency; it is common in sounds made by insects and amphibians, which are not able to vary their frequencies because their sound production mechanisms are fixed with their size.

Several groups of animals also practice *bimodal* signaling (Caldwell 2014). Bimodal acoustic communication is accomplished when an animal produces mechanical waves in two media simultaneously. For example, elephants, katydids, some spiders, and some frogs can produce airborne sounds at the same time they produce substrate-based vibrations (which pass through a solid). In some cases, the sound production and vibration production mechanisms are the same, and in other cases, two mechanical waves are produced by two independent body parts.

Masking of sounds occurs when loud "noises" prevent a sound receiver from detecting other sounds at the same frequency. Many researchers have found that masking of biological sounds by anthropogenic sounds is common in urban landscapes and in many areas of the ocean.

2.2. Visualizing a Sound Recording

There are several standard visualization tools that soundscape ecologists use. Many build on three fundamental forms of sound visualization: the *spectrogram*, which displays frequencies over time; the *waveform*, which displays amplitude over time; and the *power spectrum*, which plots sound pressure levels across all frequencies and is an excellent visualization of power spectral density. Let's see how these tools can be used to visualize the dawn chorus of the Miombo woodlands in western Tanzania right after the rainy season, in which sounds from birds, crickets, and two tropical frogs are heard (Fig. 2.3). The spectrogram shown in Figure 2.3A illustrates the frequencies of all sounds over a 2-minute recording, which range from 180 Hz to 8 kHz. Amplitude is depicted here using a gray scale, with the loudest sounds shown as black and the absence of sound as white. The waveform in Figure 2.3B displays the amplitude over time, illustrating that the sounds of the dawn chorus build for the first 30 seconds of the recording, then

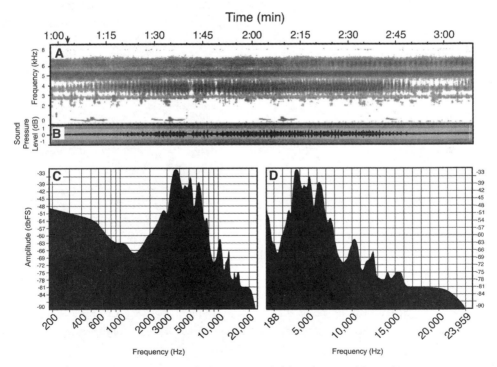

Figure 2.3 Common visualizations of sound: (A) spectrogram; (B) waveform; (C) power spectrum with logarithmic scale; (D) power spectrum with linear scale.

remain loud until about the 2:40 mark, when the chorus begins to diminish. Two power spectral density graphs are shown in Figures 2.3C and 2.3D. Figure 2.3D has a linear scale along the x-axis, whereas Figure 2.3C is plotted with a base-10 logarithmic scale on the x-axis. This logarithmic scale is the most commonly used in power spectrum plots because it aligns with the way humans perceive frequencies. Note that the dominant frequency is around 3.8 kHz.

2.3. Recording Sound with Transducers

Soundscape ecologists have principally used *passive acoustic recorders* (PARs) to create digital audio files that are then analyzed for their content. Passive acoustic recorders have several major components: a microphone, an amplifier, an analog-to-digital converter, a power source, storage media, and a programmable electronic interface that usually includes a clock. Other kinds of transducers and data loggers (i.e., parts of the PAR that support data storage) have also been deployed, including vibrometers, hydrophones, parabolic dishes with single unidirectional microphones, handheld recorders, and mid-side (M/S) microphone arrays. This section explains how microphones work, the features of microphones most useful to soundscape ecologists, and how the sounds captured by microphones are recorded and stored.

2.3.1. Microphones

Microphones are what engineers call *transducers* (i.e., converters). All microphones perform the same general function: they convert a pressure wave (sound energy) to an electrical signal (electrical energy), which is eventually stored as information on a data logger. Prior to the 1990s, most data loggers stored sound to tape, an analog form of data storage. With the advent of portable computers and devices that can be taken out to the field, data loggers eventually became digital. Devices designed for use in soundscape ecological research as passive acoustic recorders have been constructed to withstand long-term exposure to weather. (A type of instrument often used in aquatic ecosystems sends out an acoustic signal and then records how mechanical waves propagate in three dimensions. This technique is called *active acoustic monitoring*; see Stein and Edson [2016] for a review.)

Several features of microphones are important to consider when conducting research in soundscape ecology. The first is how the microphone is constructed (Fig. 2.4). There are three major kinds of microphones: A *dynamic* microphone (Fig. 2.4A) uses a diaphragm (a thin layer of material that creates a lining or partition) located next to a magnet surrounded by an internal coil, which moves when a pressure wave hits the diaphragm. A *condenser* microphone (Fig. 2.4B), on the other hand, uses a capacitor, an electronic component with two plates whose movements create voltage that varies with the distance between the plates. The voltage differences from the capacitor are the electrical signals that are eventually stored in the PAR. A *piezoelectric* microphone (Fig. 2.4C) uses the electrical properties of certain substances, like crystals, to convert pressure wave energy to

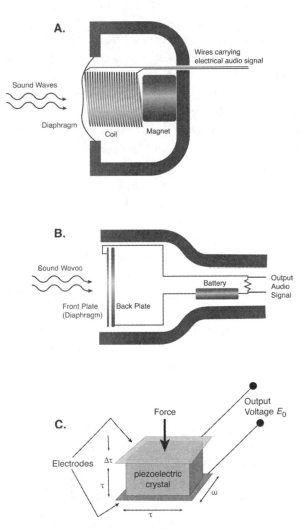

Figure 2.4 Components of three microphone types: (A) dynamic; (B) condenser; (C) piezoelectric.

electrical energy. Piezoelectric microphones create electrical current differences when the substance in use changes its form (e.g., its width or height) due to changes in sound pressure. Piezoelectric microphones are often used as *contact* microphones to measure vibrations in solids or pressure waves within short distances.

A second feature to consider is the spatial sensitivity of a microphone. Most microphones have a directional sensitivity pattern (Fig. 2.5). The sensitivity of an *omnidirectional* microphone is approximately the same in all directions (Fig. 2.5A). That of a *unidirectional* microphone is focused along a directional axis. Unidirectional microphones are generally either *cardioid* (Fig. 2.5B), with a sensitivity pattern in the shape of a heart, or *hypercardioid* (Fig. 2.5C), with a very narrow directional sensitivity pattern. A special class of microphones are those that are sensitive at their sides; these are called *bidirectional* or *figure eight* microphones (Fig. 2.5D).

A third important feature of a microphone is its sensitivity to sound at various frequencies. Many microphone manufacturers provide two measures, a frequency response curve and a measure of microphone sensitivity. The *frequency response curve* plots the microphone's typical operating frequency range on a base-10 logarithmic scale (i.e., as humans naturally experience sound) against relative response in dB. A relative response value of 0 means that the microphone detects sound at that frequency at a SPL nearly identical to that of the sound sent to it (the reference SPL); thus, the measure is called a relative response. Some microphones have the same relative response across all frequencies of interest; these microphones, called flat response microphones, are best for recording broad-range instruments such as pianos. Those that are more sensitive to certain frequencies are called shaped response microphones. Those that are not sensitive to low frequencies but are very sensitive to audible ranges, referred to as colored microphones, are best for recording in the presence of low-frequency background noise below ~500 Hz. Many microphones used with PARs are flat response microphones, but microphones that are designed for bioacoustic

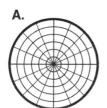
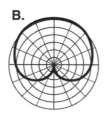
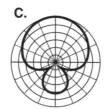
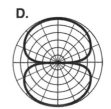

Figure 2.5 Common microphone sensitivity patterns: (A) omnidirectional; (B) cardioid; (C) hypercardioid; (D) bidirectional.

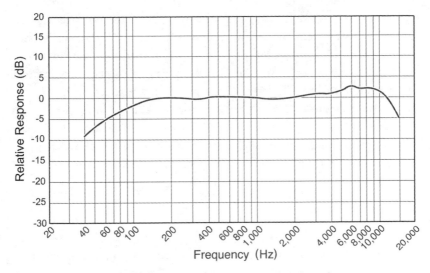

Figure 2.6 Typical frequency response curve for a microphone. Relative response is given in dB with 5 dB increments above and below reference dB of 0.

studies may be more sensitive in the 3–7 kHz range, in which where birds, frogs, and most mammals produce sound. A frequency response curve for a flat response microphone is shown in Figure 2.6. Note that this microphone has a response close to the reference SPL between 200 Hz and 2,000 Hz, is more sensitive to frequencies between 2,000 Hz and 12,000 Hz, and is less sensitive to frequencies above 12,000 Hz and below 200 Hz.

Microphone sensitivity is a measure of a microphone's efficiency of converting sound pressure levels (i.e., sound energy) into electrical energy. It is measured as the output voltage in response to a standard SPL input, set at 94 dB (which is 1 Pa) for a 1 kHz pure tone.

To understand microphone sensitivity, let's consider how this measure is typically reported. As a microphone converts a sound pressure wave to an electrical current, it is compared with a theoretically perfect microphone that would produce 1 volt of electrical energy when exposed to a pressure of 1 Pa (i.e., 94 dB SPL). A microphone's sensitivity is thus calculated as the following ratio:

$$\text{Microphone sensitivity (dBV)} = 20 \times \log_{10}\left(\frac{\text{sensitivity in mV/Pa}}{\text{output reference in mV/Pa}}\right),$$

where units of sensitivity can be either dBV (a measure of voltage in dB referenced to 1 volt) or mV/Pa. The reference value—the denominator for the right side of the equation—is set to 1,000 mV/Pa (which is also 1 V/Pa). The numerator, also called the *transfer factor*, is the voltage generated by the microphone being tested under 1 Pa of pressure. The numerator is always less than 1,000 mV, since no perfect microphone exists, so the ratio will always less than 1.0, which will produce a negative value when expressed in dBV.

To understand this calculation more fully, let's consider a microphone that produces 1.6 mV of electrical energy when presented with a 1 kHz sound at a pressure of 1 Pa. Using the equation above, 1.6 mV/1,000 mV gives 0.0016. The $\log_{10}$ of 0.0016 is −2.79. Multiplying that value by 20 gives a microphone sensitivity of −56 dBV. Microphone manufacturers will write out this entire sequence of units for a specific microphone, which will look like this:

$$-56 \text{ dBV} \pm 4 \text{ dB re } 1 \text{ V/Pa at } 1 \text{ kHz.}$$

As it is assumed that the measure is in volts, some manufactures will drop the "V" and report microphone sensitivity simply as dB. The second part of the reporting sequence (designated as "±") is the range for all individual microphones of that particular model. In other words, all microphones of that model will have a sensitivity between −60 and −52 dBV. Because sensitivity values reported in dBV are always negative, microphones with larger dBV values are more sensitive. Microphone sensitivity can also be reported in mV/Pa, as dBV and mV/Pa are interchangeable units (e.g., 10 mV/Pa = −40 dBV). Values of mV/Pa are always positive, and thus microphones with larger mV/Pa have greater sensitivity. High-sensitivity microphones are best suited for recording ambient sounds (i.e., natural soundscapes), and low-sensitivity microphones are more suitable for recording loud, isolated sound sources.

A fourth consideration is a microphone's *signal-to-noise ratio* (SNR), which is a measure of how well a microphone records wanted sounds (i.e., signals), given that microphones also create their own *background noise* (i.e., unwanted sounds, or "self-noise"). Microphones have a variety of electronic components that produce background noise, which, when amplified, make that "hissing" sound you might have experienced when listening to a poor-quality audio recording at high volume. The motion of air molecules that collide with the surface of the microphone in a quiet room can also contribute toward background noise in a recording. Background noise is normally measured in an *anechoic chamber*, a special type of room used extensively in research on noise, especially in industry. Its walls are composed of special echo-free materials (e.g., foam) with complex ridges that deaden any sound produced in the chamber and also isolate the chamber from sounds from outside. The entirety of all background noise from produced by a microphone is referred to as the "noise floor" of the microphone.

The SNR of a microphone, often reported in dB, is calculated as the simple difference between the ideal signal pressure level (in dB) and the noise pressure level (in dB):

$$SNR_{dB} = signal_{dB} - noise_{dB},$$

where SNR_{dB} is the signal-to-noise ratio in dB, $signal_{dB}$ is the SPL of the signal in dB, typically using the reference value of 94 dB, and $noise_{dB}$ is the SPL of the noise in dB. Thus, if a microphone has a self-noise SPL of 14 dB, then the SNR would be 94 dB – 14 dB, or 80 dB.

There are several practical applications of the SNR concept in soundscape ecological research in addition to measuring microphone self-noise; for example, it can be applied to the quality of recordings with respect to other contributors of noise that might exist in the recording environment. For field recordings, placing a microphone close to a sound source will allow the microphone to record the wanted sounds at SPL far above the noise floor of the microphone and any air disturbances at the microphone created by wind. The SNR of such recordings will be high, and the wanted sounds will be detectable by the human ear or by machine-learning tools. Sound engineers also consider background interior environmental sounds, such as air circulation from heating and cooling devices, as part of the measurement of noise when they evaluate SNR for their recordings.

A fifth major feature of a microphone to consider is its *dynamic range*; that is, the range of amplitudes it can replicate properly. The upper boundary of a microphone's dynamic range is the highest amplitude at which it does not distort the output signal. The lower boundary is the lowest amplitude that is above the background noise produced by the microphone itself (the noise floor) and it is sensitive to the gain settings of the microphone (the noise floor increases with increasing gain). Microphone manufacturers report dynamic ranges in dB.

A final major feature of a microphone to consider is the ability to set band filters, which make the device most sensitive at specific ranges. *Highpass band filters* (HPBFs) are typically applied at either 200 Hz or 1 kHz. They are designed to "roll" the sensitivities of the microphone so that very low frequencies are filtered out to some extent; then the frequencies are gradually "opened" until they reach the upper limit of the filter at 200 Hz or 1 kHz. HPBFs are applied where the desired focus is on sounds from ani-

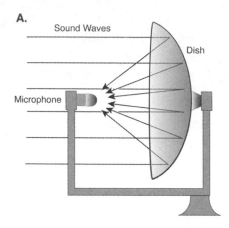

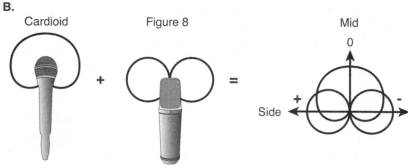

Figure 2.7 Microphones for (A) a parabolic dish and (B) a mid-side (M/S) array.

mals, which are generally above 1 kHz, but low-frequency background noise, such as highway or wind sounds, are prominent and need to be filtered out. HPBFs also have the effect of increasing the sensitivity of microphones to midrange frequencies when low-frequency sounds are intense. Low-pass band filters, which are rarely used in soundscape ecology studies, screen out higher frequencies.

Bioacousticians and soundscape ecologists often use *parabolic dishes* that "focus" sounds onto a directional microphone. The microphone is located at the center of the dish, set above and facing the dish surface (Fig. 2.7A). The dish serves to amplify the sounds coming in from one direction and reduce sounds from surrounding areas.

A *vibrometer* is a type of transducer that measures the physical displacement of an object in space. Such displacements are often measured at low frequencies. Some vibrometers can be clipped to a solid object, whose movement is then recorded as its displacement in space. Vibrometers have uses from measuring grass blade movement during the vibrational dances of spiders to measuring the vibrations of bridges to determine their structural integrity.

Hydrophones are specialized piezoelectric transducers used in aquatic systems. Most hydrophones used by soundscape ecologists are sensitive to a broad range of frequencies, since the acoustic signals used by aquatic animals range from ultrasonic to infrasonic frequencies. Many hydrophones detect sound in air, but, because of the substances used in their construction, their sensitivities in water are much greater.

Many artists who record natural sounds use a microphone array referred to as a *mid-side* (M/S) *array* (Fig. 2.7B). The classic M/S array employs a "piggyback" arrangement of two microphones with different sensitivity patterns, a cardioid (the "mid") and a figure eight (the "side"). Together, these two microphones create spatial sensitivity patterns that overlap in some areas but not in others. Thus, the M/S array generates multiple (typically three) active recording spaces and can record in a "surround sound" spatial configuration.

As mentioned above, microphones are also integrated with other electronic components. One often necessary component of a PAR is an amplifier, called a *pre-amp*, that increases the microphone's voltage in order to increase the intensity of the recording. The user can select a level of *gain*, or the "boost" in amplification provided by the pre-amp, as one would adjust the volume setting on a speaker. Gain settings are also managed to avoid *clipping*, a distortion of sound that occurs when the microphone's voltage is maximized.

2.3.2. Sound Files

There are several data formats for storing sound, but the most common format is called WAV (after "waveform"). This audio format can be converted (via a set of equations called Fourier transforms) to full-spectrum sound as the amplitude of the signal across time. A WAV file can contain one or more channels, but most have two, which produce stereo sound. The WAV file format is part of a broader class of multimedia file formats called the *resource interchange file format* (RIFF). RIFF files contain data along with a standard header that embeds other information in the metadata for the file, typically including date created, length, author or owner, and number of channels.

WAV files can be stored at different levels of digital detail, called *bit depths*. Early WAV files were 8-bit files; more recently, 16-bit and even

32-bit files have become more common. Bit depth refers to the number of 0s and 1s that store information about an event; a 16-bit file has 2^{16} levels of information (i.e., the resolution of information that can be stored is 65,536 possible values). Digital recorders that now store 32-bit depth WAV files contain so much information that clipping of microphones is, for practical purposes, not possible.

An important feature of any audio file is its *sampling rate* or *sampling frequency*: the number of measurements that the data logger makes per second to record sound. In general, the sampling rate needs to be at least twice the highest frequency being recorded by the microphone (called the *Nyquist frequency* or *folding frequency*) to adequately capture each peak and trough of the signal. Thus, for a microphone that records frequencies as high as 20 kHz (the upper range of human hearing), a sampling rate of at least 40 kHz is needed. This is why most CD-quality audio is recorded at a 44.1 kHz sampling rate: the upper limit of frequencies recorded is 22 kHz. Hydrophones, on the other hand, often have sampling rates of 98 kHz or more because they typically record a very broad range of frequencies, from ultrasonic to infrasonic. To capture frequencies at the higher end of ultrasonic communication by bats, sampling rates of 196 kHz or even greater are used.

WAV files can be *compressed*, that is, converted to a smaller file format for long-term storage to reduce storage costs. Compression can be *lossless* (i.e., involving no loss of information) or *lossy* (with only a portion of the original information saved). One of the most common compressed formats used for WAV files is PCM (pulse code modulation). Another compressed format for sound files is FLAC, which can achieve a 40%–60% reduction in file size for most recordings; larger file size reductions are possible if sounds are repetitive. A common lossy file format is the MP3 format used for portable music devices. It is lossy in that only 10%–20% of the original WAV file format is saved; an MP3 file stores only frequencies that human hearing is sensitive to, and only with the detail that humans can perceive. Soundscape ecologists always use WAV or PCM file formats for analysis, use FLAC file formats for long-term storage, and rarely analyze lossy files (e.g., MP3 files).

Most PARs contain a programmable electronic interface that allows users to specify sampling rate, bit depth, file format, and recording schedule. Many allow users to set a *duty cycle*, a schedule of recording start times and recording lengths that produces partial temporal coverage of a soundscape. For example, 10 minutes on, 50 minutes off is one common duty cycle setting, which produces 16.7% (10/60) temporal coverage. A duty cycle providing less than 100% temporal coverage is used when there are limitations on data storage or time to service the units.

2.4. Psychoacoustics

Our second way of looking at sound—as a sensory experience—requires some terms and concepts different from those we've used in discussing sound as a physical construct. The study of how humans perceive sound, called *psychoacoustics*, is an area of acoustics spanning psychology, neuroscience, and audiology (hearing). A considerable amount of research has been conducted in psychoacoustics, with much of the early work done at Bell Laboratories. Several aspects of this field are important to soundscape ecology, including how humans perceive frequency at different amplitudes, how people perceive the same frequency and amplitude when they originate from different sound sources (e.g., different musical instruments), and how factors such as age affect human sensitivity to frequency and amplitude.

2.4.1. Pitch

Human perception of frequency is different from its pure physical manifestation. This perception of frequency is referred to as *pitch*. When presented with frequencies that differ from one another in exact proportions (e.g., when comparing a pure tone of 1 kHz with a pure tone of 2 kHz), humans often do not perceive these proportions exactly (e.g., as a doubling of frequency). Thus, pitch is the human perception of frequency, a physical attribute of sound. Sounds at intensities that humans can perceive are called *audible* sounds, while those humans cannot perceive are *inaudible*. Inaudible sounds can, however, be detected by instruments such as microphones, allowing soundscape ecologists to relate an inaudible sound to its source or to an important dynamic in the ecosystem (e.g., vibrations of sand in moving dunes). Inaudible sounds may be perceived by humans in other ways, such as through the "feeling" of pressure waves moving through the body or through the tactile sensation of vibrations in an object.

2.4.2. Timbre

The term "timbre" is one of the most difficult psychoacoustic terms to define. *Timbre* relates to the ability of humans to recognize distinctions between sounds when presented with two pure frequencies at the same amplitude generated by two different sources. The classic example of such a distinction is that between sounds from two different musical instruments. Why are the sounds from two woodwinds—for example, a clarinet and a saxophone—distinct to the human ear? It is hypothesized that the difference lies in the "shape" of the start of a sound (called its attack), or of its end (called its decay), or in its sound quality, or "color," in ways that we can recognize. Timbre thus represents one way to distinguish sound sources in the environment.

2.4.3. Loudness

Human sensitivity to sound frequency is affected by amplitude. When people were presented with sounds at different frequencies and asked to adjust their amplitudes until they were equally loud, the result was a plot of perceived loudness and frequency (Fig. 2.8). This research resulted in a measure of perceived loudness called a *phon* (rhymes with John), which has been further developed and set up as an international standard (American National Standards Institute [ANSI] 1.1). A phon is a subjective measure and is reported as the median value for a group of people with normal hearing abilities. Note in Figure 2.8 how perceived loudness changes nonlinearly across frequencies. For example, notice that if a sound at 1,000 Hz is emitted at 80 dB, it is most often perceived as 80 dB, but if a sound at 10 kHz is emitted at 80 dB, it is perceived to be around 93 dB. Because subjecting people to loudness tests that would cause either discomfort or hearing loss is not advisable, and because hearing at high frequencies varies among individuals, phon curves are estimated in high-frequency ranges (shown as dashed lines in Fig. 2.8).

2.4.4. Age and Gender Differences in Sound Sensitivity

Studies (Stelmachowicz et al. 1989) on human subjects of different ages and genders have shown that the frequency threshold at which a person can detect sound decreases with age, with the largest loss occurring between the ages of 40–49 and 50–59. Sensitivities decrease across all frequencies more rapidly with age for males than for females; the largest drop in sensitivities with increasing age occurs in the 13–17 kHz range. Much of this research is applicable in urban soundscapes.

Summary

Sound is complex, and thus quite a few technical terms are required to describe it. The most common way to describe its phenomena is to describe its properties as a wave, such as frequency and amplitude. There are many

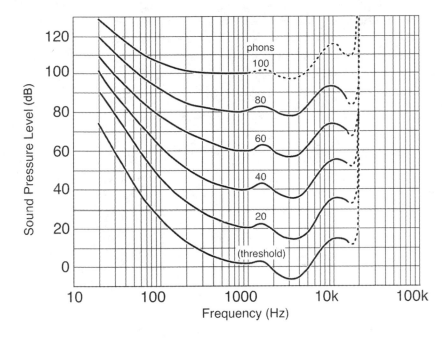

Perceived loudness contours (———)
Estimated (·····)

Figure 2.8 Human hearing response in phons (black lines)
as a function of frequency and amplitude.

ways to use frequency to describe how sound behaves in air and water and how it propagates through a solid. Amplitude is another key measure of sound, and knowing how a decibel is calculated is important to soundscape ecological research. Soundscape ecologists often measure sound using a microphone, and several forms of microphones are available for that purpose. How humans and other animals perceive sound is often different from the ways in which microphones measure it.

Discussion Questions

1. If the set of batteries attached to a passive acoustic recorder can provide it with power for 640 hours, and you would like to record sounds at a location for one year without having to return to service the unit, what duty cycle would you set to capture the most data without running out of power? Describe how you made this calculation.
2. If a 16-bit file recorded at a 48 kHz sampling rate on two channels requires

about 120 MB of storage for each 10 minutes of recording, how much storage would you need for the one-year study described above?

3. How much of an increase in resolution would be gained by shifting from an 8-bit to a 16-bit data format? From a 16-bit to a 32-bit format? Describe how you made your calculations.

Further Reading

Blake, Jared. *Field Recording for Beginners*. acousticnature.com, 2022. 136 pp.
Caldwell, Michael S. "Interactions between Airborne Sound and Substrate Vibration in Animal Communication." In *Studying Vibrational Communication*, edited by Reginald B. Cocroft, Matija Gogala, Peggy S. M. Hill, and Andreas Wessel, 65–92. Springer, 2014.

Sources of Sound in the Soundscape

OVERVIEW. This chapter provides an overview of sound sources across the biological, geophysical, and anthropogenic components of the soundscape. For each sound source, it provides a brief description of the underlying mechanisms of sound production and of the physical attributes of the sounds produced. Thus, the science of sound sources is examined across the three major sound source groups. Within the biological system, it assesses both sound production and sound perception in animals and plants. Geophysical sounds are then summarized, and their spatial-temporal-spectral variability is described. Anthropogenic sound sources are presented very generally; more details are available in Chapter 13, "Urban Soundscapes." It is abundantly clear that sound is an important component of ecosystems and that many organisms produce and/or respond to sound.

KEYWORDS: active communication space, auditory brainstem response, pulse train, stridulation, thunder, Wenz curve

---

3.1. Biological Sound Production and Perception

The kingdom Animalia is composed of nine major phyla and 1.5 to 3 million species. This section presents the current knowledge of the extent of sound production and sound perception by major animal groups.

Most animals, especially vertebrates, make sound, and it is likely that even more species of animals can sense sound waves in one or more ways. Generally speaking, our catalog of animal sounds matched to species is exceptionally poor, even for well-studied groups like the amphibians. Only for birds do we have a well-developed library of sounds. Online digital libraries of bird sounds include the Macaulay Library at Cornell University (Ranft 2004; Betancourt and McLinn 2012) and a community-based website called xeno-canto (Goeau et al. 2018).

Animal sounds can be purposeful (e.g., a bird calls to attract a mate) or produced incidentally by a behavior (e.g., feeding, locomotion). In some cases, sounds are not detectable if only one individual's behavior (e.g., feeding) makes a sound, but if hundreds to millions of individuals produce the same sound, then the accumulation of those sounds becomes detectable (e.g., sea urchins feeding on the ocean floor).

3.1.1. Animal Communication Terms

In the field of animal communication, sounds are part of a large array of animal *signals* (Bradbury and Vehrencamp 1998). The area of habitat in which the signal created by a signaler can be perceived by the intended receiver is called the *active communication space*. Nonacoustic animal signals include those that are visual, tactile (rubbing against one another, sensing temperature), and olfactory (e.g., pheromones), as well as those produced through chemoreceptive (e.g., taste) or electroreceptive (e.g., in eels) means. Animal sound production has a variety of purposes: to attract a mate, warn conspecifics, communicate status (e.g., dominance), signal behavioral intention (e.g., move away), teach skills (e.g., 1980; Janik and Slater 1997; Poole et al. 2005), establish and mark territories, identify specific individuals in a group, express individual status (e.g., hunger, thirst), detect prey (e.g., bat or dolphin echolocation), maintain flocking/swarming behavior in groups, or strengthen bonds between mating pairs. Movement in response to sound in air is called *phonotaxis*, and movement in response to sound traveling through a solid is called *vibrotaxis*. In some disciplines, such as the study of fish acoustic communication, the term *agnostic sound* is used to refer to sounds that have no known purpose.

The animal communication pathway (Fig. 3.1) is composed of a *sender*, an intended *receiver*, and the sound *propagation space*; in some cases, an unintended receiver may be an *eavesdropper*. Sounds can be generated by a sender by means of a sound production organ, and can be modified before they are transmitted. A form of communication modification occurs when sound passes through a structure that is used for another purpose (e.g., a bird's beak, which is used for feeding). In some animals, the hearing system receives all sounds at the surface of the ear, but the neurological system may process only certain signals and "filter" out others. For example, some birds process only the sounds of their own species, so that in locations where

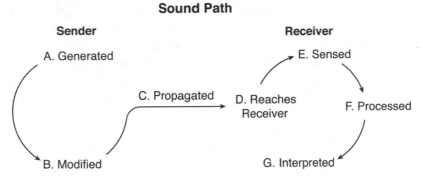

Figure 3.1 Acoustic communication pathway in vertebrates.

there are many species of birds singing (Henry and Lucas 2010), they can hear their conspecifics.

Sounds are also used by organisms as *orientation cues* (Slabbekoorn and Bouton 2008; Stanley et al. 2010; Vermeij et al. 2010). For example, coral reefs are known "hot spots" of natural resources, so many marine animals orient their movement toward reef systems. The sounds of a river can be a cue for a type of landscape preferred by an organism, where food or nesting resources might be plentiful. A special form of sound production, in which acoustic reflection is used to navigate spaces or to find objects (e.g., flying prey), is called *echolocation*. It is used by a wide range of animals, including most bats, many marine mammals, some birds (e.g., cave swiftlets), and some rodents (Sales and Pye 1974).

The term "hearing" is often used to describe how humans and vertebrates receive and process sound using anatomical structures like the ear (a major focus of Chapter 6 is perception of sound by humans). However, using the term "hearing" is restrictive if we want to consider how all organisms gather and use sound. Thus, soundscape ecologists adopt the broader term "perception" to mean "the reception of vibratory stimuli of any kind and nature, provided that the sound source is not in contact with the organism's body" (Dethier 1971; see also Pumphrey 1950). There are also many animals, mostly fish and insects, whose sensory systems perceive sound only within the near field. For example, among insects, nearly 18% of all species use some form of near-field acoustic communication, either through the air or on surfaces such as plant parts.

Research on sound perception in animals has developed considerably since the turn of the twenty-first century. Many advances have been made using an approach called *auditory brainstem response* (ABR), which is an electrophysiological far-field recording method (see Kenyon et al. 1998) that involves attaching electrodes to the body surface of an individual, sending clicks or pure tone bursts across a range of frequencies, and measuring *acoustically evoked potentials*—neurochemical responses to sound—in the individual that indicate the stimulus was received. The ABR technique has been applied to fish, crustaceans, amphibians, reptiles, and birds (Hu et al. 2009). It has also been used to confirm that some animals can detect ultrasonic sound.

3.1.2. Acoustic Communication Repertoire

Sounds produced by more than one individual are referred to with special terms. An exchange of signals between two individuals is called a *duet*. Many animals duet. Male and female mosquitoes are known to duet, as are members of gibbon families. When many individuals make sound over a short time period, this is referred to as a *chorus*. Chorusing can be *synchronous* (individuals call at the same time) or *asynchronous* (timing of calls is apparently random). Choruses can be made up of single species (e.g., frogs chorusing) or multiple species (e.g., the *dawn chorus* in terrestrial systems, the *dusk chorus* typical of marine systems).

Bioacousticians have developed a rather simple and useful nomenclature for the various components of a complex animal acoustic signal, although it is not fully standard across all animal taxa. The complex vocalizations produced by some animals, referred to as *songs*, differ from the simple vocalizations of others, most often referred to as *calls*. A string of different kinds of calls may make up a song, in which case the calls are sometimes called *syllables*. The collections of pulses that form the acoustic communication repertoires of pulsating animals, such as anurans (frogs and toads) and insects, are called *pulse trains*. Pulse trains (Fig. 3.2A) are measured by the number of pulses per train, the length of each pulse, the interpulse distance (usually in seconds or milliseconds, as indicated with a bracket and the letter B), and/or the length of the pulse train (also in seconds). Each pulse can be produced with a signal that is upswept or downswept in its spectral shape (Fig. 3.2B), and these shapes can sometimes be used to identify species. Pulse trains can also be characterized by a spectrogram (Fig. 3.2C) and/or a power spectrum (Fig. 3.2D).

3.1.3. Onomatopoeia

Onomatopoeia, the use of words that imitate the sound they describe, is common in bioacoustics. For example, many groups of animals are named after the sounds they make. Katydids are insects that produce calls in the early evening that sound like "Katy-did, Katy didn't, Katy-did, Katy-didn't."

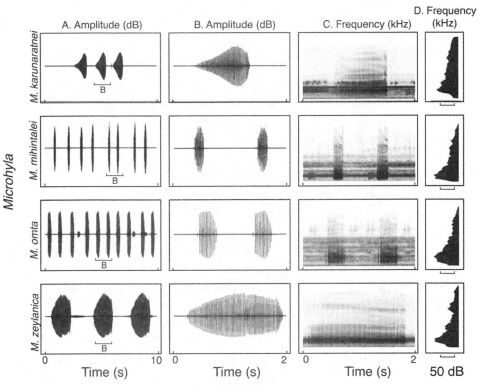

Figure 3.2 Pulse trains (A), pulse shapes (B), spectrograms (C), and power spectra (D) of four species of tree frogs from the genus *Microhyla*. The *y*-axis units are dB (A and B) and kHz (C and D). (Modified from Wijayathilaka and Meegaskumbura 2016.)

A group of fish referred to as "drummers" produce drumming sounds, a group of birds called "catbirds" make sounds like a cat's "meow," and a species of gecko in Malaysia whose common name is the "barking gecko" has a call that sounds like the bark of a dog. Sounds can be also described by the way humans perceive them. Words for sounds like "chirp," "buzz," "hiss," and "cluck" emphasize the way these sounds are modulated by animals. Finally, the complex calls of some animals have been cleverly described using phrases they resemble. The white-throated sparrow of North America says "Old Sam Peabody-Peabody," and the ovenbird, a temperate-forest thrush, says "Teacher, Teacher, Teacher."

3.1.4. Acoustic Communication by Taxonomic Group

What is known about how animals produce sound, and how they receive or perceive sound, is summarized in the remainder of this section.

Non-Arthropod Invertebrates

Simple-Bodied Animal Groups

We know that the simplest animals, such as the sponges, nematodes, cnidarians, and flatworms, lack—or have very simple—organ systems and have no known sound-producing organs. Some studies (e.g., Lynam 2006) report that a group of hundreds to thousands of jellyfish (cnidarians) produces detectable sounds in ocean recordings through their synchronous feeding behavior. Other than these ancillary reports, nothing is known about sound production or perception in this group of animals.

Mollusks

Phylum Mollusca is a very large group of animals (only the arthropods have more species) that are present in freshwater (~20% of mollusk species), marine (~70%), and terrestrial systems (~10%). Common mollusk groups include snails/limpets (gastropods), clams/oysters/mussels/scallops (bivalves), squid/octopus/cuttlefish (cephalopods), and one living group that is wormlike. Many species possess a radula (a rasping tongue), which, when many individuals are feeding, can produce "crunching" sounds that can be detected in hydrophone recordings (Kitting 1979); limpets, snails, chitons, and barnacles are commonly detected in this way. Research using the ABR technique (Hu et al. 2009) has demonstrated that the oval squid and common octopus can detect sounds in the 400–1,200 Hz range. Manipulations of acoustic soundscapes have shown that free-swimming eastern oyster larvae swim toward sounds associated with adult habitats such as coral reefs (Lillis et al. 2015).

Echinoderms

Phylum Echinodermata is composed of the sea stars, brittle stars, sea cucumbers, sea urchins, sand dollars, and a small group called the crinoids. Most live on the ocean floor, and they are found from the nearshore zone to very deep parts of the ocean. Feeding by sea urchins is known to produce sounds, which can be among the most common sounds in many nearshore zones, such as kelp forests. Feeding sea stars have also been observed to make sounds (Kitting 1979).

Arthropods

Phylum Arthropoda, with nearly a million species, is the largest group of animals, comprising the insects, arachnids (spiders, mites), and crustaceans (which are mostly marine). The most common form of sound production in arthropods is mechanical vibration (Cocroft and Rodriguez 2005). Communication by waterborne sound is also common among the arthropods (Greenfield 2002).

Forms of Sound Production

Arthropods possess an exoskeleton, which can be used in a variety of ways to produce sounds. Ewing (1989) describes five different forms of sound production in arthropods: stridulation, percussion, tymbal pulsation, vibration, and air expulsion. *Stridulation* is the act of rubbing two body parts together to produce a sound. There are many different forms of stridulation. Most typically, one body part, called the file or comb, which has teeth or ridges, is rubbed against another body part, called the scraper, which has a single peg or tooth. *Percussion* is sound production by clapping two body parts together or by striking a substrate with one body part. Flapping of forewings and snapping of mandibles are some of the ways arthropods create percussion-based sounds. Some of the loudest sounds from arthropods are produced by *tymbals*, exoskeletal membranes located on the abdomen in several groups, most famously cicadas. A pulsing sound is created as muscles compress the ridged structure of the tymbals. *Vibration*, which is different from stridulation, involves the fast movement of wings, abdominal parts, or legs to produce sounds. Finally, *air expulsion*, a form of acoustic communication similar to that of birds and mammals, which pass air through a pipelike organ, is found in several dozen species of arthropods. This broadband form of communication occurs in short-horned grasshoppers and cockroaches. Hawk moths move expelled air over a vibrating membrane, creating ultrasonic sounds as high as 90 kHz (Zagorinsky et al. 2012).

Insects

Iconic sound-producing insects include cicadas, katydids, crickets, grasshoppers, beetles, bees, wasps, flies, moths, and mosquitoes. Many sounds produced by insects are audible, but many are not, as they are emitted as ultrasonic frequencies or manifested as tiny soundscapes that would require special recording equipment (see Chapter 2) for humans to perceive, but are sensed by conspecifics. Most insects detect sound through a tympanic membrane located in the abdomen, through antennal hairs, or through mechanoreceptors in the forelegs.

- *Cicadas.* There are over 3,000 species of cicadas in the world. Their pulsating calls, produced by a tymbal, can be as loud as 106 dB (Rosales 1990). Most calls have frequencies in the range of 4–7 kHz. Several species of cicadas are periodic, emerging from an underground larval stage once every thirteen or seventeen years; others emerge each year. In places like the midwestern United States, the annual (a.k.a. "dog day") cicada generally emerges between late July and early August, depending on how hot it has been since the last freeze. It is hypothesized that a major function of cicada calls is to create spacing between individuals. Cicadas are common in tropical and temperate forests and are often found in residential areas of these ecosystems.
- *Katydids.* Katydids, or bush crickets, produce sound through stridulation. Males—and, in some species, females—stridulate by rubbing their two forewings together, one of which has a file and the other a scraper. The movement of the forewings is slow, with a stroke rate governed by air temperature, but given the high density of teeth on the file, the resulting frequencies are much greater than the stroke or pulse rate of the forewings. Some katydids produce sounds in the ultrasonic range (Montealegre-Z et al. 2006), although most call in the 4 kHz range. Resultant sounds are often harmonic in structure. The stroke rate is also governed by air temperature. Ter Hofstede et al. (2010) report that some Neotropical katydids can sense ultrasonic sounds and are therefore able to avoid predation by bats. Some katydids also produce bimodal sounds by vibrating vegetation structures. In some katydid species, males sing in a synchronous chorus.

  There are over 6,400 katydid species, of which over one-third exist in the Amazon forest (Montealegre-Z et al. 2006). Katydids are found in forested areas, residential areas (where they sing from the tops of trees), and grasslands. They sing in late summer in mid-latitude locations, commonly during the evening crepuscular period.
- *Crickets.* Crickets are mostly nocturnal insects that stridulate by rubbing their two forewings together; each forewing, or *tegmen*, is a leathery structure with a comb and a scraper. Several species of ground-dwelling crickets stridulate from the tops of their subterranean burrows (Forrest 1982).

The chirp rate of a cricket is a function of air temperature (this relationship is referred to as *Dolbear's law*). The frequencies of common crickets in North America are in the 3.8–4.6 kHz range; smaller ground (flightless) crickets produce sounds in the 7.7–8.2 kHz range. There are over 2,500 species of crickets (Alexander 1968). Crickets are found in nearly all ecosystems from 55° N to 55° S latitude.

- *Grasshoppers and locusts.* Most grasshoppers and locusts (Otte 1970) use one of two forms of stridulation: *ordinary* stridulation, which is a slow, high-amplitude sound production mechanism, or *vibratory* stridulation, which is a fast, low-amplitude mechanism. Both mechanisms involve rubbing the femur against the forewing. Some grasshoppers use ticking, a form of percussion created by the forelegs striking against the forewings. Sounds can also be produced by femur shaking, which involves striking a substrate. Most grasshopper sounds are not loud, but one species, the bladder grasshopper, is reported to produce sounds as loud as 98 dB re µPa at 1 m, at frequencies of 1.4–3.2 kHz, that can be detected 2 km away.

- *Beetles.* The largest group of insects is the beetles; there are over 400,000 described species. Few studies have focused on sound production in beetles, although several reports have centered on the well-known sound production mechanisms of bark beetles. Stridulation is common in both male and female bark beetles (Rudinsky and Michael 1972). Fleming et al. (2013) reported that one bark beetle species, the mountain pine beetle, produces sound in the 15–26 kHz range with low amplitudes (55 and 47 dB at 2 and 4 cm, respectively). Bedoya et al. (2021) examined over sixty species of beetles, mostly bark beetles along with several of the closely related ambrosia beetles, and found that over 33% produced sound, mostly by stridulation. These beetles communicate through plant tissue as they bore into trees and other woody plants. Yager and Spangler (1995) found that one species of tiger beetle responded to ultrasonic sounds (30–60 kHz), which they argued was a means to avoid bat predators.

- *Bees and wasps.* Bees and wasps produce sounds in a variety of ways; their well-known buzzing sounds, for example, are produced by beating the wings at set frequencies (e.g., 309 Hz for honey bees). They also produce sounds during the act of ventilating the hive (Ishay and Sadeh 1982), during dances, to express hunger (Ishay 1976), and as vibrations sent through burrows for orientation (Larsen et al. 1986). The cicada killer, a solitary wasp found in North America that hunts and kills cicadas, can produce sounds at high amplitudes (~70 dB re 20 µPa at 3 cm) at fundamental frequencies of 209 Hz for males and 152 Hz for females.

- *Ants.* Formally grouped with bees and wasps to form the order Hymenoptera, many of the ~12,000 species of ants stridulate by rubbing a leg against a washboard-like structure on the abdomen. Although emitting chemicals is the primary means of communication in ants, sounds (produced at frequencies of 700 Hz–1.4 kHz) are used by individuals to communicate status and by pupae to communicate with nearby worker ants (Casacci et al. 2013). These sounds are faint (~40 dB re 20 µPa at 100 mm), and it has been demonstrated that ambient sounds produced more than 1 m away do not alter ant behavior (Hickling and Brown 2000). It has also been shown that some species of leaf cutter ants in Asia use sounds to discriminate nestmates from non-nestmates (Fouks et al. 2011).
- *Moths and butterflies.* These often showy winged insects (~180,000 species) are found in nearly all terrestrial habitats on Earth. Many species generate sounds in the frequency range audible to humans by stridulation (Blest et al. 1963). There have been several reports of moths being able to detect the ultrasonic signals of bats (ter Hofstede and Ratcliffe 2016) and of some moth species using ultrasonic communication in short-range courtship (Nakano et al. 2015).
- *Flies and mosquitoes.* Perhaps one of the most elegant uses of sound during mating (see Cator et al. 2009) occurs in several species in the mosquito genus *Aedes.* Like other mosquitoes, they sense sound through their antennae (Göpfert and Robert 2001a, 2001b). During courtship, the wingbeats of male and female *Aedes* mosquitoes start at different frequencies (400 Hz for females, 600 Hz for males), and each rate increases until they duet at a shared harmonic frequency of 1.2 kHz. Fruit flies have also been shown to use acoustic communication in courtship (Tauber and Eberl 2003).

    Gasc et al. (2018), in their study of the soundscapes of the Sonoran Desert, found that sounds of flies were frequent during certain times of day and certain seasons. In what is probably a common use of sound by flies, several species use the acoustic signals of their prey or hosts (Lakes-Harlan and Lehman 2015; Bernal et al. 2006).

Spiders

Most spiders use vibrations of their substrate (a web, a grass blade, etc.) to attract mates or to detect prey. These vibrations are pure tones, often at low frequencies (typically 4–10 Hz) (Barth 1985, cited in Witt et al. 2014). Scales of vibration transmission are estimated to be 30 cm–2 m (Cocroft and Rodriguez 2005).

Crustaceans

A mostly aquatic group of arthropods, crustaceans include the crabs, crayfish, krill, lobsters, prawns, and shrimp. One group of crustaceans produces what is considered both the loudest and the most common biological sound on Earth: snapping shrimp (>1,200 species). Snapping shrimp, also called pistol shrimp, are a group of marine and freshwater crustaceans that use sound bursts created by a snapping claw to kill prey. The snaps create a high-pressure wave that forms what are called cavitating bubbles; the bursts of sound are produced by the collapse of the bubbles (Versluis et al. 2000), not by the claw itself. The sound bursts have a peak-to-peak amplitude of 189 dB re 1 µPa at 1 m (Au 1997), with a peak of lower frequencies between 2 and 5 kHz and peaks as high as 200 kHz.

Many species of spiny lobsters also make sounds (Moulton 1957; Kanciruk 1980). They use a special form of stridulation, performed by rubbing a bowstring-like structure on the plectrum, a structure at the base of the antennae, against a file located on a ridge underneath the antennae. This "stick and slip" mechanism (Patek 2001), which produces "rasps" (sounds in the 19–28 kHz range), is used when lobsters are molting and their exoskeletons are soft.

Many species of crabs (~800), some of which are terrestrial or semiterrestrial, produce sounds through stridulation, by thumping the ground, or by drumming body parts. Although it is not known whether krill produce sound, the high densities at which they commonly exist in marine environments are known to reduce sound propagation (Chu and Wiebe 2005).

Chordates

There are six major groups of chordates: the tunicates (i.e., sea squirts, marine animals with a notochord) and five vertebrate groups—fish, amphibians, reptiles, birds, and mammals. Sound production by vertebrates is common, is varied in form, and is a major component of natural soundscapes. Here, a summary of vertebrate sound production and known modes of sensation and/or perception is organized by vertebrate order, with pertinent comments on amplitude, frequencies, timing, and evolutionary "sequencing" that soundscape ecologists should consider.

Fish

The forms of sounds produced by fish are simple, as most fish lack the ability to create complex, modulated sounds (Amorim 2006). Fish sense sound with the lateral line and ears. Freshwater and marine fish can produce sound in three ways (Tavolga 1971): (1) by stridulation, accomplished through moving body parts: teeth, fins, or bones; (2) by using the swim bladder to create pulses of sound by pulsating surrounding muscles; and (3) by movement of body parts during swimming to produce hydrodynamic sounds.

Fish have the highest species richness of the vertebrate orders, with over 34,000 species. Most species of fish emit only one or two sound types. Several groups of sound-producing fish are worthy of mention. One group that possesses a rather large repertoire of sound types is the elephant fish, which emits rapid clicks in the 240 Hz–3 kHz range at the same time as high-pitched pure tones. Toadfish create courtship hums and grunts, some with harmonic properties. The benthic marine triglids (i.e., sea robins) are known to produce drumming sounds with the swim bladder, especially during breeding seasons. Other triglid species can produce clucks, croaks, barks, grunts, and growls ranging from 40 Hz to as high as 2.4 kHz, but most commonly around 400 Hz. The cichlids of the African Great Lakes produce numerous kinds of sounds, including grunts, purrs, growls, and thumps, most in the 100–700 Hz range but some as high as 8 kHz. The damselfish, commonly found in coral reefs, produce simple pops less than 1 second in length, emitted singly or in sequences of up to two dozen in a row. The gouramis emit croaks at a high frequency, often in pairs, in batches of four to eight. Catfish commonly produce drumming sounds. Other fish groups known to produce sounds include the sunfish, squirrelfish, gobies, and tigerfish.

Amphibians

Amphibians occupy nearly every natural terrestrial soundscape. This order is composed of the anurans (frogs and toads) and several non-anuran groups (e.g., salamanders, newts). There are over 5,400 species of anurans. Sounds produced by anurans are typically classified into functional categories, such as mate attraction, courtship, territoriality, aggression, and distress calls (Feng et al. 1990). Anurans can detect sound through tympanic membranes on their heads or through their lungs or mouths; in most species, sound is made only by males. Calls are often measured by call length, pulse rate, dominant frequency, call rate, pulse length, pulse amplitude, and if harmonics are present, some measure of the difference between the highest and the fundamental frequency. One of the most widespread anurans in North America is the spring peeper. This tree frog's choruses of calls are very prominent in wetlands and moist North American temperate forests in early spring. The calls, which sound like "peep, peep, peep," are produced at 2.9 kHz (Wilczynski et al. 1984).

In an analysis of ten species in the genus *Bufo* (toads) and fourteen species of hylids (tree frogs), Cocroft and Ryan (1995) found that the dominant frequency of calls ranged from 1.3 to 2.6 kHz in *Bufo* and 2.1 to 6.1 kHz in hylids. Call lengths of *Bufo* species ranged from short bursts of 690 milliseconds to long calls of 2.3 seconds. The same study also showed that average call rates were 2–3 calls per minute for *Bufo* and 1 per minute for hylids. Ultrasonic communication occurs in several species of tree frogs (Feng et al. 2006; Arch et al. 2008, 2009; Shen et al. 2011; Boonman and Kurniati 2011). Communicating in the ultrasonic range is hypothesized to be an adaptation to living near fast-flowing streams, which can produce sounds at very high frequencies.

The number of species of anurans in tropical and temperate landscapes varies, but biodiversity hot spots may be rich in anuran species: thirty-nine anuran species have been reported at a single research station in Borneo (Sah and Grafe 2019) and forty-eight species at the La Selva Biological Station in Costa Rica (Organization for Tropical Studies, n.d.).

Reptiles

Although reptiles are typically considered a relatively silent group of vertebrates, there are several interesting examples of sound production in living reptiles. The geckos are a species-rich group of lizards (~1,500 species), nearly all of which make sounds. Some produce clicking sounds; others make croaking sounds, like the tokay gecko (Brillet and Paillette 1991), which produces a long (~22 seconds) bursting call emitted at 70 dB re 1 µPa at 1 m across a 300–4,000 Hz pulse. Many geckos produce a "barking" call. One species of gecko produces sound in the ultrasonic range (Brown 1984).

Several species of turtles (~47; see Ferrara et al. [2014] for a review) are known to make sounds, either "chirps" or "clucks." A large land tortoise from South America produces chick-like clucks in a series starting at 2.5 kHz and dropping to 500 Hz in 0.12 seconds. Another large land tortoise was recorded making clucks at 1.7–2.0 kHz, producing a chorus at night within a large population. The long-necked freshwater turtle of Western Australia (Giles et al. 2009) was recorded making several sounds underwater with a dominant frequency of 1 kHz (range 100 Hz–3.5 kHz).

Crocodiles and alligators make hissing or growling sounds when threatened. A variety of snakes also produce hissing sounds, although only when threatened, not most or all of the time as is commonly imagined.

Birds

There are over 9,400 species of birds. They are found on every continent and in every terrestrial ecosystem, and they make some of the most iconic sounds in any soundscape. Many groups of birds are typically associated with specific ecosystems. For example, shorebirds are common in coastal environments; warblers, vireos, and other perching birds (called *passerines*) in temperate forests; sparrows and larks in grasslands; and penguins, cormorants, and petrels in cold subantarctic coastal climates. Most birds are *diurnal*—that is, they are active during the day—but some groups are active mostly at night, such as owls, or during the evening crepuscular period, such as tinamous and nightjars (e.g., nighthawks). Thanks to their impressive migratory abilities, the sounds of birds in landscapes are typically seasonal. In the northern temperate forests, for example, from late April through early June, the air is filled with the calls of birds that have arrived to breed and others that are passing through to breed in more northern landscapes. Birds are common to nearly all dawn choruses typical of terrestrial landscapes.

Birds produce sound using a specialized structure called the *syrinx*, located at the intersection of the trachea and bronchi. It is well developed in passerines, the group of birds that is known to have the most complex songs (and that is sometimes incorrectly equated with the songbirds). The structure of the syrinx varies across bird species, and in several species it is so highly specialized that individuals can produce two distinct calls at the same time. Bird sounds can also be produced by pecking (e.g., woodpeckers) or drumming (e.g., prairie chickens), or by using feathers during flight, as in the common snipe, which creates a loud "boom" during a deep dive.

For most birds, the greatest sensitivity to sounds occurs in the 1–5 kHz range (Dooling 1992). No birds are known to be able to detect ultrasonic sounds, although a few groups (e.g., pigeons) can detect infrasonic sounds (Beason 2004). There are a few reported cases (Pytte et al. 2004; Olson et al. 2018) of hummingbirds producing sounds in the ultrasonic range (up to 30 kHz), but further investigation found that the birds were not able to detect sounds in that range; thus they are unlikely to be using ultrasonic sounds in communication. It has been suggested (by Marler; see Dooling and Prior 2017) that birds' hearing is much like that of humans, although birds may hear temporal details of calls in the 1–4 kHz range better.

Several species of birds have served as model organisms in the field of bioacoustics. The finches, for example, have been studied extensively to understand how the avian brain processes acoustic information (Catchpole and Slater 2003). Woolley and Casseday (2004) have shown that the brain of the zebra finch (a budgerigar), is able to "tune" to spectral-temporal modulations of song, allowing the birds to filter out certain frequencies that are present in the ambient environment. The chickadee of North America has been studied to understand the structure and patterns of calls that are produced in a sequence. The birds' repertoires are often broken into specific parts and studied individually—each call is referred to as a "syllable." Work with chickadees, among a few other well-studied birds, has shown that individuals' songs can be grouped into geographic "dialects." Several species of birds have hearing so advanced that they can use calls to identify individuals in a group (Woolley al. 2005).

Mammals

Mammals are perhaps the most widely distributed animals on Earth. They live in nearly all terrestrial locations as well as in the oceans. The class Mammalia is very diverse, containing twenty-seven extant orders. Mammals probably produce the widest frequency range of sounds of any animal group (from infrasonic in elephants to ultrasonic in bats), and they use extremely specialized forms of communication that can be interpreted as language. They use two broad categories of sound production mechanisms. In *active muscular contraction* (AMC), sound is produced by contracting throat muscles to vibrate vocal cords; cats purr via AMC. In *myoelastic-aerodynamic* (MEAD) sound production, air from the lungs is used to vibrate the vocal cords; humans use MEAD to talk and sing. A summary of mammalian sound production, in taxonomic order, is provided here, with comments on special forms of acoustic communication by each group.

- *Ungulates* (e.g., swine, deer, moose, giraffes, cattle, camels). Mammals in this group are known to produce grunts, snorts, and bleats. Many species have been domesticated as livestock. Several species are known for living in large groups, or herds. A study by Kiley (1972) on several species of even-toed ungulates found that domesticated pigs made up to fourteen different kinds of sounds, most in the 2 kHz range. Horses have been reported to have five different vocalizations, including a "nicker" (at 100–150 Hz) and a "whinny" (at 400–2,000 Hz). Rhinos "grunt" or "bark." Hippos standing in water (with eyes and nostrils in the air), have been observed (Barklow 2004) to produce sounds in both air (at 214 Hz) and water (at 2,600 Hz). Peak-to-peak amplitude of these sounds was high, at 125 dB re 1 µPa at 1 m.

- *Terrestrial carnivores* (e.g., lions, hyenas, bears, civets, wolves). Terrestrial carnivores are iconic sound producers. The most well studied are the wolf and the coyote.

  The call of the wolf is often described as a "howl"; an assessment of over thirty-five howls showed that most howls were produced at 274–908 Hz, with the dominant frequency at the upper end of this range. Howls are relatively long (1–13 s). It is hypothesized (Harrington and Asa 2003) that these low frequencies are used because they carry long distances. Palacios et al. (2007) demonstrated that howls were unique enough to be statistically assigned to individual wolves.

  Coyotes are well known (Lehner 1978) for their group "yip-howl" calls produced at high amplitudes across a broad frequency range of 300–3,000 Hz and interspersed with other howl types, barks, and growls. These calls are typically 45–60 seconds long. Coyote calls are common at dusk, are seasonal, and can often be heard in suburban areas in North America (Laundré 1981).

  Hyenas, of which there are four species, are highly social carnivores (Holekamp et al. 2007) with a rich vocal repertoire, which includes a loud "whoop" that has been detected as far as 5 km away from the producer (East and Hofer 1991).

  Lion roars, produced by both males and females, are possibly the most impressive terrestrial animal sounds. Roars are produced at a fundamental frequency of 180–200 Hz with additional sounds extending to 4 kHz (Anatharksishan et al. 2011; Eklund et al. 2011); they can be heard most frequently at night, sometimes from as far as 30 km away.

- *Bats*. With over 1,300 species, the bats make up the second-largest order of mammals. Bats can be divided into two groups, one principally frugivorous and the other insectivorous. The insectivorous bats have developed highly specialized forms of echolocation (Sales and Pye 1974) using sounds most often emitted in the ultrasonic range. These bats generally use three forms of ultrasonic signals, all produced by the larynx: (1) short clicks, (2) frequency-sweep pulses, and (3) constant-frequency pulses mixed with the first two forms. The highest frequency known to occur in bats is produced by the trident bat, at 212 kHz, and the lowest by the big brown bat, at 11 kHz (i.e., within the range of human hearing). Most bats echolocate with fundamental frequencies between 20 and 60 kHz (Fenton et al. 2016).

Echolocation pulse rates increase as bats approach their prey. Bats use echolocation to build a complex image of the three-dimensional space they navigate (Sales and Pye 1974), to detect specific prey items, and to identify conspecifics and other bat species.

Both groups of bats—the frugivores and the insectivores—use acoustic communication in social contexts (Chaverri et al. 2018). Social calls tend to be emitted at lower frequencies than echolocation calls, often within the range audible to humans, because low-frequency waves travel farther. Because bats are highly gregarious animals, they use social acoustic communication to assess one another's species, sex, identity, group membership, social status, and body condition (Chaverri et al. 2018). Social acoustic communication serves several functions, including attracting a mate, displaying aggression, establishing territories, and creating group cohesion during roosting and foraging (Gilliam and Fenton 2016).

In general, the known catalog of bat echolocation calls in North America and Europe is quite broad. Several commercial firms produce instruments that can identify species' echolocation calls in the field.

- *Primates* (e.g., monkeys, chimpanzees, bush babies, gibbons, humans). Primates are extremely vocal, with very complex acoustic communication patterns (Petter and Charles-Dominiqe 1979). There are somewhere between 150 and 425 extant species. Most nonhuman primates live in tropical areas, are highly social, and are arboreal. Acoustic communication plays the same roles for primates as for other mammals and birds, namely, to attract mates, to defend territories, to establish social hierarchies, to warn of threats, and to increase social cohesion (Marler and Mitani 1988). Hearing abilities are so well developed in primates that many species use vocal communication to identify individuals.

  Several species of primates can detect sounds in the ultrasonic range (<45 kHz; species from five genera; see Ramsier et al. 2012), but primates typically communicate with dominant frequencies in the range audible to humans (<20 kHz). Recent work on tarsiers using both ultrasonic recording instruments and ABR (Ramsier et al. 2012) found that one species is capable of hearing sounds as high as 91 kHz and can vocalize in the 65 kHz range. The group of primates known as the galagos, or bush babies, is also reported to hear and produce sounds in the ultrasonic range. A detailed analysis of chimpanzee calls (e.g., pant-hoots) by Marler and Hobbett (1975) showed that most individuals vocalize with fundamental frequencies between 200 and 500 Hz and formant frequencies as high as 1.7 kHz. Durations of calls commonly range between 6 and 10 seconds.

Howler monkeys (~15 species) might produce the loudest sounds of any primate. These sounds are created at low fundamental frequencies (300–1,000 Hz), with an amplitude of 70 dB re 20 µPa at 50 m, and can be detected as far as 5 km from the producer. Howler monkeys commonly produce these calls for up to an hour.

Several species of gibbons, a group of primates found in Southeast Asia, produce some of the most complex vocalizations known, which at times have been compared with human music (Tenaza 1976; Koda et al. 2012; Geissmann 2000; Haimoff et al. 1985; Marler and Mitani 1989). Their songs are often heard after sunrise, typically as duets occurring between opposite banks of a river.

- *Elephants.* Elephants are the largest terrestrial animals. There are either two or three species, depending on how they are classified. One species lives in Southeast Asia, while the other one or two live in Africa. Elephants can make a variety of sounds, but they are well known for producing some of the lowest-frequency sounds among all terrestrial animals. An analysis of the sounds produced directly from the larynx of an African savanna elephant reported a range of fundamental frequencies from 5 to 60 Hz and amplitudes as high as 117 dB re 20 µPa at 1 m (Langbauer 2000). These sounds can be detected by another individual as far as 4 km away (Langbauer et al. 1991).

- *Rodents* (e.g., rats, mice, squirrels, beavers, porcupines, guinea pigs). The rodents constitute the largest group of mammals. It has been well known for decades that many rodents, particularly mice and rats, communicate in the ultrasonic range and use signals in this range for echolocation (Rosenzweig et al. 1955). Many ultrasonic calls are also made during copulation, in communication between newborns and adults, and during aggressive behavior between adults (Sales and Pye 1974). Many of these calls have frequencies between 40 and 70 kHz; it has been shown that newborn rodents can produce sounds as high as 120 kHz.

Many rodents also produce sounds in the range audible to humans. A study (Balph and Balph 1966) on the calls of red squirrels over a season found that they produced calls between 2.6 and 6.7 kHz, with the width of most of the calls at 1 kHz. The tree squirrels produce chirps with frequencies between 1 and 2 kHz, and emit other sounds at low amplitudes between 4 and 8 kHz (Smith 1978).

- *Marine mammals.* Marine mammals belong to several different orders: cetaceans (e.g., whales, dolphins, porpoises), carnivores (e.g., seals, sea lions, walruses), and manatees. Sound production in these animals is very advanced, as they have specialized hearing and sound production capabilities. As a group, they use communication signals in a range from ultrasonic to infrasonic, and some use echolocation for navigation. Whales can produce sounds in the 10–150 kHz range, with a sound pressure level of 120–90 dB re 1 μPa at 1 m (Kuperman 2007). There is evidence that blue whales can produce sounds as low as 9 Hz; most of their calls are at 15–20 Hz, with each repetition lasting one minute (Mellinger and Clark 2003).

- *Other mammals.* Many other mammal groups use acoustic communication, but most of them lack complex acoustic communication abilities and produce simple acoustic signals of distress or warning, or simple barks. Kangaroos and wallabies, for example, use their tails to produce a thump that communicates information to conspecifics (Blumstein et al. 2000).

Intergroup Comparisons of Animal Acoustic Communication

Allometry is the study of how any structure or function of an organism changes with body size. Several important allometric relationships related to sound production have been documented across animal groups. For mammals (Fig. 3.3), it has been shown that there is a strong negative relationship between the dominant frequency produced and body length for primates and carnivores. Examined as a function of vocal production type (MEAD vs. AMC) across mammals (Fig. 3.4), the fundamental frequency scales more strongly with body mass for MEAD vocal production than for AMC.

Among mammals, the sensitivity of hearing, which can be plotted as an *auditory response curve* or *audiogram*, is highly variable across frequencies and amplitudes. Figure 3.5 (following Kanders et al. 2017) plots the frequency-amplitude sensitivities of several mammals, showing that humans (E) are more sensitive to low sound pressure levels (i.e., low dB) than the prairie dog (A), elephant (B), lemur (C), and domestic cat (D), but not the white-beaked dolphin (F), and false killer whale (G). Note that the human ear is most sensitive to sounds at 4 kHz.

The frequency ranges of auditory sensitivity for a variety of animals are summarized in Figure 3.6. Note that several species of mammals have greater sensitivities to high (i.e., ultrasonic) frequencies than humans do, but few species have an auditory range as large as that of humans (which is exceeded only by those of bats and porpoises). The rattlesnake has the smallest range.

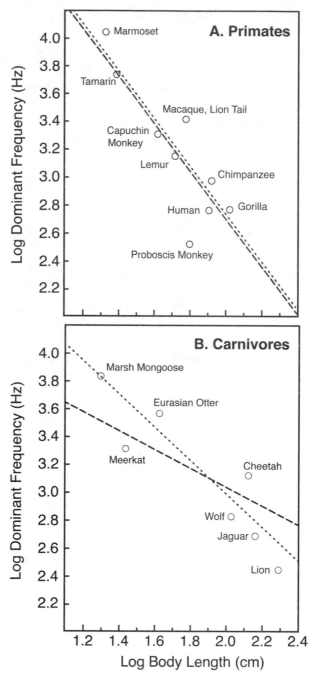

Figure 3.3 Relationship of dominant frequency of acoustic communication and body length for primates and carnivores. (Modified from Herbst et al. 2012.)

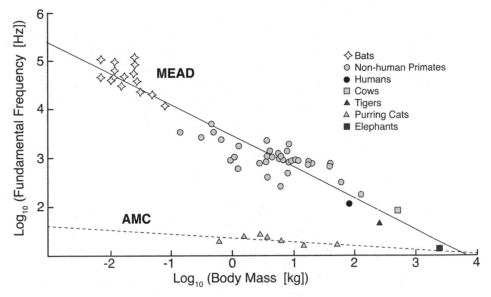

Figure 3.4 Relationship of fundamental frequency of acoustic communication and body mass for select mammals. (Modified from Bowling et al. 2016.)

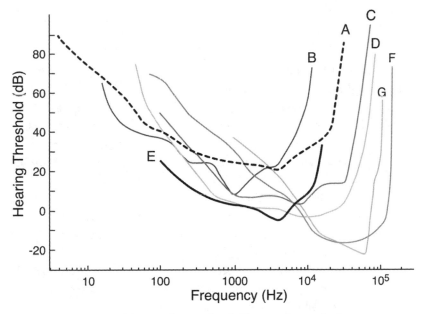

Figure 3.5 Minimum hearing thresholds across frequencies for several mammals. (Modified from Kanders et al. 2017.)

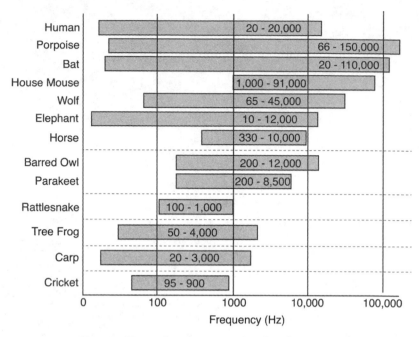

Figure 3.6 Hearing frequency ranges for selected vertebrates.

Plants

Plants respond to a broad range of stimuli, most notably temperature, light, water availability, and touch. Some respond to sound. A small branch of plant physiology, called *phytoacoustics* (Khait et al. 2019), focuses on how plants use sound to sense environmental changes (Mishra et al. 2016).

Three different kinds of plant reactions to sound have been identified (Khait et al. 2019). The first is the direct detection of vibrations on plant surfaces. Many herbivorous insects, for example, create vibrations during their movement along a plant or during feeding. Sounds produced at the frequency of insect "chewing" are known to increase plant production of glucosinolate, a chemical that repels animals. Second, the plant *Arabidopsis* has been shown to have a sonotropic response to frequencies typically associated with running water (200 Hz), growing toward these sounds as they are transmitted through soil. Finally, in a recent study in which flowering plants were exposed to artificially produced sounds of bees that pollinate them, the sugar content of their nectar increased within minutes (Veits et al. 2019).

3.2. Geophysical Sounds

Geophysical sounds were present on Earth billions of years ago, long before sounds produced by organisms. Geophysical sounds are diverse, occur in all spaces on Earth, and vary greatly across time. Examples of geophysical sound sources include wind, rain, moving water (e.g., streams, waves breaking along a beach), and intermittent thunderstorms, fire, and earthquakes. Highly focused geophysical sounds occur in solid substances, such as those produced by sand dunes (vibrations that create unique harmonics) and glaciers (sounds from calving events and colliding icebergs; sounds created during phase changes of melting, freezing, and sublimation). Many geophysical sounds are present in aquatic systems, and some, such as those caused by cavitation (i.e., collapse of a bubble that creates a wave), are common. Animals must often adapt their acoustic signals to reach the receiver without being masked by sounds from the geophysical environment; thus geophysical sounds are important to consider when studying soundscapes. Here, we summarize the variety of geophysical sounds, focusing on their acoustic characteristics as well as the processes that create them.

3.2.1. Wind in Terrestrial Landscapes

Wind, the movement of air, is everywhere, and it varies all the time. Wind is created by a difference in atmospheric pressure between two points, which results in the movement of air from the location of higher pressure to the location of lower pressure. On Earth, these pressure differences are driven by a complex set of processes that ultimately create a pattern of prevailing winds whose directions and speeds vary from the equator to the poles, as we'll see in Chapter 4.

At local spatial extents, winds can be altered by topographic features like mountains and large water bodies. Mountains can alter wind speeds in several ways. First, wind speeds typically increase with altitude on the windward side of a mountain (the side exposed to the oncoming wind) as the mountain forces the air to rise. Second, wind speeds on the leeward side of a mountain (the side sheltered from the wind) typically slow as the air descends. Third, winds moving between mountains do so with increasing speed as they become compressed into small spaces such as valleys, which create "wind tunnels." Winds influenced by topographic features like these are termed *mountain breezes*. Air currents are also created at land-water interfaces by temperature differences between land and water; these winds are called *sea breezes* and *land breezes*.

Wind speeds vary in two important time frames: hourly and seasonally. Seasonal variation is often due to changes in air temperatures over days or weeks. In mid-latitude ecosystems, wind speeds during summer and winter are often less than those in spring and fall (Fig. 3.7A). Wind speeds near the tropics (between 10° N and 10° S) are often higher in rainy seasons than in dry seasons. In nearly all locations on Earth, wind speeds at crepuscular times (dawn and dusk) are less than those at midday. Wind speeds at night are most often less than those during the day (Fig. 3.7B).

Sounds created by wind are a result of the interaction of air flow with an object whose surface slows some portions of the air more than others; this disruption of air flow, called *turbulent flow*, produces a mechanical wave. The sound we associate with wind is the result of turbulence created by the structure of the human ear. Wind is also detected by microphones, and *windscreens* or *windjammers* are often used to reduce wind turbulence at the surface of a microphone. Sounds from wind are typically without periodic waveforms and thus can be classified as a form of noise. Where wind speeds are relatively slow, sounds from wind have frequencies of less than 1 kHz, whereas wind gusts can produce higher-frequency sounds. In windy areas, and in studies where wind is not a focus, soundscape ecologists can use high-pass band filters to reduce the sensitivity of microphones to low frequencies. Some of the quietest places on Earth are those where wind does not occur, such as the insides of caves and dormant volcanoes.

3.2.2. Thunder

Thunder is created by the rapid expansion of air around a bolt of lightning. Two general forms of lightning have been identified (Few 1974). The more common form is cloud-to-cloud (or intercloud) lightning (Teer and Few 1974), which occurs in air only. The second form, ground-to-cloud lightning, occurs when electric discharge in the atmosphere produces a cloud-to-ground bolt, and then a returning ground-to-cloud bolt. Both forms of lightning heat the air around them to extremely high temperatures, creating a shock wave. The shock wave then manifests as a large pressure wave in the air, which is what is heard as thunder. Schmidt (1914) is credited with making some of the first recordings of thunder and developing a terminology of sound production by thunder. He introduced the terms "peal," "clap," "roll," and "rumble" to describe those sounds. *Peals* (Uman 2011) are sudden loud

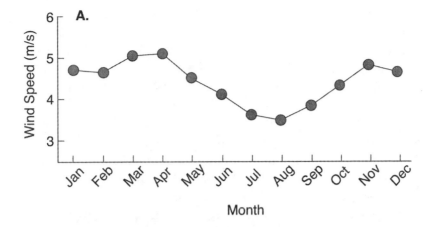

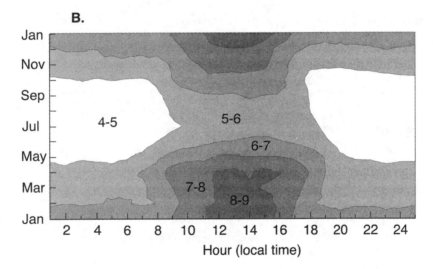

Figure 3.7 Variation in wind speeds over time in (A) Madison, Wisconsin (as monthly averages in m/s), and (B) Cheyenne, Wyoming (as hourly averages in m/s). (Modified from University of Wisconsin, Extension and University of Wyoming, Climate Atlas.)

sounds occurring in intervals of 1 second or less. A *clap* (Schmidt 1914, cited in Remillard 1961) has the same sound quality as a peal but has a shorter duration and a lower intensity. *Roll* and *rumble* are often used interchangeably to describe long, low-frequency sounds that are commonly background noise to peals and claps. The term *thunder leader* describes sounds that occur at the start of a rumble. More recent work by Few and colleagues (Few et al. 1967; Few 1969) showed that the peak in the power spectrum (i.e., the dominant frequency) for thunder is between 70 and 200 Hz. Sounds lower in the frequency range of thunder travel farther than its higher-pitched sounds, creating a typical power spectrum that falls sharply above 100 Hz (Uman 2011). Few and his colleagues also found that sounds from a thunderstorm vary over the course of the storm as the energy and height of the clouds change over time. Thunderstorms are recognized as going through three phases: growth, maturity, and dissipation (Vavrek et al. 2006), and the amount and duration of thunder varies among these phases. Few (1975) also describes rare *ripples* of thunder, which sound like a tearing of the air above; these sounds are attributable to an initial bolt of lightning not reaching the ground and the return bolt being completed as a separate event.

3.2.3. Precipitation in Terrestrial Landscapes

Precipitation occurs in three forms: rain, snow, and ice. Rain events vary considerably over space around the world (Fig. 3.8), and their distribution over time is also highly variable. In equatorial regions, rainfall is high, whereas the lowest rainfall on Earth is generally observed at about 30° N and 30° S latitudes.

Atmospheric scientists measure rainfall events in several ways. The most common is to use a "tipping bucket" instrument that records the amount of rain (in millimeters) over a fixed period (generally reported over minutes, hours, or days). Thus, a common measure of rainfall intensity is depth per unit time (e.g., mm/h). Rainfall events can be placed into three broad categories: light rainfall is defined (AMS 2012) as total rainfall of less than 2.5 mm/h; moderate rainfall as between 2.5 and 7.6 mm/h; and heavy rainfall as that which exceeds 7.6 mm/h. Other measures of rainfall (which require other instrumentation; see Cerda 1997) include rainfall duration (in minutes or seconds), return period (i.e., length of time between similar rainfall events), rainfall speed, and rainfall droplet size distribution. Rainfall droplet sizes are measured using an instrument called a *disdrometer*.

The sound of rain varies with the characteristics of a rainfall event (e.g., droplet size) and with the surface that rain strikes. Different sounds are generated by rain striking bare soil, leaves in trees or surrounding plants, open water, pavement, and even the sensor itself. Bedoya et al. (2017) and Sánchez-Giraldo et al. (2020) have discussed how the sounds of rainfall vary acoustically from low intensities, which are indistinguishable from background noise, to very high intensities, which at times clip most

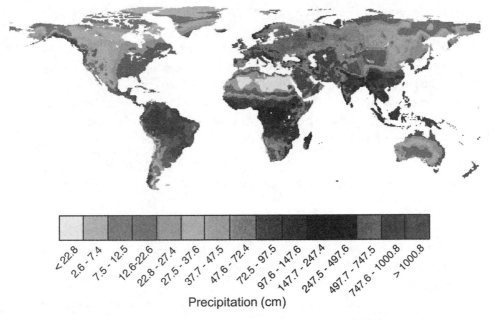

Precipitation (cm)

Figure 3.8 Average annual precipitation around the world. (From NASA.)

microphones. The frequencies of light to moderate rainfall are generally in the range of 50 Hz to around 2 kHz. Heavy rains can produce sounds above 2 kHz, reaching as high as 6 kHz, although with low intensities.

There are several methods by which soundscape ecologists can acquire rainfall data that can be associated with soundscape recordings. The first is the deployment of in situ meteorological stations, which record precipitation along with other meteorological information such as solar radiation, wind speed (and direction), air temperature, and relative humidity. A second approach is to use meteorological data in the public domain, mostly from airports. Finally, two satellite-based sensors provide information on precipitation patterns over a large spatial extent: those connected with the Tropical Rainfall Measuring Mission, which was active from 1997 through 2015, and with the Global Precipitation Measurement Mission, active from 2014 to the present.

There are two forms of frozen precipitation: snow and hail. *Hail* is the more audible and is detectable with a passive acoustic recorder. It is defined as precipitation that has a "density comparable to that of solid ice and occurs in spheroid, conical or generally irregular shape . . . with a diameter of at least 5 mm" (Punge and Kuntz 2016). The spatial extent of hailstorms is relatively small, often less than 40 km^2 (Changnon 1970, 1977). Several studies have reported on the spatial and temporal distribution of hailstorms. It has been found, for example, that hail can occur in most places on Earth and can be common or very rare. Frisby and Sansom (1967), for example, report on a "well marked" hail season that is typical in Costa Rica, occurring between April and September. Hailstorms are also common in East Africa (which has five or more per year), Thailand (in March and April), and across much of China, Europe, and the United States.

3.2.4. Flowing Water of Rivers and Streams

Sounds from the flowing water of rivers and streams are common to many landscapes and are considered relaxing by people (Agapito et al. 2014). Sounds from rivers may also be used as acoustic cues by animals that use riparian zones as feeding, roosting, breeding, or nesting sites. Flow rates, which are highly variable over time, increase with rainfall; thus, the sounds of flowing water, as measured on the terrestrial landscape, could potentially be used as an indicator of precipitation. Some streams are seasonal and are called *intermittent streams*. Those that flow all year are *perennial streams*. Flow rates are also dependent on slope, streambed substrate (i.e., whether there are boulders present in the stream), and channel configuration. For example, stream segments are often classified as pools, riffles, or runs. *Pools* tend to be deep, with slow-flowing water; *riffles* are shallow, with fast-flowing water often running over rocks or other debris (e.g., fallen trees); and *runs* are stretches of stream where water flows unimpeded. These three types of stream segments constitute different habitats. Sounds from rivers are produced in the air but also, importantly, underwater. Aquatic animals have adapted to each of these stream habitats and their unique sounds.

Perhaps the most informative studies on how physical properties of streams affect underwater sound were done by Tonolla and colleagues (2009, 2010, 2011). These researchers employed controlled laboratory experiments using an indoor flume, which is an artificial channel built to allow water to flow from one end to the other (2009). They varied water velocity and introduced flow obstructions as two hydrophones placed in the flume recorded the resulting sounds. They found that increasing flow velocity increased the width of the frequency band produced, and that sound pressure levels increased when submerged obstructions were introduced. They also showed (1) that riffles created high turbulence and air bubbles, leading to high sound pressure levels in the midrange frequencies of 125 Hz–2 kHz;

(2) that pool configurations led to low turbulence and no air bubbles, and (3) that eddies, formed when obstructions cause water to flow upstream, created high turbulence with high sound pressure levels in low-frequency bands (<63 Hz). Two outdoor in-stream studies confirmed the results of these controlled experiments and provided additional information about acoustic signatures related to water flow (Tonolla et al. 2010, 2011). In rivers and streams in Switzerland, Tonolla and colleagues (2010) found that each of the five stream habitat types they studied generated specific acoustic signatures. A broad study of streams in Europe and North America using a hydrophone floated along streams, reported in the same paper, confirmed that stream channel configuration affected midrange frequencies the most, but that transport of sediment along the streambed, in which particle collisions are common, increased sound pressure levels at high frequencies (2–16 kHz).

The underwater sounds of rivers have considerable implications for acoustic communication in aquatic organisms, which may use specific acoustic signatures to determine whether prey, predators, or conspecifics are likely to be present at a location. They may also use these sounds as indicators of stream habitats. Alterations of acoustic signatures by anthropogenic structures such as dams, docks, and erosion barriers introduce unnatural sounds; further, climate change may also affect streamflow regimes, thus altering acoustic signatures during critical times when aquatic organisms have the greatest need for acoustic communication.

3.2.5. Earth Tremors

A mechanical wave that moves through the ground is called a *seismic wave*. These waves are created by earthquakes, volcanic eruptions, landslides, and lava flows, which are common in tectonically active areas of the world (Scholz 2019). Earthquakes and other seismic events are often measured by special microphones called *geophones* or *seismometers*. Several kinds of seismic waves can be produced by an earthquake: P (primary) waves, S (secondary) waves, and L (surface) waves, which often occur in that order. P waves can sometimes be heard, while the stronger S waves are experienced as shaking of the earth. Most earthquakes create vibrations in the range of 0.01–10 Hz (Tosi et al. 2012). The dominant frequency for seismic waves is inversely proportional to the strength of the tremor (i.e., small tremors create higher-frequency seismic waves); dominant frequencies of small tremors can be in the range of 5–60 Hz. In the ocean, waves of another type, called T waves, can be caused by earthquakes. These waves occur in the SOFAR (*sound fix-ing and ranging*) channel, a naturally occurring band of ocean water that allows sounds to carry long distances.

It has been estimated that the explosive eruption of the volcano Krakatau in 1883 was the loudest sound ever to occur on Earth in recorded human history. Volcanic lightning has also been documented to cause thunder (Haney et al. 2020).

3.2.6. Sand Dunes

People have reported sounds from dunes in deserts for centuries. In some instances, the sounds are so pervasive that they can be mistaken for a low-flying jet (Androtti 2004). Some sand dunes are described as "singing dunes" (Dagois-Bohy et al. 2010). Two classes of sounds are produced by these dunes: low-frequency (60–100 Hz) booms (>110 dB re 1 μPa at 1 m; Dagois-Bohy et al. 2010) and high-frequency, harmonic "zinging" sounds. The occurrence of these sounds is influenced by several geophysical fea-tures, including time of day; sand grain size, shape, and composition; slope of the dune; and level of desiccation of sand (e.g., wet sand does not produce harmonic sounds). Interestingly, the size of the dune does not influence the kinds of sounds produced. The most important feature that influences sound frequencies is grain size (Douady et al. 2006). Recording sound from sand dunes can be accomplished using a hydrophone or contact microphone buried in the dune or a condenser microphone (e.g., a cardioid) positioned above the dune.

3.2.7. Geophysical Sounds in Oceans

Several geophysical sounds are common in ocean soundscape recordings (Wenz 1962), including sounds produced by wave motion, sprays and bub-bles created by wind and rain, and earth tremors. Biological communities vary considerably among ocean depths, and the physics of mechanical wave propagation in salt water means that low frequencies do not travel well in shallow water, as the peak-to-peak amplitudes of low-frequency waves can be greater than the depth of the water. In the SOFAR channel, which generally exists at depths of 500 m to 1 km, sounds with long wavelengths can move horizontally without attenuation by the surface or the bottom. Oceanographers sometimes study sounds in a very low frequency range, be-tween 1 and 100 Hz, referred to as the *VLF* (*very low frequency*) *band*.

Many of the sounds in the ocean are generated by *wave action*, which is created by wind (Wenz 1962). The *Beaufort scale* is a systematic scale used to classify wind speed and the resulting wave action on the ocean surface, referred to as the *sea state* (Wenz 1962). The scale assigns the wind speed and sea state a number from 1 (mirrorlike ocean surface, no wind, no waves) to 12 (hurricane conditions, winds ≥72 miles per hour, waves ≥45 feet). Marine scientists have demonstrated that the degree of wave action is correlated with wind speed and duration and is also dependent on water depth. A graph of the relationship between sound pressure levels (SPL), acoustic frequencies, and Beaufort scale categories is often referred to as the *Knudsen curves* (Knudsen et al. 1948) or, more commonly, the *Wenz curves* (Fig. 3.9). The sounds of surface wave action are produced by bubbles, water droplets from sprays, and water movement at the ocean surface. The Wenz curves illustrate that sounds produced by wave action generally have frequencies between 100 Hz and 1 kHz, with a peak in SPL between 200 and 600 Hz (note the limits of prevailing ambient ocean sounds in Fig. 3.9). As the Beaufort scale number increases, SPL increases across this frequency range and peaks at slightly lower frequencies (i.e., shifts to the left on the Wenz curves), the number of bubbles at the surface increases, and sounds with frequencies above 10 kHz are attenuated (Nystuen et al. 1993).

Underwater sounds caused by *rainfall* have been studied extensively (Nystuen et al. 1993) and have been detected even at high sea states. There have been many reports that light rainfall produces a unique 15 kHz acoustic signature that is not present in moderate or heavy rainfall events. Heavy rainfall events show a strong positive correlation between SPL and rainfall intensity at frequencies between 4 and 21 kHz (most strongly below 10 kHz). Nystuen (2001) has demonstrated that the acoustic signatures of raindrops of different sizes can be used as a surrogate for a disdrometer in the ocean.

Other sounds from the ocean include the sounds of *earthquakes*, which have frequencies below 100 Hz and extending as far down as 1 Hz (i.e., into the VLF band). *Sea ice* movements create sounds across a large range of frequencies via straining and cracking due to thermal effects, in addition to the grinding, crunching, bumping, and sliding of ice chunks (Wenz 1962).

Breaking waves are common sound sources in recordings made along coasts, whether marine or freshwater. Most soundscape studies of breaking waves, called "breakers," have occurred underwater (e.g., Deane and Stokes 2002; Loewen and Melville 1991), but a few have examined these sounds as

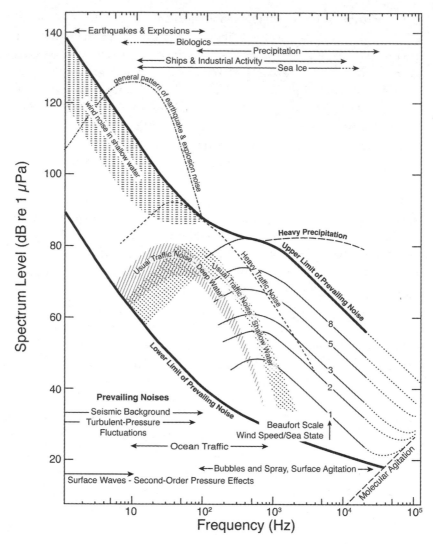

Figure 3.9 Wenz curves for ocean sounds. (Modified from Wenz 1962.)

they occur in air. The sounds of breakers are generated in an area called the *surf zone*. Three major classes of breakers are identified by coastal engineers (Galvin 1968). The first is *spilling waves*, which are common in surf zones that have gentle slopes. The second type, *plunging waves*, occurs on relatively steep beaches; these breakers have crests that plunge into the base of the wave when they become unstable. After cresting, they can transform into spilling waves. One sound from plunging waves has been referred to as a "whoomphing" or bellowing sound (Dallas and Tollefsen 2015). The third type, called *surging waves*, occurs along steep beaches where the more rapidly moving base of the wave dissipates first along the shore. An empirically derived acoustic model of airborne sound production by waves (Bolin and Åbom 2010) shows that for spilling waves, peak SPL occur at around 200 Hz for tall waves (~2 m) and that the frequencies of SPL peaks increase with decreasing wave heights (e.g., a wave 0.4 m high had a peak SPL of 1 kHz). Surging waves produce sounds with peak SPL around 100 Hz (for most wave heights 0.4–1 m). Plunging waves have peak SPL near 200 Hz. Both spilling and plunging waves produce sounds across a frequency range from 50 Hz to around 4 kHz.

3.2.8. Glaciers and Icebergs

Glaciers take two distinct forms: (1) ice caps and (2) floating ice. They are found on mountaintops, on land surfaces where temperatures rarely rise above freezing (e.g., Antarctica, northern Greenland), and along the edges of land-water interfaces in high-latitude locations. About 10% of Earth's surface is covered by glaciers (with the majority in the Antarctic, the Arctic, and the mountains of the southern Andes, the northern Alps, and the Himalayas). Several physical processes create the distinct sounds of glaciers (Jiskoot 2011), including calving (i.e., breaking apart), sliding, melting (creating flows), and sublimation. No work to date has related these glacial dynamics to acoustic properties.

Studies of the breakup of *icebergs* in the Antarctic seas have reported sounds in the VLF band ranging from 2 to 35 Hz, with fundamental frequencies of 4–10 Hz (Wilcock et al. 2014). Estimates of SPL are as high as 245 dB re 1 µPa at 1 m. Sounds from these icebergs can last hours to days. Icebergs also produce sounds when they collide or move against the ocean floor. Variable harmonic tremors from very large icebergs can last the longest. Sounds produced by small icebergs have a broader frequency range (4–80 Hz) and typically last only a few minutes to a few hours; these sounds are called *cusped pulse tremors*.

3.3. Anthropogenic Sounds

Outdoor anthropogenic sounds originate from a variety of sources, most commonly from (1) combustion engines and brakes (e.g., the "jake brakes" of trucks) of land, air, and water transport vehicles; (2) friction from vehicular contact with a surface (e.g., tires on pavement, a train on tracks, air moving through a jet engine or propeller, or a boat propeller in water); (3) construction and excavation activity (e.g., explosions, drilling, pounding); and (4) sonic objects (e.g., church bells, sirens, fireworks). Human voices, and perhaps the sounds of domestic animals (e.g., dogs, livestock), could be classified either as anthropogenic sounds or as biological sounds under their taxonomic groups.

3.3.1. Road Noise

A considerable amount of work has been done on transportation noise and its abatement (Sanberg and Descornet 1980). It is well known that the friction of tires against pavement creates sounds in the 700–1,300 Hz range; the actual frequencies of these sounds depend on vehicle type (car vs. truck) and pavement density. It has also been shown that the spectral characteristics (frequencies and dB) of road noise are influenced greatly by the textural pattern of the tire as well as the surface pattern of the road. Tires and pavement must be designed for vehicle safely and ease of control, and thus safe and quiet highways are challenging to design.

3.3.2. Doppler Effect

As moving objects like cars, planes, and trains produce sound, a physical phenomenon called the *Doppler effect* is observed in sound data obtained from stationary observers or acoustic sensors: sound originating from a moving object is perceived at a higher frequency by the observer as the object approaches and then at a lower frequency as the object moves away from the observer. This happens because the wave cycles are compressed as the object moves toward the observer and lengthened as the object moves away. Figure 3.10 shows how frequencies change as an object moves toward and then away from a stationary sensor or observer.

3.3.3. Sirens

Emergency vehicles are often equipped with loud sirens, which vary according to manufacturer and region of the world. Three different siren sounds have been identified: wail, yelp, and high-low (De Lorenzo and Eilers 1991; Supreeth et al. 2020). These sounds are emitted at frequencies between 1.5 and 3.5 kHz, which is the most sensitive range of human hearing. Some sirens produce sounds with a rapid rise in frequency and with periodic

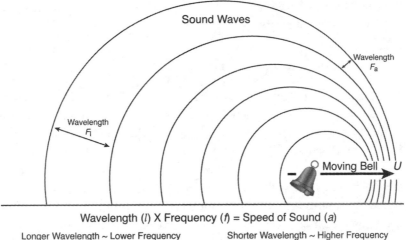

Figure 3.10 The Doppler effect.

changes in frequency and amplitude. The SPL of siren sounds are important, as these sounds need to penetrate other vehicles.

3.3.4. Horns

Horns may be mounted on motorized vehicles (e.g., cars, ships) or operated as stationary warning signals (e.g., foghorns). There are international standards for the loudness of horns (technically referred to as *audible warning devices*) in motorized vehicles: the A-weighted SPL for a horn should be more than 93 dB but not more than 112 dB at 7 meters in front of the vehicle (UN Vehicle Regulations 1984). Most car horns have a fundamental frequency of 440 Hz. Foghorns are still used today to warn boats of obstacles when fog reduces visibility. The diaphone is a type of foghorn that emits a long (3–5 s), loud, low-frequency "burst" followed by a much lower-frequency, equally long "boom." Foghorns are placed in lighthouses along marine and freshwater coasts (e.g., Great Lakes) and can be heard by residents of some cities. For example, the five foghorns located around the

Golden Gate Bridge in San Francisco, which sound, on average, for 2.5 hours per day, are part of the city's soundscape.

3.3.5. Construction Activity

As summarized by Berglund et al. (1996), sources of low-frequency sounds in cities and other places where construction activity occurs include jackhammers, blasting equipment, and pile drivers. Peak SPL occur for blasts at 50 Hz, pile drivers at 500 Hz, and jackhammers at 5 kHz. In oceans, sounds from oil rigs are commonly recorded by hydrophones in the 10–100 Hz range, often with high SPL (110–40 dB re 1 µPa at 1 m; Wilcock et al. 2014). Sounds from seismic air guns used for oil and gas extraction are considered the second most common anthropogenic sounds in the ocean, after ship propeller sounds, and are some of the loudest sounds on Earth, exceeding 220 dB re 1 µPa at 1 m (which is greater than the maximum possible decibel level in air). Air gun sounds are often within a range of 1 to over 100 Hz and can be detected several hundred kilometers away from the source.

3.3.6. Wind Turbines

Over the past few decades, the number of wind turbines in both terrestrial and marine areas has increased substantially. A considerable amount of research has been done on the sounds produced by these energy-generating devices. These sounds, created by the downward movement of the blade at each turbine rotation, are repeated about every 1 second. Most sounds from wind turbines have low frequencies, often below 250 Hz (Møller and Pedersen 2011). Some pressure waves are created below 20 Hz, even stretching down to 2 Hz. Harmonics are possible as well. Studies on sound production by wind turbines have focused on how people are affected by these low- and extremely low-frequency waves, which they perceive as "whoosh" or "swish" sounds.

3.3.7. Anthropogenic Transportation Hot Spots

There are over 40,000 airports in the world, and about a quarter of them service jet aircraft. Sounds from planes landing and taking off are considerable, even at a distance as much as 12 km away from airport landing strips, which collectively support over 100,000 landings and takeoffs per day globally. Other transportation hubs acting as "noise" hot spots include marinas and shipyards. There are over 12,000 marinas in the United States alone, supporting the docking of over a million boats.

Summary

Sounds are created by many animals and are used in some instances by plants for orientation and in response to animals nearby. Sounds from the geophysical environment are numerous and occur in many forms in the air, in water, and within the ground. Anthropogenic sounds are created mostly by machines that use combustion engines or are purposeful sounds like sirens and bells. The major types of sound sources in a soundscape are summarized in Figure 3.11.

Discussion Questions

1. Create a spectrogram that displays the frequencies of sounds made by a monkey, an insect (select one), and a bird (select one), along with the sounds of rain, wind, and thunder.
2. Discuss frequency ranges of vocalizations as a function of body size in animals.
3. Select one animal that you would like to study further. Use internet or library resources to find out how the animal uses sound to communicate (made with what part of the body), how it receives and processes sound (with what part of the body), and for what purposes it uses sound. If possible, use the internet to find a recording of the animal's call, and then use words to describe its sound to someone else.

Further Reading

Insects

Gerhardt, H. Carl, and Franz Huber. *Acoustic Communication in Insects and Anurans: Common Problems and Diverse Solutions*. University of Chicago Press, 2002.

Fish

Amorim, M. Clara P. "Diversity of Sound Production in Fish." *Communication in Fishes* 1 (2006): 71–104.
Popper, Arthur N. "Effects of Anthropogenic Sounds on Fishes." *Fisheries* 28, no. 10 (2003): 24–31.

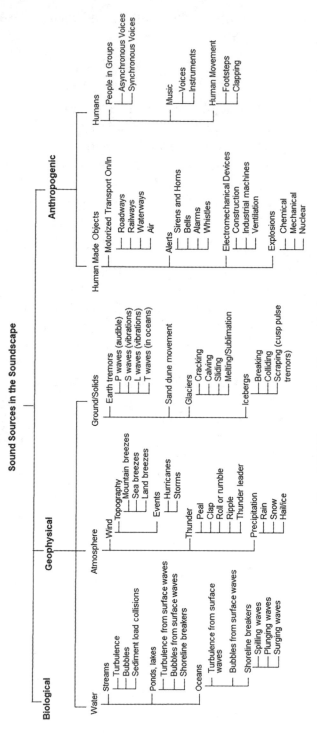

Figure 3.11 Major types of sound sources in the environment.

Amphibians

Lang, Elliott H., Carl Gerhardt, and Carlos Davidson. *The Frogs and Toads of North America: A Comprehensive Guide to Their Identification, Behavior, and Calls*. Houghton Mifflin Harcourt, 2009.

Birds

Marler, Peter R., and Hans Slabbekoorn. *Nature's Music: The Science of Birdsong*. Elsevier, 2004.

CONCEPTS

Scientific concepts help to organize information at high levels of understanding. They facilitate critical thinking and problem solving. Part II focuses on describing concepts from four main areas of study that support the new field of soundscape ecology: sensory ecology (Chapter 4), spatial ecologies (Chapter 5), sociocultural sciences (Chapter 6), and data science (Chapter 7). Those concepts that are important to the four solution spaces described in Section 1.6 are a major focus of this part of the book. Chapter 8 presents a new definition of a soundscape and recasts the field of soundscape ecology within the context of the concepts reviewed here.

4

Sensory Ecology

OVERVIEW. This chapter provides a summary of the major sensory ecology concepts that support soundscape ecology. Sensory ecology draws on bioacoustics, evolutionary biology, and animal behavior. Indeed, numerous hypotheses have been put forth to describe the underlying mechanisms that influence the environmental and neurobiological attributes of animal communication within and between species. These hypotheses, in many cases, were developed separately over time by different communities of scholars in ecology, animal behavior, and bioacoustics. Because these concepts overlap in many ways, they should not be considered mutually exclusive, but rather complementary, as they emphasize different features of acoustic communication, perception, and signal propagation through space and time. One major purpose of reviewing these concepts is to move them from their original context of focusing on particular species to the broader context of considering all animals that use sound for acoustic communication as well as nonbiological sound sources, which allows a full consideration of sensory ecology within the scope of soundscape ecology.

KEYWORDS: acoustic niche hypothesis, acoustic partitioning, acoustic space use, auditory filter, sensory drive framework

---

4.1. Sensory Drive Framework

Endler (1992) proposed a framework for the use of sensory signals by animals based on the idea that "natural selection should favor signals, receptors, and signaling behavior [i.e., where, when, etc., to signal] that maximize the received signals in relation to background noise and minimize signal degradation." Endler (1993) argued that signals, receptors, and signaling behavior are "not independent traits," but rather are traits that function together, and thus are likely to co-evolve; and that the evolution of all these traits is likely to be driven by the same process(es), which span the physics of sound propagation, neurobiology, and behavior. Endler (1993) distinguishes two areas of evolutionary effects that are important for understanding the significance of animal communication: *design efficiency* and *strategic design*. Design efficiency, he states, describes aspects of signal production, propagation, and reception, whereas strategic design focuses on the purpose of a signal (e.g., mating calls vs. territorial calls). Endler (1993) makes clear that his *sensory drive framework* (SDF) considers only design efficiency (also referred to as "sensory tuning" and "signal characteristics" in Cummings and Endler 2018) and that other evolutionary communication models are better suited for understanding and explaining the strategic design of signals.

The SDF broadly addresses all forms of communication and is not exclusive to acoustic communication modes. It considers the evolution of signals perceived through vision, hearing, electroreception, olfaction, and contact (touch and taste). The general predictions (Cummings and Endler 2018) of the sensory drive framework include the following: (1) there should be a strong correlation between the sensory system, signal properties, microhabitat choice, and behaviors associated with the signals of an individual and/or species; (2) the sensory system and the composition of the signals should exhibit predictable patterns across phylogenetic lines (i.e., closely related species should have similar signal structures and behavioral responses); (3) plasticity in sensory systems should exist in environments that have a high degree of variability across space (e.g., where habitat structure can naturally vary across a landscape) or time (e.g., where vegetation may change with seasons or times of day); and (4) in locations where a lot of background "noise" in one sensory modality (e.g., sound, vision, chemoreception) is present, a reliance on one or more alternative sensory modalities is likely.

Figure 4.1 summarizes the main components of the sensory drive framework. Two main types of processes, long-term or evolutionary (dashed lines) and immediate or functional (solid lines), connect these components. Environmental conditions can influence the reception of a signal by the sensory system, which is then perceived by the organism via the neurological system. The neurological system is employed to make decisions about predators or other signal exploiters that influence food intake, mate selection,

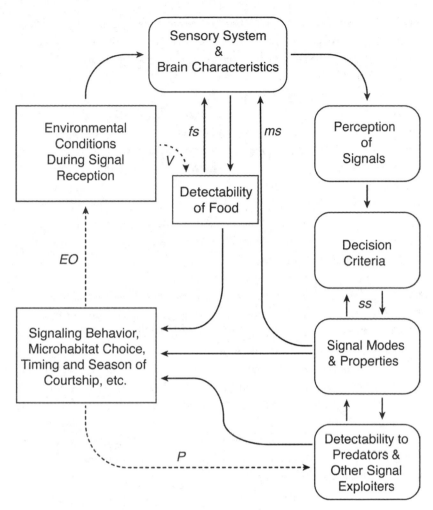

Figure 4.1 Sensory drive framework. (Modified from Endler 1992.) (A) Tuan's perception nested within experience. (B) Sentiment-sense-object conceptual framework of field. (C) Rodaway's conception of the senses, culture, and the environment. (D) Rodaway's sensuous-perception matrix. (E) Jorgensen and Stedman's tripartite sense of place model. (F) Scannell and Gifford's place attachment model with hierarchically arranged subcomponents.

or territory defense. As the SDF is meant to apply to all senses, some of its components are more applicable to non-auditory senses, especially vision; the detectability of food, for example, is often more important for the visual sensory system (e.g., note that the dashed line labeled V is specifically indicating important visual elements of the framework). Other important connections, as they affect an individual's survival, are feeding success (fs), mating success (ms), sexual selection (ss, which is the passing along of good genes to the next generation), ecological optics (EO, which is the ability of individuals to see based on conditions of the habitat) and prey visibility to predators (P).

A recent review of studies examining SDF shows some, but not universal, support for the framework (Cummings and Endler 2018). In general, it was found, after a review of over 150 studies on the topic, that the process of sensory drive (SD) had more support for visual sensory modalities than for auditory modalities when one considers both receiving and signaling components of SD, although there was clearly more support for the signaling component of SD for auditory modalities. SD was found to be most relevant for fish and birds and not so important for reptiles, and the process of SD had more support in studies that focused on aquatic systems than in those examining terrestrial systems.

Within the context of soundscape ecology, the sensory drive framework needs to be considered an important part of understanding acoustic communication in *all* animals within a landscape, as many studies of biodiversity, for example, will need to examine how habitat influences signal characteristics, and how the neurological system has adapted to the signals of conspecifics as well as background noise, including sounds produced by humans and the geophysical environment. The sensory drive framework also builds on over fifty years of bioacoustics research. This work includes studies organized around sound production (the acoustic niche hypothesis and the morphological adaptation hypothesis), sound propagation (the acoustic adaptation hypothesis), and sound reception (auditory filter research). Let's consider these hypotheses next.

4.2. Sound Production in Animals

4.2.1. Morphological Adaptation Hypothesis

It has been recognized that the physical attributes of a signal sender, such as its body size, the length of its trachea, or the structure of its beak, can directly or indirectly affect signal production. This idea has been referred to as the *morphological adaptation hypothesis* (MAH) (Pijanowski et al. 2011a). It was first described by Bennet-Clark (1998), who demonstrated, for example, that for many groups of insects, body size was inversely related to sound frequency (i.e., large-bodied insects produced the lowest frequencies).

The MAH could be extended as well to consider how body size distributions in landscapes create broad frequency ranges of biological sounds. For example, it is well known that body size distributions across ecosystems are predictable traits (Brown et al. 2007; Marquet 2002). Because nearly every ecosystem has hundreds to thousands of soniferous species, the full range of body sizes is likely to result in a broad range of frequencies that occurs naturally as a result of the relationship between body size and the acoustic frequencies of animal signals. Thus, an analysis of the soundscape could help researchers to identify "gaps" in the current community of organisms on the basis of body size and its predicted frequency.

4.2.2. Acoustic Niche Hypothesis

Krause (1987, 1993) observed that the sound produced by each "creature sits within an environmental music staff relative to frequency, amplitude, timbre, and duration of sound." Using "spectrographical mapping" of animal signals from many different kinds of habitats around the world, he argued that in older or less disturbed habitats, there are "more creatures [that] vie for acoustic space, [and] the ability to clearly articulate a voice within that space is more critical to each species' survival." He hypothesized that older habitats will have more acoustic "niches occupied." His personal observations of animal sounds at places he visited over long periods of time suggested to him that acoustic "niches remained stable across years, given time of year, day and weather patterns." He proposed an *acoustic niche hypothesis* (ANH) whereby the arrangement of the acoustic niches of all animals in a habitat represented a "fingerprint" of that habitat (Krause 1993). He further added that some animals, using the Asian paradise flycatcher as an example, have song repertoires that occupy several frequency "channels" to ensure that their signals are heard by a receiver in habitats where many species have similar "voices."

The term "niche" is an important one in ecology. A niche has been viewed in three different, but not mutually exclusive, ways. It has been described as the role or place of a species in a community (Grinnell 1917), as the range of biophysical conditions that a species can persist in in the environment (Hutchinson 1957), and as the use of resources at some point along a distribution of key ecological attributes such as prey size (MacArthur and Levins 1967). In all three definitions, the niche relates to conditions that support the ability of individuals to survive and reproduce, is quantitative, and is multidimensional; in other words, there are many aspects of the environment that contribute to the survival of a species (e.g., temperature, salinity, prey size, nutrient content). Hutchinson referred to this multidimensionality as a complex hypervolume of biotic and abiotic factors that influence birth and death rates. A well-recognized result of niche theory is the *competitive exclusion principle* (Gause 1934; Hardin 1960; Miller 1967), which states that no two species that have the same resource requirements can coexist at the same place and at the same time. Doing so would either limit both species' full use of the resource, or would result in only one species being present to use the resource, as the other would be unable to coexist with it and would become extinct.

The notion that acoustic niches exist within assemblages of animals was not entirely novel at the time Krause (1987) proposed it. In the late 1950s to early 1970s, a number of bioacousticians proposed very similar ideas. Over the years, workers in this area used a variety of terms for those ideas, including "temporal acoustic partitioning" or "broadcasting time allocation" (Cody and Brown 1969), "multispecies chorusing patterns that support acoustic social cohesion" (Alexander 1975), "acoustic interference reduction strategies" (Littlejohn 1965; Drewry and Rand 1983), the "signal interference hypothesis" and "signal confusion hypothesis" that explain mating signal partitioning (Chek et al. 2003), "animal communication networks" (Tobias et al. 2014), "acoustic resource partitioning" (Duellman and Pyles 1983), "spectral niche partitioning" (Schmidt et al. 2013), "acoustic niche partitioning" (Sinsch et al. 2012; Lima et al. 2019), "acoustic space use" (Aide et al. 2017), the "acoustic divergence framework" (Wilkins et al. 2013), and the "acoustic transmission channel competition hypothesis" (Riede 1993). Despite this diversity of terminology and perspectives over the years, all of these ideas have a few common threads. First, many, but not all, researchers consider acoustic or spectral space to be a natural resource, like food or breeding sites, that can be used by any animal. This resource is indeed vital to survival, as acoustic signaling is part of the animal's means of communicating with conspecifics (mostly males attracting a female to mate). Second, researchers have focused on three broad areas of niche or resource partitioning: spectral (i.e., frequency), temporal (i.e., usually diel, or time of day, but also seasonal), and spatial (i.e., either vertical or horizontal space use). What is important for this kind of examination is that in generalized niche theory, there is a recognition (Schoener 1974) that when more than one dimension of a resource can be partitioned (e.g., space and time), then if complementarity of species along one dimension is observed (e.g., if there is considerable overlap in space), then one or more other dimensions (e.g., time) should exhibit dissimilarity; in other words, partitioning of niches may be attained with partitioning in only one dimension. Third, many researchers have focused their work on landscapes where many closely related species are *sympatric* (i.e., co-occurring or overlapping in space), which increases their chances of being able to detect niche partitioning, as such closely related species are likely have similar signals. Fourth, a lot of the focus has been on examining signal production and signal reception together, recognizing that closely related species may have adaptations that allow them to send highly specific "carrier frequencies" and "associated side bands" (Littlejohn and Martin 1969) that are specifically "tuned" by each species' auditory system. As is traditionally done in much of the literature, we will use the term "acoustic niche hypothesis" (ANH) to refer to Krause's hypothesis, and studies that do not specifically consider spectral space as a resource will be broadly referred to as studies of "acoustic partitioning."

There have been many tests of the ANH and/or acoustic partitioning over the past fifty years. For the most part, these tests have focused on an entire group of animals (e.g., anurans, birds, fish) within the same landscape or aquatic system and/or have specifically compared the most closely related species to determine how they might avoid overlap in spectral, temporal, or spatial communication spaces. What has been found, by taxonomic group, is summarized here.

Anurans

Frogs and toads (i.e., tailless amphibians or anurans) have been examined extensively to determine whether acoustic partitioning occurs, and more specifically, whether the acoustic niche hypothesis is applicable to the diversity of sounds produced in this group of animals. In anurans, calls are used by males to attract females, often in dense vegetation or at night, when relying on visual cues is not possible (Blair 1964; Littlejohn 1965; Duellman 1967; Passmore 1977). It has been argued that mate recognition calls need to be species specific, as mating with another species is possible, but yields sterile offspring (Blair 1958; Littlejohn 1965). Thus, calls need to reduce confusion (Littlejohn and Martin 1969) in mate choice. The calls of anurans are simple (Hödl 1977), so evolutionary adjustments in calls are likely to focus on frequency, pulse rate, and so forth. Because the timing of breeding is often regulated by rainfall, and because ponds are often intermittent and behave as pulse events, temporal partitioning of calls is not possible (Littlejohn and Martin 1969).

General trends in acoustic partitioning across spectral, temporal, and spatial dimensions are not overly apparent from studies conducted over the past fifty years in anurans. In one of the first studies on the topic, Littlejohn and Martin (1969) conducted playback experiments with two co-occurring species of frogs whose calls have very similar spectral characteristics. They found that when one species' call was played, the other species would stop calling, which suggested that the calls were organized so that they would not interfere with each other at the same location and time. However, they found no evidence of other temporal partitioning (i.e., diel or seasonal). In a study of three separate locations where multiple Neotropical frog species chorused, Duellman and Pyles (1983) found that the overall structure of the acoustic patterns was similar in all three locations, but that the partitioning of calls was different in each location because the three locations had slightly different assemblages of anuran species, and that small differences in the structure of the habitat may also have influenced the call patterns of these tropical frogs. Drewry and Rand (1983) studied tropical frog assemblages in Puerto Rico and found that in groups of four to six species chorusing in several microhabitats (e.g., forest and open), species-specific calls demonstrated spectral (e.g., in dominant frequency) and temporal partitioning (e.g., duration of notes and intervals). An analysis of eleven species of frogs (Chek et al. 2003) that examined signal separation against a random, null model found little evidence of spectral partitioning in the mating calls of co-occurring species. Sinsch and colleagues (2012) examined niche partitioning in fifteen species of frogs in a wetland in Rwanda. They developed techniques that quantified the acoustic niche breadths of all fifteen species based on three measures of spectral partitioning: spectral breadth, temporal variability, and temporal overlap. They found that all three measures supported niche partitioning in this wetland, and suggested that the one species that had the most overlap may have recently dispersed to the area. They found little support for spatial and temporal (diel) niche partitioning. A study by Villanueva-Rivera (2014) of assemblages of eight species of frogs in Puerto Rico showed some evidence of spectral partitioning; they found no diel partitioning, but suggested that seasonal partitioning was possible. In montane frog species in Cuba, Bignotte-Giró and López-Iborra (2019) were able to find at least one dimension of acoustic partitioning in several assemblages of frogs, but there was no consistent trend across the spectral, temporal, and spatial dimensions of the acoustic niche. Lima and colleagues' (2019) study of thirteen species of frogs in a Brazilian pond showed that there was considerable overlap (90%) in the frequency ranges of almost half the species, but that there was spatial and perhaps temporal partitioning of the calls. Finally, a study by Coss et al. (2021) examined whether the intensity of nocturnal conspecific and heterospecific anuran calls affected females' preference for conspecific calls. Using playbacks of conspecific and heterospecific calls at three levels of intensity, they found that high intensity levels of both calls reduced the ability of females to orient toward a conspecific male.

Several conclusions can be drawn from the work done on anurans. First, acoustic partitioning across all three dimensions is generally supported, but there does not appear to be one dimension that is commonly partitioned among all anurans. Second, acoustic partitioning in mating calls serves to avoid the high cost of selecting a heterospecific to mate with. Third, a variety of environmental variables (e.g., rain) may affect the timing of calls. Finally, the intensities of multispecies acoustic signals may be a factor in determining optimal signal features (i.e., spectral and temporal).

Insects

Perhaps the first suggestion that insects partition their calls was made by Alexander (1957). Although rather qualitative in his assessment, Alexander suggested that of the more than a hundred species whose recorded calls he has examined, "all but a few can be distinguished by ear alone, and all can be separated by audio-spectrographic analysis" (1957, 110). He also noted that "when closely related species are sympatric [i.e., co-occurring in the same landscape], their calling songs usually differ radically," operating as "species isolating mechanisms" (1957, 111) to avoid heterospecific mating. Temporal partitioning was also apparent to Alexander, as singing was often separated by species into distinct diel periods. He also recognized that groups of insect species often sing together, which may function to create multispecies social cohesion that could have evolutionary significance. For example, multispecies calls could communicate the overall status of the landscape (e.g., they might signal that a predator is present). Alexander also recognized several insect species that alternate their calls to reduce overlap of the signal.

Bailey and Morris (1986) conducted a playback experiment to determine if the calls of one species of bush cricket could be masked either by noise created by the researchers and broadcast through a speaker to a female or by the calls of a sympatric bush cricket species. They found that both sounds, although at different intensity levels, caused interference and thus hindered the ability of the female to orient toward the call of a conspecific male. Schmidt et al. (2011) compared the morphological and physiological adaptations of a tropical rainforest cricket's auditory system with those of a closely related species from a temperate forest habitat in Europe. They found that the tropical cricket had more "finely tuned" auditory capabilities than the European cricket, which they argued is an adaptation of the tropical species for hearing in a high-intensity and potentially signal-masking soundscape. Sueur's (2002) study of nine species of cicadas in Mexico after the dry season found that these cicadas partitioned their calls by space (e.g., calling height) and used separate frequency channels and calling patterns that also separated them temporally.

Schmidt and Balakrishnan (2015) summarized several dozen research studies that focused on acoustic partitioning and, more specifically, the acoustic niche hypothesis in insects, and drew several conclusions. First, they found general support for acoustic partitioning in insects across spectral, temporal, and spatial dimensions, although there were no general patterns across different insect taxa (e.g., crickets, katydids, grasshoppers, cicadas). Second, they argued that an evolutionary mechanism is likely to be involved in any acoustic partitioning that acts to avoid acoustic masking when males are attempting to attract a female. However, they identified several other considerations that merited inclusion as part of any assessment of acoustic partitioning. The most important, they argued, was to understand the evolution of sound production as taxa evolved their long-distance acoustic signaling. They also argued that more studies need to consider both the distance that signals travel and the diversity of these signals across all species using acoustic communication, as well as the nature of signals; some insect species use very widely broadcast calls (grasshoppers and cicadas) whereas others use very narrow broadcasting (crickets), and the roles of these kinds of signaling strategies are still not articulated well within the ANH.

Several considerations are important in research on the ANH and acoustic partitioning in insects. The first is that among all terrestrial animals that use acoustic communication, the animals of this group were present first and thus were able to "capture" the most readily available acoustic space. Animals that evolved later, especially the birds and mammals, had to "fill" the acoustic spaces that were left open by insects. Second, most insects produce sounds in very broad frequency ranges, often with harmonics that fill more acoustic space than the sounds of other animals, such as amphibians. Third, the variety of sound production mechanisms in this group, which is far greater than in birds, mammals, or anurans, means that a study that would include all insects at a site would probably show that their sounds occupy a wide variety of frequencies, spectral patterns, and temporal patterns. Finally, given that this group of acoustic communicators is the oldest on land, the variety of ways in which they partition acoustic communication is likely to be greater than in other groups, comparatively speaking.

Birds

One of the first studies to examine acoustic partitioning in birds was by Cody and Brown (1969). They quantified the timing of songs by two of the most common bird species, the wrentit and the Bewick's wren, in a chaparral valley of California. The two species' songs are very similar in structure, so these researchers sought to determine how the species can coexist. A time series analysis of their songs showed that there was a distinct pattern to the species' periods of singing, which the researchers referred to as "broadcasting time." They also suggested that ambient sounds, such as wind, reduce the acoustic space available to these closely related, sympatric species, and that temporal partitioning of songs is one approach two or more species of birds may develop to avoid the problems of interference. The researchers also saw evidence that one species—the wrentit—patterns its song on the cues of the other species (the Bewick's wren). Ficken et al.'s (1974) study of the songs of the least flycatcher and the red-eyed vireo in Minnesota found similar patterns: the two species, which sing at the same frequencies, broadcast their songs at alternating intervals to avoid interference. One of the species, the flycatcher, was found to "strongly" insert its call into the gaps of the vireo's song. The researchers argue that the burden of the interference falls mostly on the flycatcher, which has a shorter song that can be more frequently masked by the vireo's long song. A follow-up study by Ficken and colleagues (Popp et al. 1985), used a set of audio recordings and playback experiments on four sympatric species of forest passerines. They concluded that these four common species separated their calls by waiting for the other species to rest. A similar pattern of temporal partitioning of singing was also observed in the nightingale by Brumm (2006), who used playbacks of other bird species; Brumm found that individuals started singing once the playback songs stopped. Luther's (2008) study of Amazonian bird song during the dawn chorus showed that receivers reacted preferentially to an interspecific call, which suggested that bird song production and reception are fine-tuned to very specific times of the day. In a follow-up study, Luther (2009) found that the songs of the most closely related species that sang at the same place and time were more dispersed in spectral features (eleven measures) than what would be expected of a random set of species.

A few studies have examined acoustic partitioning among many species. An analysis of twenty species of birds from fourteen families in the Peruvian rainforest by Planque and Slabbekoorn (2008) sought to determine if there were any trade-offs between temporal and spectral partitioning of songs by these sympatric species. They found no clear temporal separation of songs that had similar spectral features (e.g., same frequencies), although there were instances of small temporal partitioning in spectral spaces that were used by many species. Krishnan and Tamma's (2016) analysis of sympatric species of Asian barbets compared body size and acoustic communication features such as frequency with a null model of speciation for body size and dominant frequency. The researchers suggested that divergence in acoustic communication features such as dominant frequency among sympatric species is a "by-product of their divergent morphologies"; in other words, that partitioning of food resources by means of adaptations to foraging on different types and forms of fruits may result in different acoustic communication patterns. Chitnis et al. (2020) examined four potential dimensions of acoustic partitioning (spectral, temporal, horizontal space, and vertical space) in sympatric species of wren-warblers in a grassland landscape in India and found that these birds partition their acoustic space in all four dimensions. The spectral partitioning, they argued, was accomplished by diversifying the numbers and patterns of notes in songs.

To summarize, there is considerable evidence for acoustic partitioning in birds. Most of the evidence suggests that birds partition along the temporal dimension, and may do so in a variety of ways. The timing of their song production may be dependent on the calls of other species, and there is some evidence to suggest that some species "dominate" in multispecies calling.

Bats

Echolocation in bats is used most often to identify prey and to navigate within habitats. Siemers and Schnitzler (2004) examined five species of co-occurring bats that were of similar size and belonged to the same guild (i.e., the same feeding/habitat group). They determined that echolocation differences among these five species created separate feeding niches, and thus that echolocation differences contribute to food resource partitioning. Other work on bats (Schnitzler and Kalko 2001) has led researchers to conclude that echolocation features in guilds of bats are related to navigating within the environment. They found that acoustic partitioning resulted from species occupying habitats with different levels of complexity, or "clutter" (uncluttered habitats; background-cluttered habitats, which are typically spaces between trees and above water; and highly cluttered habitats, which are close to dense vegetation and other solid structures like cave walls and buildings), which affected their ability to navigate within these habitats and detect their prey. Other tests of acoustic partitioning by bats were conducted by Bastian and Jacobs (2015) who tested, using a playback experiment, whether bats used "private" acoustic communication channels to communicate with conspecifics. They were able determine that two closely related species were able to discriminate between each other and to do so in similarly structured habitats. They hypothesized that this was possible through improved "sensory acuity" developed by the receivers. In yet another "twist" to niche partitioning in animals, Falk et al. (2015), who studied several sympatric species of bats that preyed on over a dozen different species of katydids, found that although all the bat species were similar in body size, echolocation signal design, and prey-handling ability, they partitioned their feeding on the basis of the frequencies and other acoustic features of the katydids. The work of Jones and Teeling (2006) suggests that the evolution of echolocation in bats is influenced mostly by ecology (habitat and prey) and not by phylogenetics (i.e., genetics and taxonomic relationships). In short, work on bats has led to the recognition that acoustic partitioning may exist, but not because of the animals attempting to find an acoustic channel that is unique; instead, it results from species competing for food in the same location but specializing in different microhabitat structures, or on prey species on the basis of their size or their acoustic signals. Siemers and Schnitzler (2004) have suggested that this may be one of several ways that in which sensory ecology can structure animal communities.

Fish

Ruppé et al.'s (2015) study of fish acoustic communication examined the ANH and three dimensions of acoustic partitioning in marine canyons off the coast of South Africa. They used hydrophones to record for fifteen days and identified sixteen different fish sound types on the recordings. They examined how dominant frequencies differed among these sounds during the daytime and at night. They found that dominant frequencies did not overlap during the nocturnal period, but there was evidence of dominant frequency overlaps during the diurnal period, and the sounds co-occurred within time periods during the day. Thus, acoustic partitioning by fish in these canyons was evident at night, but not during the day, when visual cues may be more important than sound. Bertucci et al.'s (2020) study is one of the first to examine the ANH in aquatic systems—specifically, coral reef systems in French Polynesia. The researchers recorded underwater sounds for one 48-hour period, and then for another 24-hour period one month later. They identified six sound types during the day and nineteen during the night. They found evidence of temporal partitioning, as different sound types dominated sequentially throughout the day. In addition, each sound type had unique spectral features, thus suggesting that spectral partitioning occurs as well. Spectral features that were identified as serving to partition calls included dominant frequency, number of pulses in a call, and pulse duration (peak-to-peak interval). In short, the researchers were able to determine that significant acoustic partitioning did occur among the sound types that were present at this coral reef system.

Work in aquatic systems, arguably, lags behind that in terrestrial systems in terms of the number and types of habitats studied. No studies, to date, have examined how factors such as depth, the influence of the lunar cycle, and the phylogenetic determinants of sound production (i.e., how closely related species are or if body size constrains spectral features such as frequency ranges) affect acoustic partitioning.

Broad Taxonomic Research

There have been only a few studies that have examined the ANH or acoustic partitioning across diverse taxonomic groups at one location. Such studies more fully test the Krause ANH, as it considers the acoustic communication patterns of all animals at the same location. The most comprehensive in terrestrial ecosystems is the study by Aide et al. (2017), who quantified the acoustic space use of anurans, birds, and insects in rainforests across a large region in Latin America, across which biodiversity varied considerably. The researchers made 1-minute recordings (one per hour) over short durations (range of 4–13 days) at each of eight rainforest sites. They created frequency-time bins by setting a minimum intensity and a standard frequency bin size (in this case, 86.13 Hz) and then scored these bins for the presence or absence of sounds produced by animals from three main taxonomic groups (anurans, insects, and birds) in each 1-minute recording. The number of frequency-time bins by animal sound types placed in gen-

Table 4.1 Acoustic space use (ASU), regional avian species richness (ASR), and richness of acoustic morphospecies, by taxonomic group, in tropical rainforests of Latin America

| Site | ASU (%) | ASR | Acoustic morphospecies richness | | | |
			Total	Insects	Anurans	Birds
Amarakaeri, Peru	**34**	**583**	84	53	4	27
La Selva, Costa Rica	20	423	56	33	3	20
Las Cruces, Costa Rica	28	352	**91**	**55**	5	**31**
Palo Verde, Costa Rica	15	330	38	17	*0*	21
Cabo Blanco, Costa Rica	12	311	46	22	1	23
Las Alturas, Costa Rica	19	262	50	21	3	26
Cuerici, Costa Rica	13	190	*17*	3	*0*	14
El Verde, Puerto Rico	*12*	*184*	37	19	**6**	*12*

Source: Modified from Aide et al. (2017).
Note: Values in boldface are the largest values, and those in italics are the smallest. Note that insect species are the most numerous, followed by birds and then anurans.

eral categories, which they called acoustic morphospecies, was examined for each of the sites. Across all sites, they identified 419 distinct morphospecies. They then calculated *acoustic space use* (ASU)—the percentage of acoustic space occupied in the frequency-time bins by all sounds produced by animals—in each of their 1-minute recordings. They found that ASU was greatest in regions of high biodiversity (as measured independently by avian species richness maps). There was also a strong positive correlation between ASU and the number of acoustic morphospecies at each site. They concluded that ASU was greatest in areas of high avian species richness, and that ASU of insects was the greatest predictor of species richness in general (Table 4.1). They also found evidence of acoustic partitioning across the three taxonomic groups, with insects often occupying the highest frequency ranges. Overlap of acoustic spaces was common across all taxonomic groups, but the timing of their sound production reduced the amount of acoustic overlap. Furthermore, anurans possessed the smallest overall frequency range, followed by birds and then insects, suggesting that some taxonomic groups occupy more of the acoustic space than others. Sounds from the geophysical environment (geophony) and from humans (anthrophony) were excluded, as these were rare sounds in the recordings.

Putland et al. (2017) conducted perhaps the only full study of ANH in aquatic systems. They focused their research in a relatively high-biodiversity area of a shallow embayment off the coast of New Zealand. They deployed six hydrophones and recorded sounds for 54 days with a duty cycle of 2 minutes every 10 minutes. They then analyzed sounds from twelve different common species of animals, sounds from wind and rain on the ocean surface, and sounds from ships across frequency, PSD, and time. They determined that the separation of biological sounds (from fish, urchins, shrimp, and marine mammals) across frequency ranges was great, with sounds from whales occupying the lowest frequencies (10–100 Hz) and those from dolphins occupying the highest (10–100 kHz). Overall, they found support for the ANH, as there was little overlap of frequencies among many soniferous animals, and where there was overlap (in the range of 100 Hz to 1 kHz), sound production was separated temporally (into different times of the day). Both geophony (from wind, rain, and earthquakes) and anthrophony were common, and at very high SPLs, at the six sites, and the researchers reported that those sounds had the ability to mask the sounds of the aquatic animals. The animals could adjust their acoustic communication either by producing sounds when these nonbiological sounds were not present or by adjusting their frequency ranges, although many marine animals have few options to do that.

What is clear from all the studies that have examined acoustic partitioning for the purpose of understanding the role of sound in attracting conspecific mates, or have explored the relevance of the acoustic niche hypothesis, is that there are a variety of ways that acoustic signals can be partitioned in spectral, temporal, and spatial dimensions, that there is no one common means by which partitioning can occur, that most studies have examined either one group of animals or closely related species to determine the mechanisms of acoustic partitioning, and that most of the work in this area of research has focused on the terrestrial environment. Most of the work in terrestrial systems has also focused on the tropics, where species richness is high and competition for acoustic niches is likely to be high as well. More work is needed to determine if the ANH and acoustic partitioning are important in places with lower species richness, such as temperate forests and grasslands, and how all animals at a location, not just broad taxonomic groups such as frogs or birds, partition acoustic spaces. There have been few, if any, studies done on mammalian acoustic partitioning. And although they represent a small proportion of reptile species, geckos are well known to vocalize, and how they communicate within multispecies assemblages is unknown. Finally, the acoustic niche hypothesis, as proposed by Krause, needs to be updated to incorporate a variety of work in soundscape ecology and bioacoustics.

4.2.3. Extended Communication Network Hypothesis

Tobias and colleagues (2014) have suggested that multispecies communication interactions can operate in a highly specialized, complex network. Referred to as the *extended communication network hypothesis* (ECNH), their hypothesis states that the acoustic signals of co-occurring species may be optimized for habitats that they share, operating in unison to defend resources where they co-occur. Many of the species are also likely to be closely related, thus their signals are likely to be similar. According to the ECNH, multispecies acoustic communication networks operate through convergent evolution, rather than divergent evolution as suggested by the ANH. Thus, the ECNH makes predictions that differ from those of the ANH.

Tobias et al. (2014) examined their hypothesis by modeling three possible patterns of multispecies acoustic partitioning (Fig. 4.2): acoustic features might be differentiated in multivariate (i.e., spectral, temporal, or spatial dimensions) signal space at distances not differing from random (Fig. 4.2A), might be evenly spaced (Fig. 4.2B), or might be clumped or clustered (Fig. 4.2C). The distributions of distance measures for these three possible patterns are shown in Figures 4.2D, E, and F. A pattern that does not differ from random (Figs. 4.2A and D) would mean that acoustic communication features are unlikely to have been selected through evolutionary mechanisms; a pattern of evenly spaced signals (Figs. 4.2B and E) would have been created under the mechanisms described by the acoustic niche hypothesis; and a clumped/clustered pattern (Figs. 4.2C and F) would have occurred because of convergent evolution. The differences among signal features predicted by the ECNH would be smaller than those predicted by the ANH, which would show the largest departures across the multivariate signal space. The differences among signal features would vary considerably if they occurred at random—in other words, without natural selection operating to separate these acoustic signals. To test their hypothesis, the researchers recorded the dawn chorus on 47 mornings at 91 locations in an Amazonian rainforest in Peru and identified 283 bird species' songs. Each birdcall was auto-detected using Raven Pro software (see Chapter 10 on soundscape tools), and multivariate analysis was conducted to quantify the spectral and temporal differences among all calls. The researchers found strong support for the ECNH,

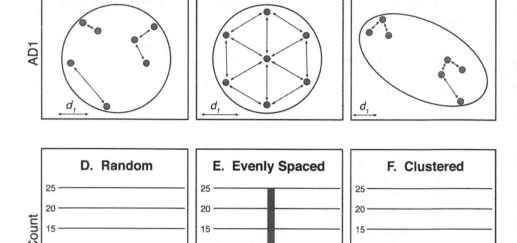

Figure 4.2 Arrangement of animal calls in a communication network in spectral, spatial, or temporal space. Each dot represents a call as characterized by two hypothetical measures (i.e., acoustic distance features AD1 and AD2). The calls may be arranged at random (A and D), equally spaced (B and E), or clustered (C and F).

as the acoustic signals showed more convergence than divergence (i.e., a pattern like that in Figs. 4.2C and F).

4.3. Propagation of Animal Communication

4.3.1. Acoustic Adaptation Hypothesis

The *acoustic adaptation hypothesis* (AAH) was proposed by Morton (1975) to describe how the physical structure of a habitat influences the spectral structure (e.g., degree of modulation) and/or temporal structure (e.g., repetition rate) of sound transmitted by an animal. The AAH is based on several important observations of sound propagation in space. It has been known for many decades that *signal degradation* occurs in two ways as a signal travels through space. First, the frequencies of a signal scatter as it interacts with solid objects in space, such as leaves, branches, and the ground. Second, the signal degrades by *attenuation*, which is the decrease in its intensity as a function of distance. According to predictions based on the AAH by Wiley and Richards (1978) and Richards and Wiley (1980), signals in forests should be structured as pure tonal whistles, while those in open landscapes, such as grasslands, should contain "trills" (i.e., components with high amplitude modulation). More recent assessments of the AAH (e.g., Ey and Fisher 2009) have led to the development of a large set of predictions that are consistent with Morton's (1975) proposal: (1) that animal signals should have longer durations in closed (i.e., forested) habitats than in open ones; (2) that call repetitions should be fewer in closed than in open habitats, as reverberations in closed habitats would create an overlap of signals; (3) that fewer frequency modulations should occur in closed than in open habitats, and that these modulations should degrade more easily in closed habitats; that the (4) minimum, (5) maximum, (6) mean, and (7) dominant frequencies of calls should be should be lower in closed habitats; and (8) that the frequency ranges of signals should be narrower in closed than in open habitats. Because the AAH places a strong emphasis on the relationship between signal characteristics and habitat structure, it is sometimes (Wilkins et al. 2014) considered a subset of concepts and theories that fall under the broader sensory drive framework, although the AAH does focus solely on acoustic communication and does not consider how sound and other sensory modalities of communication factor into the evolution of acoustic signals.

4.3.2. Tests of the AAH

One of the first rigorous tests of the AAH (Brown and Handford 1996) was done using computer-generated signals in the form of either a trill or a whistle. These synthetic signals were then exposed, using a computer simulation, to degradation conditions that produced either echo effects (i.e., reverberations like those created in closed habitats) or irregular amplitude-dampening effects (i.e., simulating the effects of wind on amplitude modulation patterns). The modified sounds were then played back, recorded, and compared with the original synthetic sounds. The differences between the original and the modified sounds were used as a measure of signal degradation. This study led the researchers to conclude that Morton's original AAH (1975) had considerable support, as signals with high amplitude modulation were less degraded in the simulated open-habitat conditions and trills propagated best in the closed-habitat conditions.

Two rather large meta-analyses that examined the relevance of the AAH to vertebrate signal structures as a function of the natural habitat have been performed; in general, they found mixed support for the AAH across all taxonomic groups, but important patterns did emerge. A literature review of studies that have addressed the AAH found partial support for the hypothesis in birds (Boncoraglio and Saino 2007), and a larger review of birds, mammals, and anurans (Ey and Fischer 2009; see also Goutte et al. 2018) yielded mixed results. Table 4.2 summarizes these meta-analyses in terms of the eight predictions of the AAH. Note that for both predictions on temporal features of acoustic signals (1 and 2), the research found mixed results for birds (the two studies came to different conclusions) and not much support for anurans and mammals. Many of the predictions of the AAH that focus on frequency (e.g., that lower frequencies should occur in closed habitats) are generally supported for birds, with mixed support for anurans and mammals. In several instances, the results even suggest an inverse trend (i.e., the data support a relationship opposite to the one predicted).

Several other multispecies assessments that have been made more recently also provide mixed support for the AAH. Peters and Peters (2010) examined the call structures of twenty-seven species of Felidae (i.e., cats). They examined whether the dominant frequencies of calls were influenced mostly by body size or by habitat type. They found that species that lived in open habitats had a lower mean of dominant frequencies than those living in closed habitats, a finding that supported the AAH. They also found no significant influence of body size on the mean of dominant frequencies once habitat type was considered, and they pointed out that body size across the species examined varied by over a hundredfold. Overall, they found considerable support for the AAH in these carnivores. Jain and Balakrishnan (2011) studied crickets and katydids in a tropical wet evergreen forest in the Western Ghats of India to determine whether vertical spatial stratification of calls followed predictions made by the AAH. They used playbacks of twelve coexisting insect species, which they broadcast from five heights, from the forest floor to the top of the canopy (0 m, 0.5 m, 2 m, 4 m, and 11 m), that were typical of the species being tested. To determine whether the location of sound production affected sound propagation properties, they measured signal attenuation, signal-to-noise ratio (SNR) as a function of distance from sound source, and envelope distortion, the difference between each call's frequency range as it was emitted and after it had passed through habitat. Five species upheld the AAH for attenuation, three species for SNR and envelope distortion, and one species upheld it for

Table 4.2 Tests of predictions generated from the AAH on signal structures in closed (e.g., forested) as compared with open habitats

Predicted signal structure	Rationale	Boncoraglio and Saino (2007), birds	Ey and Fisher (2009), birds	Ey and Fisher (2009), anurans	Ey and Fisher (2009), mammals (bats, primates, felines)
1. Longer duration	Longer signals increase likelihood that receiver detects signal given attenuation by habitat	Mostly supported	Not supported	Not supported	Mixed; in many cases, opposite of prediction
2. Lower repetition rate	Reduces overlap caused by reverberation	Not supported	Not supported	Not supported	Mixed
3. Fewer frequency-modulated elements	Structure of any modulation is reduced by dense vegetation	Mostly supported	Mostly supported	Not supported	Some supported, some opposite of prediction
4. Lower minimum frequency	Lower frequencies travel farther than higher frequencies, and more so in closed habitats	Mostly supported	Mostly supported	Not reported	Not reported
5. Lower maximum frequency	Same as above	Mostly supported	Not supported	Not reported	Some supported, some opposite of prediction
6. Lower mean frequency	Same as above	Mostly supported	Mostly supported	Not reported	Not reported
7. Lower dominant frequency	Same as above	Mostly supported	Mostly supported	Not reported	Some supported, some not supported, some evidence of opposite to prediction (fully mixed)
8. Narrower frequency range	Low frequencies transmit better, so senders don't put energy into higher frequencies	Mostly supported	Mostly supported	Not supported	Some supported, some not supported, some evidence of opposite to prediction (fully mixed)

Source: Modified from Ey and Fisher (2009).

all three spectral modifications by habitat. They concluded that the AAH was supported, but that vertical stratification of these species was determined not by acoustic properties, but rather by other aspects of their life histories (e.g., microhabitats that support food acquisition and breeding). Goutte et al. (2016) studied the spectral and temporal signal characteristics of 79 species of frogs, mostly torrent frogs, which live along loud stream banks. They used a phylogenetic approach that included morphological and environmental (i.e., known microhabitat) data to test the prediction of the AAH that calls should be of higher frequency near streams and that they should also have temporal features (e.g., short note duration) that are suited to environments with loud ambient sound. They found support for the AAH for spectral features, but not temporal patterns, of torrent frog calls. In a more recent study of birds, Sebastián-González et al. (2018) examined the relationship between signal features, principally dominant frequencies, of two groups of birds, those that are native to Hawaii and those that have been introduced. They also quantified the forest vegetation density using remotely sensed data from LiDAR (see Section 9.8). Of the seven native bird species examined in this study, only two followed the predictions of the AAH. The introduced species did not show the predicted correlations of song frequencies and vegetation density. The researchers concluded that there are probably other factors in the relationship between bird species' occurrence in the landscape and the density of vegetation, including aspects of their life histories.

In short, the AAH provides numerous predictions about how animal communication should evolve as a function of habitat characteristics. Currently, there is more support for this hypothesis among birds than among other taxa. In addition, most evidence suggests that spectral features better match the predictions of the AAH than do temporal patterns of call production.

As soundscape ecologists look to examine the important coupling of landscapes and soundscapes, propagation space needs to be characterized using the tools that landscape ecologists and terrestrial remote-sensing experts use. These tools allow us to see how the habitat is structured in two-dimensional and three-dimensional space.

4.3.3. The Lombard Effect

The French scientist Étienne Lombard published a paper that described his observation that people increased the amplitude of their voices when speaking in a location with loud background noise. This reflex effect, now called the *Lombard effect*, has been observed in other animals as well. Researchers have even found neurological auditory pathways in the brain that evoke this effect in cats and monkeys (Zollinger and Brumm 2011). The Lombard effect has been demonstrated in a variety of animals, including fish, whales, bats, tree frogs, and most bird species that have been tested. For example, it has been found that several species of birds sing at higher frequencies in cities, reducing the chance of masking where low-frequency traffic sounds are common (e.g., Fernández-Juricic et al. 2005), or increase the number of times a call is produced. It has been argued that the increase in song pitch could be an epiphenomenon of the Lombard effect (i.e., that it occurs as a result of an increase in amplitude rather than a direct purposeful shift in frequency) (Zollinger and Brumm 2011). Similarly, it has been shown that some birds, such as the European robin, sing more often at night in cities due to the presence of city noise (Fuller et al. 2007), and that other birds alter their number of calls, increase their call durations, select calls that differ in frequency, or avoid vocalizing at the loudest times (Gomes et al. 2021). There are some sounds, such as those produced by waves hitting shorelines, that are close in frequency and spectral shape to those of urban traffic, suggesting that these adjustments in vocalization are evolutionary (Gomes et al. 2021).

4.4. Sound Reception by Animals

4.4.1. Auditory Filters

There is growing evidence that the auditory processing mechanisms of animals are most sensitive to the frequency ranges, or *bandwidths*, produced by conspecifics (Dooling et al. 2000; Lucas et al. 2015). It has been proposed (Fletcher 1940) that auditory mechanisms, operating as *auditory filters*, could evolve in closely related species that produce signals that are similar in features such as frequency, modulation patterns, or timbre, to make them distinguishable by the receiver through specialized features of the neurological system. Auditory bandwidth filters could evolve to have very narrow frequency channel sensitivities such that, even if overlaps of frequencies do occur, the filters for two species communicating within a small frequency window could still allow their signals to be distinguished by receivers. Auditory bandwidth filters have been shown to vary according to habitat (as reviewed in Lucas et al. 2015); for example, forest-dwelling birds (e.g., nuthatches and titmice) have narrower auditory bandwidth filters than do non-forest birds (e.g., house sparrows). It is argued that these narrow auditory bandwidth filters are most beneficial for birds that live in forests, where modulations of signals are quickly degraded by vegetation, according to the acoustic adaptation hypothesis.

4.4.2. Auditory Filter Hypothesis

Although the auditory filter concept has not been formally posed as an organizing hypothesis in the literature, its treatment here could formulate it as complementary to the other sensory ecology concepts of this chapter. An auditory filter hypothesis (AFH), which complements the SDF and the AAH, would focus on sound reception and the anatomical and physiological mechanisms of the auditory system. A simple statement of an AFH would be that neurological adaptations should favor filters where biophonies are "crowded," species recognition is vital to survival, and closely related species co-occur in the same space. Considering how all animals perceive sound is important for a variety of reasons. Thus, researchers should consider that (1) many species may hear or perceive sounds differently, and that this diversity of auditory processing mechanisms is important, more likely so in species-rich tropical environments; (2) that human perception of sounds is often very different from that of animals, so a human-centric approach is not likely to be a valid determinant of acoustic partitioning; and (3) that habitat may influence how these auditory filters evolve, as described by the sensory drive framework.

4.5. Critique of the Current Acoustic Niche Hypothesis

The current articulation of the Krause acoustic niche hypothesis has several shortcomings that need to be addressed. The first is that Krause did not use the term "niche" in the context of the traditional and well-accepted ecological niche concept that has existed in community ecology and evolutionary biology for over a century. Thus, the ANH never makes explicit how the acoustic niche ultimately affects birth rates and death rates. The acoustic niche is also presented as unidimensional; in other words, the acoustic resource is not considered along with other environmental resources, sensory or otherwise.

A second shortcoming of the ANH is its lack of recognition that the distribution of acoustic spaces, especially across frequencies, could be an artifact of the well-documented distribution of animal body sizes in communities (Maurer et al. 1992; Woodward et al. 2005; West et al. 1997; Holling 1992). Over the past several decades, there has been speculation on how evolution and natural selection influence the distribution of body sizes within communities and across biomes around the world, but one universal pattern that exists is that many habitats support plants and animals that vary greatly in body size, by as much as magnitudes of 6 (i.e., a difference of a million between the largest and the smallest plant or animal species; Lomolino et al. 2010, 597). Animal body size is thought to vary with trophic strategy (herbivory vs. carnivory), body plan or *bauplän* (or "functionally distinct body plan," e.g., whether the animal flies or burrows, or is quadrupedal or bipedal), and the evolutionary lineages of the species present in the community (as Cope's rule states that over evolutionary time, diversification of species leads to larger-bodied species in the more recent linages). There is considerable debate about whether body size distributions are naturally discontinuous (i.e., have body size breaks or gaps; see Siemann and Brown 1999 vs. Holling 1992). Thus, if the distribution of body sizes within a community is influenced mostly by nonacoustic ecological or evolutionary pressures, then the observed variation of acoustic frequencies at a place could be an artifact of the naturally high variation of animal body sizes, since frequencies are influenced by body size.

A third shortcoming of the ANH is the assumption that the filling of acoustic niches is influenced only by what the sender produces. As described above, there is growing evidence that many species perceive sounds differently. The ANH thus needs to consider factors such as auditory filtering. Thus, a more comprehensive version of the ANH should consider the entire animal communication pathway, from sender to propagation mechanisms in the environment to the receiver.

A fourth shortcoming is that the ANH lacks a clear statement of how the properties of an acoustic signal within its niche are affected by acoustic space (*sensu* Morton and Marler 1977). Bioacousticians consider acoustic space to be the "maximum distance at which a signal can be detected against a background of noise" (Lohr et al. 2003) or, more specifically, given signal degradation over space, the space or distance "over which 'meaning' can be transmitted" through sound. This distance is especially important given that higher frequencies lose power and degrade more quickly than lower-frequency sounds. Acoustic spaces differ across frequencies, and how these differences affect the niche needs further consideration.

Finally, acoustic niches need to be considered in relation to other environmental sounds, such as those from wind, rain, and running water. These sounds, most of which occur in low-frequency bands, can mask sounds from animals that are signaling in the same range. Studies on birds, for example (see Lohr et al. 2003), have shown that for signals to be detectable, there must be a 10 dB increase in SPL over the dB of ambient noise. This critical masking ratio may be most important in the low-frequency ranges where the sound from wind is present.

4.6. Geophonic Concepts

4.6.1. Considerations

Geophonic sounds vary over space and time. There are probably important patterns of these sounds to consider at landscape to global scales, as these patterns could affect biophonic features in soundscapes.

4.6.2. Geophonies Reflect Broad Patterns of Climate Dynamics

The average amount of *precipitation* that falls each year varies considerably around the world (see Fig. 3.8). The Atacama Desert in Chile, for example, which is well known for being one of the driest places on Earth, has an average annual rainfall of less than 0.5 cm per year. In contrast, tropical rainforests can receive up to 1,100 cm per year. Global spatial patterns of precipitation are well known: the greatest precipitation occurs in equatorial regions, and the driest regions are located at about 30° N and S and 65° N and S latitude. *Rain shadows* are dry areas created along the leeward sides of mountain ranges, and thus result from topography. Water-based sound sources, such as rain, should be prevalent in areas such as the intertropical convergence zone (ITCZ), where rainfall is abundant, and should follow the seasonal wet-dry cycles that are common there. Mid-latitude locations should also have geophonic sound-source patterns that follow the typical four seasons—winter-spring-summer-fall—common to these climate zones. Soundscape ecologists could consider these questions: Are sounds produced by animals at higher frequencies, or at greater intensities, in areas of high precipitation? Does animal acoustic communication vary over time so as to be optimized for wet-dry cycles?

Wind speeds vary around the world in predictable ways. Earth's winds are oriented in "global wind belts" (Fig. 4.3), which could generate distinct patterns of wind sounds. Sailors have traditionally called the wind belt at the

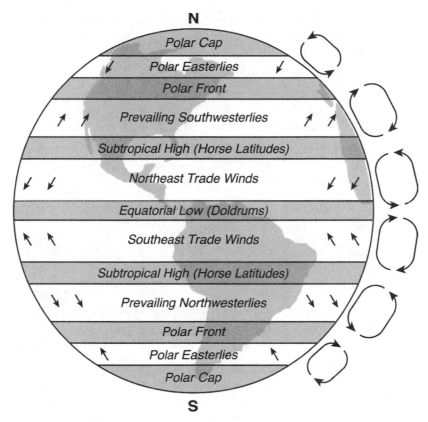

Figure 4.3 Global wind circulation patterns.

equator (5° N and 5° S) the "doldrums" because of its calmness. To the north and south of the doldrums are the northeast and southwest trade winds, which span latitudes from 5° N to 30° N and 5° S to 30° S. Another set of calm wind belts, called the horse latitudes, is found directly to the north and south of the trade winds. In the far high latitudes, polar winds are strong. In addition, nearly all terrestrial locations on Earth have a diel pattern of wind speeds, with the slowest winds typically at night or just before dawn. Winds can also vary by season and with landscape features. Some of the earliest reporting of a geophonic wind pattern is that of Eyring (1946), who described a jungle in Panama as a landscape with "wind velocity gradients so small that the sound refraction they produce may be neglected for all practical purposes" (1946, 257).

It is recognized that the sounds of *thunder* are also highly variable around the world, but that variation has been documented only by measuring lightning, by means such as NASA's satellite-based Lightning Imaging Sensor connected with the Tropical Rainfall Measuring Mission, which recorded cloud-to-cloud and cloud-to-ground lightning (Albrecht et al. 2011). These data show that lightning is most prevalent in the equatorial regions of the world, principally on land, where solar radiation heats up air and creates the turbulence characteristic of thunderstorms. Sensors from NASA, for example, have shown that the Central African Republic and Venezuela, two tropical areas with complex topographies (i.e., with mountains and lakes), are lightning "hot spots." Generally, lightning is also common in coastal areas and in mountain ranges (Andes, Alps, Himalayas).

4.6.3. Influence of Geophony on Biophonies

As described by Pijanowski et al. (2011a), geophonic patterns could influence biophony. Indeed, this possibility has been a focus of research by bioacousticians for several decades in both terrestrial and aquatic environments. Geophonic sounds tend to be in the lower frequency range (<1 kHz) when they occur at low to moderate amplitudes. These sounds include those of wind, low to moderate amounts of rain, running water, and waves hitting the shoreline in coastal environments. Within the context of the sensory drive framework and the hypotheses described above for biophonic patterns, geophonic sounds should influence the patterns of the biophonies because geophonies are likely to mask acoustic communication in low-frequency ranges.

In aquatic systems such as the oceans, geophysical sounds (see Fig. 3.9) range in frequencies from 70 Hz to as high as 20,000 Hz for precipitation, depending on rainfall amount; from 10 Hz to around 12,000 Hz for sea ice; and from 100 Hz to around 15,000 Hz for wave action at the surface, with varying intensities depending on sea state. The sounds of intermittent earthquakes, which can be common in some locations, range from 1 Hz to 100 Hz. Thus, masking of sounds in the ocean comes from natural geophysical sounds that are diverse, broadband, and at times, intense.

Summary

Soundscape ecologists have several conceptual frameworks and organizing hypotheses to draw from, and these hypotheses are likely to be a major focus over the next several years as researchers attempt to determine what mechanisms lead to specific acoustic partitioning patterns in animals.

The sensory drive framework, the morphological adaptation hypothesis, the acoustic adaptation hypothesis, the acoustic niche hypothesis, and the extended communication network hypothesis have been examined by bioacousticians over the past 50 years. Patterns of sound production by geophysical dynamics should vary due to global to local influences on wind and precipitation patterns.

Discussion Questions

1. Compare and contrast the extended communication network hypothesis and the acoustic niche hypothesis. What predictions does each make, and how might you test whether one of the two hypotheses is consistent with the soundscape you are studying?
2. How might you go about determining how wind and rain affect animal communication features in arid landscapes, or in rainforests? Discuss the types of data you would collect, the instrumentation you would use, and the analyses that you might consider.

Further Reading

Boncoraglio, Giuseppe, and Nicola Saino. "Habitat Structure and the Evolution of Bird Song: A Meta-Analysis of the Evidence for the Acoustic Adaptation Hypothesis." *Functional Ecology* (2007): 134–42.

Endler, John A. "Signals, Signal Conditions, and the Direction of Evolution." *American Naturalist* 139 (1992): S125–53.

5

Spatial Ecologies

OVERVIEW. The principal disciplines of spatial ecology that are important to soundscape ecology are summarized in this chapter. These disciplines include landscape ecology, biogeography, and conservation biology. How soundscape ecologists use concepts from these disciplines to study the variability of sound across space is emphasized here.

KEYWORDS: biodiversity, biogeography, conservation biology, ecological land ethic, endangered species, landscape ecology

---

5.1. Landscape Ecology

A soundscape is the acoustic reflection of a landscape (Pijanowski et al. 2011a). A *landscape* is defined as "an area that is spatially heterogeneous in at least one factor of interest" (Turner et al. 2001, 7). A landscape is commonly assumed to be an area a few to several hundred kilometers wide (Forman 1983, 1995); however, a strict definition based on size is often avoided, as it is agreed that the area of a landscape should relate to the problem at hand or the spatial extent of the ecosystem being managed. Landscape ecologists may consider landscapes that are terrestrial, aquatic, or urban (e.g., Ahern 2013; Wu et al. 2013). *Landscape ecology* synthesizes the approaches of geography and ecology (Naveh and Lieberman 2013; Forman and Godron 1981).

Soundscape ecology and landscape ecology share not only many thematic foci, but also a history of intellectual development (Pijanowski and Farina 2011). Landscape ecology evolved not as a "distinct discipline or simply as a branch of ecology, but rather as a synthetic intersection of many related disciplines" (Risser et al. 1984, quoted in Wu 2013). Likewise, soundscape ecology has been presented as a new paradigm that emerges from the synergy of many disciplines and interdisciplinary efforts, including ecology, animal behavior, human psychology, anthropology, and engineering (Pijanowski et al. 2011a). Landscape ecology has also benefited from the development of remote-sensing technologies that help researchers to quantify a landscape, and from tools such as geographic information systems (GIS) that enable them to store, analyze, manipulate, integrate, and visualize large amounts of data (Riitters et al. 1995). The parallels with soundscape ecology are evident, as passive acoustic recorders, acoustic analysis tools, large database technologies, and high-performance computer clusters enable soundscape ecologists to store and analyze temporally and spectrally rich, long-term, regional-scale audio files. Furthermore, the early focus of landscape ecology was tool and metric development (e.g., O'Neill et al. 1988; Turner 1990; Gustafson and Parker 1992; McGarigal 1995; Gustafson 1998; Jaeger 2000; Wu 2004). Much of the work of soundscape ecologists from the mid-2000s to the present has been the development and application of metrics, particularly acoustic indices. Finally, landscape ecologists in the early years had an applied goal of using research in natural resource management and conservation. Soundscape ecology also has many applied goals, such as noise mitigation and conservation of biodiversity (Pijanowski et al. 2002, Dumyahn and Pijanowski 2011a, 2011b).

5.1.1. Guiding Principles of Landscape Ecology

In terms of its thematic focus, the field of landscape ecology has benefited from the identification of fundamental principles that have helped to guide its further development and applications.

Guiding Principle 1: Pattern and Process Are Inextricably Linked

Since the launch of landscape ecology (see Risser et al. 1984 and Turner 1989), it has been recognized that there is an inextricable link between processes and patterns in landscapes. Landscape patterns are arrangements of one type of patch (e.g., forest) within a mosaic of other patch types. Patches are measured by their size, shape, and number and by their position in space. Within each patch are ecological processes common to that habitat; in a forest, for example, those processes might include photosynthesis, evapotranspiration, and nutrient uptake.

If we consider acoustic patterns to be combinations of sounds from biological, geophysical, and anthropogenic sources that have spectral (i.e., frequency, amplitude, duration) characteristics that we can measure at a place and over a defined time frame, then strong parallels between landscape ecology and soundscape ecology can be drawn (Pijanowski et al. 2011a). First, patterns of sounds occurring in a soundscape should reflect processes occurring in the landscape. These processes might include the flowing of a river (a geophysical process) and the songs of breeding birds (a behavioral process). Second, patterns of sound may influence organismal processes, like the evolution of hearing (e.g., auditory filters), in closely related species co-occurring in the landscape.

Guiding Principle 2: Human Activities Are Part of the Landscape

Humans are an integral part of all terrestrial ecosystems and have a unique impact on nearly all aquatic ecosystems as well (Turner 1990; McDonnell and Pickett 1990; Kates 1994; Hobbs 1997; Vitousek et al. 1997; Foley et al. 2005; Crutzen 2006). The most prominent forms of landscape transformation by humans have been the use of natural areas for food production (crops, livestock), the drainage of wetlands to reduce human disease, and the conversion of land to urban uses. Indeed, one of the major themes of landscape ecology is the role that humans play in transforming land for various uses and how these transformations affect ecological processes and spatial patterns on landscapes (Turner et al. 2001).

Landscape ecologists recognize that the surface of a terrestrial landscape serves as the template on which organisms become distributed and interact over space and time. Two key terms are used by landscape ecologists to describe surface content: "land cover" and "land use." *Land cover* (Turner et al. 1990; Lambin et al. 2001) refers to the biophysical features (e.g., vegetation type, soil type, elevation/slope/aspect) of Earth's land surface, whereas *land use* is the human purpose or intent that has shaped those biophysical features. The dual term "land use/cover" is commonly used to encompass both concepts. Currently, around 40% of the global land surface is being used by humans for urban (about 1%) or agricultural purposes (11% for crops and about 30% for pasture). About a third of Earth's terrestrial surface is forested.

What is apparent from research on the relationship of land use and biodiversity (Chapin et al. 1998; Haines-Young 2009) is that biodiversity, measured as *species richness* (number of species), is greatly influenced by land use, as well as by the intensity of the use as reflected in the magnitude of inputs to the landscape (e.g., pesticides, nutrients). Two land uses in particular—urban and agricultural—have significant negative impacts on biodiversity.

Guiding Principle 3: Heterogeneity Controls Fluxes of Animals, Materials, and Energy

Landscape ecologists recognize two ways to describe the spatial heterogeneity of landscapes with respect to land use/cover. The first (Fig. 5.1A) is to consider a landscape using a *patch-corridor-matrix model* (Forman 1995, 2014a), in which patches of natural land use/cover exist within a matrix of hostile areas (e.g., agricultural and urban). Heterogeneity is often quantified by patch size, patch shape, amount of edge and core area, number of patches or patch density, patch size variability, arrangement of patches in space (clumped, dispersed), distance between patches, and the amount of contrast between edges. These measures are used to quantify landscape fragmentation. *Corridors* may be created by long, oblong patches arranged near other patches, allowing organisms to move through the landscape. The second is the *functional heterogeneity landscape model* (Fahrig et al. 2011), which categorizes all land use/cover types by the functional roles they play for organisms (Fig. 5.1B). In this type of landscape model, functional heterogeneity is measured by compositional (number of functional land cover types, such as hedgerows, row crops, pasture) and configuration (number of patches of each functional cover type) metrics. Landscape ecologists also recognize how natural habitats function in the landscape. For example, forest patches along a river operate as riparian zones, which protect the physical land surface from erosion, serve as the major interface between land and surface water, increase the availability of food supplies or nesting sites for some animals, and protect surface water from excessive temperature shifts.

Soundscape ecologists have focused much of their work (e.g., Qi et al. 2008; Pijanowski et al. 2011a) on comparing soundscapes across land use/cover gradients, from the most natural land cover type to the most intensely used by humans. A summary of that work is provided in Chapter 11.

Guiding Principle 4: Scaling Is the Critical Lens for Understanding Landscape Dynamics

It was recognized early in the development of landscape ecology that scale was a central problem in all of ecology and, for the most part, all of science

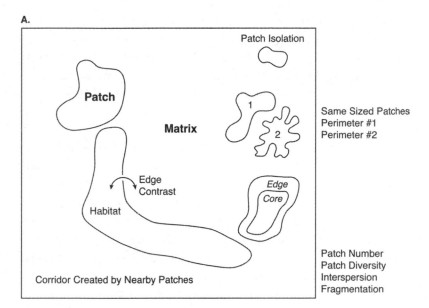

A.

Patch Isolation

Patch

Matrix

Same Sized Patches
Perimeter #1
Perimeter #2

1

2

Edge
Contrast

Edge
Core

Habitat

Patch Number
Patch Diversity
Interspersion
Fragmentation

Corridor Created by Nearby Patches

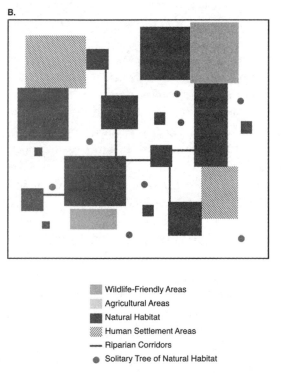

B.

Wildlife-Friendly Areas
Agricultural Areas
Natural Habitat
Human Settlement Areas
Riparian Corridors
Solitary Tree of Natural Habitat

Figure 5.1 Two conceptualizations of heterogeneity in landscapes. (A) The patch-corridor-matrix model. (B) The functional heterogeneity landscape model.

(Levin 1992). Scale has many dimensions, most notably temporal, spatial, and magnitudinal. Landscape ecologists also recognize that how the scale of a pattern affects processes and their feedbacks—that is, how the interactions of components of an ecosystem are hierarchically structured—represents another facet of scaling (see O'Neill et al. 1986; Holling 1992; Allen and Starr 2017). It is also recognized that there is no one correct scale or hierarchical level at which analyses should be performed (Levin 1992). Indeed, many landscape ecologists now recognize that "scaling laws" are rather elusive (Wiens 1989), although the need to analyze, interpret, and apply knowledge across scales is essential for the application of the research. Hierarchical schemas have been proposed, and many have been adopted by landscape ecologists. For example, a landscape can be viewed as a collection of pixels that are organized into groups with similar characteristics and thus create a patch. Patches are organized as a mosaic within a landscape, which then sits within a regional-scale area, and so on, until one reaches the largest scale of the biosphere. Temporal scaling can also be organized in broad hierarchical units of time (seconds through epochs) that are examined with summary information (e.g., average, range of values) across temporal intervals.

Soundscape ecologists are presented with parallel scaling challenges. Let us consider five. Soundscape ecologists are especially challenged by the scaling dimension of *time*. A soundscape is ultimately defined, and its composition identified, by the temporal interval that is selected by the researcher. A soundscape, for example, could be defined as the sounds in a 1-minute recording, or as all the sounds that are present over an entire year at a location. Temporal summaries of acoustic data can also be examined across locations to discover how daily, weekly, monthly, seasonal, annual, or decadal averages across several sites differ. "How long a recording or study is needed to understand how the soundscape and landscape are changing?" is often a critical question for soundscape ecologists that is related to temporal scaling. *Spatial* scaling is also an important consideration for soundscape ecologists. For example, the area that produces the sound that is recorded by a microphone is referred to as an *acoustic theater*, and how it is measured depends on several factors. Low-frequency sounds travel farther, and as such, the acoustic theater for these sounds is rather large compared with that for high-frequency sounds, which do not travel as far. Thus, the frequencies of the sounds create an uneven spatial "footprint" in any audio file. The *magnitude* (i.e., amplitude) of a sound is dependent on several physical attributes, including the amount of power produced at the sound source, the distance between the sound source and the recorder, and the medium through which the sound is traveling. *Spectral features* can also have an element of scale. Sounds can have narrow (pure tones) or broad frequency ranges; spectral scaling also depends on the sampling rate of the recording device. *Organizational scale* is the challenge when soundscape ecologists consider classifying sounds using a hierarchical taxonomy of sound sources. For example, sound sources could be very specific (e.g., down to species), could be placed into general groups (e.g., bird, insect), or could be generalized into broad categories such as anthropogenic, biological, or geophysical.

Guiding Principle 5: Landscapes Are Dynamic

Most landscapes are highly dynamic, and the study of their temporal patterns is referred to as *phenology* (Schwartz et al. 1998). Phenology is the study of those factors most responsible for the timing of plant and animal life-history events (Schwartz 2003). Key life-history events such as flowering, budding, senescence, migration, and breeding have measurable temporal patterns within each ecosystem. There have been many examples of shifts in phenologies of organisms, such as plants, in response to human activities, especially climate change (Cleland et al. 2007; Schwartz et al. 2006). In general, phenology in the mid-latitudes appears to be controlled by temperature, whereas that in the lower latitudes is controlled by precipitation patterns (Cohen- Waeber et al. 2018). In some cases, phenology patterns are controlled by photoperiod (Körner and Basler 2010). For example, the timing of bird migrations is controlled mostly by photoperiod, although minor shifts have been correlated with weather (Marra et al. 2005) or changes in the phenology of vegetation along migratory routes. Among animals, those groups most affected by global warming trends include invertebrates (e.g., insects) and amphibians. The extent to which plant and animal phenologies are coupled is currently unknown, as community interactions are often very complex, involving predator-prey relationships, timing of plant flowering and animal breeding, and the role of peak seasonal greenness in animal activity patterns across diverse ecosystems.

There are also distinct temporal patterns in the soundscape (Pijanowski et al. 2011a). In general, these temporal patterns can be referred to as *phonologies* to parallel those occurring in plant communities. Key questions that soundscape ecologists should ask when considering landscape-soundscape linkages over time include the following: (1) Is the phenology of the plant community in synchrony with either the intensity or the complexity of the soundscape? (2) If they are asynchronous, is there a predictable difference (i.e., lag) between the two temporal patterns? (3) Are the controls (e.g., daylight, temperature) on phenology and phonology the same or different? (4) Are the peaks (maxima) and valleys (minima) of phenology and phonology the same or different? (5) How do stressors and disturbances of the phenology, such as drought or wildfire, affect soundscape phonology? (6) How do intermittent events, such as desert monsoons, that might drive highly temporally defined phenological events affect the soundscape? The temporal patterns of both plants and animals are key to understanding how landscape management affects animals and their acoustic communication.

Although arguably not as commonly applied as landscape ecology has been to the terrestrial system, *seascape ecology* has had some success in applying the fundamental principles developed on the terrestrial surface to aquatic systems (Jones and Andrew 1992; Pittman et al. 2011; Pittman 2017), although mostly to coastal systems. Seascape ecology has been defined (Jones and Andrew 1992) as the field of ecology that examines "the processes that determine establishment, persistence and dynamics of patches of habitat" in aquatic systems. It focuses on diverse types of habitats found along the coast, including reefs (coral, oyster), mangroves, salt marshes, seagrass "meadows," and intertidal zones (Wedding et al. 2011). It also examines how the contributing watersheds alter the physical properties of the seawater, such as its temperature and chemistry. Human activities, such as dredging, anchoring, construction, and other hydrological alterations (e.g., damming), and their impacts on the seascape are considered part of the focus of seascape ecology (Boström et al. 2011). Jones and Andrew (1992) suggested that in seascapes, structural elements that are biological are either "habitat formers" (sedentary species that grow, such as corals), "determiners" (mobile species that control the growth of habitat formers), or "responders" (species that neither are sedentary nor control the growth of habitat formers). Spatial pattern metrics such as patch size, patch shape, and level of fragmentation have been successfully applied to seascape systems to understand the influence of seascape structure on properties such as species survival and growth rates, density, and food web structure (Boström 2011). It is also recognized that three-dimensional metrics, such as surface roughness, that have been developed by terrestrial landscape ecologists are likely to be very important measures of spatial pattern in coastal systems. Recently, seascape ecologists (e.g., Pittman 2017) have proposed that the intersection of soundscape ecology and seascape ecology could be fruitful. They have suggested, for example, that the acoustic indices developed by

Table 5.1 Key differences between landscapes and seascapes

Fundamental process	Landscapes	Seascapes
Longevity of habitat formers	Long, 10,000–50,000 year (e.g., trees)	Short, 1–50 year (e.g., corals)
Recovery from disturbance	Slow	Fast
Patch connectedness (at same extent)	Low	High
Habitat formers	Multiple species	Single species
Habitat complexity	High	Low
Resistance to disturbance	High	Low

Source: Modified from Jones and Andrew (1992).

soundscape ecologists could be used to assess the health of marine ecosystems as they are being affected by factors such as runoff from land, acidification, invasive species, and physical structures of the ocean floor. The major differences between terrestrial landscapes and seascapes are summarized in Table 5.1.

5.1.2. Measuring Landscapes

Ecologists consider several measures of biodiversity. The first is the species diversity that exists in one location or patch; this is referred to as *alpha diversity*. The second is the change in diversity that occurs as one moves between sites; this is referred to as *beta diversity*. The third is the diversity of a region, referred to as *gamma diversity*. Alpha and beta diversity are most often calculated as measures of *species richness*—the total number of species—and *relative abundances* (i.e., the proportion of all individuals that belong to each species). For example, if there are two species living in a patch, of which the first has 50 individuals and the second has 60 individuals, the relative abundance of the first species is 50/110, or 0.45. In addition to these species diversity measures, there is also functional diversity, a measure of variation in organismal traits (e.g., leaf size).

5.1.3. Landscapes and Biodiversity

Fahrig (2013) used the functional heterogeneity landscape model to describe how landscapes moderate biodiversity. This model proposes that all areas of the landscape have some role in moderating biodiversity patterns, as animals can disperse through nearly all areas of a landscape and thus be present there for some part of their life histories (Fahrig et al. 2011). Four main landscape-species interactions are considered in a *landscape-biodiversity outcomes framework* (Tscharntke et al. 2012). First, landscapes control both local (alpha) and regional (gamma) diversity, and they do so in rather simple ways. Fahrig and colleagues argue that the size of the landscape moderates species pools, which can influence alpha diversity in turn; whereas the diversity of habitats in the landscape contributes toward its beta diversity. Second, landscapes moderate *population dynamics*. Some natural habitats operate to allow spillover of populations to other habitats (resulting in spatial variation of population densities), and changes in habitat over time operate to concentrate populations in certain areas or dilute them in others. Third, landscape features select for certain kinds of traits, and thus have a role in moderating functional diversity. Finally, moderately simple landscapes are considered easier to manage for biodiversity than extremely simple (e.g., cleared) or complex landscapes. This rather holistic view of how landscapes moderate biodiversity considers nearly all aspects of biodiversity: species richness, relative abundances, and functional diversity.

5.1.4. Relevance of Landscape Ecology to Soundscape Ecology

Assessments of the relationship between landscape and soundscape patterns should consider (1) what model of landscapes should be used; (2) how landscape heterogeneity should be quantified; and (3) what core dimensions of each—for example, landscapes as areas of certain habitat patches or degrees of fragmentation, and soundscapes as biological sounds, all sounds, or the ratio of biological and anthropogenic sounds—should be examined.

5.2. Biogeography

Biogeography is the science that seeks to understand geographic variation in nature, from genes to communities, as it relates to patterns at regional to global scales (Lomolino et al. 2010). It combines ecological, evolutionary, behavioral, genetic, and abiotic factors that drive the variability of organisms across space. Its very broad and highly multidisciplinary nature has led to its being described as a "big science" (Whittaker 2014). Historically, biogeography has not considered sound as one of the features of organisms that it studies, but new efforts (Lomolino et al. 2017; Whittaker et al. 2018; Lomolino and Pijanowski 2020) are under way to describe how animal communication and organismal use of natural sounds can be part of biogeography's integrative approach.

5.2.1. Understanding All Levels of Biological Organization

Biogeography has sought to understand how all levels of biological organization, from the gene to the individual to entire regional biotas, change over space and time (Lomolino et al. 2010). One major focus has been how natural selection affects adaptations and the variability of genotypes and phenotypes at the individual to population levels. In addition, there have been major efforts to understand how variability at each of these levels of biological organization has been moderated by the spatial and temporal dynamics of the geophysical environment.

5.2.2. Biomes/Life Zones/Ecoregions

As a major focus of biogeography has been to understand the underlying dynamics of organismal communities across Earth's major biomes, soundscape ecologists should leverage the wealth of concepts, known distribution patterns of species across large spaces, and comprehensive biological perspectives that biogeography contains. Biogeographers and ecologists recognize that Earth's land cover is composed of several major forms of vegetation, which are often referred to as *biomes* or *life zones* (Merriam 1898), and most recently *ecoregions* (Bailey 1995; Olson and Dinerstein 2002). The major terrestrial biomes are forest, woodland, shrubland, grassland, scrub, and desert (Lomolino et al. 2010). Biogeographers and remote-sensing experts have developed several more detailed ecoregion classifications that span a dozen to several dozen classes. The common approach to the development of ecoregion classification systems has been to create homogeneous spatial regions that have similar vegetation cover and climate patterns, often called *bioclimatic zones*. Two of the classification systems most commonly used by biogeographers are the one presented by Olson et al. (2001) and the Köppen-Gieger classification system (Beck et al. 2018). Because ecoregions are composed of distinct plant and animal communities, studies of soundscapes need to take into consideration the types of animals that exist within each ecoregion and the geophonic sounds produced by the climate dynamics characteristic of each ecoregion.

Classifications of aquatic ecosystems are not entirely equivalent to the system of terrestrial ecoregions, as they are defined mostly by depth and are commonly separated into two principal divisions: freshwater and marine systems. Common marine ecosystems include *coral reefs* and *algal beds*, *estuaries* (some of which are also referred to as *mangroves*), *upwelling zones*, *continental shelves*, and the *open ocean*. Freshwater ecosystems are separated into two major classes on the basis of water movement: *lotic* (running water such as rivers, springs, or streams) and *lentic* (ponds and lakes). Lentic systems often contain a variety of zones based on depth, and lakes can be classified as either *eutrophic* (shallow and productive because light penetrates much of the water) or *oligotrophic* (deep and relatively low in productivity because they have few nutrient inputs and because parts of the lake are not penetrated by light). Freshwater ecosystems also include *swamps* (woody vegetation), *marshes* (herbaceous vegetation), and *peatlands* (two types are recognized, *bogs* and *fens*, where the source of water is the distinguishing feature).

5.2.3. Global Patterns of Organismal Traits

Biogeographers have long recognized several distinct global patterns of traits in both plant and animal communities, referred to as *ecogeographic patterns* (Lomolino et al. 2010). Biogeographers seek to understand the mechanisms that influence these patterns. Many latitudinal patterns of traits have also been examined along altitudinal and, for marine systems, bathymetric ranges to search for analogous patterns and identify any causal mechanisms they may have in common. Several ecogeographic patterns have been examined in considerable detail.

Bergmann's Rule

In the mid-1800s, German zoologist Carl Bergmann noticed that the largest-bodied species among taxonomic groups of birds and mammals tended to occur at high latitudes. He hypothesized that this pattern, now commonly referred to as Bergmann's rule, occurred because larger animals have smaller surface-to-volume ratios, which make those animals better able to retain heat in cold environments. Other explanations for this trend (see review in Lomolino et al. 2010) include those that focus on dispersal abilities, need for long-term storage of energy, and differences in competition or predator-prey dynamics that occur across latitudes. These trends of larger body sizes at higher latitudes have been demonstrated for mammals, birds, amphibians, reptiles, and fish. Similar trends have been observed in altitudinal or bathymetric variation in ants, gastropods, fruit flies, and flatworms. If the fundamental frequencies of acoustic communication are influenced by body size, then soundscape ecologists should find that the average frequencies of biophonies decrease as we move from the equator to the poles.

Allen's Rule

The American ornithologist Joel Asaph Allen (1877) noticed that among closely related species of endothermic vertebrates (mammals and birds), those living in hotter climates have longer appendages. Allen hypothesized that animals living in hotter climates need to dissipate heat more easily, and do so by having longer appendages in which blood vessels can carry fluids to distal areas where heat dissipates to the environment. The most common examples cited include the pinnae (external ear structures) of hares (i.e., rabbits), which are long in deserts and short in cold environments such as the tundra. How general these patterns are across animals is still debated (see Lomolino et al. 2010), but if the sizes of ears, snouts, and other body parts are determined by climate factors, that could add constraints on the animals' sound production abilities, resulting in patterns that may be evident in cross-ecosystem comparisons across latitudes.

Latitudinal Diversity Gradient

Species numbers or species densities for most taxa decrease as one moves from the equator to the poles. This latitudinal diversity gradient has fascinated biogeographers for centuries, dating back to the foundational research of Forster (1778) and Humboldt (1808). As summarized by Lomolino et al. (2010), the latitudinal diversity gradient has been reported for most major organismal groups (e.g., plants, animals, fungi, microbes), at most taxonomic levels (class, order, family), and for organisms both terrestrial and aquatic. Dozens of hypotheses have been put forth to explain this global trend. Many elegant essays have been written on the topic, most notably Hutchinson's (1959) famous essay, "Homage to Santa Rosalia or Why Are There So Many Kinds of Animals?" and James H. Brown's (2014) more recent "Why Are There So Many Species in the Tropics?" Both ecologists point to several important features of equatorial regions that make them likely to support more species than the higher latitudes: a more stable temperature throughout the year, less extreme temperatures, faster evolution, and more plant biomass.

Variation in species densities from the equator to the poles is stunning. For vascular plants, species densities exceed 5,000 species per 10,000 km^2 in the equatorial regions of South America, Asia, and parts of Africa. Higher latitudes typically support fewer than 50 species per 10,000 km^2. Mammals are also more diverse in equatorial regions of land and oceans, with as much as a 200-fold difference between equatorial and higher-latitude areas. Bird diversity, too, is greatest in tropical regions, with global maxima in the Andes region of Colombia and Peru and in central Africa. Amphibian species richness is greatest in the South American Amazon, Western Africa, northern areas of the island of Borneo, and several locations in Southeast Asia. Freshwater fish species richness is also greatest in central South America and in Southeast Asia. In short, it is estimated that over one-half to two-thirds of all global species are in the tropics (Dirzo and Raven 2003).

Soundscape ecologists are in a unique position to use their data across latitudinal gradients to document and dive deeply into the diversity of animal sounds, and hence species diversity, and to use this information to examine more closely the numerous competing hypotheses proposed to explain this global trend.

Rarity Patterns

Rare species—those with low abundances—occur around the world. Rarity is evident in two ways. First, some species have broad ranges, but their densities across those ranges are low. A second pattern of rarity is endemism, which is the occurrence of a species within a limited geographic range. Some locations around the world have relatively high *endemicity*—that is, they harbor many endemic species. These places include Réunion, an island in the Indian Ocean; ecoregions called páramos in the South American Andes; sky islands (isolated mountains) in the US southwestern desert; and large, isolated island archipelagoes such as New Zealand and the Philippines. Many species of plants and animals in tropical regions are also very rare, probably due to a "numerical" trade-off between species richness and the numbers of individuals that can be supported by the primary productivity that exists in those regions of the world.

Rarity may influence several features of the biophonic pattern at a location. First, if a species is rare in the landscape, its occurrence in the soundscape is also likely to be rare. It may not occur in any recording database, and therefore its acoustic signals will be difficult to identify. Thus, it may be challenging to create a catalog of sounds for locations such as the Amazon rainforest, where species richness is high and many species are rare. Second, rarity may also mean that the species' acoustic signal will differ only slightly from that of its ancestral species, or those of closely related species, because evolution may not be selecting for strong acoustic partitioning.

Mountain Passes

In an elegant and well-known essay, ecologist Daniel Janzen (1967) hypothesized that mountain passes in the tropics are "higher" than those in the mid-latitudes. First, he argues, temperatures at specific elevations in tropical mountain ranges remain relatively stable over the course of a year and, indeed, over many generations of the local biota. Second, because of their stable environments, tropical species tend to specialize and evolve relatively narrow physiological tolerances and geographic distributions. In the mid-latitudes, however, the seasonal and year-to-year variation in climate conditions dictate that species be more generalized in their physiological tolerances. This, in turn, allows species along mountain ranges in the mid-latitudes to occupy a larger geographic range. Placed in the context of acoustic niche evolution and, more broadly, the sensory drive framework, the community assemblages of mountain ecosystems in the tropics should be more tightly evolved and hence may have, compared with assemblages of related species in the mid-latitudes, more distinct acoustic niches.

5.2.4. Geological History of Land

Biogeographers often consider the geological history of the area they are studying to understand the ecological processes occurring there. Indeed, many ecologists have recognized that it is challenging to distinguish between the effects of historical and current ecological factors on the distribution and interactions of species (Endler 1992; Jetz et al. 2004). This challenge might be taken up by soundscape ecologists, too, as they attempt to understand the distributions of the animals, and thus the acoustic signals, that occur at a place. For example, some areas of the world were glaciated only 10,000–40,000 years ago. Much of the central portion of North America north of 40° latitude was covered by glaciers, and thus the current distribution of animals in this region is a recent phenomenon. Thus the species co-occurring in today's ecosystems are often different from those that coexisted during previous glacial and interglacial periods. Likewise, the drop in sea level caused by the formation of glaciers transformed shallow seas into land bridges that enabled dispersal and mixing of faunas across otherwise isolated regions (e.g., Southeast Asia and the islands of Java, Sumatra, and Borneo). As glaciers melted and sea levels rose again, new mixes of species were isolated on these islands. How might these historical events affect the way acoustic niches are partitioned to reduce overlap by the multiple species (and often novel species combinations) that might have colonized the same area? How do these geological histories affect the way that biophonies have evolved over evolutionary time? For example, are younger ecosystems more "finely tuned" across all species in their adaptations to the habitat, and to the acoustic niches that have resulted from colonization, than older ecosystems that have not undergone such changes? Are younger ecosystems more closely biophonically arranged than a random model? In other words, are climatically stable places such as the rainforests of the Amazon more optimally adjusted to acoustic partitioning than younger ecosystems?

5.2.5. Intersection of Biogeography and Soundscape Ecology

The intersection of soundscape ecology and biogeography has been labeled *sonoric biogeography* (Lomolino and Pijanowski 2020). Surprisingly, biogeographers have been relatively "silent" on the role that acoustic communication might play in shaping the distributions of species globally. But soundscape ecologists can conduct comparative studies across Earth's ecoregions to address many of the spectral hypotheses described in Chapter 4 in the context of the rich intellectual history of biogeography. We can examine these hypotheses within the fundamental geographic dimensions of biogeography: area, isolation, elevation/depth, and latitude. Table 5.2 summarizes the four persistent themes of biogeography and lists their possible parallels in soundscapes that could drive an assessment of the relationship between these themes and the variability of soundscapes globally.

5.3. Conservation Biology

Major conservation organizations around the world have begun to use passive acoustic recorders (PARs), acoustic indices, and soundscape ecological principles to help them assess and manage protected areas. This section presents a summary of the origins of conservation biology in modern Western culture, standard definitions of protected areas developed by international organizations, the scope and practice of conservation biology today, approaches to monitoring and assessment of management

Table 5.2 Persistent themes of biogeography and those that relate to soundscape ecology

Biogeographic theme	Soundscape ecology parallel
Classifying regions by their biotas	Can soundscapes be similarly classified by biota, and is there homogeneity within, or are there characteristics that distinguish between, global regions such as biomes?
Reconstructing historical development of lineages	Are there clear patterns of acoustic communication across lineages—for example, do closely related species have the most similar acoustic signals? Are there parallel developments of nonacoustic sensory capabilities among closely related species?
Explaining the differences in abundances and species richness among geographic regions	Are the biophonies across geographic regions influenced by the same abundance and species richness trends that are found regionally and globally, especially across latitudes?
Explaining the differences in geographic variation between closely related species, including trends in morphology, physiology, behavior, genetics, and demography	Does biophony vary in locations where there are many closely related species, and/or are the trends in body size, demography, etc., reflected in the composition of biophony?

Source: Modified from Lomolino and Pijanowski (2020).

practices, and principles of conservation planning as well as its sociocultural dimensions.

5.3.1. Origins of Modern Western Conservation Thinking

There have been three distinct historical origins for the concept of conservation of nature in Western culture (Callicott 1990). The first historical strand originated with eighteenth- and nineteenth-century Enlightenment scholars who articulated a *romantic-transcendental preservation ethic*. These individuals, most notably Ralph Waldo Emerson, Henry David Thoreau, and John Muir, argued that nature had great aesthetic and spiritual value, which was counter to the contemporary view that nature existed purely for human consumption. In the view of Emerson and others, nature should be treated as a temple. A second strand came from scholars who saw a need to sustain natural resources in the present and into the future. Often referred to as the *resource conservation ethic*, it was a widely held view in the late nineteenth and early twentieth centuries in North America and Europe.

The third view, originally presented by nineteenth-century naturalists from Europe (such as botanist Alexander Gibson) and later by wildlife ecologist Aldo Leopold, is referred to as the *ecological land ethic*. It posits that nature is a system of complex processes that are highly integrated and that no one component can be preserved without the others. In *A Sand County Almanac* (Leopold 1949), Leopold argues eloquently that people have an ethical responsibility to conserve nature and that, because humans are part of nature, conserving it will ultimately have a positive effect on human society. This third view of conservation is driven by science, as opposed to aesthetics (like the romantic-transcendental preservation ethic) or economics (like the resource conservation ethic). The ecological land ethic argues for use of land and water resources in a way that protects all the elements and processes of a place based on an integrated understanding of the ecosystem.

As we shall see in the next chapter, natural soundscapes can have spiritual value to local communities. In addition, the natural sounds of a place have high recreational value and positive effects on mental well-being (see Chapters 6 and 13). And soundscapes can be used to assess the ecological health of a location (see Chapter 11). Although all three of these values have relevance to soundscape ecological work, the last—the role of soundscapes in assessing the ecological health of a landscape—is emphasized by most soundscape ecologists.

5.3.2. Twenty-First-Century Motivation for the Conservation of Nature

Several assessments in the past twenty years have summarized the status and trends of the Earth system since industrialization. The most comprehensive assessment was the United Nations' Millennium Ecosystem Assessment (MEA) (2005), which involved over 1,500 scientists and policy analysts who synthesized current knowledge of the status of ecosystems and human well-being. To accomplish this, they developed a framework that categorized *ecosystem services*—the ecological elements and processes that human life depends on—into four classes: *provisioning* (e.g., food, water, fiber), *regulating* (e.g., air and water filtering), *cultural* (e.g., spiritual, creation), and *supporting* (e.g., soil formation) services. They used this framework to assess those ecological services as they are being affected by drivers of change (e.g., land use change, chemical inputs, climate change). They also reported that current rates of extinction for amphibians, birds, and mammals are 100 to 1,000 times greater than rates calculated from the fossil record, and that projections made by models showed that future rates would be more than 10 times greater than current rates. Following the MEA, many of its lead scientists (Carpenter et al. 2006) assessed research gaps. Their findings included the need to advance monitoring technologies that can go uninterrupted for a sufficient length of time to determine species richness trends, to link ecological and sociocultural indicators of change and well-being, and to develop more transdisciplinary approaches that can integrate the ecological and social disciplines that address the complexity of the Earth system (Carpenter et al. 2009).

Another widely discussed framework for assessing the status of Earth's ecosystems is the concept of planetary boundaries (Rockstrom et al. 2009; Steffan et al. 2015). *Planetary boundaries* are a set of global threshold values that the total societal use of the Earth system must not cross if we are to maintain natural and social well-being. Nine broad classes of planetary boundaries were identified, and each was placed into one of three status zones: safe, critical, or exceeded. The nine boundary processes are land use change, freshwater use, biogeochemical flows (of nitrogen and phosphorus), ocean acidification, atmospheric loading, ozone depletion, novel entities (e.g., releasing human-made substances into the environment), climate change, and biologic integrity (i.e., biodiversity). Current assessments show that we have exceeded the planetary boundaries for biologic integrity and nitrogen flows, and are within the critical zone for climate change.

More recent assessments of global biodiversity trends are alarming. The International Union for Conservation of Nature (IUCN) reported in 2019 that one in seven of the species that are known to scientists are classified as threatened or endangered. An assessment of North American bird abundances (Rosenberg et al. 2019) has shown that over the past fifty years, populations of 529 species have decreased by nearly 30%, or by a total of three billion individuals. Birds in some ecosystems have had more declines than those in others; for example, grassland bird populations have declined by as much as 75% over this fifty-year period. The percentages of species in major taxonomic groups of animals (Table 5.3) classified as endangered or threatened range from 12% for mammals to 42% for amphibians.

5.3.3. Protected Area Typologies and Major Conservation Organizations

Conservation practices focus on a variety of goals to ensure that nature is preserved. Arguably, two of the most common approaches to its preservation can be distinguished by the different spatial scales at which they focus. The first approach is to use a global *biodiversity hot spot* analysis that identifies those regions of the world—terrestrial and marine—that harbor

Table 5.3 Percentage of species in each taxonomic group considered endangered or threatened

Group	Percentage	Comments
Amphibians	42	Most species use acoustic communication. Species are known to have natural "boom-and-bust" population cycles, and habitats are being lost and threatened by climate change; thus long-term acoustic monitoring is required. Catalog of calls is poor for most groups.
Grasshoppers	39	Many create sounds through stridulation and through wingbeats. Monitoring in natural habitat of grasslands shows they are changing due to the presence of wind.
Freshwater fish	34	Catalog of calls is poor.
Freshwater shrimp	32	Catalog of calls is poor.
Tiger beetles	23	Most species make audible sounds through stridulation, but there is no catalog of these sounds.
Mammals	17	A majority of species make sounds. Catalogs of sounds for some groups (bats, marine mammals) are very good, but those for many groups are poor. Catalogs of sounds for some species (primates) are extensive, but vocalizations are complex.
Birds	12	Catalogs of calls across the entire group are very good and excellent for some.

Source: Data from Kareiva and Marvier (2012); see also Noss et al. (2013).

the greatest numbers of species. Developed first for tropical rainforests (Myers 1988), this approach uses mapping technologies such as GIS to identify the spatial boundaries of "exceptional concentrations of endemic species which are experiencing rapid loss of habitat" (Myers et al. 2000, 853; see also Reid 1998). In one of the most comprehensive biodiversity hot spot analyses conducted, which illustrates the potential of this approach for conservation, Myers and colleagues (2000) used spatial distributions of plants and animals to delineate twenty-five geographic hot spots that harbored 44% of the known species in their global database within an area that represented 1.4% of Earth's terrestrial surface.

A second approach has focused on managing existing *protected areas*. The IUCN defines a protected area as "a clearly defined geographical space, recognized, dedicated and managed, through legal or other effective means, to achieve the long-term conservation of nature with associated ecosystem services and cultural values" (Dudley 2014, 8). The IUCN has worked across a va-

Table 5.4 Protected area management categories

Category	Category name	Description
Ia	Strict nature reserve	Strictly protected for biodiversity and also possibly geological/ geomorphological features, where human visitation, use, and impacts are controlled and limited to ensure protection of the conservation values
Ib	Wilderness area	Usually large unmodified or slightly modified areas, retaining their natural character and influence, without permanent or significant human habitation, protected and managed to preserve their natural condition
II	National park	Large natural or near-natural areas protecting large-scale ecological processes with characteristic species and ecosystems, which also offer environmentally and culturally compatible spiritual, scientific, educational, recreational, and visitor opportunities
III	Natural monument or feature	Areas set aside to protect a specific natural monument, which can be a landform, seamount, marine cavern, geological feature such as a cave, or a living feature such as an ancient grove
IV	Habitat/species management area	Areas set up to protect particular species or habitats, where management reflects this priority. Many will need regular, active interventions to meet the needs of particular species or habitats, but this is not a requirement of the category
V	Protected landscape or seascape	Areas where the interaction of people and nature over time has produced a distinct character with significant ecological, biological, cultural, and scenic value, and where safeguarding the integrity of this interaction is vital to protecting and sustaining the area and its associated nature conservation and other values
VI	Protected area with sustainable use of natural resources	Areas that conserve ecosystems, together with associated cultural values and traditional natural resource management systems. Generally large, mainly in a natural condition, with a proportion under sustainable natural resource management and where low-level, non-industrial natural resource use compatible with nature conservation is seen as one of the main aims

Source: Modified from Dudley (2014).

riety of conservation organizations to develop a generalized categorization of protected area definitions (Table 5.4) and governance types (Table 5.5).

Currently, over 190 governments have signed the *Convention on Biological Diversity* (CBD), which pledges all governments to conserve biological diversity, promote sustainable use of the components of biological diversity,

Table 5.5 Protected area governance types

Type	Description
Governance by governments	Federal/national/ministry organization in charge, subnational ministry or agency in charge, or government-delegated (such as to a nongovernment agency, or NGO).
Shared governance	Collaborative management, joint management, or transboundary management
Private governance	By individual owner, nonprofit organization, or for-profit organization
Local community governance	By indigenous or local communities; declared and run by local communities.

Source: Modified from Dudley (2014).

and support the fair and equitable sharing of the benefits of the biosphere's genetic resources (CBD 2021; cbd.int/intro). In 2010, the CBD established a set of targets called the *Aichi Biodiversity Targets*, which state that by 2020, at least 17% of Earth's land surface and at least 10% of coastal and marine ecosystems will be part of protected areas with the goal of conservation of nature. Currently, the World Database on Protected Areas (WDPA), which is updated monthly by governments, lists over 258,000 protected areas, which cover 15.1% of the terrestrial surface and 7.7% of the marine area (protectedplanet.net/en, accessed Oct 5, 2021). The total footprint of protected areas has increased since the IUCN was founded, with a fourfold increase for terrestrial protected areas and a twentyfold increase for marine and coastal protected areas. Despite these achievements, many scientists and conservation managers believe that the total footprint of protected areas is far short of what is needed to protect biodiversity and the ecological and evolutionary processes that underpin it.

Although it is well recognized that protected areas are helping to support efforts to maintain biodiversity at local, regional, and global scales, the effectiveness of management of these areas, and how well the existing network of terrestrial and marine protected areas preserves global biodiversity patterns, is still under debate (Rodrigues et al. 2004; Gaston et al. 2008; Watson et al. 2014; Gray et al. 2016; Hanson et al. 2020; Ward et al. 2020). In general, it has been found that biodiversity tends to be greater inside terrestrial protected areas than outside them (Gray et al. 2016). What is often lacking, however, are adequate monitoring programs, research that examines how well management practices meet conservation goals, and a determination of whether low to moderate levels of human activities in protected areas contribute toward their degradation (Jones et al. 2018). In addition, other factors that can decrease biodiversity, including illegal hunting, disturbance by recreational activities, and natural disturbance by wildfire or the retention of natural fire regimes, are not well monitored (Schulze et al. 2018).

Several conservation organizations (World Wildlife Fund, The Nature Conservancy, Conservation International) have used PARs and soundscape ecological principles to help monitor and assess large areas of terrestrial and marine ecosystems for changes in biodiversity. Two of the earliest such efforts were those of the US National Park Service, coordinated in the Natural Sounds and Night Skies (NSNS) office in Fort Collins, Colorado (discussed in Chapter 11), and Conservational International's multi-sensor, multi-site Tropical Ecology Assessment and Monitoring (TEAM) program.

5.3.4. Characteristics of Modern Western Conservation Biology

Soulé (1985) has argued that conservation biology's goal is to understand the principles of biodiversity and to develop plans and tools for preserving this biodiversity on Earth. Modern conservation biology has several important features. First, conservation biology is a *synthetic* science, spanning not only ecology, evolution, behavior, and geophysical sciences, but also social sciences, planning and management, economics, ethics, education, and law (e.g., environmental justice). Second, it has been recognized as a *crisis discipline*, much like cancer biology, as it often requires action before all data are available and analyzed and conclusions are drawn. In other words, decisions are often made with considerable uncertainty. Third, it is *holistic*, in the sense that focusing on a single species or a short timeline is not considered a valuable goal; rather, it seeks to understand the ecological and evolutionary processes (Moritz 1994) of whole ecosystems at macroscopic levels. Fourth, it is inherently *spatial* (Pressey et al. 2007), as it uses the concepts of landscape ecology and biogeography and tools such as GIS and remote sensing.

5.3.5. Monitoring and Assessment in Conservation Science

To determine whether protected area management is effective at local, regional, and even global scales, monitoring programs are often put into place. These programs integrate a variety of data collection and analysis approaches to create indicators of biodiversity and habitat alteration. Noss (1990) argued that monitoring of biodiversity should entail the *hierarchical use of indicators* focused at four scales: (1) regional landscape, (2) community-ecosystem, (3) species-population, and (4) genetic. This top-down, hierarchical approach moves from a coarse-scale assessment of the landscape using maps of satellite data, to tracking animal communities, assessing vegetation diversity across space, and assessing ecosystem dynamics (e.g., water availability), to assessing species-population variability (e.g., abundances and species spatial-temporal variation) and genetic variability. At each of these scales, monitoring should seek to quantify the composition, structure, and function of the area being studied.

Monitoring for biodiversity change should focus on (1) an entire group of organisms (e.g., birds) so that biodiversity trends can be assessed (Butchart et al. 2010); (2) one or more endangered or threatened species that exist in the protected area (Noss 1990); or (3) one or a limited set of "flagship," "umbrella," or "keystone" species that may reflect change across ecosystems (Simberloff 1998). Simberloff (1998) recognized that landscape-level assessments were needed for monitoring biodiversity, but that assessments focused on one or a few species could be cost-effective. Monitoring of *umbrella* species, which are species that are common across a management unit, would track organisms with a large footprint and would thus assess those management practices that have the largest spatial impacts. Monitoring of *flagship* species, which are those with charismatic appeal, would be useful to generate public enthusiasm and, potentially, funding. *Keystone* species could also be monitored, as they are the species that have the broadest impacts on the ecosystem. Simberloff argued that conservation managers should select the one of these three species-specific approaches that connects best with landscape-scale patterns and processes. He adds that monitoring of keystone species, if only one or a few species must be chosen, might be the most consistent with a landscape-scale management paradigm. Keystone species may not, however, be a panacea as indicators of biodiversity, as they are not likely to be present in all ecosystems (Simberloff 1998).

The application of networks of PARs, acoustic indices, and soundscape ecological principles could be well integrated with the hierarchical indicator approach of Noss, along with some of the species-specific ideas suggested by Simberloff. For example, a *keynote* species, to borrow a term used by Schafer to refer to an "acoustic keystone species," is a species that has an important role in the ecosystem and is common enough to be an indicator of habitat quality at the landscape scale. As Simberloff suggests, relying on a single species does not provide adequate information about the status and trends of a management area; however, a *multi–keynote species indicator*, in the form of a keynote species acoustic index, could capture the spatial and organizational taxonomic scale necessary to assess the status and trends of management units. Acoustic monitoring could also provide soundscape recordings that could be played back to the public (via kiosks, exhibits, websites, etc.), bringing the charismatic appeal of a place to one of the most emotionally important human senses.

5.3.6. Conservation Planning Approaches

Several commonly adopted conservation planning procedures use an approach referred to as *systematic conservation planning* (Margules and Pressey 2001; Groves et al. 2002; Sarkar and Illoldi-Rangel 2010; Pressey et al. 2007). This approach (Table 5.6) involves developing measures or surrogates for biodiversity, developing conservation goals or targets, identifying high-priority conservation areas and potential spatial gaps in protection, and implementing a monitoring program. Consistent with the hierarchical monitoring approach of Noss (1990), systematic conservation planning involves a series of steps that are fully integrated and often repeated on a regular basis, such as monitoring → intervention → assessment → adoption of new management practices. This sequence of recursive steps employs what are known as *adaptive management principles* (Holling 1978; Walters 1986; Walters and Hilborn 1978), which emphasize an iterative process of examining and reexamining conservation goals, experimenting with new approaches, analyzing data from experiments and monitoring, and altering plans on the basis of new evidence, analyses, and research results. In addition, conservation planning should leverage traditional software tools, such as GIS and satellite remote sensing, and integrate them with new hardware and software tools that reduce costs, increase reliability, and allow for a reduction in time from data analysis to decision making (Sarkar and Illoldi-Rangel 2010).

5.3.7. Sociocultural Dimensions of Conservation

Conservation managers recognize that landscape management for conservation must address several key sociocultural issues. They recognize that

Table 5.6 Stages of systematic conservation planning

Stage	Examples of activities
1. Measure and map biodiversity	Measure status and locations of individuals, populations, communities, and ecosystems across the heterogeneous landscapes. If resources are available, focus intensely on rare and endangered species
2. Set conservation goals	Establish quantitative targets, such as number of species present, vegetation type, sizes and features of habitats, and the kinds and intensities of disturbances that may occur (if involving human activities)
3. Review existing conservation areas	Assess how current goals are being met by existing conservation areas and identify threats to those areas
4. Select additional conservation areas	Determine if other areas could help support conservation goals and how they can do so (e.g., increase connectivity, add more critical habitat)
5. Implement conservation practices	Add more protected areas, change restrictions
6. Manage and monitor conservation areas	Use PARs and other monitoring technologies to track changes in biodiversity over space and time

Source: Modified from Margulis and Pressey (2000).
Note: Implementation is not stepwise, but rather iterative, with all stages explored during management.

public support needs to be engendered—locally, regionally, or globally—to reach conservation goals. These sociocultural dimensions of conservation are more acutely challenging in developing countries, where there are many biodiversity hot spots, but excluding people from these areas is culturally problematic, since many local indigenous communities have used these lands for centuries. To address this issue, sometimes referred to as the *conservation-development mix debate* (Salafsky 2011), trade-offs must be examined and goals adjusted to allow for use of protected area resources by local communities. Conservation managers must also consider the kinds and levels of human activities that occur *outside* managed areas, as these activities can affect conservation goals inside the boundaries of the area that is being managed.

Engaging the Public in Conservation

There is considerable debate on how to engage the public in conservation efforts, as engagement requires addressing both facts (pursued through science) and values (based on ethics). Indeed, promoting conservation practices is often considered *advocacy* (Brussard and Tull 2007). As noted by Cai et al. (2008), advocacy for biodiversity requires a careful set of considerations that should place a high priority on science and policies. In addition, conservation management requires public education, so that people are aware of the problems, potential solutions, and management goals (Brussard and Tull 2007).

Conservation-Development Interplay

The relationships between poverty and the environment are very complex and poorly understood. An assessment of global biodiversity hot spots and poverty levels (Sachs et al. 2009) has shown that over 20% of the land surface is composed of locations with both high numbers of threatened and endangered species and high levels of poverty; many of these areas are in sub-Saharan Africa, Southeast Asia, and the central portions of South America (see also Naughton-Treves et al. 2005). The co-occurrence of biodiversity and poverty requires managers to balance protection and use of the natural resources in these landscapes (Salafsky 2011). A more recent assessment of protected areas showed that globally, nearly one-third of all protected land is under intense human development pressure (Jones et al. 2018). At the local level, some conversation scientists and managers have drawn on traditional ecological knowledge, developed over centuries by local communities that have managed landscapes sustainably through generations, to help support conservation goals. Chapter 6 presents a more thorough overview of TEK and its potential to interface with soundscape monitoring. Some scholars have suggested (e.g., Fisher et al. 2021) that conservation also needs to occur outside protected areas, particularly in commodity production landscapes such as those that support forestry (i.e., wood products) and agriculture. If production landscapes are to continue to support ecosystem services such as healthy soils, clean water, natural nutrient cycling, and waste decomposition, they argue, then biodiversity will need to be maintained in those landscapes. High levels of biodiversity will also mean that pollination, seed dispersal, and pest species regulation—all of which are required for commodity production—will continue to occur in those landscapes. To ensure high levels of biodiversity in production landscapes, researchers have suggested that they be managed by increasing the number of patches of natural vegetation, creating corridors to connect them, and buffering sensitive areas within the landscape matrix of production and natural vegetation cover.

Summary

This chapter, along with Chapter 4, describes the core concepts in ecology that frame how soundscape ecologists can examine the dynamics of the soundscapes they study. The fields of landscape ecology, biogeography, and conservation biology have rich histories. The concepts from these fields presented here should serve as a starting point for more research that helps scientists to understand the variability of the soundscape given the ecological, evolutionary, and behavioral variability of animal life on Earth. Conservation biology can benefit from knowledge produced by soundscape ecology and sensory ecology as well.

Discussion Questions

1. Pick two locations on Earth. Using Google Earth and a screen grab utility, capture the landscapes of these places. Find information for each of these two locations to place them in their Köppen climate zones. Describe the spatial features of these locations in terms of land cover patterns.

2. Use the internet to find out how the geophysical environment in each of the locations you chose in Question 1 changes over time. How variable are daylight per day throughout the year, monthly average temperature, and monthly average precipitation? Describe these climate patterns in relationship to the ecosystem for these two locations.

3. Using online resources, find a conservation plan and critique its public outreach plans. How extensive are they? How do they address knowledge gaps and cultural values?

Further Reading

Groom, Martha J., Gary K. Meffe, and C. Ronald Carroll. *Principles of Conservation Biology*. 3rd ed. Sinauer Associates, 2006.

Lomolino, Mark V., Brett R. Riddle, and Robert J. Whittaker. *Biogeography: Biological Diversity across Space and Time*. 5th ed. Sinauer Associates, 2017.

Turner, Monica G., Robert H. Gardner, and Robert V. O'Neill. *Landscape Ecology in Theory and Practice: Pattern and Process*. Springer, 2001.

Sociocultural Concepts

6

OVERVIEW. Just as landscape ecology has incorporated people—their behaviors, attitudes, and management actions in space—into its disciplinary focus, soundscape ecology must include people in its inquiries, and must do so in a way that builds on advances in the social sciences and humanities (see Guyette and Post 2016). This chapter provides an introduction to and overview of relevant concepts and theories that either have emerged outside the ecological sciences or have recently been integrated into them, but not specifically through sound. Concepts and theories from the social sciences (e.g., anthropology, psychology) and humanities (e.g., ethnomusicology) will support a more holistic understanding of the role that sound plays in people's perceptions of their environment. We review the fields of ethnomusicology, sensory ethnography, ecomusicology, sensuous geography, sense of place, traditional ecological knowledge (TEK), nature connectedness, psychoacoustics, and soundscape values. We begin by exploring how these different disciplines have defined perception. The areas of study summarized here are not mutually exclusive, and the labels for their disciplines, and even the specific terms they use, are currently evolving and at times represent lines of scholarship that have become blurred due to the variety of emphases chosen by scholars in these fields.

KEYWORDS: ecomusicology, emotions, ethnomusicology, nature connectedness, perception, place attachment, sense of place, sensorium, sensory ethnography

---

6.1. The Key Concept of Perception

Gibson (1966) is often cited as the first scholar to develop a multidisciplinary concept of perception. *Perception*, to Gibson, is the act of gathering important, survival-based information about the environment through an interrelated set of senses as an organism moves through space collecting environmental information over time. Senses are used not only for passive information gathering, but also for seeking information actively. Perception is therefore a dynamic process for gathering information about the environment, in contrast to the earlier paradigm of sensation-based perception. Many organisms, including humans, perceive their environment through five major senses possessed by most vertebrates: vision, hearing, touch, smell, and taste. Some scholars suggest that other senses may exist as well, such as a sense of direction and a sense of time, and that not all animals have the same level of development across sensory systems, as there are a variety of specialized cells, tissues, organs, and neurological capacities across all animals that receive, process, and store information. Organisms perceive their environment by using all their senses working together as the *sensorium*.

Human geographer Tuan (1977), who is credited with articulating the concept of perception from a more cognition-centered view, suggested that information about our environment and about people around us is gained through *experience*, which is sequentially integrated with *sensation* (bodily contact with the sensory modality, such as a chemical for taste, a physical object for touch, or waves moving through space for light and sound), *perception*, and *conception* (*cognition*) (Fig. 6.1A). Experience is the mode by which an individual gains knowledge of and constructs the reality of the environment (Tuan 1975). Gibson and Tuan offer similar views of perception, both of which emphasize that perception occurs as an organism immerses itself in, and moves through, its environment. In Tuan's (1977) view, *place* provides an individual with a "center of meaning." Tuan's seminal book about perception in space and place, *Topophilia* (which translates as "love of place"), details how sensory systems act to reinforce cultural norms and environmental information through *emotions*. Tuan also agrees with the Gibsonian view that perception is "both the response of the senses to external stimuli and purposeful activity in which certain phenomena are clearly registered while others recede in the shade or are blocked out."

Both Gibson and Tuan emphasize that perception also needs to focus on various aspects of the environment that an organism is perceiving, including

A.

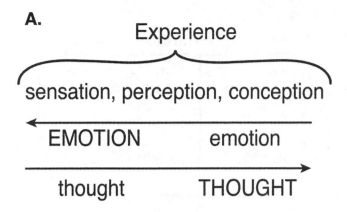

B.

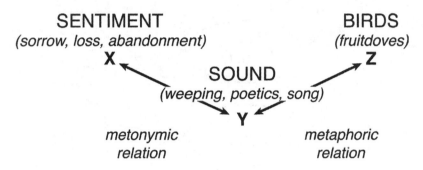

C.

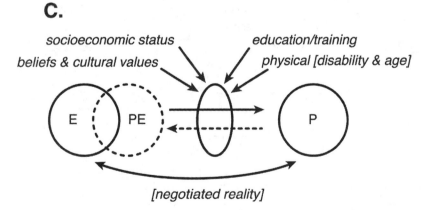

Figure 6.1 Six different conceptualizations of human perception.

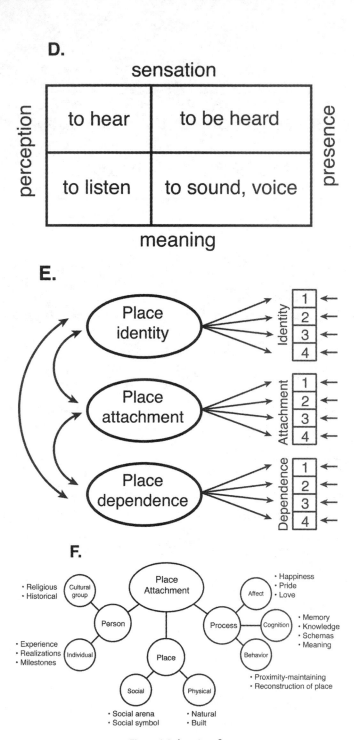

Figure 6.1 *(continued)*.

the *uniqueness* of a place (across space and the experiences of the organism; is it "home" or new?), its level of *boundedness* (the spatial-temporal scale at which the organism is sensing its surroundings), the organism's *movement* and thus position in space, and the amount of *change* that occurs in the interval during which the organism experiences the environment. To Tuan (1977), these aspects combine to form a "world view" of place and the individual's relationship with it. Finally, many scholars (e.g., Tuan 1974; Ingold 2002) who study perception argue that perception occurs at two levels: at the individual level, and at the level of the group or community.

6.2. Ethnomusicology and the Anthropology of Music

6.2.1. Fundamentals of Ethnomusicology

The term "ethnomusicology" (*ethno* = "group," *music* = "music," *ology* = "study of") was coined by Dutch scholar Jaap Kunst. There is no single well-standardized definition of this field (Merriam and Merriam 1964; Nettl 1983; Rice 1987; Post 2014), but the variation among the definitions that have been proposed reflects its breadth and depth (cf. Nettl 2013). Merriam and Merriam are often credited with the simplest and most frequently cited definition: ethnomusicology is "the study of music in culture, as concept, behavior and sound" (1964, 7). In his *Anthropology of Music*, Merriam and Merriam further suggest that ethnomusicology involves "welding together aspects of the social sciences and aspects of the humanities in such a way that each complements the other and leads to a fuller understanding of both" (1964, 7). Many other definitions of ethnomusicology emphasize dependence on data collection, analysis, and the application of new knowledge (i.e., how new knowledge is used to solve problems). Similarly, in *The Study of Ethnomusicology*, which has classically been called "The Red Book" by ethnomusicology students, Nettl suggests that ethnomusicology is "the study of all of the world's music from a comparative perspective, and it is also the anthropological study of music" (1983, 204).

Over the past five or more decades, some of the seminal work in ethnomusicology has focused on how sound and soundmaking by communities (Feld 1986) link environmental perception, experiences of language and music, and human expressions, especially those that are vocal (Feld 1994). Feld's work, originally conceived as a study of "sound as a cultural system" (Feld 1984), used sounds produced by people and other sound-producing agents to understand how shared feelings and beliefs exist and evolve in a group. A well-known starting point of this inquiry is Feld's (1986) classic study of the Kaluli society, an indigenous group of people in New Guinea living in a rainforest rich in sounds. Feld quickly learned of "the myth of the boy who became a muni bird" and saw that this one myth represented a significant link between people and place through the auditory senses. From this myth, Feld's (1986) work carefully dissects the role of natural sounds, particularly those of birds, in reflecting specific forms of *sentiments* (i.e., expressions) and *ethos* (i.e., beliefs). For example, the Kaluli people incorporate the mournful sounds of doves into songs and poetics that express a shared feeling of loss, abandonment, or sadness. These sonic expressions use the same tonal structure as the "weeping" sounds of the local dove. The calls and songs of weeping announce and recall spirits and texts that are sung as poems, called "bird song words," and relate environmental features such as plants, animals, light, and sound to place and event. For the Kaluli, these poetic songs are a "path" between people and their environment; they become a "series of place names that link the cartography of the rainforest to the movement of its past and present" (Feld 1994, 2) and link "place to that of the spirit world of birds." Sounds are also part of the Kaluli taxonomy of birds, which consists of groups of birds organized around their sounds: some groups of birds "say their name"; others "weep," "whistle," or express the sentiment of "cheer." Thus, classification and metaphor are intertwined. Feld's research thus showed that indigenous cultures may have visual-auditory-sensate relationships that link people and nature tightly. How people and nature co-evolve, and the role that the senses, especially hearing, play in this process, provided him insights into society's reliance on the materials and objects in the environment. Feld developed a simple tripartite diagram, shown in Figure 6.1B, that illustrates the Kaluli's relationships of *sounds*, *sentiments*, and *birds*. Here, Feld identifies a *metonymic* relationship—which reflects the contiguity of sound and sentiment—and a *metaphoric* relationship of birds and sounds—which emphasizes the analogous similarity of place and society.

Feld (see Rice 2018 for an excellent summary) eventually advanced his original ideas into an area of inquiry he calls *acoustemology*, a term that combines "acoustic" and "epistemology" (theory of knowing). This broader view of the application of sound, people, and place emphasizes the role that sound plays in forming an understanding of the environment and its dynamics, and how sound and other senses are integrated to form a perception of the environment as people move about and interact with the environment using all of their senses. Rice (2018) argues that this new conceptual paradigm was made to address what Feld argued was a rather "static" and "audiocentric" view espoused by Schafer (1994, 2012) in his definition of soundscapes.

6.2.2. Sensory Ethnography

More recently, several ethnographers have built on Feld's work by recasting the more traditional ethnographic field of study into one that some have labeled *sensory ethnography* (Pink 2009; see also Ingold 2003; Howes 2014). Sensory ethnography advances the fields of ethnomusicology and cultural anthropology by emphasizing the importance of the close relationship of theory, methods, and analysis of ethnographic "data" to sound production and sound perception in the environment. It emphasizes methods and theories of *ethnography*, which involves direct observations of people with the purpose of documenting what they do, how they do it, and why they may do it; placing these observations in the context of place and situation; and integrating these accounts with theory of human behavior and culture (Hammersley 2018). This approach forms a practice of inquiry that is "iterative" and "inductive," "drawing on a family of methods" in which the researcher is immersed in place, people are agents of the embodiment of knowledge, and culture is examined in detail. Thus, sensory ethnography contrasts with the earlier, more traditional twentieth-century paradigm of ethnography, which Delamont (2004) describes as "participant observation during fieldwork." Pink (2015) emphasizes the need for the researcher to be reflexive, which entails using a collaborative approach to the production of knowledge (see Section 9.9.3) whereby the researcher and observant co-create knowledge through "shared sensory perception of experiences" (Pink 2015, 6).

Ingold (2000) argues that sensory ethnography should focus attention on the role of sensory experiences in *purposeful* activities that integrate people, place, and culture. To understand how humans perceive the environment, ethnographers must examine human sensory experiences of the environment (i.e., place) and culture through the lens of three major personal or group activities: livelihoods, dwelling, and skill. Procurement of sustenance through hunting or gathering practices is a *livelihood* (i.e., a practice that sustains life), which Ingold argues is separate from perceiving the environment through the worldview of "myth, religion and ceremony"; rather, livelihoods elucidate how people and nature are integrated consciously as people learn how to hunt animals and find fruits and vegetables in their landscape. *Dwelling*, on the other hand, encompasses how individuals construct an understanding of their surroundings "through their perception of the landscape, the idea of environmental change, the practice of wayfinding and the properties of vision and hearing" (Ingold 2000, 3). Its emphasis on movement (of the body) and change is important for shaping the human-environment experience through the senses. Finally, *skill* encompasses the interactions of people and technology and how tool development influences the ways in which humans perceive their environment through the senses. Ingold also states that sensory ethnography should consider humans to be animals functioning in their environment, using these three lifeway activities as models; for example, he argues that ethnographers should consider borrowing from ecologists who study animal foraging behavior. Indeed, most of the arguments that Ingold (2000) makes about perception of the environment focus on the use of the word "organism" rather than "person" or "animal"; research thus becomes a multispecies endeavor. Taking this perspective, he contends, begins to move the field beyond the traditional Western way of thinking that focuses on the duality of people and nature.

Sensory ethnographers (Pink 2015) also recognize that it is important to understand individual variability in terms of sensory abilities, experiences, and how sensory systems are used. Recognizing this also means that researchers need to characterize this variability as well and should not over-generalize. Finally, as many sensory ethnographers (e.g., Howes 1991) have argued, the Western five-sense taxonomy (seeing, hearing, feeling, smelling, and tasting) should not be an a priori set of fundamental senses for everyone. Thus, researchers must consider their assumptions and how the senses might operate in combinations apart from their own (see Pink 2013 for an excellent overview of this argument).

6.2.3. Acoustic Ecology and Schafer's Soundscape

Acoustic ecology, or as some have called it, *soundscape studies* (Truax 1974), is the study of the interrelationship of sound, nature, and society (Westerkamp 2002). This area of study grew out of the work of F. Murray Schafer and his colleagues at Simon Fraser University in the late 1960s. This work eventually evolved into the World Soundscape Project (WSP) (Westerkamp 1991), which had a more global focus. By the 1990s, the field of acoustic ecology had emerged from the humanities and, to a smaller extent, the social sciences, and it has grown to encompass a large community of scholars who work under the umbrella of the World Forum for Acoustic Ecology (WFAE).

The field of acoustic ecology has made three significant advances in our understanding and appreciation of soundscapes. The first is its develop-

ment of a set of terms and concepts that allow people to discuss and compare sounds and soundscapes across space and time and to relate them to
culture (Truax and Barrett 2011). The first set of terms, introduced by Schafer, relate to the features of a soundscape. To Schafer, various sounds of a
place have different importance and thus should be named accordingly. *Keynote sounds* are those sounds created by a landscape's geography of climate,
water, vegetation, and animals that imprint themselves so deeply on the
people hearing them that "life without them would be sensed as a distinct
impoverishment" (Schafer 1994, 10). *Soundmarks*, on the other hand (the
term was developed to parallel "landmarks"), are sounds whose unique qualities are regarded by people of a community as valuable and thus "deserve
to be protected" (Schafer 1994, 10). The term *signal* is used to described
those sounds in the foreground that are consciously listened to. Sounds can
also be considered either *hi-fi* (i.e., high-fidelity) sounds, in which signal-
to-noise ratios are high and the sounds are clearly heard, or *lo-fi* sounds, in
which signals are "overcrowded" and the sounds cannot be heard well.

The second significant advance of acoustic ecology is its activism on be-
half of the protection of natural sounds and the reduction of *noise*. Schafer distinguishes several different interpretations of noise, as unwanted
sounds, loud sounds, unmusical sounds, or disturbing sounds, but con-
cludes that noise should be considered sound that is unwanted. Since the
1960s, researchers have argued that many soundscapes are becoming dominated by noise. This change, argued Schafer and colleagues, reduces our desire to listen. Noise devalues the act of listening and demotes it as part of
our overall sensorium as compared with other senses, particularly vision.
Conscious listening, and thus the status of humans as major "sound makers," is minimized if we do not listen. Schafer also introduces the term *earwitnesses* to describe people who testify to what they hear, *ear cleaning* as a
process to train oneself in listening discriminately to sounds, and *clairau-
dience* to describe those individuals who have exceptional hearing abilities
with regard to environmental sounds. *Sacred noise* refers to sounds that are
socially acceptable and may not be banned. Schafer states that traditionally,
sacred noise referred to sounds like thunder, volcanic eruptions, and others
that represented divine displeasure with people, but in modern society it
consists of unnatural sounds that are culturally acceptable, such as church
bells, amplified music, or loud vehicles.

Finally, many scholars, such as Westerkamp (2002), have focused on the
area of artistic expression called *soundscape composition* (see Section 14.2).
To soundscape composers, the artistic expression of sonic spaces that are
composed of environmental sounds is meant to be the "sonic transition of

meanings about place, time, environment and listening perception." She argues that soundscape composers create their work to "speak back" to the problematic "voices" in the soundscape. Their compositions should inspire, reinforce a sense of place, and communicate purpose to the listener and composer who share the sonic expression. Another dimension of her argument (Westerkamp 1988) is that music-as-environment (i.e., monotonous corporate music with no meaning) produces a "schizophonic" effect on the listener by "dislocating him/her from the physical present and self."

6.2.4. Ecomusicology

One of the earliest expressions of interest by ethnomusicologists in ecology (as a disciplinary field, as opposed to the study of music and nature) is contained in the work of ethnomusicologist Jeff Todd Titon. Titon's work attempts to draw from discoveries within ecology, conservation biology, and sustainability science and to purposely develop numerous parallels to those fields. For example, Titon (2009) argues that to sustain music that is endangered or even threatened with "extinction," we should follow guidelines developed by conservation biologists as they use knowledge about what constitutes healthy and sustained ecosystems. He points to four principles of sustainability science and conservation biology and argues for them to be applied to the sustainability of music around the world: that diversity must be maintained (i.e., we need to main a diverse world music heritage); that growth is unsustainable (e.g., that exploitation of music practices solely for economic gain will serve to create staged authenticities); that all parts of the ecosystem are connected (e.g., that many parts of a music culture, not just the music's sound, need to be sustained together, including lifeways); and that ecosystems require stewardship (e.g., music practices need caretakers). Titon (2015) has also drawn on the parallels between sound production by humans, in the form of music, and nonhuman animal sound communication. First, he proposes that the main purpose of sound for humans and other living beings is to *"announce presence"* (Titon 2015, 26). Sound production locates a person or organism in space and time, both for itself and for other beings who may experience the sound. His second proposition is that sound *establishes co-presence*. Sound production is the act in which, by means of simultaneous vibration, two or more individuals become mutually aware. Communication of co-presence by organisms is common, as sounds produced by individuals of the same species have a function (attracting a mate, establishing territory), as do those produced by individuals of different species (e.g., predators and prey). His third proposition is a *sound community of co-present beings*, which constitutes a group of individuals that share sonic communication. Titon (2015) argues that this third level of co-presence is characteristic of animal sound communities that exist in nearly every natural area of the world. These sonic connections establish direct relations among living beings. These interconnections among beings and the environment are the basis of Titon's "Appeal for a Sound Commons for All Living Creatures" (2012) and form the structure of a *sound ecology* (2020) that is useful for sustaining the integrity of global sonic communicative and aesthetic practices as they evolve.

The field of ecomusicology was initiated during the time Titon (2013, 2015) was exploring how the knowledge of ecosystems and the scientific study of ecology can help ethnomusicology create a new perspective on how humans interact with the environment and how social and natural systems are similar given the focus on sound. Ecomusicology emerged as "the study of music, culture, and nature in all the complexities of those terms" (Allen and Dawe 2016, 1). It considers music and sonic issues or practices as they relate to nature and the environment (Allen 2013). Titon (2013) argued that ecomusicology is the study of music, culture, and nature in the context of ecological crises. These authors (Allen 2013; Allen and Dawe 2016; Titon 2013) all point to early forms of this area of scholarship, most notably the work of Feld (1993) in his coining of the term "eco-muse-ecology" and Schafer's concept of the soundscape (1969, 1994).

Allen and Dawe (2016) emphasize that ecomusicology is a "field," not a "discipline," because it needs to build across many disciplines, including ecology, ethnography, musicology, psychology, and anthropology. Key features of ecomusicology include the role of activism (including "well-founded scholarship that argues a point based on facts and evidence" [Allen 2011b, 417]) and *ecocriticism*. Ecocriticism is part of a larger field that considers literature, film, and other media that portray human-environment relationships and engage activist viewpoints. In a way, then, ecomusicology borrows from the Schafer WSP view that human-nature relationships, as connected through sound and the soundscape, are deteriorating, and new knowledge is needed to address these problems.

6.2.5. Ecoethnomusicology

Ecoethnomusicology, as presented by Guyette and Post (2016, 42), is "eco-musicological research that engages ecological events and issues and their relationship with musical expression." Ecoethnomusicologists describe human behavior as it relates to the ecological features of landscapes to demonstrate that "human and non-human sounds and sound-making play equally important roles in providing ecological knowledge about a sound landscape" (2016, 43). Ecoethnomusicology also emphasizes the strong role of ethnography in discovery. The connection with human behavior evolved from the early anthropological work of Feld (1984) and Seeger (1987), among others, who acknowledge the important work of Schafer and colleagues. Ecoethnomusicology embraces the scientific treatment of ecosystems, culture, and what its practitioners call *sonic practices*. Sonic practices include all forms of sonic expression (songs, poetics, calls, mimesis of natural sounds) that are cultural, as well as the use and interpretative meaning of environmental sounds (e.g., thunder, ice cracking) that reflect place. Several expressive forms of sound production or music making that reflect the cultural meanings of the natural system have been identified by ecoethnomusicologists. These expressive forms could be classified as follows, although these classes are not formal, nor are they mutually exclusive:

> *Sonic expressions.* This class includes sounds produced by individuals or groups who mimic the sounds of nature or tune their throats (see Levin and Edgerton 1999) to the resonance patterns of wind or running water. Studying the timing, use, purpose, and meanings of these sounds could allow ecologists, anthropologists, and ethnographers to understand the connections people have with their environmental surroundings.
>
> *Multispecies communication.* Included in this category are yells and whistles made to livestock and other animals for the purpose of information exchange between people and animals. Herders also listen and respond to calls of the herd in many cases, using these as clues to livestock health. Herder birthing songs are another example of this form of communication; herders sing to mothers giving birth to relax them and make birthing of young less stressful (Post 2020).
>
> *Ecological songs.* Some indigenous communities sing songs that tell stories about place and the sounds of nature or of animals and plants that exist there, or that serve as a guide to or map of a landscape. Examples of these songs are presented in Chapter 14.
>
> *Acoustic cues.* Some sounds from the environment are used to indicate the status of a place (e.g., thunder indicates storms approaching) or how a place is changing. Just as scientists may research sound source, tim-

ing, magnitude, and location to determine how a place is changing, the people of a place may use these cues to understand environmental dynamics as well, and may incorporate them into sonic expressions or ecological songs.

Research on these sonic expressive forms focuses on who learns them and when, how they are learned (in what conditions and/or situations), and how adaptable are they over time. What is the status (i.e., social power) of individuals who learn and use them? Are they used by individuals, or by a group? Is the expression considered performance, and is it symbolic? Can a topology like that proposed by Feld (1986) be developed that begins to organize these expressions for each community?

6.2.6. Extending Ethnography to Include Soundscapes and Sonic Practices

Post and Pijanowski (2018) outline a "way forward" for integrating the work of ethnomusicological and cultural anthropological scholars with that of biophysical scientists, including remote-sensing experts, climate scientists, hydrologists, data scientists, landscape ecologists. Their approaches emphasize (1) that the social sciences and humanities can play a role in directing ecological research (design of studies, integration of data across qualitative and quantitative domains); (2) that disciplines need to "meet in the middle" to ensure that research is well integrated; (3) that research goals should be developed using co-production approaches that integrate the people who are being studied into new discoveries; and (4) that research should focus on an outcome that is shared across the disciplines of all team members (e.g., it should address the biodiversity crisis as well as how communities perceive their environment and the changes that are occurring). They also suggest that *knowledge co-production* is necessary to understand how humans perceive their environment, and that the subjects of study should be recognized as scientists as well as artists. They argue that Western scientists can learn from local scientists about how the local natural system changes, why it changes, and what elements in the system are important for its sustainability. Finally, they recognize that music itself is a form of soundscape—one that may be isolated (as in a concert hall) or participatory (e.g., outdoors and with human music integrated with natural soundscapes). Approaches for integrating this kind of research are presented in Chapter 9, "Measuring the Soundscape."

6.3. Sensuous Geography

Following the work of Feld (1986), many scholars in the fields of anthropology, ethnography, and geography began to detail how human senses affect the way individuals and communities perceive their environment in space and time, and the ways in which the senses also reinforce cultural norms about human-environment interactions. Perhaps the most detailed advance following Feld was the work of Rodaway (1994), who outlines a theory of *sensuous geography* that is based "not solely on perception geography, not an experiential geography, but a geography of the senses in a duality" that encompasses "sense as meaning" and "sense as sensation or feeling." Rodaway (1994) credits the work of Gibson (1954, 1968), who described the ecological theory of the senses. Gibson (1968) argued that this duality is best thought of as *perceptual systems*—in which the senses have (1) interrelationships that are used by the body and mental processes to sense structure and changes in the environment, and (2) a set of *ecological optics* that describes how the environment structures optical, tactile, auditory, and other sensory stimulation. Gibson's (1968) *ecological theory of perception* emphasizes that sensory stimuli are modified as they pass through the environment (e.g., sound is degraded or reflected) and are then "read by" the sensory organ systems, so that the environment provides "information, not merely raw data." Sensing the environment in such as way provides individuals and groups with "space perception of the total environment." Perception is also a social experience shared by individuals or groups. Rodaway also recognizes that individuals will vary in their perception of the environment due to differences in age, gender, social position, and the effects of education, training, and social norms. This human-environment interaction as viewed by Rodaway, as influenced by Gibson (1968) and the human geographer Yuan (1974; 1977, 18), is depicted in Figure 6.1C. Note how the environment (E) and the individual (P) are separated by the perceived environment (PE) and a host of cultural "filters." A person's full understanding of the structure of the environment and the nature of its dynamics is a "negotiated" reality by which environmental stimuli, cultural filters, and an individual's integrated sensory system form what the individual views as the environment around them. Rodaway further emphasizes that the sensory systems are also "active," not passive, as they "explore, collect, respond to, and participate in" the complexity of environmental stimuli, and that the ways in which these systems respond to these inputs evolve over time.

Rodaway also suggests that the senses work together in cooperation, are hierarchical (e.g., seeing dominates listening), are sequential (i.e., they are often ordered in how they are sensing the environment), and operate in reciprocity; "reciprocity" refers to the fact that the senses are used to reinforce the human-environment experience. An additional aspect of sensing the environment is movement, which introduces the importance of the body and the way in which it orients itself in the environment and its sensory organs (eyes, ears) in space. Movement is also an aspect of sensing that allows the individual the ability to continually monitor space through exploration and continuous evaluation. Finally, each sense can be understood through a matrix of sending and receiving and sensation and meaning (Fig. 6.1D). Rodaway's sensory matrix emphasizes, in the vertical direction, integration of physical and mental processes (sensation vs. meaning), and in the horizontal direction, the gathering of information and the projection of one's self into the environmental space (perception vs. presence).

Rodaway's more specific focus on sound is termed *auditory geography*. His assessment of the role that sound plays in sensing the environment also considers Schafer's concept of the soundscape (1994, 84–88). He suggests that Schafer's concept provides several useful perspectives for those interested in sensuous geography. First, he points out that the vocabulary introduced by Schafer has a degree of "simplicity" (1994, 84), and is strongly based on "visual metaphors" (1994, 84) that allow scholars and the public to understand how various types of sound influence our ability to understand our surroundings. Notions of soundmarks, hi-fi and low-fi soundscapes, and keynotes are helpful. The most useful concept is the application of the term "soundscape" to "the geographical space of a particular sonic characteristic" (1994, 86), but Rodaway argues that Schafer's use of the term lacks a detailed description of "soundscapes as auditory experience" (1994, 86).

Advances made by Rodaway in auditory geography extend our understanding of how sounds are presented to an individual as unique stimuli as well as how the auditory sense is used in coordination and integration with the other senses. For example, Rodaway argues that the "auditory world not only surrounds us but we seem to be within it and be participants" (1994, 91). We can hear sounds all around us, and the experience differs from seeing. In addition, the physical form of sound is modified considerably by the environment. Following Gibson's (1968) ecological theory of perception, factors such as distance, environmental composition (forests, tall buildings), and the presence or absence of wind modify the sound from the original sound source, and that modified stimulus is interpreted as part of the overall sensing. "Sounds fill spaces" and "gives character" to places, which is indeed what creates a soundscape. Finally, Rodaway suggests that hearing may not be considered as important to human sensing of the environment as seeing, as our hearing abilities have evolved not to include a broad range of sound sources, but rather are specifically tuned to the sounds that represent human speech, as people possess the ability to recognize individuals by the sound of their voices.

Rodaway also emphasizes that sound is not merely physical, but also *emotional* (see Section 6.7 below), following the arguments of Wyburn et al. (1964) and Yuan (1974), who also recognize that sound creates cultural expressive forms, such as music. Auditory experiences not only play a role in how an individual anticipates and encounters activities occurring in a place (e.g., hearing a highway and walking toward it), but also serve as one form of "memory of a place." Sound sensing can also be a tactile stimulus, as vibrations of objects felt through the hands or through an object like a walking stick add another dimension to auditory geography, although one that is close to, rather than far from, the sound-source experience.

6.4. Sense of Place Studies

Scholars in environmental psychology (Gibson 1979; Cantor et al. 1991; Jorgensen and Stedman 2001), cultural anthropology (Feld and Basso 1996), planning and design (Relph 1976), and human geography (Tuan 1977) have made significant contributions toward describing how animals and humans perceive "place" through their senses and experiences and have labeled this phenomenon *sense of place* (SoP). Arguably some of the most prolific work (see Lewicka 2011 for a review) has occurred in environmental psychology, which has focused on developing frameworks that outline the major components of sense of place along with methods that can be used to measure it in humans. Most of the recent research has focused on assessing the attachment of people or communities to urban, suburban, or rural places in developed countries, or to national parks they visit. As such, most research in this area emphasizes places that are built up or are experienced by individuals for a short time.

The first widely cited framework, proposed by Jorgensen and Stedman (2001), considers sense of place (Fig. 6.1E) as having three somewhat independent components: *place identity*, *place dependence*, and *place attachment*. Place identity, they argue, encompasses the attributes of self that "define an individuals' personal identity in relation to the psychological environment, by means of a complex pattern of conscious and unconscious ideas, beliefs, preferences, feelings, values, goals and behavioral tendencies and skills relevant to this environment" (Proshansky 1978). Place dependence, on the other hand, is the strength of an individual's association between themselves and specific places. Finally, place attachment is the affective (i.e., emotional) relationship between an individual and the environment that transcends cognition. Note, however, that scholars in environmental psychology who study SoP do not agree on its terminology (see Jorgensen and Stedman 2001), on precise definitions of its components, or on how they are measured, as there are so many disciplines that have conducted research on this topic.

A second, well-recognized SoP conceptual framework is the one offered by Scannell and Gifford (2010). A cursory examination of this framework and comparison with Jorgensen and Stedman's framework also illustrates that the terms in this field are not well established, as they are arranged differently. Scannell and Gifford's (2010) framework builds on the work of Tuan and Relph and is consistent with that of Feld, Rodaway, and Ingold. In their "PPP model," place attachment is separated into three dimensions (Fig. 6.1F): people, place, and process. A person's attachment to place is one that is formed by the individual's experiences and reinforced through cultural and social interactions. Place is defined as the natural and built components of the local environment as well as the social "arena" of that place (e.g., family, religion, learning, play). Finally, place attachment is also influenced by a variety of processes involving the individual that occur at the place; these processes may be behavioral (moving, building), cognitive (developing knowledge, creating meaning), or emotional (developing emotional connections to place, which might include happiness, pride, and love).

6.5. Traditional Ecological Knowledge

6.5.1. The TEK Complex

A group of researchers in the social sciences has focused on traditional ecological knowledge to understand how indigenous people live and use resources within their environments. *Traditional ecological knowledge*, frequently abbreviated as TEK, is the "cumulative body of environmental *knowledge*, resource management *practices* and worldview *beliefs* of the local community evolving from the adaptive processes handed down through generations by cultural transmission" (Berkes 2017, 9). The study of TEK focuses on the relationship of people to the environment and relationships between people through this knowledge-practice-belief complex. It seeks to understand indigenous "ways of knowing and ways of doing" and, as such, emphasizes the process of knowledge development and transmission rather than a cataloging of facts about the environment. Traditional ecological knowledge is often contrasted with knowing through science. Much of the work conducted by scholars examining TEK across indigenous groups has found that they often construct knowledge using local rules of practice that are reinforced culturally; that their resource use is flexible, they monitor resources, and they rely on a diversity of practices to ensure that they are resilient to change and surprise; that traditional ecological knowledge is holistic; and that the local environment, with its soils, water, plants, and animals, shapes personal and cultural identity (nativescience.org/html/traditional_knowlege.html, accessed October 19, 2020).

6.5.2. TEK and Science

Berkes (2009) has argued that many scholars studying TEK have focused too much on the differences between TEK and Western scientific knowledge. He argues that a practical solution is to use both TEK and deductive Western scientific approaches to co-produce knowledge; using insights from both perspectives provides a broad understanding of people and place. Berkes (2017) also argues that TEK can provide keen insights about biodiversity (e.g., what species are present, or were present, and important aspects of their life histories) for natural resource management (e.g., how resources vary across time and space), for the conservation and preservation of protected areas, for stewardship of biodiversity, for environmental monitoring approaches and priorities, and ways of responding to extreme weather events and disasters, as well as leading to insights into environmental ethics.

Another important perspective on the relationship between TEK and Western science is the view expressed by Chief Robert Wavey (1993) of the Fox Lake Cree Nation located in Manitoba, Canada. Wavey argues against fully accepting a complementary local view of TEK and science, since science "is based on discovery and has provided the foundation for the industrialization of wealth in the hands of those nations with the greatest scientific capacity. Traditional ecological knowledge is not another frontier for science to discover" (1993, 16). If one "contemplates linking traditional ecological knowledge and science to support the healing of Mother Earth, I ask you to resist seeking to discover. I urge you to accept what is obvious" (1993, 16). If science is to be integrated with TEK, he states, then there must be "respect, understanding, the recognition of traditional rights, and the recognition of existing indigenous stewardship of many regions of the earth" (1993, 16). Thus, scientists, in their efforts to understand the role that TEK plays in indigenous communities, need to develop shared visions and strong, lasting partnerships that build a relationship of trust and respect with these communities. In addition, Johannes (1989) argues that scientists who have not become accustomed to living in the ecosystem that they study need to be trained to seek their new information from local community knowledge holders and not to rely solely on "books" and other conventional scientific sources of knowledge to frame their discovery of how people interact with their environment. A similar view of the relationship of TEK and science is espoused by Agrawal (1995), who advocates that management of natural resources should not merely move beyond a simple dichotomy of indigenous versus scientific and into a complementary effort, but rather into one where there is "greater autonomy for indigenous peoples" to manage their own land. Berkes et al. (1995) expressed yet another view of the relationship between TEK and science (see Table 6.1 for a summary of their differences); that is, that the "gap" between these two perspectives is *decreasing* as Western science begins to develop broad conceptual ideas such as chaos theory, resilience, and the science of surprise, which are less about facts and more about holism, which they argue is more usefully integrated with TEK.

6.5.3. TEK Transmission

An important aspect of studying TEK is understanding how ecological knowledge is transmitted between people and across generations. It is well accepted (Ruddle 1993; Grenier 1998) that TEK is held by elders, children, men, and women, and that knowledge holders will vary according to specific age, education, daily experiences, roles and responsibilities in the community holding that knowledge, and the level of "outside influence" that each person is exposed to in life (e.g., Western education). The transmission of knowledge is specifically designed to be age specific, systematic, and often very site specific; to construct knowledge and practice from simple to complex; and to be taught and learned in fixed periods (e.g., harvesting of specific food). It may also involve a form of reward or punishment associated with learned tasks. Knowledge is often encoded in stories, songs/poetics, dances, myths, cultural values, beliefs, laws, language, rituals and ceremonies, the taxonomy of local life, and the use of tools and equipment (Grenier

1998). Finally, TEK works to help to shape an individual's relationship to the environment by building a local knowledge of land and animals, land and resource systems, and social institutions, all within a larger worldview (Berkes 1999).

TEK is designed to be passed along to subsequent generations as objective truth, but also internalized as objective reality (Ruddle 1995). Indigenous communities often codify this knowledge by creating specific rules, formalizing practices, and integrating the knowledge into the language. Most of the literature on the transmission of traditional knowledge, including ecological knowledge, stresses the role of understanding values, individual attributes, and attitudes rather than the cataloging of the knowledge. Transmission of knowledge, either as information or skill (i.e., practice), is systematic; one transmission framework includes eight learning steps: from familiarization, observation, simple steps with assistance, entire task complex with assistance, entire task complex under supervision, entire task complex as assistant to instructor, individual experimentation, to equal partnership with instructor. Finally, modes of transmission vary according to sedentary or mobile lifeways. Many indigenous groups that live in grassland and savanna ecosystems have adopted a nomadic way of life in which they move from place to place depending on expectations of rainfall, seasonal variation in natural resource conditions, and specific wildlife behaviors (e.g., migration of ungulates, spawning by fish).

6.5.4. TEK and the Future

There has also been a great deal of interest in the TEK of indigenous communities from the standpoint of its potential loss, as it is viewed as important not only to the sustainability of any ecosystem, but also to how it might be most prudently managed as natural resources globally become scarcer. Two aspects of this threat have been identified (Wavey 1993). The first has been the historical disruption of social order within indigenous communities by outside government and economic interests. Changes in local cultural governance systems, educational systems (e.g., requiring youth to attend schools not traditionally supported by local customs), and cultural norms (e.g., belief systems) are introduced into these indigenous communities. This disrupts the transmission of traditional knowledge (ecological and non-ecological). A second threat is the modification of the local environment by these outside interests. Changes such as hydroelectric dams, mining, and harvesting of forests have the effect of "obliterating the reference points and actual resources that (oral) maps are intended to share."

Table 6.1 Comparison of TEK and Western scientific approaches

Dimension of knowledge generation	TEK	Science
Scale of observations	Local	Global
Form of inquiry	Mostly qualitative	Mostly quantitative
Drive to collect facts	Small	Large
Speed of accumulation of facts	Slow	Fast
Longevity of observations	Long (diachronic)	Short (synchronic)
Reliance on verification	Limited	High
Level of interest in theory building	Low	High

Source: Modified from Berkes et al. (1995.)

In recognition that TEK can play a critical role in sustaining ecosystems around the world, national and international governments have established specific initiatives that support indigenous communities in their protection of TEK and its use in providing information that is needed for sustaining these land resources for future generations. Noteworthy efforts that have created TEK initiatives include the documentation and study of TEK by a variety of international environmental policy forums, including the Commission on Environmental, Economic and Social Policy (CEESP) of the International Union for Conservation of Nature (IUCN), and various working groups in US federal government agencies, including the Fish and Wildlife Service, National Park Service, Forest Service (Vinyeta and Lynn 2013), Geological Survey, and Bureau of Land Management (BLM), to name a few. The Canadian government has also supported a variety of First Nations TEK initiatives. A thematic section of the Ecological Society of America was formed over a decade ago in response to the need to study and understand TEK from the standpoint of ecosystem management and ecosystem understanding.

6.6. Nature Connectedness and Relatedness

Another strand of research in environmental psychology that is relevant to soundscape ecology has been the measurement of *nature connectedness* (e.g., Mayer and Frantz 2004), *nature relatedness* (Nisbet et al. 2009), *inclusion of nature in self* (e.g., Schultz 2002), and/or *nature deficit disorder* (Louv 2005) in individuals and communities. These areas of environmental psychology research have been motivated by scholars who have desired to understand the factors that influence a person's positive environmental behavior (Fabio and Kenny 2018). Much of this focus is also placed in the context of modern society and how its interactions with the natural environment differ from those of thousands of years ago. First, it is recognized that most people in the world today do not have a high degree of exposure to nature, as more than 50% of the human population now lives in cities and many people spend most of their time indoors. It is also true that this lack of exposure to nature is growing over time. Second, researchers have recognized that people who state that nature is close to them, or is part of them, often express a high level of respect for, and interest in, nature. This recognition builds from Aldo Leopold's original notion, as outlined in his *Sand County Almanac*, that for people to care about and be stewards of nature and the land that supports it, they must feel connected to it (Leopold 1949). This "ecological land ethic" also implies that people will not do harm to nature if they feel connected to it. Finally, the noted biologist E. O. Wilson (1986, but see also Fromm 1964; Kellert and Wilson 1995) has even suggested as part of his *biophilia hypothesis* (*bio* = "living systems," *philia* = "love") that there has been genetic-cultural co-evolution of love for nature in humans, and that over their evolutionary history, they have developed an "innate" love of nature, life, and living things. Environmental psychologists and many environmental activists have argued that for human society to have empathy (i.e., care) for nature, it must have a psychological "connectedness" to it. Another part of this argument is that for human society to care about nonhuman systems on Earth, it should have a worldview that it is part of the natural system and thus reflect a high level of "relatedness" to nature. Although nature connectedness and nature relatedness are very similar terms, their use in the literature has remained mostly independent since both concepts emerged in the late 1990s.

Schultz (2002) is credited with one of the key definitions in this field of study. In his view, nature connectedness is "the extent to which an individual includes nature within his/her representation of self." Other researchers have investigated the effects of nature connectedness on well-being and emotional state: does being connected to nature, and thereby having empathy for it, improve a person's well-being and happiness when they have experiences with it (Di Fabio and Kenny 2018)? Several studies (Mayer et al. 2009) strongly support a relationship between nature connectedness and psychological well-being, including positive effects on mood, cognitive processing, physical health, and ability to reflect. Nature connectedness differs from any measure of simple enjoyment of being outdoors or environmental activism expressed by an individual. It is instead a measure of a person's appreciation for and understanding of their own, and their society's, interconnectedness with all living organisms on Earth (Nisbet et al. 2009).

6.7. Affective Qualities of Soundscapes

The field that integrates the scientific study of sound and how it is perceived by humans is called psychoacoustics (Zwicker and Fastl 2013). This field includes the study of how sound affects human hearing in quiet, loud, or noisy (i.e., masking) environments and across frequency ranges, and how age, gender, and cultural differences affect human sensitivity to sound. Human health applications, such as treatment of hearing deficiencies and the physiology of sound perception (i.e., how stress may occur in response to different acoustic intensities or frequencies) are also part of this field.

In psychology, a variety of terms are used to describe the psychological response to events or situations, with each representing a different level of psychological awareness (Munezero et al. 2014). Affect, or *affective* quality, is the set of non-conscious expressions of the mind and body in response to changes in the environment and others. *Feelings* are unreasoned opinions or beliefs. *Emotions* are conscious mental reactions to an experience; they can be physiological or behavioral. And *sentiment* is the attitude or thought expressed by a person. Psychoacousticians have studied emotions and sentiment for decades, and they have developed many conceptual models of sensory processing, one of which is shown in Figure 6.2. Here, sound that is created in the environment passes through three main processing

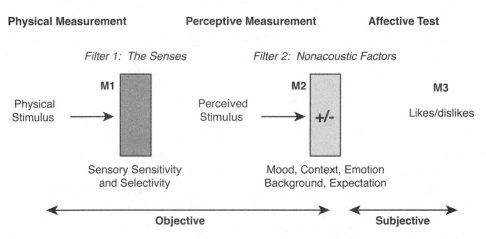

Figure 6.2 A conceptual model of sensory information processing.

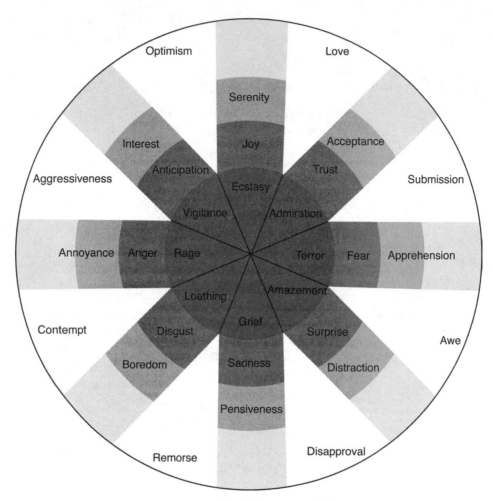

Figure 6.3 Plutchik's emotion wheel.

domains: *physical*, *perceptive*, and *affective*. In this conceptualization of sound processing, two filters are recognized: one that reflects sensory selectivity and sensitivity, and another that filters sound within a context that elicits emotions, but at the highest level of generality, such as like or dislike. There is considerable debate about what constitutes an emotion (see Smith and Lazarus 1990), but some of the most widely accepted human emotions are anger, anticipation, joy, trust, fear, surprise, sadness, and disgust. Plutchik (2001) has arranged these into an *emotion wheel* (Fig. 6.3) that places opposing emotions on opposite sides and those that are most similar next to one another. The wheel also organizes these emotions by intensity (e.g., annoyance < anger < rage). Mixtures of the primary emotions (e.g., awe = apprehension plus distraction/amazement) are labeled in the white spaces between their two components. Other emotions that are not described on the wheel are also possible, and are often aligned with one or more emotions on the wheel (e.g., curiosity = anticipation and surprise).

It is also argued (e.g., Fredrickson 2001) that positive emotions can strengthen personal worldviews, social interactions, and social-environmental connections that are beneficial to the individual or to a community. Plutchik (1982) also points out that emotions are generally composed of a sequence of events in the following order: a stimulus event (e.g., a threat by an enemy) that triggers inputs from the sensory system; a cognitive component, which involves associating meaning with the event (e.g., danger); a feeling (e.g., fear); an overt behavior (e.g., run); and a function or purpose (e.g., protection). Regarding the last component, Plutchik and others (e.g., Ekman 1992; Smith and Lazarus 1990) have argued that emotions have evolutionary significance for people and some animals, as they support the fundamental needs of the individual or community (e.g., find food, find a mate, flee from danger) and facilitate action as well as reinforcing knowledge about the surroundings.

6.8. Attention Restoration Theory and Related Concepts

Kaplan and Kaplan (1989) and Kaplan (1995) outline a theory they call *attention restoration theory* (ART) to explain the positive effects (i.e., pleasure and positive physiological changes) of nature on human well-being. ART assumes that human cognitive effort is needed to direct one's attention toward an object, event, or phenomenon. Kaplan and Kaplan hypothesize that humans have evolved to live within natural systems (i.e., nature) and thus do not require effort to direct their attention when existing in these surroundings. On the other hand, when humans are within their unnatural contemporary surroundings (e.g., in a city, in a building), they need to focus their attention and therefore develop *directed attention fatigue*. This condition results in irritability, decreased social interactions, decreased helping behaviors, negative emotions, and decreased performance on tasks. In natural systems, human attention is directed by fascination, which involves involuntary attention and "a replenishment of the cognitive resources" (Ratcliffe et al. 2013). There are four main elements to attention restoration. The first is "being-away," a state of having distance from the aspects of one's routine, contemporary surroundings that are unnatural. The second is "extent," which is the degree of isolation or distance from contemporary settings—whether one is in a distant wilderness or a nearby nature center. The third is "compatibility," which is the ease of human function within the natural system; it is proposed that humans function more easily in natural systems than in unnatural settings. The fourth is "fascination," the experience of being in nature with features such as clouds, sunsets, breezes, and animals moving, which provide cognitive relaxation that is important to cognitive restoration and/or psychological escape that nature provides.

Often considered in contrast to ART is the stress recovery theory (SRT) of Ulrich (1983), who asserts that being present in nature provides people with exposure to certain environmental qualities that are lacking in contemporary settings. These qualities include those that are aesthetic (e.g., complex patterns in surfaces such as trees, landscapes, land-water interfaces, waves), those that demonstrate that resources are available (e.g., water, trees, soils), and those that demonstrate the absence of the threats that are common in modern built environments. These environmental qualities, they argue, reduce stress.

6.9. Soundscapes as Coupled Human and Natural Systems

Many scholars in geography recognize that all places on Earth are composed of complex interactions between people and the environment that operate over multiple spatial-temporal-magnitude scales. These interactions are viewed as systems, often called *coupled human and natural systems* (CHANS) (Liu et al. 2007), human-environment interactions (Moran 2010), or socio-ecological systems (SES) (Berkes et al. 2000). In these systems, individuals, communities, economies, societies, and cultures operate within, and depend on, the biosphere and shape it. The biosphere, in turn, limits what society can do and how it operates, and society must evolve as the biosphere changes.

Research in this area, which is often referred to as CHANS or SES research, is based on the systems theories of ecologist Eugene Odum (1983). Odum's work was ultimately derived from the early developments of systems dynamic modeling by Forrester (1961) and from systems theory work applied to physics and biological systems by von Bertalanffy (1950). These researchers considered *systems* to be complex mutual interactions of elements that are connected through a variety of feedbacks. All systems are thought to be driven by exogenous variables such as climate, and most can be considered open systems, in which inputs, such as energy, ultimately control how the system evolves. All connections between elements in a system have either a *positive polarity* (if one element increases, the connected elements also increase) or a *negative polarity* (if one element increases, the connected elements decrease). Feedback loops can be either of two types. In *reinforcing feedback* loops, positive feedbacks generate an additive measure to all elements or maintain an equilibrium. *Balancing feedback* loops are composed of a negative and also a positive polarity effect.

CHANS (Liu et al. 2007) research emerged mainly from the geographic sciences, spatial studies in which social and natural scientists collaborate on questions that focus on human-dominated landscapes and aquatic systems. Coupled human and natural systems are examined through the study of a system's components and interactions, which have distinct behaviors. These interactions include couplings across spatial scales, the degree of heterogeneity of system components in space, couplings within and outside system boundaries, cascades (the operation of a behavior across multiple components or layers), the presence of positive and/or negative feedback loops, time lags, and legacy effects (past events that affect current system dynamics). These systems may operate either top-down or bottom-up, may have *emergent properties* (new properties that emerge at higher levels that are more than the sum of their parts), and/or may change states (go from good to bad). Systems may also transition between states by passing a tipping point. One major focus of CHANS is to study these systems using a variety of methods—mostly mixed methods, which borrow and integrate methods from the natural and the social sciences. A major focus of CHANS research is to understand system dynamics in the context of sustainability.

There are several benefits to applying a systems way of thinking that integrates humans and natural systems (i.e., as CHANS). First, a systems approach provides a holistic view of all the important elements that exist at a place. And it allows us to depict dependencies among the elements graphically. A systems view also allows researchers and policy analysts to understand how environments and social systems change over space and time. Systems approaches have been used over the years to help us understand the ways in which humans affect the environment, how the environment affects people, how both natural and human elements can be repaired if altered, and how they can be sustained over time. Finally, systems approaches provide a means for collaboration across disciplines, as many disciplines are now using systems perspectives to examine and document the complex connections between people and the environment.

Summary

There is considerable research on the sociocultural dimensions of sound and on how the sensorium is used by people to perceive their environment. This research has generated a lot of conceptual theories that integrate human experiences, cognition, emotions, and other behavioral factors to understand how people use their senses to live in their environment. These rich theories have come from anthropology, psychology, geography, and ethnomusicology. And many advances have been made by scholars who work in rich interdisciplinary fields where knowledge from two or more disciplines is used to understand complex topics.

Discussion Questions

1. Compare and contrast the ways in which Gibson, Tuan, and Feld conceptualize human perception.
2. Sketch a coupled human and natural system that has sound as a component and is of a place that is very familiar to you. This place can be your home or campus or another place you have been to many times. In your CHANS diagram, include all the components that interact, but highlight those components that are manifested as sound.
3. Do you possess any ecological knowledge that is valuable for the next generation? If you do, how does this knowledge get passed to them? Which aspects of this knowledge are sonic, or have acoustic properties that are important?

Further Reading

Berkes, Fikret. *Sacred Ecology*. Routledge, 2017.
Feld, Steven. *Sound and Sentiment: Birds, Weeping, Poetics, and Song in Kaluli Expression*. With a new introduction by the author. Duke University Press, 2012.
Gibson, James Jerome, and Leonard Carmichael. *The Senses Considered as Perceptual Systems*. Houghton Mifflin, 1966.
Ingold, Tim. *The Perception of the Environment: Essays on Livelihood, Dwelling and Skill*. Routledge, 2021.
Kaplan, Stephen. "The Restorative Benefits of Nature: Toward an Integrative Framework." *Journal of Environmental Psychology* 15, no. 3 (1995): 169–82.
Louv, Richard. *Last Child in the Woods: Saving Our Children from Nature-Deficit Disorder*. Algonquin, 2008.
Pink, Sarah. *Doing Sensory Ethnography*. Sage, 2015.
Relph, Edward. *Place and Placelessness*. Pion, 1976.
Tuan, Yi-Fu. *Space and Place: The Perspective of Experience*. University of Minnesota Press, 1977.

Data Science Concepts

7

OVERVIEW. The advent of passive acoustic recorders has meant that soundscape ecologists are now presented with the challenges of managing and analyzing big data. These challenges are not trivial, and they often require a different kind of thinking. They are forcing ecologists either to collaborate closely with data scientists or to become experts in the areas of machine learning, software development, and data management. One of the fastest-growing areas of STEM (science, technology, engineering, and mathematics) is data science. This chapter summarizes big data concepts, data management, and approaches to computational thinking that differ from those used in ecological research that relies on traditional statistical analysis.

KEYWORDS: artificial neural network, big data, data analysis pipeline, deep learning, fourth paradigm of science, machine learning

7.1. Big Data and Soundscape Ecology

7.1.1. Key Terms of Data Science

The data types that are useful for conducting soundscape ecological research are diverse and large, and these two characteristics distinguish soundscape ecology from other kinds of ecological and sociocultural research. For example, soundscape ecology studies often require the analysis of landscape data (i.e., spatial data, some of it from satellite-based sensors), weather data (i.e., highly temporal, nonacoustic data), animal and plant survey data (e.g., multivariate data), and socioeconomic data (i.e., survey data and ethnographic

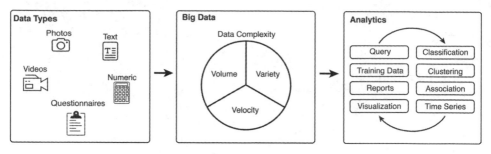

Figure 7.1 The cornerstone components of big data.

interviews). As multiple PARs are often deployed in landscape and regional studies, the amount of data being brought into research laboratories is unprecedented. Soundscape ecologists, who may collect hundreds to millions of audio recordings, are often challenged with trying to manage, analyze, and interpret this massive volume of audio data (Servick 2014). Such large and diverse databases constitute what is called *big data* (Fig. 7.1). "Big data" is a relatively new term, coined only a few decades ago, to describe the data that are being generated in fields like health care (e.g., genomics, imagery), astronomy and physics, and business (e.g., financial transactions and information associated with purchases). Big data are being generated from the need to combine different *data types*. Managing, analyzing, and interpreting big data requires specific approaches, which are broadly called *analytics*.

A growing number of ecologists and social scientists are now working with *data scientists* (Provost and Fawcett 2013). *Data science* is the "set of fundamental principles that support and guide the principled extraction of information and knowledge from data" (Provost and Fawcett 2013, 52). Data science is a data-driven exercise that draws from a variety of other fields, most importantly statistics, visualization, and computer science. The use of the term "science" implies that this field is one where "knowledge is gained through a systematic study" (Dhar 2013). By extension, then, "data science" is "the study of generalizable extraction of knowledge from data" (Dhar 2013). Data science requires knowledge in database management, data formats, computer programming, computer engineering (i.e., hardware), artificial intelligence, and statistics. It is considered a modern evolution of traditional statistics. Given the massive "deluge of data" in scientific fields such as medicine, physics, and genomics, some argue that the traditional scientific method is obsolete (Anderson 2008).

Kitchin (2014) made a statement about big data that is central to the challenges faced by soundscape ecologists as they analyze their data: "The challenge of analyzing Big Data is coping with abundance, exhaustivity and variety, timeliness and dynamism, messiness and uncertainty, high relationality, and the fact that much of what is generated has no specific question in mind or is a by-product of another activity." PARs can generate massive databases; these acoustic data contain a large variety of information and have many dimensions, they come at fast rates, and the technologies are always advancing. Integrating soundscape data with landscape, atmospheric/ocean, and sociocultural data represents an enormous challenge, especially considering the need to apply these data to immediate problems.

7.1.2. Key Statistical Terms

To understand why analyzing big data requires new perspectives, one must first consider what traditional statistics was designed to do for scientists. Traditional statistics was developed in an era (1880s–1980s) when big data were not present. As such, many of the traditional statistical tests are based on the theoretical tenets that guided the development of statistics during that time. To understand these tenets, let us consider a few key terms in statistics: variables, samples, populations, and inference. A *variable* is any characteristic that can be measured and that can be represented with different values. Examples of variables include age, temperature, and income. A *sample* in statistics is a collection of individual values (Sokal and Rohlf 1987) that represent a subset of the entire population. A *population* consists of all possible values of a variable (Steel and Torrie 1980); it is the *universe* of all potential values. Thus, traditional statistics assumes that sample data represent a subset of the population. It also assumes that this subset is randomly pulled from the population. The core tenet of traditional statistics is that data from samples are used to understand the values across the entire population for which estimates such as *mean* (i.e., average) are not known. This understanding is accomplished through *inference*. Statistical inference has generated several important parameters that are used often, most notably p-values and confidence intervals, which are applied to the samples to test hypotheses about certain estimates (e.g., mean) of the population. In statistical inference, a p-value is the probability that any observed difference between two samples is due to chance (Leek and Peng 2015), and a *confidence interval* is the expected range of values for a particular estimate, such as the mean, based on all values in the population. Many of these statistical inference parameters are affected by the size of a sample. If the sample is *very* large, then using these parameters to test hypotheses about estimates of values in the population is not valid. In many cases, the number of recordings in a soundscape study exceeds the "comfortable" limits of sample size that traditional statistics is based on. Thus, many big data analytics focus on data exploration and discovery of knowledge through inference that builds on discovering patterns in data, which can then be used to create new knowledge.

Some important assumptions were made in the development of statistical inference tests. The two most often cited are that any two samples from a population are *independent* (i.e., one sample does not influence the other) and that samples are *normally distributed*. The second assumption requires more explanation. A normal distribution follows a bell-shaped curve when the range of values is plotted along the *x*-axis and the number of cases for each value is represented along the *y*-axis. The central maximum value represents the mean. If the data plotted are normally distributed (see Chapter 10 for tests of normal distributions), then many of the traditional statistical tests can be used to test a variety of hypotheses about the data. These tests are called *parametric tests*. If the data are not normally distributed, another set of traditional statistical tests can be used; the most common among them are categorized as *nonparametric tests*. In these cases, data are often transformed into ranked values and then used in statistical tests.

Two more issues need to be discussed relative to traditional statistics and their use in analyzing big data. First, if analysis requires addressing how two or more variables are related, most of the traditional statistical methods available are based the assumption that their relationship is *linear*. As such, many traditional statistical models estimate parameters for a line that quantifies these relationships. In these cases, too, the data must be normally distributed if the parameters (e.g., slope and *y*-intercept) for a linear relationship are to be valid. Within traditional statistics, some "workarounds" are possible if this assumption is not met. If the data are not normally distributed, then they may be able to be transformed by taking a log base 10 of the original value, so that the transformed data are distributed normally. If the data cannot be transformed, then *nonlinear* techniques are applied to them. Numerous nonlinear techniques have been proposed to address the shortcomings of linear approaches.

The second issue is the fact that big data are often considered *noisy*. This means that the data may have inherent errors within each of the variables. Traditional statistics do not perform well on noisy data.

7.2. Characteristics of Big Data

The three characteristics of big data—that is, their *volume*, *variety*, and *velocity*—are known to data scientists as the *big data 3 core Vs* (Laney 2001;

see also Kitchin 2014). In this section, these three characteristics, and three other "Vs" that are sometimes added (Gandomi and Haider 2015), are described as they relate to conducting soundscape ecological research.

7.2.1. Volume

Volume refers to the amount of data (Orgaz et al. 2016). Although no set size distinguishes big data from non-big data, some ranges have been suggested. However, it is first necessary to understand how data are stored on computer storage devices and how the amount of data is reported before database size can be understood. Data are coded as strings of values of 0s and 1s. One position in this string is referred to as a *bit* (which means it can be coded as a 0 or a 1). A string of 8 bits (a sequence of eight 0s and/or 1s in a row) creates a storage form called a *byte*, which is the smallest unit of information that encodes a character of text. A bit is designated by a lowercase b, and a byte is designated by a capital B. A *kilobyte* (KB) is 1,024 bytes (a base-2 measure equal to 2^{10}), a *megabyte* (MB) is 1,024 KB, and a *gigabyte* (GB) is 1,024 MB. Many computers today have hard drives with capacities around 500 GB. Soundscape ecological research, however, often relies on the greater capacities of advanced storage technologies in the form of university/government disk "farms," which also back up data to robotic tape drives or move data to fast flash memory devices (SSDs), co-located with supercomputers possessing hundreds to thousands of CPUs (central processing units). More recently, cloud services such as Amazon Web Services (AWS), which offer storage, CPUs, and even analytical tools accessible through the internet, are serving as platforms for processing scientific data like the kinds generated by soundscape ecologists.

Recent surveys of information technology (IT) specialists (de Oliveira et al. 2003) have suggested that big data is often thought of as data measured in *terabytes* (TB, which is 1,024 GB) or *petabytes* (PB, which is 1,024 TB or 1 million GB). Many social media sites, such as Facebook and YouTube, are cited as examples of big data sites; these platforms store data in the *exabyte* (EB or 1,024 PB) to *zettabyte* (ZB, or 1,024 EB) range. Many soundscape ecological research studies generate data in the low to high TB range (e.g., a few to over 100 TB), and laboratory and organizational data archives are now exceeding 1 PB.

Because research in soundscape ecology requires quantifying the heterogeneity of the landscape, soundscape ecologists often use geospatial data, especially data from satellite-based sensors that provide information on land surface properties, which are collected at every location, in some cases (e.g., MODIS) every day. A sound file can be used to create a variety of other data, such as (1) all sound sources labeled within the recording along with measures of the labeled data (e.g., sound-source name, mean frequency, peak amplitude), which could number hundreds to thousands of values in a several-minute-long recording in a tropical rainforest; (2) acoustic indices (e.g., ACI; see Section 9.4), which can also number in the dozens, and which could be calculated for small subsets of the files, creating hundreds to thousands of values; and (3) cultural meanings of individual sounds or of all sounds, which could be coded for many individuals who listen to the recording. In other words, the high level of content and interpretative meaning in acoustic files means that *big data could beget more big data.*

7.2.2. Velocity

The second V of big data is velocity: the speed at which data are generated. It also refers to the speed at which data must be transferred across a computer network to maintain a near real-time ability to move them from the site of data collection to a long-term storage facility. Another facet of big data velocity is the speed at which data can be analyzed, as many applications of big data—health care, environmental catastrophes, business planning—allow for only a brief window of time between data collection and decision making (Gandomi et al. 2015). Similarly, use of acoustic data, combined with other ecological and sociocultural data, for conservation management allows little time for analysis. Indeed, new sensors are being developed that are connected to the internet so that the information they collect can be displayed in near real-time (Kitchin 2014); this is called *edge computing.*

7.2.3. Variety

The third V of big data, variety, has two major dimensions. First, estimates are that 75% to 95% of most big data, such as health care data, is unstructured (Gandomi et al. 2015; Das and Kumar 2013). Structured data are often stored either as flat files (e.g., an Excel spreadsheet with rows and columns) or as relational tables (e.g., as a Microsoft Access table consisting of multiple tables that are related to one another through structured linkages). The term *unstructured data* refers to the fact that the files, although they are often organized by topic, are not related through data tables, and that their analysis requires very complex approaches customized to each file type.

The second major dimension of variety in big data is the variety of file *formats*. Generally, big data formats include tables (with numbers), audio, video, and text (Gandomi et al. 2012). Another type of data often considered important is transactional data, which store information in the form of logs, reports of what occurs, and so forth, along with data and analyses. Finally, nearly all forms of data are paired with *metadata*, which is standard information about the data being stored. Metadata often include the date, time, and location of data collection, the database custodian (i.e., the database contact person), coding information (if necessary to interpret the data), data format, and sensor type or sensor ID. In many fields, organizations have developed metadata standards and have distributed templates for their use.

7.2.4. Veracity

Because big data are so voluminous and complex, it is impossible to know if every single data point is 100% valid, or if all are equally reliable. For social media, it is also unknown whether data can be "trusted." This feature of big data has been referred to as *veracity*. It is very likely that some data could be corrupted by a failed or faulty sensor. As data are now often streamed directly from the sensor to a data archive, automated data cleansing (i.e., error checking) routines should be developed (Jin et al. 2015) to improve the reliability of the data. Perhaps surprisingly, some big data are now considered more reliable than traditionally collected data. For example, light detection and ranging (LiDAR) sensors can provide more accurate measurements of canopy height and foliar density than an observer on the ground using traditional handheld measuring tools. And a digital recording of bird-calls can be played, replayed, and compared with reference calls, so that the bias that is common in field identification using binoculars and listening is reduced. In short, the veracity of big data varies according to how it is collected and used for decision making.

7.2.5. Variability

Because big data can come from sensors that have periodic failures, variability is another characteristic of big data. *Variability* refers to the fact that there are often peaks and troughs of data production (Gandomi and Haider 2015). Gaps in data collection are common. Sometimes, long-term monitoring will evolve through several versions of sensors, each of which may have improved sensitivities, improved power use, and increased resolution, which requires researchers to match, and perhaps even transform, the data that are created across sensor versions. All these characteristics contribute toward big data variability.

7.2.6. Value

One other "V" term used to describe big data is *value* (Gandomi and Haider 2015). For most big data, the value of a single data point is very low compared with its value as one member of an entire dataset. In other words, the value of big data comes from the ability to analyze a dataset in its entirety, rather than in small portions. Big data are also considered to be of high value when they are distributed to large communities of users so that they can be used in large numbers of analyses. In theory, as big data grows across all the other Vs, its value should increase as well.

7.3. Soundscape Analytics

7.3.1. Basic Terms

The analysis of soundscape ecological data often requires the adoption of approaches used to analyze big data, particularly those referred to as data mining, machine learning, data perceptualization, and data fusion. Together, these approaches form the basis of *analytics*. Another important concept is the *data analysis pipeline*, the sequence of steps from data collection to the final step of knowledge generation. Finally, all these approaches, used singly or in combination, fall within the scope of a growing theoretical area in data science (Frawley et al. 1992; Piatetsky-Shapiro 2000) described by many scholars as *knowledge discovery in databases* (KDD). KDD emphasizes the fact that new knowledge is the intended outcome of the data-driven discovery process. More specifically, it is defined as the "nontrivial extraction of implicit, previously unknown, and potentially useful information from data" (Frawley et al. 1992).

The remainder of this section describes the conceptual foundations of data formats, aspects of big data, and the key components of data mining. Although the focus will be on analyzing large audio file collections, there is also a summary of how soundscape and landscape data can be analyzed together along with the generalized approaches of mixed methods (see Section 9.10), which are common in sociocultural, educational, and geographic research, in which quantitative and qualitative data are analyzed together.

Before we examine the main features of each step in the typical data analysis pipeline, it will be helpful to define other key terms in big data analytics. *Data mining* is "the application of specific algorithms for extracting patterns from data" (Fayyad et al. 1996, 39). Data mining draws on both the traditional statistics that is commonly used in ecology, engineering, business, and the social sciences and advanced computer-based approaches broadly known as machine learning, artificial intelligence, and pattern recognition (Tan et al. 2016). Data mining can be used either for *descriptive* tasks—that is, to describe patterns (correlations, clusters, trends) within data—or for *predictive* purposes—that is, to predict one value from other values in the dataset. Data mining differs from the related activity of *information retrieval*, which is the task of recovering stored information from a database. *Machine learning* (Jordan and Mitchell 2015), on the other hand, is the process of "teaching" a computer to learn about patterns in a dataset; machine learning is necessary when datasets are exceptionally large and patterns are difficult to detect with simple plots. Machine learning is often applied to classification problems. For example, a machine-learning algorithm could be used to detect the call of the dink frog in many audio recordings from a tropical rainforest. Many machine-learning algorithms, referred to as *pattern recognition* algorithms, are based on our understanding of how living systems learn about patterns in the environment. The term "pattern" has been defined (Watanabe 1985) as "opposite of chaos; it is an entity, vaguely defined, that could be given a name." Machine-learning algorithms are derived from the broader body of theory called *artificial intelligence* (AI). This field of study attempts to understand different kinds of learning that occur in living systems and how they can be replicated using computer algorithms. AI has been called "brain-inspired computing." General AI tools include artificial neural networks (ANN), genetic algorithms (GA), and artificial bee colonies (ABC). AI and machine-learning tools have been successfully employed in business, human health care, and geography, and with audio, visual, text, and geospatial data, all of which are used for soundscape ecological research. A form of AI in which computers learn without much human specification of rules, composed of multiple processing layers of learning algorithms, is called *deep learning* (LeCun et al. 2015). Many deep learning tools have been developed and are being used in business, medicine, and astronomy.

There is also a subfield of machine learning, called *active learning* (Settles 2009), that uses algorithms to determine which instances (e.g., audio files) in the database need to be labeled by an "oracle" (a human annotator). Labeling of soundscape data is an approach commonly used by soundscape ecologists and bioacousticians to build call detection tools. Active learning for labeling (Druck et al. 2009) increases the accuracy of labeling and can reduce the cost of labeling data compared with selecting instances for labeling using a random approach.

Data fusion is the process of combining data from different sources to create one database (Mitchell 2007). Data fusion can occur across sensors, variables, locations and time, and is done for the purposes of creating greater quality data (Wald 1999). Data fusion often occurs with different databases in numerical and text data formats.

One of the important activities commonly performed along the data analysis pipeline is *visualization*. It is considered good practice to visualize data at each step, as data may require cleansing, subsetting, transformation, or fusion with other data to generate the knowledge necessary to address a problem. For acoustic data analysis, the use of visuals while listening becomes *data perceptualization*. For example, soundscape ecologists can view a spectrogram alongside a set of acoustic indices while listening to a recording using audio file software at any point along the data analysis pipeline.

Finally, as many studies in ecology and other scientific disciplines are now finding, the number of variables in many studies is equal to or greater than the number of samples per variable. In the past, scientists collected many samples (e.g., hundreds) for only a handful of variables (e.g., temperature, rainfall). However, the era of big data has created a situation in which many hundreds to thousands of variables can be collected. This abundance of variables creates a situation that data scientists call *high dimensionality*. Bellman (1957) coined the term "curse of dimensionality" to describe the multiple mathematical challenges presented by databases that have many variables. These attributes may also covary, creating a situation of *collinearity*, which may inflate one or a host of closely related variables and thus reduce the ability to detect important relationships in complex data. A variety of techniques have been developed in statistics and big data analytics to reduce dimensionality in highly dimensional data. The aim of these techniques is to reduce the number of dimensions to a manageable number (often three). In statistics, this process is called *ordination*. Analysis is then performed on the ordinated data.

7.3.2. Data Types

At the highest level of consideration, data can be divided into three types: categorical, numerical, and binary. *Categorical* data (Agresti 2002) have a measurement scale that consists of classes. A common categorical variable in soundscape ecology is land use/cover, which can be classified as urban, agriculture, shrubland, forest, open water, wetlands, or barren. Other common categorical data include taxonomic categories for plants and animals (e.g., birds, amphibians, mammals). These kinds of categorical variables that are not ordered are called *nominal*. Nominal data are often labeled using nouns. In contrast, some categorical data can be ordered. Example of such *ordinal* variables are size (small, medium, large), condition (good, fair, poor), height (understory, midstory, canopy), and date (in various formats).

Numerical data come in two main subtypes. *Discrete* data are represented as integers that are often used in counting. Age is an example of a discrete numerical variable. The other subtype is *continuous* data, which are data that can take on any value (e.g., weight, height). Numerical data are reported as real numbers, such as 10.1 or 10.1294783.

A third data type is *binary* data, which is placed in one of only two categories. Such categories include true or false, yes or no, success or failure, heads or tails (on a coin), male or female, alive or dead. These data are most often represented as 0s or 1s. In many cases, 1s represent true, yes, success, and so forth, and 0s represent false, no, or failure.

Conversions between data types are common. For example, binary data can be created by employing a rule that is applied as a Boolean condition to discrete or continuous data. For example, we might consider only sounds that are above a noise floor defined as –50 dBFS. Sounds above the noise floor would be considered "true," and as such, they would be assigned a value of 1. Sounds below the noise floor would be assigned a value of 0. Categorical data can be created from numerical data using a process called *binning*. Ranges of values can be used to convert data into ordinal data. For example, numerical measurements of height above sea level could be placed in one of three ranges: low, medium, and high elevations.

7.3.3. An Archetypal Data Analysis Pipeline

Big data analysis requires a set of steps that involve a set of general rules or best practices. These steps are typically performed along a pathway that often involves "looping" back to redo an earlier step. Let us examine some of the tasks or approaches that are associated with each point along an archetypal data analysis pipeline (Fig. 7.2) using the example of audio files as they are processed by soundscape ecologists.

> *Sensor in the field.* Data are collected in the field on storage devices, commonly as one file per recording. Between researcher visits to the PAR,

dozens to hundreds of files are collected on the storage device. When the PAR is serviced, an empty SD card is placed in the PAR and the SD card with the collected data is brought back to the lab.

Offload to data disks. SD cards are placed in an SD card reader that is physically connected to a computer through a cable (e.g., a USB 3.X cable), and the collected data are copied to hard drives on a local computer, to drives in a data center, or to a data cloud service such as Amazon Web Services. A large data repository that stores highly diverse data (acoustic, tabular, text) together is often called a data lake.

Create metadata. A software program is used to create a metadata file, usually from the filename. For example, PARs manufactured by Wildlife Acoustics name WAV files using a standard naming convention of sensor ID + date + recording start time and create a metadata file for each WAV file. The metadata file may be organized either as a record in a database or as an XML file.

Data cleansing. Data ingestion programs typically examine files from the field for potential errors. Files that are shorter in length than expected or have unusual acoustic features (e.g., amplitudes that suggest microphones were not working) should be flagged or not entered into the database.

Appending. New field data are often appended to existing databases. With big data, they are typically appended frequently, and in some cases, where sensor data (e.g., satellite images) are sent directly to a server, they are appended continuously. A log file is typically generated that includes the amount of newly appended data, date and time, and the name of the person performing the operation.

Data characterization. In this step, a summary of the database is created, which might include the following: number of records, maximum and minimum values for each variable, the amount of missing data or number of error flags, and the sizes of all files in the database.

Data selection/data labeling. Researchers often create a subset of their data using sampling procedures (described in more detail in Chapter 10); this process is called *data selection* or *subsetting*. The subset may then be labeled using any one of several approaches developed by bioacousticians and soundscape ecologists. One option is to use only a portion of a recording for analysis. For example, an hour-long recording would be difficult to listen to and label, so a 1-minute portion might be selected from that file and labeled.

Data transformation. Transformation of acoustic data can be done in a few different ways. One common approach is to calculate a variety of acoustic indices for each recording and store these in the database. These indices transform the data from a WAV file to a numerical value that is more easily analyzed. A second form of transformation might be used if data are to be analyzed using traditional statistical models; in that case, data distributions that depart greatly from a normal distribution" may have to be transformed using a log, sine, cosine, or other transformation model. Finally, researchers may be required to use binning or encoding approaches to convert numerical data to categorical (e.g., names of soniferous animals) or binary (e.g., presence, absence) formats.

Reduction of dimensionality. When many variables are contained in the data, a variety of approaches can be used by soundscape ecologists to reduce these dimensions, typically from dozens or more to three.

Core analytics. Data scientists recognize four kinds of analyses that can be performed on big data: *classification, association, clustering,* and *time series analyses.* Indeed, many of the advances in big data analytics have focused on these four core analytical approaches, and generalized machine-learning algorithms based on these approaches have been developed and used by soundscape ecologists to discover patterns in data. A summary of these approaches, with examples of their use by soundscape ecologists, is provided in Chapter 10.

Pattern discovery. The patterns generated from big data analyses can be discovered through a host of standard visualization techniques as well as new ones that have been developed to complement big data analytics. The discovery process here needs to be driven by the kinds of questions being asked along with the design of the field study. For example, asking how the soundscape differs within and outside protected areas, how a soundscape changes after an event such as a hurricane, or how the soundscape varies across an ecological gradient (e.g., soil moisture) or a gradient of human activity (e.g., levels of urbanization) would be one way of looking for patterns in big data.

Evaluation and presentation/dashboards. A summary of the new knowledge gained can be shared with others (e.g., natural resource managers) using web-based tools that provide visuals such as charts. These summaries can be configured as dashboards to allow users to query the results using interactive tools or to visualize the soundscape data along with landscape (i.e., as maps) and/or other data (i.e., weather, human population, economic activities).

Visualization or perceptualization of data can occur along the entire data analysis pipeline, at the points marked in Figure 7.2 with a circled V. For example, the dates or times of missing data should be visualized, perhaps

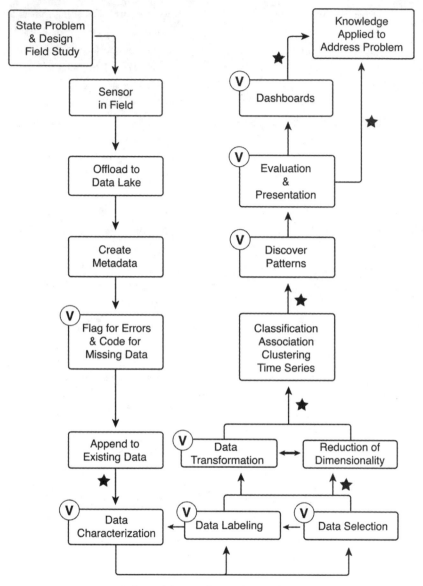

Figure 7.2 Archetypal data pipeline analysis pathway. A circled V represents an opportunity for perceptualization; a star represents an opportunity for data fusion of non-acoustic with acoustic data.

as a timeline. Any type of errors could also be viewed along a visual time-line. Analyses and visualizations commonly used in big data analytics are described in Chapter 10.

Nonacoustic data can also be fused with acoustic data along several points in the data analysis pipeline (indicated by a star). For example, data from weather stations, satellite data on surface characteristics such as plant stress, or data on the spatial configuration of vegetation within the PAR's acoustic theater can be integrated at various steps.

7.4. Data Management

7.4.1. Basic Terms and Concepts

Research in soundscape ecology will invariably involve the need to manage data. As an example, let's consider how the big data analytics approaches described in the previous section can be used on the database that is presented in Figure 7.3. This database contains data from sound files from a study in Borneo, recorded with a Wildlife Acoustics PAR, sensor ID 015088, on a 10-minutes-on, 20-minutes-off duty cycle during the month of February, 2014. Two error flags were created, one for unusual recording lengths (i.e., not 10 minutes) and another for recordings made during PAR servicing. Wildlife Acoustics's Kaleidoscope software program was used to calculate a variety of acoustic indices (e.g., intensity, spectral, and SNR). Each column created in the database represents a *field*, or variable; the term "field" is more common among data scientists and "variable" among statisticians. An important type of field used in database management is the *record ID*, *primary key*, or *index field*. Fields in database systems can be of several types: numeric, integer, string (alphanumeric, such as A1), or character (letters or text only, such as "*Passer domesticus*"). The rows in databases are referred to as *records*.

Special fields, called *key fields*, link two or more data tables together. Let us consider why these fields are important. *Relational databases* are common to the types of data structures that soundscape ecologists use. A relational database is often needed when data for one sensor, such as a PAR, is collected at different time intervals. Consider that the latitude and longitude of a sensor is collected, and thus stored, only once. Sensor IDs and locations, and perhaps the date when each sensor started collecting data, are in one table. In another table are landscape characteristics, such as habitat type, soil type, and slope. Another table could include daily remote-sensing information such as plant condition at, and perhaps within, the acoustic theater of the sensor. Hourly air temperatures, wind speeds, and so forth could be recorded but associated with all sensors at a study site, or, if weather stations are placed near PARs, could be used as additional data for soundscape analysis. All of these relational tables could be linked through their key fields. An illustration of how relational tables are linked and what variables (i.e., fields) are contained in them is often referred to as a *data model*.

Summarizing Fields							Flags		Intensity Acoustic Indices												Spectral Acoustic Indices					SNR Acoustic Indices							
Sensor	Day	Channel	Duration	Date	Time	Hour	Flag 1	Flag 2	Mean	SD	SEM	Median	Mode	Q25	Q75	IQR	Skew	Kurt	SFM	SH	NDSI	ACI	ADI	AEI	BI	BGN	SNR	ACT	EVN	LFC	MFC	HFC	CENT
015088	2	0	577	2/8/2014	30:25.0	11	0	0	7632.76	4777.33	211.13	8613.28	11800	3402.3	11757	8355	6.2567	58.714	0.4911	0.8928	0.9228	317.19	1.373	0.851	28.196	-41.56	22.39	0.063	0.324	0.194	0.109	0.044	0.436
015088	1	0	600	2/8/2014	00:00.0	12	1	1	7272.81	4214.62	186.26	6373.83	11972	3789.8	11025	7235	2.748	17.037	0.4956	0.9114	0.9843	291.77	2.609	0.496	119	-37.09	9.74	0.035	0.222	0.044	0.107	0.054	0.483
015088	1	0	600	2/7/2014	00:00.0	19	1	1	4887.45	4069.31	179.84	3962.11	3962.1	3531.5	5297.2	1766	5.487	35.829	0.7959	0.8419	0.9927	294.71	2.65	0.48	68.023	-19.29	4.83	0	0	0	0.022	0.061	0.275
015088	1	0	600	2/7/2014	30:00.0	18	1	1	4255.89	4860.15	214.79	2110.25	43.07	344.53	6718.4	6374	10.723	143.31	0.4188	0.8419	0.0843	294.83	2.153	0.692	60.083	-27.2	7.6	0.003	0.025	0.031	0.04	0.035	0.264
015088	1	0	600	2/8/2014	00:00.0	13	1	1	5509.88	2933.75	129.65	4392.77	4177.4	3746.8	6632.2	2885	3.0357	12.369	0.2827	0.8404	0.9979	301.38	2.325	0.622	193.06	-23.73	5.63	3E-04	0.002	0.035	0.048	0.054	0.477
015088	1	0	600	2/7/2014	00:00.0	20	1	1	5062.27	4771.23	210.86	3962.11	43.07	689.06	7752	7063	10.397	137.82	0.8675	0.9143	0.9143	293.18	2.691	0.452	52.095	-26.7	5.06	3E-04	0.002	0.021	0.038	0.099	0.238
015088	1	0	600	2/8/2014	30:00.0	12	1	1	8419.32	4367.23	193.01	9388.48	0	5297.2	11886	6589	4.6815	35.857	0.4652	0.8884	0.9856	286.62	2.395	0.613	78.336	-35.28	8.15	0.012	0.12	0.045	0.087	0.063	0.45
015088	1	0	600	2/7/2014	00:00.0	19	1	1	5696.97	5213.74	230.42	4005.18	43.07	1550.4	8096.5	6546	7.2995	71.633	0.4903	0.8692	0.9661	295.48	2.815	0.363	57.427	-24.81	4.35	3E-04	0	0	0.027	0.086	0.255
015088	1	0	600	2/7/2014	30:00.0	18	1	1	3868.63	4649.53	205.48	1421.19	43.07	129.2	6847.6	6718	12.009	170.19	0.3678	0.8057	0.4044	303.04	2.394	0.596	53.234	-28.76	4.74	4E-04	0.003	0.02	0.028	0.025	0.261
015088	1	0	600	2/8/2014	00:00.0	13	1	1	5553.03	3232.42	142.85	4392.77	3617.6	3617.6	6675.3	3058	2.6437	9.7731	0.3463	0.8628	0.9933	299.39	2.531	0.534	165.91	-29.86	6.35	0.002	0.01	0.034	0.049	0.08	0.484
015088	1	0	600	2/8/2014	30:00.0	14	1	1	5412.78	3015.81	133.28	4392.77	3617.6	3703.7	6330.8	2627	2.8442	10.959	0.3091	0.8426	0.9955	301.31	2.367	0.61	188.23	-25.2	5.95	2E-04	0	0.039	0.047	0.047	0.489
015088	2	0	600	2/8/2014	00:00.0	15	1	1	5440.11	3174.5	140.29	4392.77	4306.6	3617.6	6416.9	2799	2.6903	9.8519	0.3352	0.8507	0.9967	300.89	2.441	0.587	170.2	-26.39	6.68	0.002	0.013	0.045	0.05	0.048	0.477
015088	2	0	600	2/8/2014	00:00.0	14	1	1	5856.51	3635.1	160.65	4435.84	3832.9	3574.5	7235.2	3661	2.5598	8.8917	0.4264	0.8775	0.9936	298.96	2.504	0.557	145.94	-32.61	6.41	0.003	0.025	0.048	0.048	0.074	0.487
015088	2	0	600	2/8/2014	00:00.0	16	1	1	5454.91	3446.33	152.31	4263.57	4091.3	3531.5	6373.8	2842	2.7554	10.324	0.3828	0.8605	0.987	299.26	2.556	0.528	141.46	-29.13	6.61	9E-04	0.003	0.04	0.055	0.073	0.487
015088	2	0	600	2/8/2014	30:00.0	15	1	1	6191.94	4021.68	177.73	4521.97	4177.4	3574.5	7450.5	3876	2.5043	8.7509	0.4638	0.8884	0.9842	297.35	2.366	0.629	152.95	-31.46	7.8	0.023	0.142	0.023	0.057	0.064	0.483
015088	2	0	600	2/8/2014	30:00.0	17	1	1	5311.96	4282.22	189.25	4349.71	646	2019.8	7019.8	4953	0.7661	1.962	0.5187	0.9236	0.5591	307.82	2.923	0.26	52.4	-23.57	14.71	0.055	0.157	0.041	0.044	0.044	0.475
015088	2	0	600	2/8/2014	30:00.0	16	1	1	7028.4	5789.88	255.88	5813.96	0	2756.3	11326	8570	8.299	98.4	0.7435	0.937	0.9391	296.7	2.193	0.688	70.902	-43.98	16.35	0.141	0.365	0.035	0.074	0.039	0.47
015088	2	0	600	2/8/2014	30:00.0	17	1	1	5245.69	4429.05	195.74	4349.71	0	1894.9	6847.6	4953	4.0804	41.202	0.5292	0.9173	0.6938	323.59	2.628	0.485	57.866	-32.23	24.27	0.102	0.488	0.058	0.046	0.051	0.444
015088	2	0	600	2/8/2014	30:00.0	18	1	1	6155.63	4787.98	211.6	5254.1	2971.6	2928.5	7278.2	4350	2.3542	11.413	0.6237	0.927	0.9588	321.58	2.077	0.693	72.629	-38.82	28.92	0.209	0.958	0.055	0.092	0.051	0.489
015088	2	0	600	2/8/2014	00:00.0	18	1	1	5913.41	3427.72	151.49	5684.77	7192.1	4220.5	7192.1	2972	3.9135	21.2	0.3646	0.8432	0.9748	309.35	2.264	0.631	106.51	-30.69	17.15	0.124	0.707	0.073	0.088	0.044	0.488
015088	2	0	600	2/8/2014	00:00.0	19	1	1	6179.88	3387.93	149.73	5555.57	5512.5	4694.2	6804.5	2110	3.7334	18.118	0.384	0.845	0.9935	310.41	2.97	0.267	74.497	-32.9	12.37	0.002	0	0.07	0.063	0.091	0.483
015088	2	0	600	2/8/2014	00:00.0	19	1	1	5698.04	3348.52	147.98	5167.97	5770.9	4349.7	5943.2	1593	3.3616	14.973	0.3636	0.8436	0.9866	315.19	2.626	0.489	59.828	-32.99	15.79	0.008	0.015	0.078	0.061	0.048	0.474
015088	2	0	600	2/8/2014	00:00.0	21	1	1	6261.52	3307.33	146.16	5684.77	5900.1	4995.7	6890.6	1895	3.8217	18.202	0.3732	0.8351	0.996	306.46	1.999	0.745	100.8	-33.36	8.64	9E-04	0.003	0.078	0.072	0.111	0.481
015088	2	0	600	2/8/2014	30:00.0	20	1	1	6052.7	3394.66	150.02	5555.57	5469.4	4780.4	6761.4	1981	3.486	15.547	0.3678	0.8393	0.9958	303.3	3.091	0	91.417	-34.02	10.61	0.001	0.002	0.087	0.072	0.083	0.464
015088	2	0	600	2/8/2014	00:00.0	20	1	1	6437.16	3488.29	154.16	5684.77	5814	5038.8	6847.6	1809	4.2425	22.643	0.3823	0.8314	0.9969	326.86	2.773	0.407	84.123	-31.75	10.12	0.001	0.071	0.068	0.104	0.083	0.487
015088	2	0	600	2/8/2014	30:00.0	21	1	1	5337.25	3918.93	173.19	4866.5	5426.4	2627.1	6546.1	3919	1.4137	4.4961	0.4606	0.9082	0.7994	324.29	2.442	0.569	58.456	-31.51	22.06	0.052	0.145	0.03	0.041	0.053	0.477

Figure 7.3 A sample database table structure for a sound file with acoustic features and other key variables.

When bivariate or multivariate data are plotted in two-dimensional or three-dimensional space and groups of values cluster together in a portion of the plot, that area is called a *feature space*. At a minimum, there should be two feature spaces in a bivariate plot. Often, there are more than two, and in highly dimensional data, there could be numerous feature spaces. The objective of many data exploration approaches is to determine the nature of feature spaces by quantifying boundaries between them. These boundaries can be linear or nonlinear and can be applied to two or more dimensions.

7.4.2. Querying

If we consider the sample database in Figure 7.3, we will see that several of the summarizing fields can be used to explore the data for patterns. For example, researchers may be interested in characterizing the data across different groupings. They could then explore the means or ranges of values for each sensor in the database. Other types of groupings could explore how means change by day, between studies, or between treatments. The common way to do this is to perform a query using a *group by* phrase in a query-based language like SQL (Standard Query Language).

7.5. Textual Analysis

7.5.1. Forms of Textual Analysis

For soundscape ecologists, there are two main forms of textual analysis that can be undertaken in research. The first is the use of *textual data mining* (TDM) tools (Hearst 1999) that support *knowledge discovery in textual databases* (KDT) (Feldman and Dagan 1995; Tan 1999). Many of the same kinds of data mining operations that are used for numerical data are also used for textual data; these include the use of tools and algorithms that do classification, association, clustering, and time series analysis. For the most part, TDM is a semi-automated means of discovering patterns and trends across many documents, in which one should "take the mining-for-nuggets metaphor seriously" (Hearst 1999). One of the first applications of TDM was the mining of text from thousands of journal articles (Swanson 1986) on two distinct topics in medical research—specifically, fish oil and Raynaud's syndrome—when a new hypothesis proposed that there was a relationship between the two. Researchers assembled over 3,000 articles published before this hypothesis was proposed and used a textual data mining routine to look for similar patterns in these two independent sets of literature. They found that the literature in both areas supported two premises: that fish oil affects a person's blood chemistry, and that it does so in a way that benefits patients with Raynaud's syndrome.

The second form of textual analysis is the use of software and approaches employed in the social sciences and the humanities to do transcription analysis, content analysis, discourse analysis, and coding and text interpretation. Much of this work follows from the application of grounded theory methodology (see Chapter 9).

7.5.2. Textual Features

TDM and KDT can help to discover patterns in four levels of textual features that are contained in documents:

Characters: The individual component letters, special characters, numerals, and spaces that form the building blocks of words, terms, and concepts. Character-based representations are often not useful unless they contain what KDT data scientists call positional information.

Words: The basic semantic (i.e., meaning) level contained in documents. Generating word lists is often an extensive task and by itself is not a useful activity if there are many documents to analyze.

Terms: Single words and multi-word phrases. Processing of terms using automated tools often requires that words be converted to normalized terms (e.g., past tense to present, plural to singular) so that there is less redundancy at the start of an analysis of all words in a document. Users can specify a set of candidate terms that will act to subset those most relevant to the analyses.

Concepts: Concept-level textual features can be generated by manual or automated means via statistical, rule-based, or hybrid approaches. Many TDM tools allow users to search for single words, multi-word phrases, and whole clauses that can then be used to create concept identifiers.

Term-level and concept-level features can be created using TDM tools to develop hierarchies, term and concept correlations, and term and concept frequencies, distributions, and hierarchical graphs. Patterns of terms and concepts are often sought across documents at an inter-document level of analysis. Terms and concepts are often derived from *domains*, specific areas of study, each of which has its own ontologies, lexicons, and taxonomies.

Domain knowledge informs TDM with specific hypotheses and text exploration pathways. Another common analysis of documents is an examination of trends over time: Do terms and concepts emerge or disappear? Are there concepts that remain stable over time in their frequency of occurrence?

7.6. Software Tools

This section is a brief, but not exhaustive, summary of the software tools commonly used by soundscape ecologists. The objective here is to showcase the breadth and depth of the software that soundscape ecologists might use.

7.6.1. Database Management Tools

Nearly all soundscape ecologists will need to store data in a database management software package. To date, there is no database management system that directly stores audio files for use in research. Rather, many soundscape ecologists use analytical tools such as R (described later in this section) to calculate a variety of acoustic indices or other features, which are then stored as either numerical or text fields in the database system. The database management software packages most commonly used by soundscape ecologists include Microsoft Access, Oracle, MongoDB, and MySQL. In some cases, soundscape ecologists create their own relational tables that allow them to query the data. Purdue's Center for Global Soundscapes (CGS) has a MySQL database system called Pumilio (Villanueva-Rivera and Pijanowski 2012) that stores pathnames (file names and directory locations) for WAV or FLAC files. Some other, more advanced database management systems have been developed specifically for heterogeneous big data. Such systems include those that are called *NoSQL* (or sometimes "Not Only SQL"), as they offer the features of SQL relational tables and SQL queries, but also more flexible query options for unstructured data. Two big data NoSQL tools are known by the clever names of Hive and Pig. In some instances, soundscape ecologists use spreadsheet software packages such as Microsoft Excel for analysis and visualization of smaller subsets of data.

7.6.2. Operating Systems

Soundscape ecologists, as well as most data scientists, need to consider a variety of operating systems (OS) for data management and analysis (note that some data scientists call an OS a *platform* or *engine*). Most scientists are familiar with the standard business operating systems, Microsoft's Windows and Apple's MacOS. Many of the database management software packages listed above, as well as common analytical tools are available for these operating systems. Another type of operating system that is less common, but popular with data scientists, is UNIX-based systems, such as Linux. Many university supercomputer facilities run some version of UNIX. There are a few big data operating systems as well. The best known is Hadoop (White 2012), whose Hadoop Distributed File System (HDFS) (Shvachko et al. 2010) is designed specifically to store and analyze big data where data storage and CPUs are distributed across many servers, commonly located in corporate data centers around the world. HDFS has been optimized for use by businesses that have extremely large numbers of users and a variety of services (e.g., Google, Facebook, Amazon). In fact, these businesses initiated the development of Hadoop around the mid-2000s and made it *open source* software, which allows users to modify its source code (i.e., program) for their own specific uses. A modified version of Hadoop is Spark (Zaharia et al. 2012), which has advantages for soundscape ecologists and data scientists because it is designed to place data in memory and keep them there for a long time so that analyses can be conducted across the entire data analysis pipeline without having to access data servers multiple times. Because using memory—or more specifically, *random access memory* (RAM)—is much faster than reading and writing from data drives (either SSDs or HDDs), Spark is far faster (Zaharia et al. [2012] estimate it is 40 times faster) than Hadoop, especially for using R analysis tools and, more generally, machine-learning algorithms.

7.6.3. Analysis Tools

There are five categories of analysis tools that soundscape ecologists use. The first is tools that focus on *machine learning* and are not specifically designed for any one data type, although most are for use with numerical data. The second is tools specifically designed to use *audio files*, such as WAV files, as inputs for analysis. A majority of these tools are limited in that they allow users to work with only one file at a time. A third category of tools is those designed for *scripting*. Scripts command the computer to carry out small tasks, unlike computer programs, which involve many commands that often have to be compiled to create an executable program. The most common scripting language is Python, which is a general-purpose scripting language used across a variety of software tools (i.e., it allows for the integration of two or more tools). The fourth category of tools is those used by social scientists and scholars in the humanities for whom *text* contains important information from their work, particularly interviews. Finally, there are software tools designed specially to manage, manipulate, integrate, analyze, and visualize *spatial data*. Let us consider each of these categories and provide some examples of the software packages most commonly used by soundscape ecologists.

Machine Learning

Arguably, the R suite of tools (www.r-project.org), with two million users, is the "standard" for ecologists and most scientists interested in analyzing numerical data (Wickham and Grolemund 2016). R is an open source (and free) software language composed of a core set of more than twenty-five packages, and many more are created and distributed by users through the Comprehensive R Archive Network (CRAN). R is often referred to as a collection of tools (packaged as toolkits, or *libraries*) that support a computational environment (Venables et al. 2021). Users of R work in one of two environments. The first, and probably the most common, is the command window. This environment has a prompt, and the user provides a command or list of commands in a file. The second environment employs a graphical user interface that allows users to interact with R using visual objects and menus. R packages commonly used by soundscape ecologists include Seewave, Spectro, TuneR, and SoundEcologyR. A special implementation of R for Spark, called SparkR, is used especially for large machine-learning tasks and for data stored in the Hive database management system.

Audio File Processing

Often, it is necessary to visualize a sound file using a spectrogram, waveform, or frequency/amplitude plot. Several software packages exist that allow researchers to view, listen to, and manipulate a sound file. Several of these packages also allow users to label or annotate specific areas of a recording. The sound file processing packages used most often by soundscape ecologists include Audacity (Audacity Team 2021), Adobe's Audition (https://www.adobe.com/products/audition.html), SonicVisualizer (https://www.sonicvisualiser.org/), Kaleidoscope (https://www.wildlifeacoustics.com/products/kaleidoscope-pro), Avisoft (http://www.avisoft.com/), and Raven Lite and Raven Pro from Cornell University (https://ravensoundsoftware

.com/). The last three software packages are designed for soundscape ecologists and bioacousticians who are interested in labeling sound files for use in machine learning (covered in more detail in Chapter 10) and in calculating a variety of acoustic features of interest to their fields. The first three packages are designed for a wider audience, which includes many users in the entertainment industry, so they are best used for visualizing, manipulating, and applying specific audio filters (e.g., high-pass band filters) to sound files.

Scripting

Scripting languages differ from *programming* languages in that scripting languages operate through an interpreter, whereas programming languages use a compiler to create a program. Perl, JavaScript, and Python are some of the most common scripting languages used by data scientists and soundscape ecologists. Of these three, the most widely used is probably Python, as it interfaces well with the R programming environment (Oliphant 2007). Two major Python-based toolkits (also sometimes called *libraries*) are used by scientists: NumPy (Oliphant 2006) and Scikit-learn (Pedregosa et al. 2011; Garreta and Moncecchi 2013). NumPy provides users with the ability to work with data in arrays and matrices. Scikit-learn contains a set of common machine-learning algorithms, such as those for clustering, classification, and dimension reduction, that use NumPy and a few other custom tools to do machine learning. Other popular Python-based machine-learning libraries include MLPy (Albanese et al. 2012), PyBrain (Schaul et al. 2010), SHOGUN (Sonnenburg et al. 2010), and PyMVPA (Hanke et al. 2009). JavaScript is most commonly used by data scientists when data and analysis need to be interfaced with a web browser. Some tools, such as the set of deep learning tools in TensorFlow (Abadi et al. 2016), are implemented in Python and JavaScript.

Textual Analysis

Soundscape ecologists, commonly in collaboration with social scientists and scholars in the humanities, may also conduct research that involves interviews with people about how they perceive their soundscape and surrounding environment (see Chapter 6). Interviews require qualitative analysis, and the software packages that have been developed for that purpose are referred to as *computer assisted qualitative data analysis* (CAQDA) platforms (Lewins and Silver 2010). Many CAQDA platforms manage and analyze text, photos, audio, and video, and some can incorporate numerical data as well. Lewins and Silver (2010) describe two kinds of CAQDAs: (1) code-based theory building software and (2) text retrievers/textbase managers. Code-based theory building software assists researchers in analyzing qualitative data using thematic coding of large portions of data, sometimes called "chunks," which allows users to identify thematic lines (this is the retrieval aspect). Tools in these packages allow users to examine relationships across issues, concepts, themes, individuals, and groups, so that higher-order patterns emerge from the narrative databases. Text retriever/textbase manager software tools allow users to create keyword co-occurrences, draw "proximity plots" for showing relationships between keywords, and produce visualizations of words and associations across the narrative database. The CAQDA Networking Project at the University of Surrey provides a summary of the most commonly used CAQDA software platforms along with tutorials (see also Lewins and Silver 2010). The most widely used commercial CAQDA software platforms include NVivo (Hilal and Alabri 2013; Wong 2008), Atlas.ti, Leximancer, and The Ethnograph. Free CAQDAs include Aquad, Coding Analysis Toolkit, RQDA, and qcoder. Aquad, RQDA, and qcoder are based on the R statistical language platform. Many of the CAQDAs can be used to discover patterns of concepts in text and are thus excellent tools for conducting reviews of big literature (Nunez et al. 2016) in addition to interview text.

Spatial Data Analysis

Software tools that manage, analyze, manipulate, integrate, and visualize spatial data are called *geographic information systems* (GIS). These tools have helped facilitate research in landscape ecology since the late 1980s. Most GIS software is aligned with what some geographers call a "science of space," and therefore the term "GIS" is used simultaneously with "*GISci*" (Goodchild 1992; Goodchild et al. 2007; Longley et al. 2005; Goodchild and Haining 2004). Thus, the structure of spatial analysis aligns closely with spatial theory. One of the most widely used GIS software packages is the ArcGIS platform produced by Esri (Environmental Systems Research Institute). Point, line, polygon, and raster data can be managed, integrated, analyzed, and visualized on this platform, mostly through maps as the core visual tool. GIS software that focuses on satellite imagery is referred to as remote-sensing software; the most commonly used is Erdas (Bolstad 2002). Noncommercial software packages are also commonly used; these packages include QGIS and GRASS (Geographic Resources Analysis Support System). An array of R libraries is available, too (see Bivand 2003 for an early review and Bivand et al. 2013 for most recent), including *sp* and its successor, *sf*; *gdal* and associated R libraries of *rgdal* (Bivand 2016); and *raster* (Hijmans 2015).

7.6.4. Visualization Tools

Soundscape ecologists use a variety of visualization tools, many developed by data scientists. Probably one of the most widely used visualization packages is *ggplot2*, developed for the R environment (Wickham 2011). This package can create graphics in eight categories (Prabhakaran 2019): correlation (e.g., scatterplot, jitter plot), deviation (e.g., diverging bars), ranking (ordered bar chart, dumbbell plot), distribution (e.g., histogram, box pot, violin plot), composition (pie chart, waffle chart), change (e.g., time series plots, calendar heat maps), groups (e.g., dendrograms, clusters), and spatial (using Open Street Map or Google Road Maps). The *ggplot2* package is built from the well-known graphics theory for data visualization by Wilkinson (2012), referred to as the "Grammar of Graphics." The Grammar of Graphics is a rubric of data structures and visualizations that together help inform the researcher at different levels of detail. A set of JavaScript-based tools, called D3 (or Data-Driven Documents) (www.d3js.org), provides an array of unique visualizations of data that can be easily transported to the web, as the scripts work within a web browser (see D3 overview at https://github .com/d3/d3/wiki). Finally, Tableau (Murray 2013) is a commercial software package that is designed to connect to large databases on data servers, where users can then visualize the data in a variety of ways. It is used mostly in business.

7.7. Client-Server-Cloud Technologies

Given the glut of data in many areas of science and business, the ways in which researchers work with data have evolved over the past several years. Prior to the generation of these massive databases, many researchers analyzed data on their own computers. With the advent of big data during the early 1980s, new data management approaches had to be developed. These new approaches included the development of large data centers, called *servers*, to store and allow access to data. Many of these servers were managed by large organizations such as national governments. Retrieving these data eventually required the use of web-based interfaces that would allow a user to query the data and then extract them for download to their desktop computers, which, to distinguish them from the server, are called *clients*. As machine-learning tools matured, a need for large supercomputer facilities emerged. These facilities were developed as collections of rack-mounted computers, often numbering in the thousands, that are co-located with data servers. Researchers connect to these supercomputer data centers through special programs that allow for command line communication between the client and the data center. Moving data between locations is often a challenge, especially if the database is large, and the transfer may take hours to weeks to complete. A variety of tools are available to facilitate such transfers. One common transfer tool is Globus (Foster 2006). It allows the user to set up connection endpoints between clients and servers and manages data transfers simply and securely; if a connection is lost during the transfer, Globus can reconnect and resume the transfer where it left off.

More recently, many large IT firms such as Google, Microsoft, and Amazon have developed remote data storage and generic tool repositories that allow researchers to upload their data to what are called *cloud services*. Users are charged for the use of tools, CPUs, and storage. Connections to these cloud services are made through a web browser.

7.8. A Fourth Paradigm of Science?

A well-known essay by Microsoft's chief data scientist, Jim Gray (Hey et al. 2009; Hey and Trefethen 2003), suggested that, with the enormous ongoing deluge of data, science is now entering its fourth paradigm. Gray suggested that the first paradigm of science, begun by Aristotle, focused on objective observation. The second paradigm, best exemplified by Isaac Newton, focused on experimentation and attempting to understand cause-and-effect relationships. The third paradigm was created by mathematician and computer scientist John Von Neumann, who argued that a computer and its associated software can be used to develop models that represent processes and patterns of the real world, and as such, these simulations represent our understanding of how the world works. Gray argued that the *fourth paradigm of science* is a data-driven exploratory process. In many cases, data are not collected by the researcher who uses them, but have been collected over time for many purposes and are readily available online. The task of the scientist is to know what data to use, how to integrate them, and how to infer patterns that generate knowledge. This approach, Gray argues, shifts research and scientific discovery away from experimentation and more toward an exploratory mode that relies on tools such as machine learning and visualization techniques to generate the knowledge needed to answer specific questions. Gray also argued that domain scientists, such as ecologists, will need to collaborate with data scientists who understand how data are stored, analyzed, and visualized.

What is not clear from Gray's description of his fourth paradigm is whether there is a "hybrid" option, in which a soundscape ecologist who establishes a traditional field research project that is designed to examine causal relationships not only generates big data from that field research, but also uses data from satellite-based and other ground-based (weather) sensors that are integrated with audio files and other data such as interviews. Does this model of data use through careful experimental design (see Chapter 9) constitute this new fourth paradigm? It is indeed data-intensive, and it is likely to involve collaboration with data scientists.

Summary

In the early days of soundscape ecology, researchers often had to become engineers, trying to figure out how to deploy microphones that could withstand the hardships of outdoor environments and record audio data for long periods. Once the data were brought back to the laboratory, they needed to become data managers. Eventually, small companies started to produce reliable and robust PARs, and ecologists started working with data scientists; These changes allowed ecologists to work on other tasks, such as data mining and developing new tools for analysis. In short, understanding the fundamental concepts of engineering and data science as they relate to soundscape ecology is important, both for knowing how to collaborate with engineers and data scientists and for helping to innovate within those collaborative relationships. Gray's fourth paradigm is an interesting perspective on the progression of science. Whether it is true remains to be seen, but any transition to a highly data-driven science needs to consider what this new paradigm of discovery means.

Discussion Questions

1. Consider a social media platform like Facebook. What elements of Facebook illustrate all six Vs of big data summarized above?

2. If you had to select one of the analysis tools listed above (e.g., GIS, Scikit-learn, a CAQDA, or Python), which would you choose? Why?

3. Compare and contrast knowledge discovery in databases and knowledge discovery in textual databases as these two types of analytics are described above. How might you combine these tools (in a general sense) in any research project that involves the use of PARs and interviews with people who are listening and providing meaning to soundscapes?

Further Reading

Tan, Pang-Ning, Michael Steinbach, and Vipin Kumar. "Data Mining Cluster Analysis: Basic Concepts and Algorithms." In *Introduction to Data Mining* (2013), 487–533. Pearson Education India.

Hey, T., and A. Trefethen. 2020. The Fourth Paradigm 10 Years On. *Informatik Spektrum* 42 (6): 441–47.

8

Soundscape Ecology as a Nexus

OVERVIEW. This chapter focuses on synthesizing the previous chapters of Part II and the literature into a more holistic and comprehensive articulation of the concept of soundscapes and the principles of soundscape ecology. It is argued here that the soundscape represents a multilayered *nexus*: it is a nexus of all sounds of a place, a nexus of discovery that requires many disciplines whose theories and methods require integration, and a nexus of science and policy where synergies between research and application can be created (see Fig. 1.5). This chapter is composed of several sections: (1) an expansion of the definition of the soundscape based on its inherent duality; (2) a presentation of the guiding principles of soundscape ecology; (3) a reformulation of existing organizing hypotheses and a presentation of new hypotheses; and (4) an expansion of existing theories and concepts from ecology, social sciences, and humanities that could be enhanced through the consideration of the soundscape.

KEYWORDS: revised acoustic niche hypothesis, soundscape definition, soundscape nexus, soundscape synergy

8.1. Expanding the Definition of the Soundscape

8.1.1. Definition

For the remainder of this book, a new definition of "soundscape" will be used. It follows the one originally proposed by Pijanowski et al. (2011a, 2011b), which was developed from the previous definitions of Southworth and Schafer. This new definition also incorporates elements of the more recent ISO standard definition of the term, which emphasizes that sound-scapes consist of sounds perceived by humans. However, the new definition presented here recognizes that soundscapes are perceived by humans *and* other organisms, and thus has a multispecies focus:

> A soundscape is the collection of all sounds—biological, geophysical, and anthropogenic—that occurs at a place and within a given time frame, which is created from ecological events, processes, and human activities, at landscape scales and larger, and is perceived by organisms, including animals and people.

This definition recognizes the important duality of the soundscape that Granö, Southworth, Schafer, and Kang—and to some extent, Tuan, Gibson, and Feld—emphasize: that it has both a *physical* and an *experiential* dimension. As we have seen in this book, mostly through the early chapters (Chapters 2–7), there are many aspects to each dimension.

8.1.2. Physical Dimension

- A soundscape is composed of a *holistic* acoustic mixing of all processes that create distinct sounds (*sensu* Pijanowski et al. 2011a).
- There is a *fluidity* to all soundscapes; in other words, soundscapes possess an impermanence or transience (*sensu* Schafer 1976).
- All soundscapes are *continually dynamic*, and as such, no two soundscapes are precisely the same.
- Soundscapes characterize a *place* over a defined time frame (Dumyahn and Pijanowski 2011a).
- Soundscapes have *temporal patterns* (i.e., rhythms) that follow the typical temporal trends found in ecosystems; these rhythms are measurable at minute, hourly, and seasonal time scales (in other words, soundscapes are *rhythms of nature*; see Pijanowski et al. 2011b).
- There is a tight *coupling* between soundscapes and landscapes, and changes to landscapes can be measured through the soundscape (Pijanowski et al. 2011a).
- There should be meaningful correlations between *silent* variables (measures that are not sound) and *sonic* variables (Pijanowski et al. 2011a).
- All soundscapes are *spatially variable*, creating a complex sound field with sounds that originate from different locations. The timing of these sounds is also highly variable, and their movement across space is unique to each sound within the soundscape.

- Audio recordings are *digital representations* of a soundscape that do not capture precisely how it occurs in the real world (*sensu* Truax 2019).
- A digital recording of a soundscape, if preserved, represents a *digital fossil* of the sound of a place at the specified time frame.

8.1.3. Experiential Dimension

- Perception of a soundscape is based on how sounds are processed by an organism's *auditory system*, which includes mechanisms for sensation and, in many organisms, cognition (Tuan 1974; Gibson 1960).
- Perception of the soundscape is based, in part, on how hearing is *combined with other senses*. The environment is perceived through each organism's *sensorium* (Gibson 1960; Tuan 1974).
- Perception of the soundscape is an *experience* (Tuan 1974).
- All animals perceive the same soundscape differently because their sensoria differ in ways that depend on evolutionary and ecological factors (Endler 1992).
- Within a species or a group/community, there is *individual variation* of perception based on age, gender, education, skills, and experiences (Ingold 2000; Pink 2009).
- Perception of the environment through sounds is influenced by social interactions and context; in other words, *culture* acts as a lens to an individual's use of, and interpretation of, sound (Feld 1986).
- Soundscapes *characterize a place* for an individual based on how that place is perceived (Tuan 1974).
- Soundscapes are often *segmented* by organisms into individual sounds or small combinations of sounds and are thus not used holistically (Tuan 1974).
- The perception of the environment through soundscapes changes as an organism *moves* through space (Ingold 2000).
- The information contained in sounds and soundscapes can be transmitted between people and across generations as *knowledge about the environment and society* (*sensu* Berkes 1993).
- Soundscapes contain information about how a *place is changing* (Relph 1976).
- Soundscapes are audible *at scales that matter* to the organism that perceives and draws information from them (Tuan 1974; Relph 1976).
- Animals *optimize their acoustic communication* through sound production and sound perception mechanisms so that information contained in their signals is received (Endler 1992).

- Because sounds are part of the natural system as well as the human system, and because they are perceived by people and animals, the soundscape represents a *coupled human and natural system* (*sensu* Liu et al. 2007).
- Soundscapes are composed of sounds that can elicit *emotions* and *mood* and thus provide context for organisms capable of such neurological and sociocultural processing (Gibson 1960; Tuan 1974).

8.2. Guiding Principles of Soundscape Ecology

The following guiding principles, based on past and current research in soundscape ecology, can be considered core to advancing soundscape ecological research and its applications. These guiding principles can be grouped into three categories.

8.2.1. Sound Sources

Soundscapes are composed of the complex integration, over space and time, of sounds from three major sound sources:

A1: *Biophony* is shaped by many ecological, behavioral, and evolutionary factors. It occurs because acoustic communication is important for nearly all groups of animals, but it varies according to the full sensory capabilities of organisms.

A2: *Geophony* is shaped by topographic and habitat features at the landscape/seascape scale and by climate at regional to global scales.

A3: *Anthrophony* is shaped by human activities at the landscape/seascape scale.

8.2.2. Forces That Drive the Dynamics of Soundscapes

As recognized by a variety of scholars over the years, soundscapes are highly dynamic. Understanding the characteristics of these dynamics can be accomplished by keeping the following principles in mind:

B1: Soundscapes reflect the patterns and processes (human and ecological) occurring in *landscapes*.

B2: Soundscapes are also a reflection of the large-scale patterns and processes occurring at regional to global scales consistent with the principles of *biogeography*.

B3: *Perception* of soundscapes is complex and highly variable among individuals and cultures.

8.2.3. Doing Soundscape Ecological Research

Conducting ecological or social research on soundscapes requires broad expertise. Because of this, soundscape ecological research has the following characteristics:

C1. Comprehensive studies of soundscapes are likely to be highly *transdisciplinary*, requiring multiple methods that are integrated in complex ways.

C2. Conducting soundscape ecological research is likely to require new approaches that are being developed as part of the contemporary era of *big data*, including data mining, data fusion, machine learning, and perceptualization of data.

C3. Soundscape ecological research may require a shift to the *fourth paradigm of science*, which is data-driven, discovery-focused, based on advanced computational technologies, and possibly hypothesis-poor.

8.3. Organizing Hypotheses for Soundscape Ecology

Many areas of scholarship contain a set of organizing hypotheses that aids researchers in the testing of guiding principles and even in reconsideration of core definitions. The organizing hypotheses of soundscape ecology, as presented in Pijanowski et al. (2011a), are summarized in this section, and new ones are proposed as well.

8.3.1. Biologically Focused Hypotheses

Considerable research has occurred over the thirty years since the *acoustic niche hypothesis* (ANH) and others described in Chapters 4 and 5, such as the acoustic adaptation hypothesis (AAH) and the morphological adaptation hypothesis (MAH), were first proposed, so a recasting of these hypotheses is arguably in order.

Revised Acoustic Niche Hypothesis

In light of the considerable work in acoustic partitioning, the extensive research that has focused on the sensory drive framework and related hypotheses, the shortcomings of the original ANH (all described in Chapter 4), and the fact that the original ANH was not posed by Krause as a scientific hypothesis (although it was treated as such in Pijanowski et al. 2011a), a revised acoustic niche hypothesis (RANH), with the following hypotheses and predictions, is in order:

> *Hypothesis*: All acoustic signals produced by all animals in a natural eco-system can be separated across spectral frequencies, timings, lengths, modulations, timbres, and amplitudes; there are few or no overlaps within these dimensions; natural selection is operating on senders, receivers, or both; and signals are also selected on the basis of habitat structure and background noise.

The following predictions of the RANH are based on the forms of acoustic partitioning (Fig. 8.1):

> *Prediction 1.* Acoustic signaling should be driven by competition for acoustic space and should not be an artifact of the natural distribution of animal body sizes that occur in the ecosystem (Fig. 8.1A).

> *Prediction 2.* Closely related species should show distinct acoustic characteristics that reduce the overlap of their signals in any of the dimensions of acoustic signaling. These distinctions may include shifts in frequencies (Fig. 8.1B) or development of special auditory filters (Fig. 8.1C).

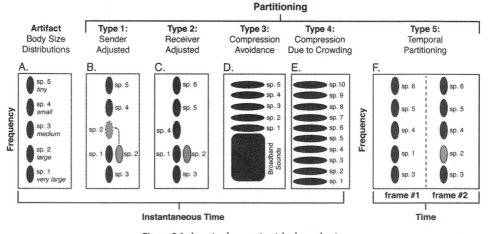

Figure 8.1 A revised acoustic niche hypothesis.

Prediction 3. There should be greater competitive exclusion from acoustic space in ecosystems that have a lot of broadband noise (background noise), which compresses the available acoustic space (Fig. 8.1D).

Prediction 4. In locations where species richness and abundances are great, such as the tropics, acoustic partitioning should be evident as spectral frequencies that are narrower, timing that is highly specialized, or acoustic signaling positions that are highly specific to microhabitats (distributed vertically or horizontally in space) (Fig. 8.1E).

Prediction 5. Among closely related species or species that have very similar spectral features, temporal partitioning is likely (Fig. 8.1F).

Prediction 6. Among species that are not closely related, those species that appeared earliest in the history of the ecosystem should exhibit the greatest departures from expected signaling characteristics. In other words, if the frequency of an animal's signal based on its body size should be X and it is observed to be Y, the difference between X and Y should relate to avoiding masking by another species that was present earlier than the species being examined.

The three main dimensions of acoustic partitioning (spectral, temporal, and spatial) are illustrated in a simple way in Figure 8.2. Here, each of five species is shown in multidimensional acoustic space, in which we assume there is no additional nonbiological background noise (e.g., sounds from wind or rain). The distances among species in inter-dimensional space represent niche distances, and the size of each sphere represents that species' niche breadth. Each dimension could have dimensions of its own (i.e., could be hierarchical), with *space* divided into microhabitats that are horizontally and vertically distributed; with *time* divided at the scale of inter-signal distances, diel (i.e., hourly), diurnal (day vs. night), or seasonal; and with *spectral* divisions by frequency, amplitude, or timbre. Additional factors to consider would be phylogenetic differences, body size, and habitat constraints on signal design.

Integrating the Acoustic Adaptation Hypothesis (AAH), the Morphological Adaptation Hypothesis (MAH), and the Sensory Drive Framework (SDF)

A more comprehensive view of how current research in sensory ecology plays into our understanding of soundscapes, and in particular, biophonies,

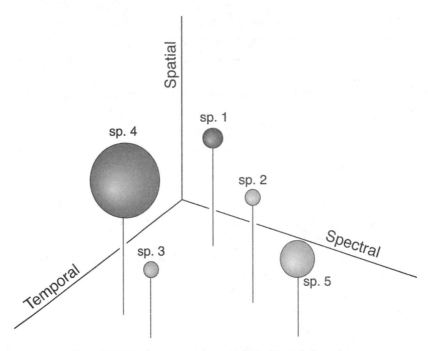

Figure 8.2 Major dimensions of a revised acoustic niche hypothesis.

requires researchers to consider Endler's (1992) sensory drive framework and the other major sensory ecology concepts within the context of animal communication at the landscape scale (Fig. 8.3). This means that sound production and reception by a species should be placed within the context of ambient sounds (i.e., geophonies) and communication patterns of other species located within the same landscape. Note that the ANH and AAH drive diversification of animals' acoustic features and that the habitat drives homogenization through the optimization of those features for habitat characteristics. Existing morphological constraints (a form of phylogenetic constraints) also influence the breadth and diversity of the sounds that can be produced by a species. As we have seen from our reviews of the ANH and AAH in Chapter 4, no single adaptive pathway appears to apply to all species or to any major taxonomic group in particular.

8.3.2. Geophonically Focused Hypotheses

Because wind and other geophysical dynamics (such as flowing water in rivers) are common to landscapes, seascapes, and riverscapes, how animals adapt to these patterns in natural areas should also be explored by soundscape ecologists. Here we present a few geophonically based hypotheses that soundscape ecologists may want to consider in the future.

Comparable Animal Acoustic Adaptations to Similar Geophonic Sounds Hypothesis

There are several geophonic sounds that have similar spectral features (e.g., low frequencies, lack of spectral form), including the sounds of waves breaking along beaches, the sounds of running rivers, and even the sounds of wind. All of these ambient sounds could potentially drive acoustic communication adaptations. This observation suggests the following:

> *Hypothesis*: Given the similar spectral features of most geophonic sounds, animals that live in the presence of these sounds exhibit similar adjustments to their acoustic communication signals to reduce chances of masking.

If this hypothesis is true, then the following predictions can be made:

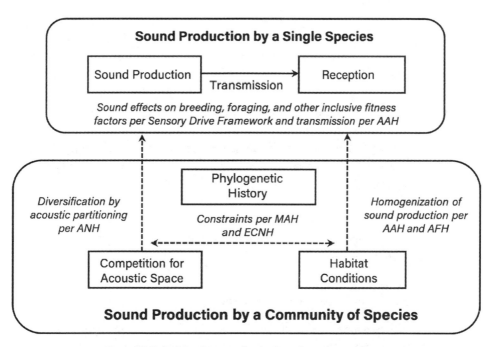

Figure 8.3 Roles of major contributing hypotheses to soundscape ecology operating at the species and community levels.

Prediction 1. All animals that use acoustic communication where geophonic sounds are present should produce sounds that are above the frequency of geophonic sounds.

Prediction 2. Animals that live in windy areas and animals that live in areas where loud water sounds exist should exhibit similar adaptations to avoid their signals being masked. Close relatives of those species that do not live in these highly geophonic areas should not have those adaptations.

Acoustic Signal Adaptations in Quiet versus Loud Geophonic Landscapes Hypothesis

In areas where sounds from rain and wind are prevalent, animals could adapt the timing and spectral features of their acoustic communication signals to avoid masking by these low-frequency sounds. Likewise, in areas that have low wind speeds and are arid, acoustic communication could occur within these low-frequency ranges, since masking would not occur. Comparisons among ecosystems that vary in the intensity of geophonic sounds could provide information about selection for spectral features that facilitate animal communication.

Hypothesis: Biophonic patterns should be adjusted to the occurrence of geophony in terms of its magnitude and timing throughout the day or season.

If this hypothesis is true, then we can make the following predictions:

Prediction 1. In landscapes that have high-intensity geophonic sounds, vocalizations by individuals that exemplify the Lombard effect should be common when these sounds are present.

Prediction 2. Closely related species that evolve separately in quiet versus loud geophonic landscapes should exhibit different adaptations in the timing or spectral features of their calls.

8.3.3. Anthrophonically Focused Hypotheses

No hypotheses have yet been presented that describe important relationships among patterns and processes that influence human-produced sounds. Here are two that might be considered:

Urban Sounds as Proxies for Energy Use Hypothesis

Most sounds produced by humans come from machines, and in particular, from combustion engines. Therefore, we can posit that the amount (i.e., as measured across time and space) and amplitude of anthropogenic sounds are measures of energy use by humans in the landscape.

> *Hypothesis*: Given that most sounds from anthropogenic sources are from combustion engines, the spatial extent and intensity of those sounds will be positively correlated with energy use and with the number of people using these machines.

If this hypothesis is true, then we can make the following predictions about the occurrence of anthropogenic sounds in space and time:

> *Prediction 1.* Measures of anthropogenic sounds should be greatest during times of greatest energy use, such as times when people are going to and returning from work, and during days of the week when travel is most frequent (e.g., Mondays through Fridays in Western societies). Seasonal trends may also be evident if travel occurs during specific times of the year (e.g., summer in temperate climates).

> *Prediction 2.* Spatial variability in measures of anthropogenic sounds should exist at a variety of scales and should correlate with the spatial variability of energy use.

> *Prediction 3.* Spatial variability in measures of anthropogenic sounds should also be positively correlated with human population densities.

Spatial-Temporal Patterns of Urban Sounds Affect Housing Choices through Economics Hypothesis

Many sounds produced by humans, such as road noise, are considered unwanted. As these sounds vary in space and time, it can be argued that people who have the necessary resources to avoid them will do so.

> *Hypothesis*: Levels of noise (intensity and duration) in urban landscapes will be inversely correlated with the resources of human populations in those landscapes.

If this hypothesis is true, then we can make the following predictions about the relationship between noise and the economics and demographics of local communities:

> *Prediction 1.* Household income should be inversely correlated with noise levels (intensity and duration).

> *Prediction 2.* Home values should be inversely correlated with noise levels (intensity and duration).

It is also argued that as income and housing values are also affected by race in many areas of the world, research in this area could be enhanced by understanding noise and exposure to it as an environmental justice issue (Bullard 2005).

8.4. Extensions to Existing Ecological and Social Science Theories

8.4.1. Conservation Biology and Soundscape Ecology

Conservation biology has been presented as a field that possesses both functional (i.e., data-driven) and normative (i.e., value-driven) postulates. Several parallels between the postulates of conservation biology and those of soundscape ecology are listed in Table 8.1.

8.4.2. Synergies between Spatial Ecologies and Biophonic Patterns

Numerous conceptual bridges exist between the spatial ecologies summarized in Chapter 5 and soundscape ecology. For landscape ecology and soundscape ecology (Fig. 8.4), species richness, population levels, functional traits, and management constraints and practices give rise to patterns of plant and animal biodiversity. For animals using acoustic communication, acoustic niches, functional constraints such as body size and sound frequency relationships, and habitats constraints that control sound propagation all give rise to the biophonic patterns at a place. Similarly, soundscapes are the acoustic representation of both pattern (composition of sound sources) and process (movement or change at a place); their composition is heterogeneous, as it is always a mixture of many sound sources across ecosystem components that are biological, geophysical, and anthropogenic; the soundscape can be viewed across a multitude of temporal, spectral (frequency and magnitude),

Table 8.1 Postulates of conservation biology and soundscape ecology

	Conservation biology	Soundscape ecology
Scientific	Species are products of evolution	Species' acoustic communication signals are a result of the evolution of all signals from animals and the presence of geophonic sounds
	Species are interdependent	Sounds in an ecosystem are one dimension of how species in the ecosystem function interdependently
	Extinctions of keystone species can have long-term consequences	The loss of a keynote organism may have cascading effects in acoustic communities as predator-prey relationships that are based on sound may be altered
	Introductions of species can alter diversity	Introductions of species could mask endemic species' communication signals and/or change predator-prey dynamics that are based on sound
Normative	Diversity is good	Highly diverse biophonies are good
	Biological diversity has value	Natural soundscapes have social value for humans

Source: Modified from Dumyahn and Pijanowski (2011b).

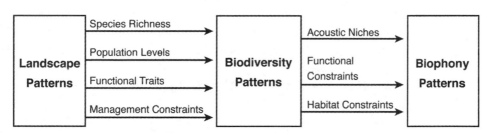

Figure 8.4 Conceptual bridges between landscape, biodiversity, and biophonic patterns.

and spatial scales; soundscapes are inherently dynamic, possessing characteristic rhythms of nature; and sounds from humans, as well as how humans perceive them, are a critical part of soundscape ecology.

Biogeography has been a discipline that seeks to understand the broad patterns that exist, regionally to globally, in the Earth system, but particularly in the biosphere. Soundscape ecologists need to understand how broad Earth system processes shape the biological patterns that exist across important gradients, such as latitudinal and altitudinal variation. These patterns include those driven by latitude; by body size and body form trends;

by regional climate variability and trends, often referred to as the climate regime; and by impacts on animal behavior, as well as the influence of long-term Earth dynamics on local species distributions.

Finally, the tools and techniques of soundscape ecology can play important roles for conservation biology, especially given the great need for more information on the relationship between landscapes and biodiversity trends, and on how conservation management can influence local to global patterns of biodiversity. Soundscape ecologists can play a large role in conservation management by monitoring animal populations and by developing a better understanding of the interaction of landscape and soundscape.

8.4.3. Extending Sense of Place Frameworks to Include Soundscapes

Sounds—and more precisely, soundscapes—are missing from nearly all SoP conceptual frameworks. This is an area that needs considerably more research, especially by those in the social sciences who are interested in the impact of sound on sense of place, place attachment, nature connectedness, and attention restoration potential. But enhancements of these frameworks to include soundscapes are possible. For example, Dumyahn and Pijanowski (2011b) have realigned some of the survey questions that Jorgensen and Stedman (2001) used to measure SoP (Table 8.2) to include the importance of sounds and the soundscape in the development of a sense of place rubric by an individual or community. These questions could be answered using Likert scale options (e.g., 1 = strongly disagree, 5 = strongly agree, 3 = don't know).

Other possible advances could build on the traditional work in topophilia. According to the theory of Tuan, Gibson, Rodaway, and others, the sensorium, sounds, and soundscapes could be introduced to Scannell and Gifford's (2010) hierarchical, tripartite PPP model as one additional layer (Fig. 8.5). Here, the sensorium of an individual or community is added as an umbrella that informs all three processes, each in a different way. The soundscape informs cultural attachment to place through symbolic sounds that occur in the landscape (e.g., church bells in Christian communities, evening chants in Muslim communities). The roar of crowds at sporting events could form that "social arena" component of place.

8.4.4. CHANS, TEK, and Soundscapes

Let us consider a coupled human and natural system (CHANS) that includes human and natural acoustic elements (Fig. 8.6). This system contains

Table 8.2 Adding a soundscape dimension to measures of place identity, place attachment, and place dependence

	Place identity	*Place attachment*	*Place dependence*
Item # 1	Everything about this location is a reflection of me.	I feel relaxed when I am at this location.	This location is the best place for doing the things that I enjoy most.
	Everything about these sounds is a reflection of me.	I feel relaxed when I'm surrounded by these sounds.	This location is the best place for hearing the sounds that I enjoy most.
Item # 2	This location says very little about who I am.	I feel happiest when I am at this location.	For doing the things that I enjoy most, no other place can compare to this location.
	This soundscape says very little about who I am.	I feel happiest when I hear these sounds.	For hearing the sounds that I enjoy most, no other place can compare.
Item # 3	I feel that I can really be myself at this location.	This location is my favorite place to be.	This location is not a good place to do the things I most like to do.
	I feel that I can really be myself while surrounded by these sounds.	This is my favorite soundscape	This location is not a good place to hear the sounds I most like
Item # 4	This location reflects the type of person I am.	I really miss this location when I'm away from it too long.	As far as I am concerned, there are better places to be than at this location.
	This soundscape reflects the type of person I am.	I really miss these sounds when I'm away from them for too long.	As far as I am concerned, there are better soundscapes than at this location.

Source: From Dumyahn (2012).

biophysical and human components all linked through either silent or sonic elements. Note that the atmosphere generates several climate elements, including temperature, precipitation, and wind. Some of these elements create sound. Precipitation and temperature, along with soils, support the kinds of vegetation that exist on the landscape. Humans, due to their need for food, may transform that land from its natural state (e.g., forest) to cropland. Forested landscapes would support the natural composition of the animal community, which disperses seeds, consumes vegetation, pollinates plants,

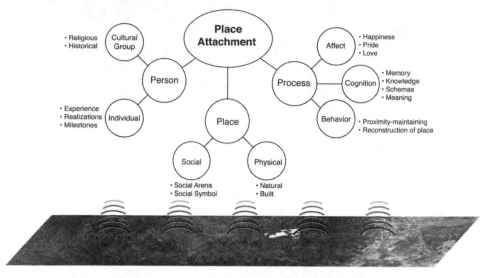

Figure 8.5 One way to integrate soundscapes into a sense of place model.

and so forth. The forest vegetation–animal community operates as a *reinforcing feedback loop* in this system. The presence of animals in agricultural landscapes can be problematic, however, as they could destroy crops in a variety of ways. A high abundance and diversity of animals in the forest would result in a high diversity of animal sounds. The coupling of species richness and diversity of biological sounds represents one of the *silent-sonic couplings* of this generalized system. The sounds from the geophysical environment, as presented in Chapter 4, represent a frequency range with the potential to mask communication in low-frequency ranges; this interaction could be considered a *negative effect* of these elements. The sounds of rain, running water, and wind, as well as sounds from animals, could be used by people as cues to the status of the environment. Howling wind, for example, could be a good predictor of an impending storm. If people have high levels of nature connectedness, they may use these environmental cues to make decisions, which, according to the research summarized above, tends to result in pro-environmental behaviors. As described by Feld (1986) and others who study indigenous communities, many of these communities have traditional sonic expressive practices that include songs, poetics, mimesis of natural sounds, and dances that reinforce their connection to nature. The use of sounds to understand environmental change and determine environmental status, the level of nature connectedness, and the use of sonic practices could all constitute traditional ecological knowledge. TEK is known to create and

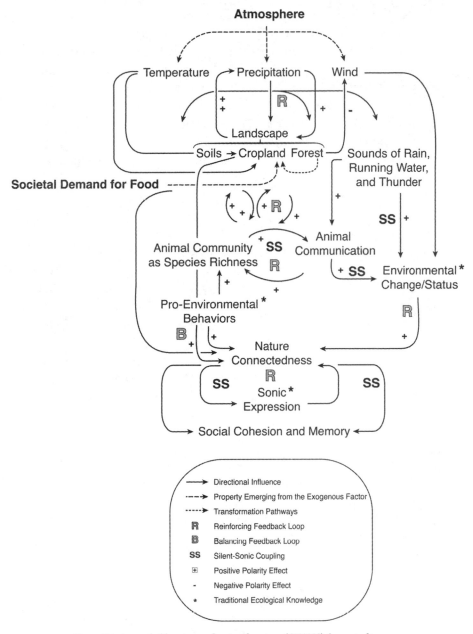

Figure 8.6 A coupled human and natural system (CHANS) for an indigenous community of herders whose members use sonic practices in an environment (a grassland ecosystem) that is undergoing transformations.

establish social traditions, which create cohesion within the community and serve as its memory of how it exists within the environment.

CHANS research is often focused on systems behavior, the nature of the connections between elements, and the consequences of decouplings. For example, does a decoupling result in a simple change to the overall system behavior, or could it lead to collapse of the system and its transitioning to another kind of system that, for example, may not use sounds from the environment as a source of information about how the environment is changing? What is the role of silent-sonic couplings and decouplings in these systems?

8.4.5. Derived Benefits of Natural Soundscapes

Soundscapes are proving to be excellent proxies for biodiversity, and thus for the ecological health of a landscape (Table 8.3). Sound is one of the critical ways in which animals function within the ecosystem (Endler 1992). If natural soundscapes are protected, then the biological diversity that exists in the landscape will be also preserved. Soundscapes also have societal value (Dumyahn and Pijanowski 2011a). They provide a sense of place (see Feld 2004; Schafer 1994) for many people, serve as sources of information on environmental change (Post and Pijanowski 2018), and have cultural value (Feld 1986).

Table 8.3 Soundscape values and derived benefits of quality soundscapes

Soundscape value	Associated benefits from soundscapes
Human well-being	Improved human health, reduced stress, improved quality of life, and equitable access to high-quality soundscapes
Wildlife well-being	Ability to detect prey, avoid predators, find a mate; decreased stress, reduced need for modified acoustic communication (elevated intensities)
Sense of place	Cultural, historical, and natural places provide unique sounds that meaningfully connect people to a place
Landscape perceptualization	The coupling of visual and sonic increases the aesthetic appreciation and value of place and experience
Ecological integrity	Soundscape composition reflects natural processes and organismal diversity

Source: Dumyahn and Pijanowski (2011a).

8.4.6. Soundscapes and Principles of Natural Resource Management

To assist natural resource managers in their efforts toward the conservation of soundscapes, Dumyahn and Pijanowski (2011a) introduced several types of soundscapes that might be considered priorities for conservation. These soundscapes are listed in Table 8.4, along with their value, potential threats, the goals for their conservation, and the form of monitoring needed to ensure that those conservation goals are met.

8.4.7. Soundscapes as Common Pool Resources

A variety of natural resource management organizations, such as the US National Park Service, specifically recognize soundscapes as a resource worth protecting (NPS 2006). The NPS defines a soundscape as "all the natural sounds that occur in parks, including the physical capacity for transmitting those natural sounds and the interrelations among park natural sounds of different frequencies and volumes," whose condition is "protected under the Organic Act of 1916 of the United States" (NPS 2006, 2020). Pavan (2017) argues that soundscapes can be modified in several ways: through the loss of species, the modification of habitat, and the intrusion of non-natural sounds. Other alterations to the soundscape include the addition of invasive species and changes in climate. The natural sounds of protected areas such as national parks are threatened directly by the introduction of noise (Buxton et al. 2017) and indirectly by the alteration of the natural landscape (Buxton et al. 2017) through the introduction of roads and modification of habitat (e.g., selective logging).

It has been argued that one approach to managing soundscapes is to consider them common pool resources. Ostrom (1990) argued that natural resources can be considered *common pool resources* (1) when they are used by multiple users; (2) where their use by one user can either result in the "substractability" of use or be of benefit to others; and (3) where the exclusion of others is difficult or costly. *Substractability* is the degree to which the use of a natural resource by one person reduces its value to others (Ostrom 1990, 2009). Common pool resources differ from resources that are considered public, private, or toll goods. *Public goods* are those resources that have low substractability of use and are difficult to exclude people from; one example could be a city park situated near a loud traffic corridor. *Private goods* are those that are privately held, for which rules of exclusion are clear and easy to enforce, such as the large landholding of a rancher who may enjoy

Table 8.4 Priorities for soundscape conservation

Soundscape type	Examples	Value(s)	Threat(s)	Management goals	Monitoring
Natural quiet soundscapes	Wilderness areas, national parks	Solitude, sense of place for people, experience of remoteness	Intrusive sounds from humans	Protect areas from low-frequency noise	Monitor for low-frequency sounds from humans
Sensitive soundscapes	Breeding habitats for rare, endangered, threatened species	Ecological integrity	Intrusive sounds from humans	Protect areas from low-frequency noise	Monitor for sounds produced by rare, endangered, or threatened species, and for low-frequency sounds from humans
Threatened soundscapes	Places where intrusive non-natural sounds predominate and threaten the integrity of animal communication in biodiverse areas	Ecological integrity	Intrusive sounds from humans, habitat alteration that changes sound propagation	Mitigate excessive noise	Monitor for non-natural sounds; compare natural sounds of focal place with a reference landscape
Unique soundscapes	Unique places of the world that harbor globally unique sounds	Global acoustic heritage	Intrusive sounds from humans, activities at or near the sites	Mitigate excessive use of sites, reduce or prohibit intrusive sounds	Monitor for the unique sounds and also for threatening sounds

Soundscape type	Examples	Value	Threats	Protection	Monitoring
Recreational soundscapes	Lakes (with motorboats) and rangelands; city parks	Human connection to nature and sounds that relate to the activities at these places, escape from urban sounds	Sounds other than those from the recreational activity and the natural sounds	Mitigate non-recreational sounds	Monitor for non-recreational sounds
Representative natural soundscapes	Representative ecosystems such as deserts, grasslands, coral reefs	Last vestiges of natural areas with unmodified soundscapes, global ecological heritage	Sounds from non-natural sources and human activities	Protect from non-natural sounds and from human activities	Monitor for all sounds for long periods, compare with altered soundscapes in same biome
Cultural soundscapes	City markets, church bells, coastal areas with foghorns, evening Muslim prayers	Sense of place, cultural acoustic definitions, historical values	Sounds from non-cultural sources being introduced and masking cultural sounds	Protect from non-cultural sounds	Monitor culturally significant soundmarks
Everyday soundscapes	City soundscapes, residential soundscapes, rural soundscapes where humans are common	Sense of place	High levels of low-frequency sounds, continuous low-frequency sounds	Protect from harmful levels of sounds mostly from human-produced machines	Focus on dB and the amount of exposure (levels, time) to high levels of low-frequency sounds

Source: Modified from Dumyahn (2012).

natural sounds but can exclude others from that enjoyment. *Toll goods* are those resources whose use is paid for directly by users and which have low substractability of use when used by one person.

Three approaches have been proposed for managing common pool resources: privatization (Demsetz 2000); the government-directed "command-and-control" approach presented in the classic "tragedies of the commons" work of Hardin (1968); and the application of collective action (Ostrom 1990) to self-government of the resource. The privatization approach would be a market-based way of achieving solutions. The command-and-control approach would require strict rules of use for all users, with consequences issued as fines or other forms of penalties. The collective action approach would require users to self-organize, and to monitor resource use and quality over space and time to ensure that no users experience reduced access to or enjoyment of the resource. Dumyahn and Pijanowski (2011b) argue that for soundscapes to be managed properly as common pool resources in places like national parks and wilderness areas, users' resource use rights need to be well documented, as proposed by Schlager and Ostrom (2011).

Summary

This chapter argues that soundscapes, and the field of soundscape ecology, serve as a nexus for research and applications that can use sound to understand the natural world, and its interactions with humans, at landscape scales. The study of soundscapes is multilayered, requiring concepts and methods from diverse disciplines, a duality of understanding (of physical presence and perception), and application to a variety of natural management problems, such as biodiversity loss, climate change, and noise in urban settings.

Discussion Questions

1. Critique the new definition of soundscapes presented in Section 8.1. Is it an improvement over the earlier definitions of Southworth, Schafer, and Pijanowski and the one created as an ISO standard? What might be some challenges of using this definition in scientific research?
2. Are there any physical or experimental characteristics missing from this chapter that should be also considered in this "duality of the soundscape" understanding, especially considering the information presented in Chapters 4 through 6?

3. Hypotheses that focus on geophonic and anthrophonic sound sources are few. Propose new ones for consideration by soundscape ecologists. Do so using the structure used in this chapter, in which the hypothesis (a new idea about how something works) is articulated before the predictions (the data you would need to find if the hypothesis were true). Make sure you make at least two predictions that you could use to test your hypothesis.

PART III

METHODS

9

Measuring the Soundscape

OVERVIEW. Soundscape ecologists use many tools and approaches to measure the soundscape, including the use of acoustic features and acoustic metrics, labeling of audio files, deployment of sensor networks, quantifying features of landscapes, and visualizing sounds to understand the content and dynamics of the soundscape.

KEYWORDS: acoustic index, passive acoustic recorder, soundscape trigger, soundscape phase, sensor network, sentinel sensors, entropy, false color spectrogram, aural annotation, soundwalk, knowledge co-production, grounded theory

9.1. Passive Acoustic Recorders

9.1.1. Standard Architecture

Numerous types of passive acoustic recorders are manufactured today. For terrestrial systems, those available include the Song Meter (e.g., SM4) series from Wildlife Acoustics, the Swift from the Cornell Laboratory for Ornithology, and the Bioacoustic Audio Recorder from Frontier Labs, Australia. PARs suitable for ultrasonic recordings include the Song Meter Bat (SM4Bat) series from Wildlife Acoustics, the Anabat from Titley Scientific, and the RPA2 from Peersonic. Several units record in both the audible and ultrasonic ranges; these include the AudioMoth and the uMoth (openacousticdevices.info), and AURITA (reviewed by Beason et al. 2019). For aquatic, particularly marine, soundscape studies, many PARs have been developed that are suitable for recording animal, geophysical (e.g., underwater earthquakes), and anthropogenic (e.g., ship noise, blasts) sounds. An excellent review of marine PARs is provided by Sousa-Lima et al. (2013).

Nearly all PARs are equipped with power and storage capabilities (e.g., SD cards). Several PARs can be attached to solar panels. Several battery types are used. Alkaline batteries have more power, and weigh about 20% less, than rechargeable batteries and last longer in cold weather. Additional features of PARs include a programming interface (either on the unit or through a software program that produces a parameter file) and an onboard clock. Some PARs include a GPS device, which stores location information and synchronizes the onboard clock to a global clock source.

Some PARS allow researchers to program them. Programs can be built to detect the calls of specific species at the site and alert the researcher by broadcasting a simple text message to a remote server using a wireless connection. This form of sensor design, called *edge computing*, is probably the future of soundscape ecology technologies. Edge computing applications include detecting destructive or invasive animals, measuring storm energy from thunder, and detecting the presence of rare or endangered species.

9.1.2. Deployment Considerations

PARs are often mounted on trees at eye level. Some researchers have also arranged PARs vertically in forests, from eye level (in the understory) to midstory to the canopy. Mounting terrestrial PARs in non-forest landscapes is challenging. Mounting them less than 1 m off the ground will mean that very low frequencies will not be recorded, as those frequencies are attenuated by the ground. Placing PARs on a post is common; however, a post can serve as a perch for an animal to sing from, creating an artificial soundscape. Thus these mounts may require a perching deterrent, such as metal spikes.

Marine PARs often require divers to attach the units to a mooring device; they are designed to have buoyancy so as to float above the seafloor. Divers service the units by hauling them to a boat or to the surface, where they swap storage devices and batteries.

Windscreens are important for terrestrial PAR microphones. Usually designed to clip on at the base and fully cover the microphone, these devices can become brittle over time and thus lose their ability to reduce wind sounds at the microphone. Many soundscape ecologists also note that they are favorite targets of mammals that are attracted to their scent. At Purdue's Center for Global Soundscapes (CGS), we have found that soaking windscreens in a bucket of local water for 24 hours introduces the local background scent to the material. Some PARs, especially those placed in areas with large carnivores or in large colonies, may need to be placed within a protective cage. Finally, many researchers orient all their network PARs in the same direction so that they are all exposed to wind in the same way.

9.2. Recording Parameters

Soundscape ecologists must consider several recording parameters, which include *sampling rate*, *duty cycle*, and *frequency band filters*. The sampling rates most commonly used to store sound as WAV files are 44,100 Hz or 48,000 Hz for terrestrial studies and 96,000 Hz to 182,000 Hz for aquatic (marine) studies. Sampling rates of 186,000–384,000 Hz are used for ultrasonic recordings. Duty cycles, the amounts of time the PAR is set to record and not to record, vary by study, with many studies using 10 minutes on, 50 minutes off; 10 minutes on, 20 minutes off; 1 minute on, 59 minutes off; or 59 minutes on, 1 minute off. The last of these duty cycles can be implemented to capture the full soundscape during a study, but provides the sensor the time to write large sound files to storage.

A second consideration in setting PAR recording parameters is the length of the study. Many soundscape ecologists focus their recording efforts on critical animal life-history events for which communication is vital, such as breeding, territory defense, and migration. Studies can be as short at two weeks, but some have extended over years or, in some marine studies, decades.

At CGS, pre-study recording sensitivity analyses are common. These analyses involve placing two or more sensors of the same model next to one another, each set with a different *high-pass band filter* (none, at 200 Hz, and at 1 kHz) and with a different microphone *gain setting*. We set these PARs to record nearly continuously for 2–7 days. We then evaluate the recordings made during peak winds (mid-day), at the loudest time of day (dawn chorus in terrestrial environments or dusk chorus in aquatic ones), and at the quietest time of day (generally 2–3 hours before sunrise in terrestrial systems). We select those parameters that produce the best sound quality for labeling and acoustic index calculations.

9.3. Sensor Networks and Sensor Arrays

To measure soundscapes across landscapes, soundscape ecologists often deploy several, or even hundreds of, PARs to quantify the spatial variability of soundscapes across large areas (Towsey et al. 2014). These groups of PARs are often referred to as *sensor networks*. Towsey et al. (2014) argue that sensor networks allow researchers to address landscape-level or community-level concerns such as animal diversity across space and time, interactions between animal species within communities, and effects of noise on animal communication. Sensor networks can also be used to study modifications to the terrestrial landscape through a variety of human activities such as minor (e.g., selective logging) to major (e.g., clear-cutting a forest) habitat alterations; impacts of wildfires, livestock grazing, and agricultural management practices; changes to regional climate that modify precipitation and/or temperature regimes; and the impacts of invasive species on habitats or animal species interactions. Effects of human activities are also manifested in marine environments: changes to water temperature and chemistry have the potential to modify marine ecosystems such as coral reefs, and rising ocean levels have the potential to alter coastal ecosystems such as mangroves.

A general sensor network design (Fig. 9.1A) used at CGS (Pijanowski et al. 2018) places sensors simultaneously across ecological and human activity gradients. Measurements across the gradients are then compared with a reference condition (see Hughes et al. 1986; Hawkins et al. 2010). Researchers should consider recording at least five disturbance levels to assess trends across these gradients. Replicates of sensors at each level—a minimum of three—allow for an assessment of variability at each level. Recording along natural ecological gradients provides researchers with information about how those gradients differ from human disturbance gradients. Natural ecological gradients (Vitousek and Matson 1991) include those across elevation, moisture, temperature, salinity/biogeochemistry, foliar age/height/density, plant species composition, and soil type. Quantifying the effects of disturbances or stressors across natural and anthropogenic gradients is important if we are to quantify the shape (e.g., linear or nonlinear) of the relationship between disturbance and response (see Larned and Schallenberg 2019).

Purdue's CGS also considers three types of sensors with different *deployment schedules*: (1) midrange, (2) sentinel, and (3) synoptic. *Midrange sensors* are used during a critical life-history event for an animal community, generally spanning 2 weeks to 3 months. Midrange sensors often have nearly continuous duty cycles, as many PARs can record for 4–6 weeks without servicing. *Sentinel* sensors are used to assess seasonal changes or to monitor soundscapes in relation to slow changes in the environment (e.g., climate

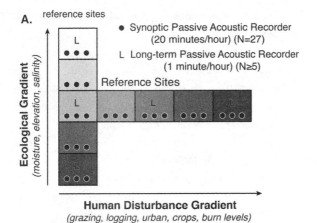

Figure 9.1 Considerations for deploying acoustic sensor networks. (A) A general sensor network design that shows how sensors can be deployed in groups (i.e., replicates) of classes (illustrated as boxes) organized along a human disturbance and an ecological gradient. (B) A typical deployment of sensors for broad-spectrum studies that shows a diversity of acoustic sensors operating at one site.

change). Sentinel duty cycles of 1 minute per hour can provide a temporal window of at least one year for many terrestrial and marine PARs without servicing, but often are run for several years so that annual climate patterns can also be measured. *Synoptic sensors* use a short recording length (viz., 1 minute) but make a large number of recordings within a limited time period (viz., 2–3 hours). Other, highly portable recording devices, such as Handy Zoom F3s or hydrophones connected to a portable data logger that are dropped over the side of a boat, are manually turned on and off and thus do not have duty-cycle schedules. Sentinels also serve to place intensive mid-range studies in the context of long-term changes in soundscapes (i.e., providing temporal rigor), and synoptic sampling can increase spatial sample sizes to assist in quantifying spatial variability (e.g., providing spatial rigor).

Further considerations for the deployment of sensor networks include the need to ensure that all sensor clocks are synchronized, that all sensors have the same recording parameters, and that sensors are placed in locations where they are measuring sound independently from one another.

Purdue's CGS has conducted *broad-spectrum studies* that attempt to measure all sounds across the audible (e.g., SM4s) and ultrasonic (SM4Bats) ranges, as well as aquatic sounds (using hydrophones) along the edges of land-water interfaces. Figure 9.1B illustrates a typical deployment. A full spectrum–wide set of studies has also been implemented by Purdue CGS that uses sound/vibration sensors, such as vibrometers attached to trees to record wood boring insects, ultrasonic recorders to detect bats, and hydrophones in ponds to measure activities of fish and crustaceans, all co-located at a long-term earthquake station. These studies are designed to use sound to measure as many dynamics of the environment as possible at one location.

Sensor arrays, in which non-acoustic sensors are added to an acoustic sensor network, are often used to determine how the environment is changing. For example, meteorological stations can be co-located with PARs to provide critical information on wind speed, air temperature, air chemistry, precipitation, solar radiation, relative humidity, and barometric pressure. Aquatic sensors in freshwater and marine environments typically measure many of these same variables, but also salinity, turbidity, and conductivity. Other sensors can provide additional information about the changes of the environment, including soil moisture and leaf wetness, among others, and tree bands can be used in long-term soundscape studies to monitor the growth of trees. Streams within landscapes whose soundscapes are being studied can be continuously measured with stream gages, which provide information on water velocity, water flow (i.e., volume), and water levels.

Camera traps have also been co-located with PARs to quantify activity patterns of non-vocal animals. These studies are designed to determine how changes in environmental factors, such as barometric pressure changes over short (e.g., minutes to hours) or long periods (e.g., seasonal trends), correlate with larger-scale environmental changes, such as climate change.

9.4. Soundscape Indices

Sueur et al. (2014) presented one of the first summaries of soundscape indices used by soundscape ecologists. Soundscape indices are grouped into five broad categories: intensity indices, acoustic event and feature statistics, acoustic indices, sound type statistics, and call detection statistics. (Table 9.1 contains a list of common soundscape indices.) As Figure 9.2 shows, these soundscape indices fall into a range from those that can be calculated automatically, but provide little specific information regarding the nature of sound sources, to those that provide a detailed account of the number, intensity, and duration of sound sources in every recording. The level of effort required for the researcher to perform the calculations varies from high to low. The uses of these soundscape indices fall within three broad paradigms (i.e., approaches to problems by the research community), which are also

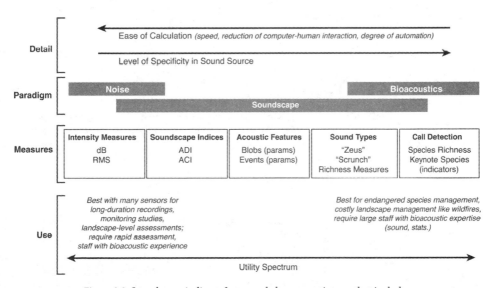

Figure 9.2 Soundscape indices of arranged along a continuum that includes intensity measures, acoustic event and feature statistics, acoustic indices, sound-type statistics, and call detection statistics. Also shown are use, level of specificity of sound sources, ease of calculation, and the research paradigm that is most often aligned with these soundscape indices.

Table 9.1 Summary of the major categories of soundscape indices used by soundscape ecologists

Index name	Abbreviation	Description	Reference
1. Intensity Indices			
Intensity levels	L_{eq}, L_{10}, LA_{eq}	Often called statistical noise levels, these indices use small time "slices" of a sound file to calculate dB, and the summary statistic is reported. L_{eq} is the equivalent sound level, which is average sound level. L_{10} is the level exceeded 10% of the time. LA_{eq} is A-weighted; L_{day} is an A-weighted equivalent noise level for a 12-hour day (also denoted as $LA_{eq(12h)}$)	—
Root mean square of the SPL	RMS	Root mean square of SPL of the raw signals	Rodriguez et al. 2014
Background noise	BG-dB or BGN	Average wave envelope, often used to remove acoustic activity that is not from a distinct sound source	Lamel 1981, cited in Towsey et al. 2014; see also Burivalova et al. 2017
Acoustic activity	—	Fraction of frames in a 1 min segment where signal activity is 3 dB above the BGN	Towsey et al. 2014
Mid-band activity	—	The fraction (0.0–1.0) of the mid-bands (defined by user; e.g., 483–3,500 Hz) that exceeds a spectral amplitude threshold	Towsey et al. 2014
Signal-to-noise ratio	SNR-dB	The SPL (in dB) between the BGN and maximum envelope amplitude	Towsey et al. 2014
Band levels	—	A windowing of dominant frequency bands reported as SPL	Staaterman et al. 2014
2. Acoustic Event and Feature Statistics			
Count	—	Number of acoustic events that exist as >3 dB above BGN	Towsey et al. 2014
AV event duration	—	Average duration of acoustic events that exist as >3 dB (start and stop) above BGN	Towsey et al. 2014
Average signal amplitude	ASA	Mean amplitude of the wave envelope	Towsey et al. 2014

Index name	Abbreviation	Description	Reference
Pulse train automatic detector (counts)	—	Use of software designed to window within frequency bands that count pulses or snaps; based on amplitude threshold	McWilliams and Hawkins 2013
Spectral statistics	—	Average, median, peak, range (high and low), delta (difference between high and low), frequency per recording or sampling window after adjustments (e.g., dB above BGN).	Gottesman et al. 2020a
Spectral centroid	SC	Weighted mean of the frequency in a recording with weights calculated as magnitudes for each frequency bin	Peeters 2004 and see also Peeters and Rodet 2002
Ridge features	verRidge, horRidge	Local maxima of an acoustic feature called a ridge	Dong et al. 2013 (with Towsey and Rowe, technical paper)
Low-level acoustic features that are either timbral or temporal based	MFCC, SC, ZCR, etc.	Includes a long list such as MFCC (Mel-frequency Cepstrum Coefficients) and zero crossing ratio (ZCR)	Fu et al. 2011; Bellisario and Pijanowski 2019; see also Lutter 2014
Mid-level acoustic features that are rhythm, pitch and harmony	BH among several others	BH is beat histogram, that could be used for pulse train analysis, many others have been developed	Fu et al. 2011; Bellisario and Pijanowski 2019; see also Lutter 2014
Acoustic features of sound types	—	Distinct acoustic signals that have measurable (e.g., mean and frequency range, spectral shape, repetition patterns)	Gottesman et al. 2020a; Aalbers 2008
Spectrogram color descriptors (a.k.a. blobs)	GCH, ACC, CCV, and BIC	Used in image classification, these are measures of color histograms (GCH), auto color correlation (ACC), color coherence vector (CCV) and border/interior pixel classification (BIC). Values are calculated from PNG imagery of spectrograms.	Dias et al. 2021; also Penatti et al. 2009

(continued)

Table 9.1 (*continued*)

Index name	Abbreviation	Description	Reference
Spectrogram texture descriptors	GLCM, LBP, Fourier texture	Gray Level Co-occurrence Matrix (GLCM) and Local Binary Patterns (LBP) are among descriptors calculate. Values are derived from PNG imagery of spectrograms.	Dias et al. 2021; also Penatti et al. 2009

3. Acoustic Indices

Index name	Abbreviation	Description	Reference
Acoustic complexity index	ACI	Degree of frequency and amplitude modulation calculated across a spectrogram. There are several variations of ACI	Pieretti et al. 2011
Temporal entropy	H_t	Temporal complexity	Sueur et al. 2008
Spectral entropy	H_f	Spectrum complexity	Sueur et al. 2008
Acoustic entropy index	H	Envelope and spectrum entropy	Sueur et al. 2008
Bioacoustic index	BI or BIO	Created as a proxy measure of bird diversity, amplitude is calculated in 1 kHz bands between 2 and 11 kHz are each divided by the quietest band in this range	Bolman et al. 2007
Standard deviation of SPL	—	Standard deviation of frequency bands within a recording	Gottesman et al. 2020a
Median of amplitude envelope	M	Used as part of AR calculation but also reported separately	Depraetere et al. 2012
Acoustic richness	AR	Combines M and H_t	Depraetere et al. 2012
Acoustic dissimilarity	D_f, D_t, D		Sueur et al. 2008
Acoustic diversity index	ADI	Entropy applied to activity above a threshold dB for each frequency band	Villanueva-Rivera et al. 2011
Acoustic evenness index—Gini	AEI, AIEve, AEI-Gini	Gini coefficient–based index applied to activity above a threshold dB for each frequency band	Villanueva-Rivera et al. 2011
Normalized difference spectral index	NDSI	Ratio of biophony to anthropony (typically sounds occurring in the biologically active frequency bands >2 kHz and those below which are from human produced noise)	Kasten et al. 2012

Index name	Abbreviation	Description	Reference
Normalized power spectral density	nPSD or PSD	Converts N frequency bands to watts/kHz using Welch (1967); can be normalized (nPSD)	Gage et al. 2014
Soundscape saturation	S_m	The amount of sound above BGN that occurs in temporal-spectral segments of the spectrogram	Burivalova et al. 2017
Number of spectral peaks	NP	The number of amplitude peaks in a recording	Gasc et al. 2013
Soundscape power	POW	Calculated for each frequency band by difference of max intensity and background noise (BGN)	Towsey et al. 2014
Average intensity per frequency band	—	The average intensity for a frequency band, split into 1 kHz bands or centered across frequency bandwidths of active animals	Pekin et al. 2014
Composite urban quietness index	CUQI	Uses a combination of NDSI and ACI and their measures at edges and cores of urban parks to assess relative quietness of the core area	Tsaligopoulos et al. 2021
Green soundscape index	GSI	Combines measures of the perceived extent of natural sounds (PNS) and perceived extent of traffic noise (PTN) derived from survey data, from listening and/or soundwalk surveys	Kogan et al. 2018
Overall soundscape assessment using Swedish soundscape-quality protocol	OSA	Developed using a statistical analysis of dozens of soundscape perception descriptors from people who listened to acoustic environments; two major axes are commonly generated: pleasantness and eventfulness	Axelsson et al. 2010

4. Sound-Type Statistics

General statistics of sound types summarized by time of day, season or sites; statistics can include mean, diversity or richness (i.e., counts)	—	Developed using a statistical analysis of dozens of soundscape perception descriptors from people who listened to acoustic environments and annotated sounds based on general sound sources	Gottesman et al. 2020a

(continued)

Table 9.1 (*continued*)

Index name	Abbreviation	Description	Reference
Sound type and/ or morphospecies mapped to acoustic space use	ASU	Percentage or number of occurrences over a time interval by sound source as acoustic space use (ASU)	Aide et al. 2013, 2017
5. Call detection statistics			
Use of sound-source information derived from automated, semi-automated, and manual interpretation of sounds placed into sound-source categories	—	Examples include sound-source richness, dominance, or diversity per unit time compared across sites (meant to parallel this use in community ecology for species and population abundances)	Aide et al. 2017

Source: Modified from Sueur et al. (2014) and Towsey et al. (2014).

summarized in Figure 9.2. The use of intensity indices has been focused on studies of noise in urban environments (i.e., noise paradigm). Soundscape ecologists are more interested in sound sources and how acoustical patterns can be measured as proxies for these sources (i.e., soundscape paradigm), and in the use of automated call detection tools that employ statistical and/ or machine-learning techniques (see Chapter 10) to determine the presence or absence of species' calls in a recording (i.e., bioacoustics paradigm).

9.4.1. Intensity Indices

Intensity indices quantify sound pressure levels (SPL) of a recording. They can be grouped across ranges of intensity or for defined frequency bands. The most commonly used are those calculated as the dB exceeded in a recording for a certain percentage of the time (e.g., L_{10}); the root mean square (RMS) of the SPL of the raw signals; and the fraction of a band that exceeds a certain dB level.

9.4.2. Acoustic Events and Features

Acousticians have identified many measures of *acoustic events and features* that quantify the patterns of acoustic evens and the frequency and amplitude of acoustic patterns within a time interval. The number of these measures that have been introduced by this community is impressive (see Fu et al. 2011 for an excellent summary), and audio processing software now generates dozens to hundreds of measures per recording. Acoustic features can also be extracted from spectrograms using software that calculates image features using the same principles as facial recognition software. These acoustic features have been referred to as *blobs* by those in the machine-learning image classification community. Another group of acoustic features focuses on counts of *acoustic events*, some of which may be placed in specific categories. Acoustic events can be derived with automated routines (Farina et al. 2018), semi-automated routines (using automated routines that are then manually updated, corrected, or enhanced), or manual routines (see Section 9.5 on aural annotation).

9.4.3. Acoustic Indices

Acoustic indices (AIs) quantify the complexity, diversity, and/or breadth of sound sources in a soundscape. The most widely used acoustic indices are described here. A subset of these indices is referred to as *classic AIs* (ACI, H, D, ADI, AEI, NDSI, and BI); many are included in software programs such as Seewave and SoundEcologyR.

The acoustic complexity index (ACI; Pieretti et al. 2011) is the average absolute proportional change of spectral amplitude for frequency bins in adjacent spectra summed over all frequency bins and the entire recording. ACI is used because biological sounds are complex compared with sounds from the geophysical environment (e.g., wind, running water, rain) and from machines, and that complexity is measured as variable amplitude over time. ACI is thus designed to measure biological sounds. Its calculation requires several steps, in which the sum of one step is subsequently input into the next. First, intensity values I_k for frequency bins f_l are divided into temporal steps k in the time interval j, and differences in intensities d are calculated for the adjacent time bins as such:

$$d_k = |I_k - I_{(k+1)}|,$$

where I_k and I_k+_1 are acoustic intensities of neighboring time bins for defined frequency bins, and the d_k values for all the frequency bins are added for the temporal step of the recording such that

$$D = \sum_{k=1}^{N} d_k \text{ for } j = \sum_{k=1}^{N} \Delta t_k.$$

Here, N = the number of Δt_k in j, and D is the sum of all d_k in time interval j. Note that j should be set to 1 in Seewave and 60 in SoundEcologyR.

ACI is also one of the acoustic indices that uses a short-time Fourier transform (STFT) set of algorithms to determine frequencies. Specific window functions (usually either Hamming or Hanning) are used for these calculations; the most common method for calculating acoustic indices is to use a fast Fourier transform value of 512 samples in both Seewave and SoundEcologyR. The next calculation is to make this a relative intensity value, which gives

$$\mathrm{ACI}_j = \frac{D}{\sum_{k=1}^{N} I_k}.$$

This calculation is done for each frequency bin Δf_l across all j in the recording such that

$$\mathrm{ACI}_{\Delta fl} = \sum_{j=1}^{m} \mathrm{ACI}_j,$$

where m = the number of j in the entire recording. The final $\mathrm{ACI}_{\mathrm{tot}}$, the value used for analysis, is the sum of all $\mathrm{ACI}_{\Delta fl}$:

$$\mathrm{ACI}_{\mathrm{tot}} = \sum_{l=1}^{q} \mathrm{ACI}_{\Delta f_l} \text{ for which } \Delta f = \sum_{l=1}^{q} \Delta f_l,$$

where q = number of Δf_l. $\mathrm{ACI}_{\mathrm{tot}}$ can also be reported with subscripts that denote the range of frequencies that are being calculated.

The acoustic diversity index (ADI; Villanueva-Rivera et al. 2011) is a measure of the diversity of acoustic energy as it is distributed across frequency bands. ADI is calculated by creating N frequency bins ($N = 10$ is the most common) and calculating the proportion of energy in each frequency bin that is above an amplitude threshold (e.g., –50 dBFS). ADI is calculated in a manner analogous to its use in organismal diversity based on Shannon's entropy measure:

$$\mathrm{ADI}_{N,\Delta AT} = -\sum_{i=1}^{N} p_i \times \ln(p_i),$$

where p_i is the proportion of the frequency band i that is above the amplitude threshold (AT) relative to the total amount of energy above the amplitude threshold for each frequency band, N is the number of frequency bands ($N = 10$ is the default in SoundEcologyR), and Δ is the frequency band limit if it is not the default in SoundEcologyR.

The acoustic evenness index (AEI; Villanueva-Rivera et al. 2011) uses the same frequency bands to calculate the Gini coefficient, which measures the inequality of values among frequency bands. As Shannon's diversity and Pielou's evenness are inversely related (e.g., high diversity gives low evenness values), researchers should use one or the other. ADI uses natural logarithms, whereas AEI calculates the area under the curve of p_i values aligned in rank order.

It should be mentioned that both ADI and AEI require sensitivity analyses and will vary with PAR settings, such as gain, as well as with the sensitivity of the microphones. ADI and AEI should not be used across PAR models (e.g., when a study integrates SM2, SM3, SM4, and/or AudioMoths). ADI and AEI were first developed for and applied to terrestrial systems, so default AT values are likely to be different for aquatic, and perhaps urban, applications. As such, researchers should adjust AT values across a range of, say, –50 to –30 DBFS to determine the best amplitude threshold for their recordings.

Acoustic entropies (H_f and H_t; Sueur et al. 2008) are based on measures of probability mass function (PMF) of spectral and temporal recording sample lengths. A PMF is calculated from discrete values (e.g., amplitude), which are normalized so that they sum to 1. Using probability mass functions for amplitude, a measure of temporal entropy, based on time (also called the amplitude envelope), becomes

$$H_t = -\sum_{t=1}^{n} A(t) \times \ln A(t) \times \ln(n)^{-1},$$

where n is the size of time series computing amplitudes across the PMF, measured as the number of digitized points. Likewise, spectral entropy H_f is calculated from the mean spectrum based on a short-term Fourier transform (STFT) and then converted to a probability mass function, and then computed as such:

$$H_f = -\sum_{f=1}^{N} S(t) \times \ln S(t) \times \ln(N)^{-1},$$

where S is the spectral probability mass function of length N and H_f is the spectral entropy. A final acoustic entropy index H can be calculated as the product of both:

$$H = H_t \times H_f.$$

All values of acoustic entropy, H_f, H_t, and H, range from 0.0 to 1.0. Values near 0.0 are sounds that are pure tones, and values close to 1.0 are white noise (i.e., recordings containing many frequencies with equal intensities).

The bioacoustic index (BI; Boelman et al. 2007) is calculated as the area under a decibel-frequency curve for sounds occurring between 2 and 8 kHz, which is the range in which avian vocal activity occurs.

An *acoustic dissimilarity index* (D; Sueur et al. 2008), is a measure of the differences in temporal and spectral entropy between two or more sensors (PARs). Computing differences in each type of entropy between two sensors recording at the same time with the identical settings (e.g., sampling rate) can be accomplished thus:

$$D_t = \frac{1}{2} \sum_{t=1}^{n} |A_1(t) - A_2(t)|,$$

$$D_f = \frac{1}{2} \sum_{f=1}^{N} |S_1(t) - S_2(t)|.$$

The full acoustic dissimilarity index D is calculated as

$$D = D_f \times D_t.$$

The values of D_f, D_t, and D also range from 0.0 to 1.0 (Sueur et al. 2008). D is designed to estimate beta acoustic diversity.

Depraetere et al. (2012) developed two additional metrics based on Sueur and colleagues' (2008) probability mass function for amplitude. One of those metrics, the *median of the amplitude envelope*, or M, is

$$M = \text{median}[A(t)] \times 2^{(1-\text{depth})} \text{ with } 0 \leq M \leq 1,$$

where $A(t)$ is the amplitude probability mass function and depth is the bit depth of the recording (typically 16). As M values are typically small, M is standardized by dividing by the maximum M value of all recordings. Using M and H_t, an acoustic richness index (AR) can be derived (Depraetere et al. 2012) as follows:

$$AR = \frac{\text{rank}(H_t) \times \text{rank}(M)}{n^2}, \text{with } 0 \leq M \leq 1.0,$$

which combines ranks rather than values.

The *number of spectral peaks* (NP; Gasc et al. 2013), which is the number of amplitude peaks in a recording, is calculated after two simple rules are applied. First, all peaks across the spectrum are selected using amplitude and frequency thresholds, and then only those peaks that exceed an amplitude slope are selected. Second, peaks that exceed a frequency separation (e.g.,

200 Hz) are selected; for those falling within this window, only the largest amplitude is selected. After these two rules are applied, the number of peaks for each recording is used as the acoustic index.

The BGN index is a measure of background noise (i.e., the most common continuous baseline level of acoustic energy). It is calculated as follows (see Towsey et al. 2014): First, a spectrogram is divided into frequency bins, f, and temporal bins, c. Intensity values are saved for each of the temporal bins within each frequency bin. The mode for each frequency bin, across all c temporal bins, is then used as a measure of background noise:

$$BGN_f = \text{mode}(dB_{cf}),$$

where the mode is the most common value across all c bins in frequency bin f.

Soundscape power (POW; Towsey et al. 2014) is computed using BGN:

$$POW_f = \max(dB_{cf}) - BGN_f,$$

which is the maximum intensity, in dB, for each frequency bin f and temporal bin c, calculated for each frequency bin. POW_f represents a measure of signal-to-noise ratio, as background noise is subtracted from the other sounds that are more intense and/or closer to the microphone (Towsey et al. 2014).

Soundscape saturation (S_m; Burivalova et al. 2017) is the proportion of frequency bins that are active in m minutes. Two steps are required to calculate S_m. The first is to set all cf bins from the spectrogram used above for POW and BGN to either active or inactive status, using the following conditional Boolean test:

$$a_{mf} = 1 \text{ if } (BGN_{mf} > \theta_1) \text{ or } (POW_{mf} > \theta_2); \text{ otherwise, } a_{mf} = 0.$$

In other words, if BGN or POW is above a certain amplitude threshold, then the cf bin is set to 1 (i.e., active); otherwise, it is set to 0 (inactive). Burivalova et al. (2017) suggest a sensitivity analysis varying θ_1 and θ_2 across a range of values that would yield a "near-normal" distribution of a_{mf} values for all recordings in the study. Once the amplitude thresholds are determined, a soundscape saturation measure can be obtained by summing all a_{mf} values across all N frequency bins in a recording that is m minutes long:

$$S_m = \frac{\sum_{f=1}^{N} a_{mf}}{N}.$$

Power spectral density (PSD; Kasten et al. 2012) is a measure of the amount of energy within a specified frequency range over a defined time interval. One version of this index developed by Gage and colleagues (2014), nPSD, normalizes across all frequency bands by scaling a value from 0.0 to 1.0 for each frequency band (typically ten bands between 1 and 11 kHz). These values are then used to calculate the normalized difference spectral index (NDSI; Kasten et al. 2012). This index calculates the nPSD for the most biologically (β) active frequency band in the range of 2–11 kHz (i.e., the largest nPSD for the eight 1 kHz frequency bands) and for the anthrophonic (α) range (1–2 kHz) such that

$$\text{NDSI} = \frac{\beta - \alpha}{\beta + \alpha}.$$

Values for NDSI range from –1 to +1; +1 is the condition in which all sounds are in the biologically active frequency bands and –1 is the condition in which all sounds are anthrophonic. An alternative implementation of NDSI is also used (Villanueva-Rivera and Pijanowski 2018) whereby the nPSD for β is calculated as the sum of all frequency bands.

The composite urban quietness index (CUQI; Tsaligopoulos et al. 2021) compares values of ACI and NDSI calculated at patch edges (e.g., four corners and four locations at midpoints along edges) with their values calculated at the center of the patch, representing the core (i.e., a total of nine sampling points). ACI is assumed to be measuring biological sounds and NDSI (especially when negative) to be measuring anthropogenic sounds. Tsaligopoulos and colleagues present CUQI as

$$\text{CUQI} = \frac{\text{NDSI}}{|\text{NDSI}|} \times (\text{ACI}_{\text{max}} - \text{ACI}_{\text{min}}) \times \frac{\text{ave}(\text{ACI}_{\text{edge}})}{\text{ACI}_{\text{core}}},$$

where NDSI is calculated following Kasten et al. (2012), ACI_{max} is the maximum ACI value derived from all nine sampling locations (eight edges and one core), ACI_{min} is the minimum ACI value derived from all sampling locations, $\text{ave}(\text{ACI}_{\text{edge}})$ is the average of the eight ACI measures at the edges, and ACI_{core} is the acoustic complexity of the one core site. A "large, positive CUQI score indicates an equally balanced acoustic environment of high complexity" [whereas] . . . a negative or low CUQI score indicates an ununiformed acoustic environment of low complexity where anthropogenic and masking [of] sounds prevails (Tsaligopoulos et al. 2021, 3151).

The Green Soundscape Index (GSI; Kogan et al. 2018) uses several perception scoring metrics (e.g., Zamba perception methodology; Kogan et al.

2017) to ask human subjects who are in specific landscapes (e.g., urban park landscapes) to score the perceived extent of sounds that are urban (e.g., traffic sounds) and those that are natural (e.g., birds, wind, moving water). Scores range from 1 to 5, with 5 meaning "dominates the soundscape." The average scores for the two sound types are then used to calculate GSI:

$$\text{GSI} = \frac{\text{ave(PNS)}}{\text{ave(PTN)}},$$

where ave(PNS) is the average of the perceived natural sounds score and ave(PTN) is the average perceived traffic noise score. Soundscapes with GSI values less than 1.0 are dominated by traffic, whereas those with GSI values greater than 1.0 are dominated by natural sounds. Soundscapes with GSI values at about 1.0 are considered balanced (i.e., have major anthropogenic and natural sounds with equal intensities).

The *overall soundscape assessment* (OSA; Axelsson et al. 2010) is a simple metric that relates perception of a soundscape across two major axes of soundscape quality. Axelsson and colleagues (2010) reported that human subjects found sounds in their living areas to align across two dimensions: pleasantness and eventfulness. To develop these soundscape quality axes, a set of soundscape quality descriptors is used to collect information from subjects through a soundwalk, a playback experiment, or an outdoor survey (see the descriptions of methods for studying perception later in this chapter). The survey focuses on human perception across several major perception classes: noise annoyance, pleasantness, quietness or tranquility, music-likeness, perceived affective quality, restorativeness (see Section 6.8), soundscape quality, and appropriateness.

9.4.4. Sound Types

Sound types characterize distinct sounds in recordings by describing a sound's source using onomatopoeia (e.g., thud, boom) or its spectral structure (e.g., pulse). Sound types are most appropriately used when the sound source is not known and ways of referring to individual, distinct sounds are necessary. Some researchers (Aide et al. 2017) have used the term *acoustic morphospecies*, or just *morphospecies*, to refer to the distinct sounds in a recording. In other fields, such as speech recognition, the term *phoneme* (distinctive phonetic unit) is used to describe a unique sound type. Sound types can be identified in recordings using manual or semi-automated approaches, such as the one outlined by Gottesman and colleagues (2020a), in which simple rules guide the identification of sound types. The sound-type taxonomy in Figure 9.3 shows a hierarchical classification of eighteen sound types that occur in a tropical pond located in Costa Rica. They are distinguished from one another using rules that focus on number of pulses per second, frequency properties, amplitudes, and period lengths.

9.5. Aural Annotation

9.5.1. Aims of Annotation Methods

Bioacousticians for years have used a "listen and label" approach to identifying sound sources, mostly at the species level. In psychology, such approaches are called *sensory evaluation methods* (SEMs). These approaches are being used in soundscape ecology to summarize data from which call detection and sound-type measures (i.e., counts of specific sound sources) can be calculated. Some soundscape indices are often developed from summary data (i.e., counts of specific sound sources). Several methods that have been applied by soundscape ecologists to label sound data are summarized here.

9.5.2. Annotation Settings

Information can be attached to audio recordings in several kinds of research settings. A common setting is an indoor room where a listener (annotator) is provided with a computer, sound files, software that allows the annotator to view spectrograms of the recordings and play back the sound, headphones, and something (e.g., a pad of paper) with which to document the annotation. Additional information such as photos, videos, or maps may also be provided. Many soundscape ecologists will also have available species-specific recordings to ensure that their annotation is correct; sound libraries such as xeno-canto and the Macaulay Library at Cornell University are commonly used. Techniques such as audio selection (i.e., windowing of calls) may be used to ensure that only those sounds of interest can be heard.

A second type of annotation setting in common use is a large room that accommodates a group of listeners, where recordings are played to the group using speakers. This approach allows several listeners either to score a recording independently or to discuss the sounds as a group before annotating the file. A third approach is to use a *playback instrument* in the field to

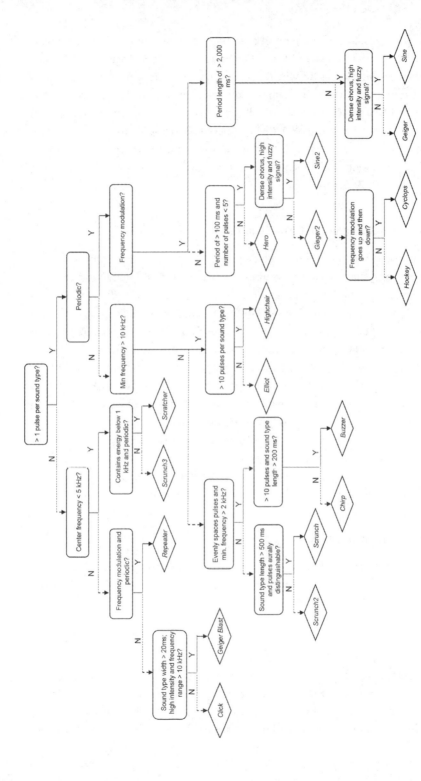

Figure 9.3 Sound taxonomy for aquatic animals in a tropical pond. (From Gottesman et al. 2019)

play the sounds to a listener and then have them annotate the recording. A fourth approach, often used in urban soundscapes and national parks, has been *intercept surveys*, in which researchers stop and question people who are experiencing the soundscape and have them complete a survey that is attached to a recording that was obtained at approximately the same time. Finally, some acoustic ecologists have asked participants to do a soundwalk and then complete a survey on the soundscape. Recordings may be made during the soundwalk using a handheld recorder capable of calculating acoustic indices for the location.

9.5.3. Visualizing Long-Term Recordings

Soundscape ecologists have also developed specialized spectrograms to visualize long-term acoustic events, such as storms, ships moving through the seascape, or long animal choruses. *Long-term spectral average* plots (LTSA; see Kaplan and Mooney 2015) and *false color spectrograms* (Towsey et al. 2014) have been developed for these purposes. False color spectrograms are made using a software program (AP.exe) that combines dozens to thousands of short (e.g., 1-minute) audio samples that are then stitched together. As many duty cycles have long recording times (e.g., 10 minutes on per hour), audio software is used to extract a 1-minute sample at a random point in the longer recording window. Small frequency bins are then used to calculate several AIs (e.g., H, ACI, BGN), and their relative intensity is scaled (in a range of 1 to 100) to one of three colors (hence the term "false color spectrogram") and set as bright (dominant) or off (not a large value) for each of the AIs.

9.5.4. Soundscape Content

Numerous methods have been employed by soundscape ecologists to annotate their soundscape recordings. This section summarizes some of the most common approaches.

Presence

The first, and perhaps the least labor intensive, approach is to listen to a recording and score it for the *presence* of a sound source. Many of the applications of annotation in soundscape ecology have focused on specific taxa, such as birds or fish, and so experts that are knowledgeable about distinct sounds that particular animals make can score a recording for their presence. Depraetere et al. (2015) used this method to score the presence of bird species in several locations. This information was then used to quantify avian species richness at each site to determine the quality of the habitat independently of acoustic indices, which were also being used.

Presence can also be scored using *high-level* annotation. For example, Bellisario et al. (2019b) listened to short (<6-second) recordings to annotate them for the presence of sounds in the three major sound-source categories: biological, geophysical, and anthropogenic. A set of these labeled recordings containing nine different combinations of these sound-source categories was scored to determine how often the categories co-occur. This kind of labeling is considered high-level because only major sound-source types are documented; low-level annotation would involve annotation to very specific sound sources, such as animal species (e.g., dink frog) or type of anthropogenic sound source (e.g., train whistle).

Counts and Acoustic Space

Another approach to labeling is to *count* the number of acoustic events produced by each sound-source type. If a particular bird species calls ten times in a 1-minute listening interval, for example, this count can be noted along with the sound source. Some soundscape ecologists have also measured the *duration* of each sound in a recording, generally by viewing a spectrogram after applying an amplitude threshold to reduce the presence of background noise. These scores may be presented as a function of time (e.g., number of events per minute) or as a proportion of the length of the recording, placed into bins of 5% or 10% of recording length. Durations of temporal *overlap* of sound sources can be measured as well, as these are important for studies of masking or for examining dimensions of acoustic partitioning. Another method of annotation is to score the amount of spectral-temporal *space* that is occupied by the sound source. Aide and his colleagues (2013, 2017), for example, developed a standardized technique with his ARBIMON system that divides the spectral-temporal space into small frequency-time bins. The sound source for which there is *acoustic space use* (ASU) is then associated with these frequency-time bins. Finally, some researchers have assessed the percentage of a spectrogram occupied by a sound by listening to a recording and viewing a spectrogram in detail, then calculating the sound source's relative acoustic space occupancy in the spectrogram.

9.5.5. Perceptions

In any of the annotation settings described above, people can be asked a variety of questions related to how they perceive the acoustic environment, using a playback of a recording or an actual soundscape with an in situ survey. Axelsson et al. (2012) have used several techniques to gather perception information. Listeners can be asked to select from long lists of perception descriptors that have been randomly shuffled so that the listener does not always select the first one on the list, thus introducing bias (see Section 9.5.7). Perception descriptors can also be provided with Likert-scale scoring options so that quantitative information is obtained for the qualitative soundscape descriptors (see Section 9.9.6). For some perception research, audio recordings are played at the same time a photo is provided. Tests of sensory perceptive differences between the auditory and the visual senses will require the use of recordings, photos, or both as part of the study. Many perception studies have been done with university students (considered a convenient sample) or by intercepting visitors at research locations (such as urban parks).

A variety of other information is sometimes gathered by researchers during perception studies, including nature relatedness or sense of place scores (i.e., how respondents relate to nature or the place that is being studied), as well as respondents' age, education, and gender affiliation.

9.5.6. Qualitative Information

A listener may also be asked to provide qualitative scores when annotating soundscape content. For example, listeners could assign each of their labels a confidence score (1 = not fully certain but it is my best guess, 2 = moderate level of confidence in my label, or 3 = high level of confidence in my label). In addition, the listener could provide a purity score (0 = lots of background noise, 1 = some background noise but signal is somewhat clear, 2 = signal is fairly clear, 3 = signal is very clear and no other sounds occur within the spectral space).

9.5.7. Bias

As with all human-identified data, user bias can be quantified following the approach of Gasc et al. (2018), in which two sets of listening comparisons are made. First, a set of labeled audio files is presented, but without the labels, to another listener, who labels the information independently from the first listener. Second, the original listener is provided with a second presentation of the files they labeled, but without information on how they labeled those files in the first pass. The two tests allow researchers to determine the number of within-listener and between-listener differences, which may reflect bias. Assessments of this kind are referred to as inter-rater error assessments or inter-rater reliability assessments (Landis and Koch 1977). In the event that there are discrepancies between the labels in these two tests, a third listener (generally a more experienced listener) may be asked to break the tie. By combining the results of these bias tests with confidence scores and purity scores, the difficulty of classifying the sound types in the recording can be quantified and error rates at different levels of classification can be assessed.

9.6. Phases, Transitions, Triggers, and Cycles of Soundscapes

Soundscape ecologists have recognized that soundscapes are composed of consistent sounds over distinct time intervals. Time intervals such as the dawn chorus and dusk chorus have been well established by bioacousticians and soundscape ecologists. Seasonal time intervals, such as spring bird migrations in the mid-latitudes, are characterized by distinct sounds (e.g., bird vocalizations). Distinct time intervals that have a regular composition of sounds are referred to as *soundscape phases* (Fig. 9.4). These

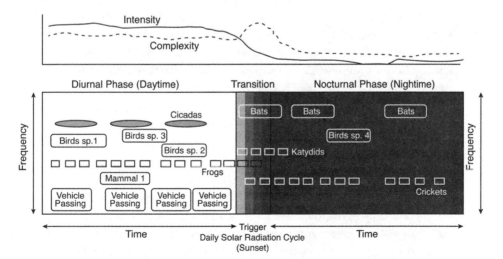

Figure 9.4 Soundscape triggers, phases, transitions, and cycles.

phases are regulated by a variety of environmental factors, such as the diel (i.e., daily) or seasonal solar radiation cycles. In marine systems, it is well known that lunar cycles regulate dusk chorusing by fish and other marine animals (e.g., Aalbers 2008), with peak acoustic activity occurring during the evening crepuscular periods but also during the new moon (Staaterman et al. 2014). Environmental factors that regulate soundscape phases are referred to as *soundscape triggers*. An interval between soundscape phases, referred to as the *soundscape transition* period, may occur over a period of seconds (e.g., transition from cicadas to katydids in the evening can occur within a 30–90-second interval; Pijanowski, pers. obs.) or last much longer (e.g., hours to days). Finally, because soundscape phases often exhibit regular periodicity—diel cycles, for example, are very common (see the section "Rhythms of Nature" in Pijanowski et al. 2011b)—these *soundscape cycles* can be measured over longer periods that include multiple iterations of a cycle, which can be compared across space (e.g., an ecological gradient such as those found across elevation or moisture) or across levels of human activity (e.g., different land use intensities). Figure 9.4 shows how soundscape phases, transitions, and triggers are organized over time.

When we more broadly consider soundscape phases across hourly to seasonal time intervals, they can be identified by the factors that create district time periods. Schafer's (1994) generalized observational plots of sound activity patterns throughout a year were some of the first attempts to examine these long-term trends (Fig. 9.5). As another example, consider the hourly traffic patterns that exist in many areas of the world. Morning traffic sounds, which peak when people travel from home to work, are then repeated when people return home from work. Weekends and holidays will result in soundscapes with compositions and hourly patterns different from those on weekdays. Some areas may also experience distinct seasonal patterns because activities occur during certain times of the year; college campuses are bustling during spring and fall semesters but are often quiet during summers. Likewise, recreational and vacation locations will experience increases in traffic and other sounds (e.g., many people talking or shouting at markets and shopping venues during the summer).

Geophonic sounds many vary over distinct time intervals as well. Wind often peaks during the middle of the day and diminishes during crepuscular periods. Wind speeds often vary over seasons as well. In the equatorial regions of the world, one or two rainy season–dry season cycles occur (depending on latitude), bringing with them differences in wind and rain.

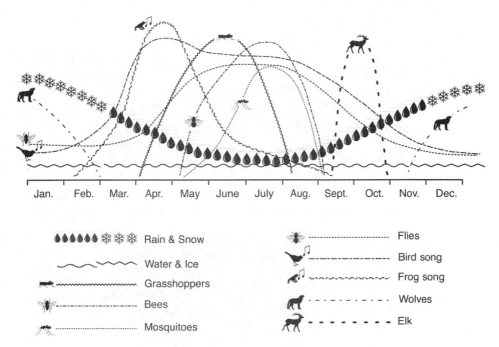

Figure 9.5 Temporal variability of sound sources in a mid-latitude landscape. (Modified from Schafer 1995.)

One of the most illustrative examples (Fig. 9.6) of how environmental factors trigger soundscape phases and transitions comes from the research on soundscape composition just before, during, and after a full solar eclipse (Buckley et al. 2018). The sound intensities across nearly all frequencies decreased at the height of the solar eclipse; air temperatures also decreased during this period, as did light values. The calling activity of insects exhibited a large shift during the eclipse, as cicadas decreased their activity and three species of crickets increased theirs. Interestingly, the frequencies produced by orthopterans (e.g., crickets, katydids) decreased during the eclipse as well.

9.7. Supplemental In Situ Survey Data

Ecologists have long conducted plant and animal surveys to monitor trends in species composition over space and time. Many of these techniques have become standardized, and best practices have been documented. Common techniques used for plant and animal surveys, for both terrestrial and marine environments, are summarized here.

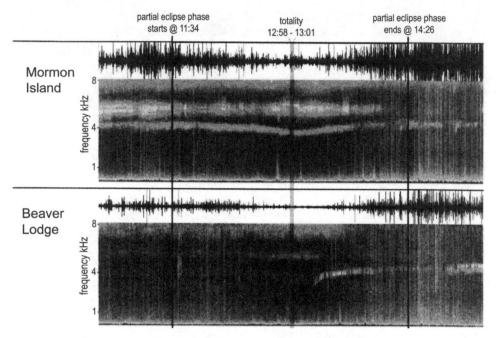

Figure 9.6 Changes in sound and atmospheric conditions during a solar eclipse at two study sites in Nebraska. The eclipse occurred at 1:00 p.m. local time on August 21, 2017. (Modified from Buckley et al. 2020.)

9.7.1. Animal Surveys

Royal and Nichols (2003) provide a comprehensive summary of best practices for animal survey work that includes techniques such as *line transects*, *point counts*, and other approaches employed to estimate species richness and densities as well as user bias. Monitoring of insects has been accomplished by ecologists in a variety of ways (Muirhead-Thompson 2012). These include the use of *pitfall traps* (Laub et al. 2009), *sweep net sampling* (e.g., McCoy 1990; Janzen 1973), *light traps* (e.g., NightLife traps; see Price and Baker 2016), *suction traps* (e.g., Juillet 1963), *sticky traps* (e.g., Heinz et al. 1992), *suspended nets* (Taylor 1962), *heat attractor* and *wind-funnel traps* (Pijanowski 1991), and *pheromone/bait traps* (Vick al. 1990). Mammal surveys most frequently use live traps (often baited) and *camera traps* (O'Connell 2010). Camera traps are placed near a PAR and are operational during soundscape recording, as they are noninvasive and silent. The movement of an animal within range of the camera triggers it to take a photo. A time stamp and sensor ID are burned into the image, which is stored with a filename associated with the sensor and time of the photo. A variety of automated techniques (Yu et al. 2013) have been developed to identify the animals in these photos.

Surveys of fish and other aquatic organisms in marine and freshwater environments have also been standardized. These surveys are often conducted in situ with endangered species or in marine protected areas. Underwater surveys of aquatic animals (Harvey et al. 2020) follow some of the same generalized approaches as terrestrial surveys, but are conducted using scuba gear. Counts of fish and other aquatic animal species and estimates of their relative sizes are made within a rectangular path by divers who spend a fixed amount of time at each of the sampling spaces. Recently, diver surveys have been supplemented with underwater video cameras (e.g., Willis and Babcock 2000; Watson et al. 2005) and surveys made from videos.

9.7.2. Vegetation Surveys

Several vegetation survey standards have been developed, many of which have been implemented as common quantitative approaches to assess the structure and diversity of plant communities. One of the most widely used approaches is that of the USDA Forest Service's Forestry Inventory and Analysis (FIA) Program. Some of the first FIA plots in the United States have been in operation since the 1920s (Smith 2002), although the measurements used have changed over time. FIA-like plots also exist in other countries around the world (Liang et al. 2016). FIA plots involve the use of four concentric inventory plots (one central and three in the twelve o'clock, four o'clock, and eight o'clock directions proximal to the central plot), which also have three circular microplots that are designated for high-quality structural and/or species-level data collection. Data collected at these plots include diameter at breast height (DBH), species, tree height (a.k.a. canopy height), and canopy cover. As these methods are well developed and documented, soundscape ecologists could benefit from adopting them to measure forests in their study sites.

Grasslands are often assessed using the *line-point intercept method* (Floyd and Anderson 1987), which standardizes measures of grassland cover (percentages of foliar cover, bare ground, standing dead vegetation, litter, and rock fragments), vegetation heights (in height ranges, reported as number of stems that extend to that height), soil condition, and landscape properties such as site stability and erodibility. A cloth tape measure is placed on the ground and points are randomly selected along the taut tape (which is anchored); grassland cover characteristics are then recorded at each point from the top of the vegetation to the ground. In grassland and savanna landscapes, soundscape ecologists may consider this or other similar techniques for doing plant surveys, as doing so will allow their data to be compared with those from other locations where these methods have been used.

9.8. Quantifying the Landscape for Soundscape Research

There are several ways that landscape ecologists quantify landscape structure. These methods involve the use of remotely sensed data, such as those from satellites, planes, and drones (also called unmanned aerial vehicles, or UAVs), and those developed using geographic information systems (GIS), in which vector data, such as lines for roads, rivers, and boundaries and polygons for lakes, are stored, integrated, and analyzed with other spatial data.

9.8.1. Quantifying Landscape Structure and Dynamics

One of the most significant types of spatial databases used to characterize the landscape is *land use/cover maps*, often distributed as national databases (e.g., Roy et al. 2015), although a few are global (e.g., Bartholome et al. 2005). These maps store information about land use/cover types using a hierarchical classification scheme (Loveland et al. 2000). Standard land use/cover types at the highest level of classification (Anderson et al. 1976) are urban, agriculture, shrubland, forest, open water, wetland, and barren. Subclasses of each are also typically included in these spatial databases; for example, the urban class often includes subclasses such as residential, retail/commercial, industrial, transportation, and city park. Agricultural land use/cover is often broken into food production types such as row crops, orchards, pastures, and combined animal feed operations. Data are stored in vector format (e.g., as polygons of similar use/cover) or as a raster, which is a grid of square pixels with a specific resolution (often at pixel sizes of 30 m, 100 m, or 1 km).

Another type of spatial database used by landscape ecologists to quantify landscape structure is topography maps. Referred to as *digital elevation models* (DEMs), these maps, some with global coverage, are distributed by a variety of organizations (e.g., NASA) at typical pixel sizes of 30 m × 30 m for those with global coverage, although local DEMs can be as fine as 50 cm × 50 cm. DEMs can be used to quantify elevation (in meters above sea level), slope (expressed as a measure of rise per run or height over distance, also called grade), and aspect (the orientation of the slope using a 360-degree measure with 0 being north).

Linear landscape features, such as streams and roads, are also used by landscape ecologists to quantify landscape structure (Forman et al. 1984). These data, which are stored as lines, are typically divided into classes, such as transportation, and subclasses, such as highways, rural roads, residential streets, on-ramps, and off-ramps. These subclasses could be useful to associate with different types of sounds (e.g., trucks, buses, cars) and their amplitudes. Classification of river segments by stream order, such as primary, secondary, or tertiary, can be used to infer water volume, occupancy by certain kinds of animals, and so forth.

Mapping of habitat is often done through forms of remote sensing based on electromagnetic energy sensors. Electromagnetic energy sources that are measured include gamma rays, X-rays, ultraviolet light, visible light, near-infrared light, microwaves, and radio waves; each of these terms refers to a portion of the electromagnetic spectrum. Many kinds of electromagnetic energy sensors exist, and their products are referred to as *imagery*. These sensors are often mounted on satellites, planes, or drones. Land and water surfaces are often mapped using electromagnetic imagery with data stored as slices (e.g., spectra) of electromagnetic wavelength bands. Land use/cover is often mapped using multispectral imagery (MSI), which refers to imagery that has between 3 and 15 spectral bands. One component of NASA's Landsat program is its Thematic Mapper (TM), a satellite-based multispectral sensor that has been collecting land surface imagery since 1972 and is primarily used to map land use/cover. Landsat TM data are stored in about a half dozen bands and at a pixel size ranging from 90m × 90m to 30m × 30m. Landsat is also used (Cohen and Goward 2014; Roy et al. 2014) to map wildfire, drought, crop type, biomass (carbon) change, density of harmful algal blooms in water, and land and water surface temperature. TM time series are used to measure changes in coastlines, glaciers, and dunes.

NASA's Moderate Resolution Imaging Spectroradiometer (MODIS), also a multispectral sensor, is a satellite-based remote-sensing platform on two satellites, Terra and Aqua, that maps land surface conditions on a near-daily basis (Justice et al. 2002). Distributed as multi-day (most commonly 16-day) composites at pixel sizes of 250 m × 250 m to 1 km × 1 km, these landscape condition maps can be used to assess changes in vegetation condition over long time intervals. Wildfire is also mapped using MODIS (Giglio et al. 2003). Since the early 2000s, MODIS imagery has been distributed for free to the scientific community as a series of over two dozen data products. Because they are created as a time series, maps from MODIS have been used to quantify landscape dynamics and phenology over time. Other multispectral sensor platforms include Advanced Very High-Resolution Radiometer (AVHRR); Sentinel-1 and Sentinel-2; ECOsystem and Spaceborne Thermal Radiometer Experiment on Space Station (ECOSTRESS) on the International Space Station; and the series of Satellite pour l'Observation de la Terra (SPOT) sensors.

Both Landsat TM and MODIS have been used to create greenness indices (Carlson and Ripley 1997). One of the most common is the Normalized Difference Vegetation Index (NDVI), which is given as

$$NDVI = \frac{NIR - Red}{NIR + Red},$$

where NIR is the near-infrared (light wavelength of 400–700 nm) value for a pixel and Red is the red (light wavelength of 700–1,000 nm) value for the same pixel. NDVI values, which represent the photosynthetic capacity of plants at each pixel, range from –1 to +1: values of less than 0.0 indicate dead plants, water, or rocks; values of 0–0.33, unhealthy plants; values of 0.33–0.66, moderately healthy plants; and values of 0.66–1.0, very healthy plants.

Some sensors measure reflectance of wavelengths across the entire electromagnetic spectrum; this remote sensing technology is often called *hyperspectral imagery* (HSI) to distinguish it from multispectral imagery or, more correctly, to reflect the way the technology operates, as *imaging spectroscopy* (Goetz et al. 1985). Some of the applications of imaging spectroscopy (Bioucas-Dias et al. 2013) include detecting the chemical composition of plants, water, and soils. Satellite-based hyperspectral sensors include the Advanced Land Imager (ALI) on the EO-1 satellite that was operational from 2000 to 2017 (NASA), the DLR Earth Sensing Imaging Spectrometer (DESIS) on the International Space Station, and the Environmental Mapping and Analysis Program (EnMAP).

Finally, several forms of optical remote-sensing data from Light Detection and Ranging (LiDAR) technologies (Dong and Chen 2017) have been used to characterize the *three-dimensional structure* of vegetation and/or buildings across a landscape. LiDAR works by sending out hundreds to billions of laser pulses and measuring the time it takes them to return (reflect back), then using that measure to determine distance. The data are then used to create a three-dimensional point cloud that represents the land surface as well as the vertical structure of vegetation. LiDAR data are often collected by UAVs or by plane, although NASA recently (in February 2019) mounted a LiDAR instrument on the International Space Station (called GEDI), which will collect continuous data for about 80% of the land surface for three to five years.

9.8.2. Associating Landscapes with Soundscapes

Data on landscape structure and condition can be associated with soundscape data in a variety of ways. A typical approach (Fig. 9.7) is to use a GIS to store all spatial data, including the positions of PARs. The GIS can be used to extract the spatial extent of the landscape for which quantification is needed by using a *buffer* command. Pijanowski et al. (2011a) used 100 m, 300 m, and 1,000 m buffers around PARs to associate land use/cover amounts with each PAR, thus relating summary statistics of soundscapes to the landscape's acoustic theaters. Larger buffers would be necessary for low-frequency sounds. These spatial calculations, called *areal amounts*, can

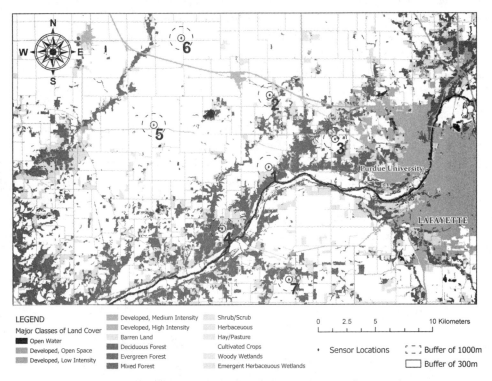

LEGEND

Major Classes of Land Cover

- Open Water
- Developed, Open Space
- Developed, Low Intensity
- Developed, Medium Intensity
- Developed, High Intensity
- Barren Land
- Deciduous Forest
- Evergreen Forest
- Mixed Forest
- Shrub/Scrub
- Herbaceuous
- Hay/Pasture
- Cultivated Crops
- Woody Wetlands
- Emergent Herbaceuous Wetlands

0   2.5   5   10 Kilometers

• Sensor Locations   Buffer of 1000m

Buffer of 300m

Figure 9.7 Use of geographic buffers around sensor locations illustrates the diversity of land use types and landscape features in each acoustic theater.

be expressed in units of square meters or as proportions (e.g., urban = 0.1; agriculture = 0.4; forest = 0.5). Topographic information for the location of the sensor can also be used as a spatial value; if DEMs are available for the acoustic theater, then summary statistics, such as average elevation and average slope, can also be calculated using the GIS.

Landscape ecologists are often interested in the *spatial patterns* of landscapes. Tools such as FRAGSTATS (McGarigal 2015) are used with raster data to calculate patch, class (e.g., forest), and landscape indices. Standard patch statistics include patch size, patch edge, and patch shape. Class statistics summarize these values (e.g., patch size) for each patch in a class, but also summarize these as a group of all patches of the same class. Class statistics include the total area of a patch and the total amount of edge for all patches of that class. Statistics that summarize groups (i.e., landscape indices) include the average patch size, the size of the largest patch, the range of patch sizes, and a variety of summary statistics such as the variation in patch size. Spatial arrangement indices can also be calculated, including interpatch distances, measures of clumpiness or aggregation, and the amount of edge and core habitat. Landscape indices make the same calculations that class indices do, but do not consider classes as separate entities. Landscape measures are used to quantify the level of fragmentation of a land use or land cover type. Separate measures for a PAR may include values associated with the patch that the sensor is in (e.g., the size and shape of the patch where the PAR is located).

A second spatial statistical summary that can be made of landscapes is the total *length of any linear network*. For example, if the GIS is used to create buffers for the acoustic theater, a summary of the lengths of all roads within that buffer is a measure of the potential road traffic sounds. These statistics can be summarized by subclass, too.

Another set of common measures of landscape structure is *distance from features* measures. The distance from a PAR to the nearest road or stream provides a measure of the likely influence of that feature on the sounds in its recordings. The distances from airports or city centers could be additional measures of how far the PAR is located from an anthropogenic transportation hot spot (see Section 3.3.7).

Finally, some landscape ecologists use voxels (*volumetric measures*) that may contain LiDAR point data on the density of vegetation in vertical zones and the height of the vegetation as an additional three-dimensional measure of the landscape. Several forest three-dimensional metrics have been developed by landscape ecologists and remote-sensing experts (see Jung and Pijanowski 2015). Derived products from MODIS provide another common measure that ecologists use frequently, which is leaf area index (LAI), a measure of the area of leaves per unit ground area that exists between the ground and the top of the canopy (Breda 2003). Tropical rainforests have the greatest LAI, often more than 6, and dense coniferous forests can have values that reach 10. LAI at the PAR location, or a summary statistic of LAI within the acoustic theater, can be used to quantify leaf density across a landscape.

9.8.3. Acoustic Phenology Models

A generalized landscape-soundscape-disturbance model of *acoustic phenology*, the change in soundscape composition over time, is shown in Figure 9.8. Two time scales are displayed, seasonal (part A) and diel (part B).

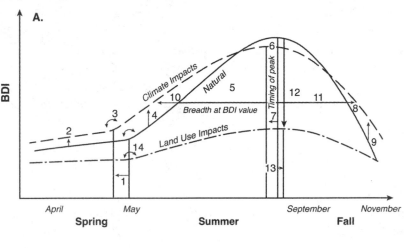

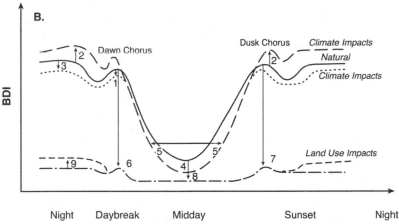

Figure 9.8 Soundscape-landscape-disturbance model. See text for details.

In both parts, the solid line shows the natural phenological trend in animal acoustic diversity (labeled here as BDI, or biological diversity index, which could refer to any of the soundscape indices mentioned above that quantify biological sounds) for a typical temperate forest in Indiana. Peaks occur late in the summer (August), when birds, frogs, and insects all contribute to the acoustic diversity. Early in the year, most vocalizations come from birds; then migratory birds arrive, and anurans emerge, followed by insects (e.g., crickets, cicadas, katydids). However, as we know from current literature on animal phenology, as atmospheric temperatures rise with climate change (dashed line), many ectothermic animals will emerge earlier, shifting the peak of this phenology earlier and creating earlier spring "inflection points." Alterations to the habitat (dashed and dotted line) are known to reduce animal species richness, which has been shown to decrease animal acoustic diversity. The habitat structure and composition models will help to elucidate trends in land use change (comparing them with natural conditions) and the dynamic landscape model will help to characterize the shifts due to changing climate. Measures of phenological change (numbers 1–14 in the figure) characterize the degree to which climate change and habitat alteration will affect animal diversity as reflected in the proxy of animal acoustic diversity.

9.9. Qualitative Methods: Measuring Soundscape Perception

9.9.1. Key Terms

Scholars in the sociocultural sciences (e.g., political science, anthropology, and psychology), humanities, journalism, oral history, education, and medicine have employed a variety of qualitative approaches to study the human dimensions of their fields. Combined, these fields represent an area of scholarship often referred to as the *human sciences*, which has advanced several major shared methodological approaches. These approaches have a rich history, are especially well developed and mature in several major areas, and have been applied even by ecologists and others in the physical sciences. They have involved collaborative efforts across the biophysical and human sciences to understand human behavior, perceptions, cognition, and sentiments as they relate to the environment.

One important type of qualitative research applicable to studying human perceptions of soundscapes is called *phenomenological* research. Phenomenological research is characterized by "close attention to the details of participants' lived experience as well as an emphasis on participants' interpretation of their experience" (Frost et al. 2014, 121). Researchers who study *lived experience* are interested in studying people's everyday life experiences in order to understand the unique nature of each human situation (van Manen 2016). The phenomenological perspective has many supporters (see Serres 2008; van Manen 2016; Moustakas 1994; and Pink 2015) as well as some detractors (see Howes and Classen 2013), but the term is widely used in psychological and ethnographic research.

This section is by no means a complete "how-to" guide on each qualitative methodological approach. There are extensive guides and handbooks available that provide details on these approaches at introductory to advanced levels. This section describes the most common qualitative approaches; high-level "best practices" and reference guides should be consulted for more in-depth pursuits. Prior to a categorical description of methods, two broad methodological perspectives are described: grounded theory and knowledge co-production. A presentation of mixed-methods approaches, which combine quantitative and qualitative approaches, follows in Section 9.10.

9.9.2. Grounded Theory

One very broad theory of research in the human sciences is called grounded research theory, or simply *grounded theory* (Glaser and Strauss 1967). In this approach to qualitative data collection and analysis, the focus "initially is on unraveling the elements of [human] experiences . . . [that] enables the researcher to understand the nature and meaning of an experience for a particular group of people in a particular setting" (Moustakas 1994, 4). Theory is generated from data collected, and "hypotheses and concepts are worked out in the course of conducting the study and from an analysis of the data" (Moustakas 1994, 4). According to the originators of the concept, grounded theory is a level of theory that is considered "middle range," falling between the "minor working hypotheses" of everyday life and the "all inclusive" grand theories found in many biophysical sciences (Glaser and Strauss 1967, terms in quotes are as used on pp. 32–33). The aim of grounded theory is to develop theory from the data through a careful examination of the various forms of data collected that is iterative, thorough, and alert for gaps in knowledge or inconsistencies, and which requires frequent tasks of coding (i.e., labeling) the data during both data collection and analysis, all culminating in new knowledge through mostly inductive means, but also revisiting data through deductive reasoning and eventual verification. Grounded theory emphasizes a fluidity in the process of discovery; there are no pre-defined sequential steps. Grounded theory (Strauss 1987) emphasizes a "grounding in data," which has also been interpreted as a "grounded theory of analysis." Grounded theory can apply to the use of one of the methods described below or to a collection of methods used in a study. There are also interesting parallels between grounded theory and knowledge discovery in databases in that both focus on data to derive hypotheses, which are then tested by developing general themes from patterns in the data.

Glaser and Strauss (1967) recognize two forms of grounded theory. The first is *substantive grounded theory*, where research focuses on empirical areas of human science such as professional education, cancer care, or noise in cities. The second form, which they call *formal grounded theory*, aims to develop ideas from data collected to examine conceptual areas of human science, such as authority and power or nature connectedness. They also emphasize that grounded theory is theory as a process rather than theory as a product; any product in theory development is "momentary" (1967, 12). They add that the grounded theory approach allows for the construction of multiple theories that can be used to contrast with one another and drive reasoning that is more inductive across the data. This approach to theory development differs from the more traditional view that theories are grand, singular, and apply across a large array of contexts, individuals, and groups.

Theory development within the context of grounded theory has three main elements (Glaser and Strauss 1967): *categories*, *properties* of categories, and *hypotheses*. Categories and properties are concepts that are inferred from the data through inductive reasoning. Categories are the conceptual *elements* of the theory derived from the data; a property is a conceptual *aspect* or *element* of a category. A category developed through substantive grounded theory might be the loss of natural sounds due to human activities like mining, and its properties might include the relationship between the loss of natural sounds and the demographic properties, such as age, of the individuals who used to experience those sounds. The third element is *hypotheses*—ideas suggested by the data that relate categories and properties—that provide explanatory generalizations at any one of many conceptual levels. Thus, a hypothesis might be that older individuals experience greater emotional and physiological harm due to the loss of natural sounds due to mining activities than do younger individuals. An excellent summary of what grounded theory is not is provided in Suddaby (2006); it is instructive for novices in this area of research.

9.9.3. Knowledge Co-production

Knowledge co-production grew out of epistemological research (i.e., re-search on how knowledge is developed) involving multiple disciplines and diverse actors (e.g., researchers, study subjects, decision makers, and the public). The term was first used by Gibbons and Nowotny (e.g., Gibbons et al. 1994; Nowotny et al. 2013), who were interested in describing the features of what they called *Mode 2.0 research* or *new knowledge production* (NPK), a transdisciplinary way of working that is socially distributed and differs from traditional singular disciplinary ways of generating knowledge (which they label as Mode 1.0). By "socially distributed," they mean that the new knowledge produced is *credible*, *salient*, and *legitimate* for the users of that knowledge (Cash et al. 2003). The knowledge co-production perspec-tive derives its foundations from Guston (2007), who describes two ways of co-producing knowledge at the interface of science and policy. In the first, researchers and the policy community work together, and research informs policy. Guston (2007) argues that many instances of scientific misconduct have eroded the relationship between these communities. In the second, the scientific (a.k.a. research) community works with boundary organiza-tions that can assist them in communicating with the policy community. *Boundary organizations* are important groups that create new knowledge in-teractively and, in some instances, in real time (see Cash et al. 2003). Exam-ples of boundary organizations are university extension groups that work closely with communities and individuals to support "effective information flows" (Guston 2001, 404).

Mode 2.0 and boundary organization research approaches have merged into a much broader concept of knowledge co-production. *Knowledge co-production* has been defined as consisting of "loosely linked and evolving clusters of participatory and transdisciplinary research approaches" that have emerged in recent decades (Norström et al. 2020). This form of knowl-edge production has five main characteristics (Hessels and Lente 2008):

> *Context-based.* The social, economic, political, and ecological circumstances are embedded in the research activities and questions. "Context" refers to place, problem, or social group. The context should also involve the understanding of relevant language(s) and a shared understanding of concepts, terms, and how knowledge is used.
> *Pluralistic.* Knowledge co-production acknowledges the diversity of knowl-edge sources across disciplines as well as all actors in the knowledge

production activity (e.g., ecologists, subjects whose lived experiences are being studied, policy analysts, social scientists, people in business, journalists, and data scientists).

Goal-oriented. Work by all actors has clearly articulated goals that are meaningful, shared, and related to solving the problem that is being addressed. This dimension of knowledge co-production involves setting realistic milestones, deliverables, and metrics for success that are obtainable and valuable for all actors in the research enterprise.

Interactive. The research method supports many interactions that span a variety of activities, including the development or refinement of research questions, the design of data collection, the collection of diverse data, and the involvement of all actors in the research enterprise. The number of, timing of, and intervals between activities, the number of participants, and the distribution of roles and responsibilities are only some of the considerations in this important element of knowledge co-production.

Reflexive. Research that combines qualitative and quantitative data requires a dialogue that constantly reexamines the data and how they are being collected to address the problem. Researchers become more conscious of the role that their work plays in improving the well-being of nature and/ or of people.

9.9.4. Interviews

Interviews are one of the most "firmly embedded research methods in contemporary qualitative research" (Pink 2015, 73). The purpose of an interview is to gain insights into, or understanding of, opinions, attitudes, experiences, processes, behaviors, or predictions (Rowley 2012). It is especially useful to understand the "lived experiences of people and the meaning they make of these experiences" (Seidman 2019, 9). It is not, as Seidman (2019) explains, an approach specifically designed to test hypotheses directly. Rather, it is a technique to explore and understand how people live by interpreting experiences important to the research questions being analyzed. It is work focusing on the "interest in other individuals' stories because they are of worth" (Seidman 2019, 9).

Interviews can take several forms, ranging from preset, standardized, and closed-ended, to semi-structured, where some questions are asked in a particular sequence but others are asked after part of the interview elicits an idea about a gap or inconsistency in knowledge for the interviewer, to free-form conversations (a.k.a. dialogues; see Post and Pijanowski 2018). Pink (2015, 75) considers the interview to be "a social, sensorial and affective encounter." Interviews can be one-on-one or in a group setting; the latter types are commonly referred to as focus groups (Rowley 2012). The interview can be conducted at the place of the lived experiences (such as outdoors, next to a river, where sounds are being experienced and their meaning described, or indoors, in a schoolroom, office, or community center where decision making is typically discussed); over the phone, through chat sessions on the internet or by video conferencing, or in a neutral location (where both interviewer and interviewee are new to the location). Interviews can also be "mobile," in which the interviewer and interviewee walk together (also called go-along interviews) or visit an important place together. Interviews can be recorded using handheld digital audio recorders, video recorders, or notes written on paper for later transcription to a digital format. Interviews can be conducted once with each interviewee, or as a series that attempts to cover several kinds of topics (e.g., a person's life history; detailed descriptions of lived experiences, possibly in situ; and discourse on the meaning and consequences of change). Interviews can be conducted with people in power (sometimes also called key informants, such as elected officials or CEOs), with people who have extensive lived experiences and who can tell a life story based on those experiences (e.g., a Mongolian herder who is raising livestock in rural grassland landscapes), with elders (i.e., people who can provide information about how lived experiences have changed over time), or with people whose lived experiences have changed due to the circumstances under consideration (e.g., people whose soundscape has been changed by a global pandemic). Interviews can include the use of playbacks (e.g., playing a sound and eliciting comments), visuals (e.g., showing photos of animals and asking about sounds and meaning), or videos (e.g., performances of songs about nature). Some interviews are conducted using cards or statements; when the topic or term on a card is used, it is picked up and then associated with the narrative, or eventually even discarded to ensure that all topics for the interview are covered. Finally, interviews may require live translators. The additional time required for the translation of questions and responses must be considered in the planning of the questions ahead of time and in considering the length of the interview. In many cases, translation is not just a literal translation of words, but may also require a cultural description, by the translator, of a question or response. Translation may also require the interviewer and the translator to work together to determine how technical terms used by the researcher need to be conveyed during the interview. For example, the term "soundscape" is a rather unique term in English, and its direct translation into other languages might not be strictly literal.

Interviews are labor intensive. They require many months of preplanning to ensure that the selected interviewees, topic, and setting are adequate for the interviewer to develop an understanding of lived experiences and their meaning. They also require that notes or audio recordings be transcribed, and that all of the interviews be analyzed as text documents. The length and spacing of interviews are also important considerations. Seidman (2019) has suggested that 90-minute interviews provide, in most cases, an adequate amount of time to ask questions, but not enough to exhaust either the interviewee or interviewer.

Pink's (2015) sensory ethnographic studies emphasize an alternative viewpoint on the contemporary interview: that the interview is a phenomenological event in itself. Pink (2015, 79) points out that as the "researcher and interviewee move through their [interview] route, they unavoidably verbalize, engage with and draw together a score of ideas, sensed embodied experiences, emotions, material objects and more." In other words, the interview can be an emplaced way of knowing through the shared experiences of the interviewer and interviewee. The interviewer becomes an apprentice to the sensory experience of others.

Finally, elicitation techniques are often used in interviews, in which participants are asked to focus on a topic or sensory modality with specific instructions (e.g., "listen to the rhythm of the frogs") and to document their experiences, meanings, attitudes, and beliefs. One common sonic elicitation technique, used by Feld (2001) in his New Guinea work, is to record natural sounds, play them back to participants, and have conversations that allow the researcher to create a body of sensate information to explore and analyze.

9.9.5. Soundwalks

The *soundwalk* is a special form of data collection first used by acoustic ecologists (Shafer, Truax and Westerkamp) in the 1970s as part of the World Soundscape Project. It is an "excursion" with the purpose of listening to the environment, and for some, an intense introduction to experiencing uncompromised listening (Westerkamp 2010). A more contemporary version of the soundwalk as a research approach is to conduct an interview during the walk (see Adams et al. 2008 and Vokes 2021 for a review of these methods). The observer/interviewer may ask questions, take photos, or make recordings along the path as well (Semidor 2006). Some researchers have focused on explicitly connecting the landscape to the soundscape (e.g., Berglund and Nilsson 2006) so that the visual and sonic experiences can be integrated and understood. Berglund and Nilsson (2006) also used a set of descriptors (i.e., statements) that participants could select from to describe soundscapes and landscapes along a predefined soundwalk path. A follow-up interview focused on the experience was conducted to document meanings, attitudes, and beliefs. In some instances, researchers have soundwalk participants complete a questionnaire after the walk.

9.9.6. Surveys and Questionnaires

Surveys are approaches to collecting information, usually from a large sample of individuals, in the form of questionnaires or interviews (sometimes called *survey instruments*). Questionnaires can be issued orally during interviews or over the phone, in writing through mail or email surveys, or built into mobile apps. Surveys are "well suited for descriptive studies, but can also be used to explore aspects of a situation, or to seek explanations and provide data for testing hypotheses"; they are "an approach and not a method" (Kelley et al. 2003, 261). Surveys sample the larger population, so sample size and sampling approaches should attempt to avoid bias. For example, researchers need to ensure there is an even distribution of surveys across demographic categories, such as age, gender, and wealth, that are important to the understanding of the topic. Questions on surveys can be open-ended (allowing for long answers) or closed-ended (only a few options can be selected); many surveys are semi-structured, which means that they have a combination of open-ended and closed-ended questions. Survey experts will test a pilot questionnaire on a small sample of the population of interest to ensure that the questions are understood and the responses are useful.

Many survey questions are set up to allow for *Likert-scale* answers. Surveys of this type provide a question or statement that the participant responds to by choosing a ranked option from a rating scale (Table 9.2). Likert scales are often designed to have symmetry around a neutral response by providing an odd number of options (so that the neutral response is the center point). A common set of Likert-scale options, with their values, is *strongly disagree* (–2), *disagree* (–1), *neither agree or disagree* (0), *agree* (1), or *strongly agree* (2). An even number of options is also possible, generally with no neutral option available.

Over the past few decades, many survey procedures have been improved,

Table 9.2 Some common Likert-scale options for hypothetical survey questions related to sound and soundscapes

This soundscape is a reflection of me. (seven-point scale)	How satisfied are you with the sounds of your community? (five-point scale)	The sounds from this bird in this recording are beautiful. (two-point scale)	How often do you hear these kinds of soundscapes? (four-point scale)
Very strongly agree			
Strongly agree	Highly satisfied		Never
Agree	Satisfied	Agree	Almost never
Not relevant	Neutral	Disagree	Sometimes
Disagree	Dissatisfied		All the time
Strongly disagree	Highly dissatisfied		
Very strongly disagree			

and best practices have been well described. For example, one challenge with sending emails asking people to complete online or mobile app–based surveys is the need for a good response rate. Dillman (2000) has developed the *tailored design method* to address this challenge; it outlines best practices for formatting and wording questions, approaches that make surveys and correspondence more personal, and procedures that create the largest response rates (e.g., a four-wave approach, in which surveys are sent out to those who have not responded up to four times, each time with a specific note). Other best practices include comparing those participants who responded with those who did not so as to understand the potential bias that might occur in the survey. Survey response rates should always be reported in research publications. Many professional organizations have now published guidelines on the ethical conduct of survey research (Kelley et al. 2003), including the American Psychological Association.

There have been several fields, particularly psychology, in which standard perception and human state *psychological scales* have been developed and used to address a broad range of topics. One example is the *nature relatedness* (NR) scales of Nisbet and Zelenski (2013). Several forms of these scales have been created, used, and compared. For example, Nisbet and Zelenski (2009) present a variety of statements about nature relatedness that survey respondents then score on a seven-point Likert scale. Their NR-21, a twenty-one-item questionnaire, presents statements such as, "I presently recognize and appreciate the intelligence of other living organisms," which participants score in terms of how they feel at present. A large number of responses are then used to assess patterns in the data, perhaps as they relate to the context (e.g., where a respondent is located) or to the respondent's lived experiences (e.g., comparing urban and rural groups). A short form, NR-6 (Nisbet and Zelenski 2013), has been shown to be as informative as the lengthy NR-21 form. Other psychological scales that have been used include the New Environmental Paradigm (NEP) scale (Stern and Dietz 1995), which measures attitudes toward environmental protection; the *Connectedness to Nature Scale* (CNS; Mayer and Franz 2004), which measures a person's opinions of their relatedness to nature; the Inclusion of Nature in Self (INS) scale (Schultz 2002), a unique survey that involves using rings labeled "Me" and "Nature" that are arranged as Venn diagrams by respondents; the New Ecological Consciousness Scale (Ellis and Thompson 1997), and several happiness indicators, scales, or standard statements, including the Positive Affect Negative Affect Schedule (PANAS; Watson et al. 1988); a fifty-four-statement Psychological Well-Being Inventory developed by Ryff (1989); and the Satisfaction with Life Scale (SWLS; Diener et al. 1985), a measure of subjective well-being. Many of the scales are set up to explore three main psychological dimensions: self-identity, affect (i.e., emotions), and behavior (i.e., what kinds of actions have been, or will be, taken). Many of the studies that use these scales employ a form of correlation analysis called exploratory factor analysis (R- and Q-mode) or a more hypothesis-driven form called confirmatory factor analysis.

9.9.7. Participant Observation

The *participant observation* method "explicitly incorporates the collection and recording of information gained from participating in a social setting and observing what is happening in the setting explicitly in the analysis" (Bernard 2011, 252). It is often regarded as originating from anthropology in the early twentieth century (Malinowski 1922; see Firth 1985 for interpretations of Malinowski's work). Contemporary practices have evolved into a set of observational and documenting techniques, theory of knowledge discovery, and models of participant observation that, for example, identify different roles for researchers and participants (Zahle 2012). For some, participant observation is a means by which a researcher attempts to understand "practical knowledge" (Zahle 2012); for others (Bernard 2011), the approach serves as a means to learn the tacit aspects of culture that are not directly observable. Observers (i.e., researchers) live through daily activities with participants and gain familiarity with their "lived experiences." Unconscious behaviors such as how to sit, how to stand, or where to walk are learned by doing through shared activities in which the participants provide feedback on the actions of the researcher. Best practices also require that researchers be unobtrusive (i.e., not be the focus of attention), embedded (i.e., live with the individuals or communities being observed), and subtle (e.g., not request input on their actions).

There are many ways to observe. Emerson et al. (2011) recognize five kinds of observations: (1) initial impressions, (2) personal opinions on what is significant and unique, (3) observations of how others react to the observer's actions, (4) observations of how routine actions are organized and enacted, and (5) documentation of patterns of events, actions, and settings that begin to suggest emergent patterns in how the lived experience exists within the context and how it varies (e.g., according to environmental conditions). Techniques for documenting participant observations, broadly described as reflexive reflection, have been well researched and documented, and best practices have been outlined in guides such as *Writing Ethnographic Fieldnotes* (Emerson et al. 2011).

There have been several different perspectives on how sensory information should be collected during the course of participant observation. The first considers all senses, exploring how each is involved in the perception of the environment by the corporal self; this perspective takes a very phenomenological approach and often generalizes across all populations (Serres 1985, as described in Howes and Classen 2013). A second perspective, espoused by Pink (2015), also uses a phenomenological approach, but does so with the assumption that the researcher needs to become an objective participant observer of the subject, who is likely to have a different worldview of the senses. In Pink's view, sensory participant observation is a form of auto-ethnography of the senses that is practiced "by visiting other people's sensory environments" (Pink 2015, 101); that is, by sharing an environment with the research subject(s). Another important aspect of sensory participant observation is to explore experiences through the lens of the researcher's own past lived experiences. For example, a researcher from a Western megacity can generate knowledge about how an indigenous community uses natural sounds for understanding environmental change by addressing how these sounds may be familiar in other contexts, and what other meanings can be inferred by their presence or absence in the non-megacity landscape.

9.9.8. Mapping Cultural Places

Cultural mapping is the identification and recording of the spaces that are cultural resources for communities and people. Maps can be made (Duxbury et al. 2015) of culturally relevant places (e.g., churches, synagogues), stewardship boundaries (i.e., areas that a group or groups occupy permanently or temporarily where they affect the composition of the landscape), the social uses of an area (e.g., learning, walking), and built areas and their uses. Recently, efforts have been made to introduce cultural intangibles into traditional maps (Longley and Duxbury 2016) by including symbolism (e.g., religious, historical), locations of information (e.g., stories or sources of environmental status or change), and places that create identities for individuals or groups, all of which have value (e.g., recreational), provide sense of place, and affect quality of life. Post and Pijanowski (2018) argue that culturally important sounds can be mapped and associated with meanings so that a better understanding of landscape, soundscape, and human perceptions can be achieved. Cultural mapping is often a participatory exercise; in other words, it is a process that involves individuals from a community that embodies the knowledge about cultural meanings.

9.9.9. Ethical Considerations for Qualitative Research

A number of ethical considerations are important in qualitative research. In the social sciences, many ethical considerations have now become part of a formal process in which research protocols are reviewed and approved by an Institutional Review Board (IRB).

Informed Consent

One of the most important considerations in the IRB review is *informed consent*: the principle that research subjects should understand the risks and benefits of, and alternatives to, the research activity in which they agree to participate. Informed consent requires a description of, and an assessment of the subjects' understanding of, these risks, benefits, and alternatives. The review also covers the way in which information about a subject's identity is recorded, stored, processed, and reported (most approaches detach a person's identity from the data). Informed consent is thus an agreement between researcher and research subject. The agreement is often presented to the subject orally and in written form, and the subject is often asked to sign a form that is kept by the researcher, who is thereafter guided by the terms set forth in the protocols approved by the IRB.

Trust

A researcher who works with human subjects must recognize that to ensure that unbiased information is being presented to them, an investment in the kinds of activities that build trust is important. Trust building requires researchers to spend time with participants in activities other than the research, to return to research sites several times during a project, and to recognize the local norms and practices that are important to individuals and communities. Trust is also built when researchers return to research sites and provide results of the study to the community that contributed to the research.

Extractive versus Proactive Approaches

Researchers that come to a community and collect data for use in data analysis, but never report their findings to the community that was the subject of the research, are considered to employ an *extractive* approach to data gathering. Communities that are the subject of research are often very interested in knowing how their investment of time and effort in the research project was used, and thus are interested in the results of the research. If a researcher wants to return for follow-up work, but has not shared the previous results, the community may not be as willing to continue to invest time and effort in that researcher's work, or that of any future researchers who may approach them. Researchers should avoid being extractive and find ways to return their research results to the community in a form that is usable by its members.

9.10. Mixed Methods

Mixed-methods research is an approach to data collection, analysis, and generation of discoveries in which both quantitative and qualitative methods are integrated into one study (Creswell 1999). Mixed-methods research has a rich history in sociocultural sciences, education, and medicine.

Integrating the ecological and sociocultural components and interactions of a soundscape most often requires the use of mixed-methods research (Tashakkori et al. 1998; Greene 2007; Mayring 2007; Denscombe

2008; Snelson 2016). Mixed-methods research, by definition, requires the integration of at least one qualitative and at least one quantitative method. Advances in mixed-methods applications in education and medical research over the past thirty years have generated a set of broad mixed-method typologies. Mixed-methods research designs may be *fixed, emergent,* or "blended." Fixed designs use approaches predetermined at the start of a study, whereas emergent designs are developed during the discovery process (Creswell and Plano Clark 2018). Mixed-methods research designs can also be interactive (or *convergent;* i.e., quantitative and qualitative data are collected simultaneously) or based on a typological ordering (or *sequential;* i.e., sequential data collection "strands" are planned). Sequential designs (Fig. 9.9A) are considered either *exploratory* or *explanatory* depending on which strand of research, qualitative or quantitative, drives the study and is thus conducted first. Convergent studies (Fig. 9.9B) can be conducted with the strands in parallel, in full isolation from each other, or with the strands fully integrated (i.e., at the same time and place and with the same subjects).

What is important to consider with mixed-methods research is that there is probably no one correct way to arrange the methods; that the integration of qualitative and quantitative data is very challenging; that qualitative research can be used to help frame the design and analysis of quantitative research, and vice versa; and that in most cases, this kind of work involves researchers who are experts in each qualitative and quantitative research method, as well as those who have done mixed-methods research.

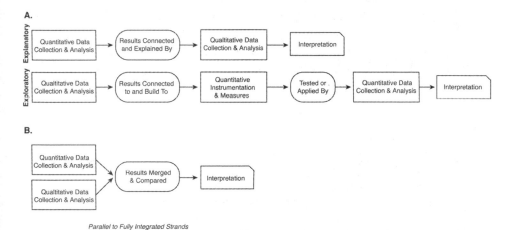

Figure 9.9 Mixed-methods data collection and analysis pathways. (Modified from Creswell 2002.)

Summary

Soundscape ecologists have numerous ways to measure the soundscape and its associated landscape, in terms of both the physical acoustic environment and how people perceive it. Recent advances in other fields in which quantitative and qualitative information needs to be integrated have great potential to serve as models for ways that soundscape ecologists might integrate the diverse data that can be collected in any study.

Discussion Questions

1. Design a study that would examine how acoustic indices could be used for a rapid assessment of the effects of selective logging in an area, given that the area also supports the presence of a rare bird species. Consider for this study the use of AIs, annotated recordings, and measurements of the vegetation in this forested landscape.

2. A city planner would like you to determine whether an existing park is meeting its intended purpose of providing restorative affect to residents of the city. Using a set of PARs and a small team of social science experts (who have IRB approval for their informed consent protocols), determine how you might conduct a study of this park. What data would you compile, and how might they be integrated?

3. Develop a survey instrument that would allow you to play back sound recordings from four biomes—a Paleotropical rainforest, a grassland, a desert, and a temperate wetland—to university students in a wildlife biology class and assess their perceptions of these places. The survey could explore any one of the perception dimensions presented in this chapter, or examine a combination of them. Describe how you would compile the information and then use it to compare the students' perceptions of the different biomes.

Further Reading

Sueur, Jérôme, Almo Farina, Amandine Gasc, Nadia Pieretti, and Sandrine Pavoine. "Acoustic Indices for Biodiversity Assessment and Landscape Investigation." *Acta Acustica united with Acustica* 100, no. 4 (2014): 772–81.

Creswell, John W., and Vicki L. Plano Clark. *Designing and Conducting Mixed Methods Research*. Sage, 2017.

Kelley, Kate, Belinda Clark, Vivienne Brown, and John Sitzia. "Good Practice in the Conduct and Reporting of Survey Research." *International Journal for Quality in Health Care* 15, no. 3 (2003): 261–66.

10

Analyzing Soundscape Data

OVERVIEW. Central to the analysis of soundscapes is the need to use approaches employed by data scientists. These approaches, broadly termed *big data analytics*, use tools that build on traditional statistical approaches but also include machine learning. The fundamental principles of each of the techniques along the data analysis pipeline are presented here. Specific details of these quantitative analytical approaches are presented in excellent textbooks on the topic of data mining, such as *Introduction to Data Mining* (Tan et al. 2016), and *Data Mining: Concepts, Models, Methods and Algorithms* (Kantardzic 2011). Qualitative approaches used by sociocultural researchers are also summarized here, with references provided for those interested in exploring these kinds of analyses further. This chapter is by no means exhaustive, but it does present a diversity of analytical approaches that can be used by soundscape ecologists and those with whom they collaborate in other disciplines.

KEYWORDS: accuracy metrics, association analysis, classification trees, clustering, sampling, textual analysis, time series

10.1. Data Cleansing

Data cleansing, or data cleaning, is the process of exploring data for "possible problems [to] endeavor to correct the errors" (Maletic and Marcus 2000). Doing this by hand is not practical, or even feasible, when data volumes are large and data are heterogeneous. Data cleansing invariably involves removing duplicate records and records that are not relevant to an analysis or to maintaining the data archive, identifying structural errors such as improper labels, flagging or coding missing data, and validating data. Data validation requires researchers to identify values that do not make sense or are obvious errors (e.g., if one channel of a PAR has malfunctioned). The ultimate goal of data cleansing is to produce data of the best quality possible that will support knowledge discovery in databases. Data cleansing should aim to support the five characteristics of *data integrity* in the following ways (Tableau 2021):

Validity: by ensuring that values fall within the expected range for the data type

Accuracy: by ensuring that values are close to true values

Completeness: by knowing if all required data are stored within the database

Consistency: by ensuring that the data are consistent across the same database and, if related to others, that the linkages are consistent as well

Uniformity: by ensuring that the data are consistent in units of measure across all variables and databases

10.2. Data Characterization

There are several important considerations when researchers begin to explore big data. Data characterization focuses mainly on central tendency (e.g., mean of values) and dispersion (e.g., range of values) using statistical measures and visualizations.

10.2.1. Univariate, Bivariate, and Multivariate Analyses

Data can be explored by focusing on one variable at a time (called univariate analysis), on pairs of variables (bivariate analysis), or on groups of three or more variables (multivariate analysis). Bivariate and multivariate analyses are guided by how data types are combined. For example, bivariate analysis could be conducted on the following pairings: two numerical variables, two categorical variables, and one categorical and one numerical variable.

Univariate Analysis

Common univariate numerical analyses include calculating metrics of central tendency, dispersion, and the shapes of distributions. Measures of *central tendency* include the mean, median, and mode. The *mean* can be calculated as

$$\bar{x} = \frac{\sum_{i=1}^{n} x_i}{n},$$

which is the sum of all values divided by the number of values. This form of the mean is the *arithmetic mean*. Another form, the *geometric mean*, is used by ecologists and social scientists in situations where there is a very large range across a small number of values, or where values are skewed, with many small or many large values. The geometric mean is the *n*th root of the product of all values, as summarized by the equation

$$\left(\prod_{i=1}^{n} a_i\right)^{\frac{1}{n}} = \sqrt[n]{a_1 \times a_2 \times a_3 \times ... \times a_n},$$

where the symbol $\prod$ (pi) is the cross product, or the product of all values *a* that range from 1 to *n* (i.e., all values, moving through the list of *i* items).

The *median* is the value in the dataset that divides it into two equal parts. In other words, it is the value where half of the values are below it and the other half are above it. Finally, the *mode* is the most common value in the dataset.

The measures of *dispersion* used in statistics include the standard deviation, standard error, variance, and range. The *standard deviation* (SD) is the average difference of all values from the mean. For the SD, the mean, $\bar{x}$, is calculated, and then each value, x_i, is subtracted from the mean. Each difference is then squared, and all the squared differences are summed. This last result is called the *sums of squares* (SS). The SS is then divided by the total number of values (i.e., the sample size, *N*) and then its square root is calculated:

$$\sigma = \sqrt{\sum_{i=1}^{N} \frac{(x_i - \bar{x})^2}{N}},$$

where σ (the Greek letter sigma) is the symbol for SD.

The *standard error* (SE) is the standard deviation adjusted by the sample size. The SE is calculated as

$$\sigma_{\bar{x}} = \frac{\sigma}{\sqrt{n}}.$$

Variance is closely related to SD. Variance is calculated just as SD is, but the final value is not transformed by the square root, and the denominator is a bit different. Variance is designated as σ^2, and its equation is

$$\sigma^2 = \sum\nolimits_{i=1}^{N} \frac{(x_i - \mu)^2}{N},$$

Range is also reported, although not commonly. Range is the difference between the smallest and the largest value for a variable.

Data can also be transformed using means and variances. One common form of data transformation is to calculate a *z-score*, also called the standard score, for each value. A z-score is the proportional difference of a value from the mean. A z-score of 0 means that the value is the same as the mean. A z-score of 1 means that the value is exactly one SD from the mean. Z-scores can be negative (smaller than the mean) or positive (larger than the mean). The equation for a z-score is

$$Z = \frac{x - \bar{x}}{\sigma},$$

where Z is the z-score for the value x, $\bar{x}$ is the mean, and σ is the standard deviation. Z-scores are useful when researchers need to compare values from two distributions that have different means and/or standard errors. For example, if a soundscape ecologist wanted to compare values from two acoustic features recorded at one site, Z-scores could be calculated for each value of each acoustic feature in the recordings.

Finally, the shape of a plot of the distribution of values is an important consideration for some statistical analyses. A histogram plot of a distribution, like that in Figure 10.1A, can help researchers to visualize not only the central tendency and dispersion of the values, but also the shape of the distribution, which can be symmetrical about the mean, narrow, or wide. Measures of symmetry and dispersion are commonly employed in data exploration. *Skewness* is a measure of symmetry about the mean. Distributions that have more values to the right have positive skewness, and those that have more values to the left have negative skewness. Those that have a "bell-shaped" curve have no skewness. *Kurtosis* is a measure of the narrowness or broadness of the distribution. Positive kurtotic data have a narrow bell-shaped curve with many values close to the mean. Negative kurtotic data have a nearly flat distribution across all values, although there are still more values close to the mean than values far from the mean.

Univariate analysis of categorical and binary data is rather limited in comparison to its usefulness for analyzing numerical data. A majority of univariate analyses focus on creating histograms of counts, percentages, or proportions (count of category *i* divided by sample size for each category).

Bivariate Analysis

Examining how two variables relate to each other is a very common data exploration approach. One of the most common methods of doing so is the *scatterplot* (Fig. 10.1B). Let us consider a scatterplot of the relationship between ACI and ADI for 273 recordings made in Borneo at the same location. The plot shows several areas where ACI has a positive (upward) slope for many low ADI values, but for the most part, ACI is not strongly related to ADI for these recordings.

If more than one pairwise comparison is needed, a set of scatterplots can be made that aligns the variables in a way that allows researchers to scan for patterns across a diverse set of variables. As there are often relationships between each variable pair, these relationships can be quantified using simple statistics. The most common of these statistics is *covariance*, a measure of the direction of the relationship between two variables. It is calculated much like variance, but for two variables:

$$\text{cov}(X, Y) = \frac{\sum_{i=1}^{N} (X_i - \overline{X}) (Y_i - \overline{Y})}{N},$$

where X and Y are the two variables, $\overline{X}$ is the average of observations of variable X, and $\overline{Y}$ is the average of observations of variable Y. Note that if there are more than two variables, a set of pairwise covariance values can be calculated. The range of possible values for covariance is $-\infty$ to $+\infty$.

Another common measure of the relationship between variables is *correlation*, which measures the strength and direction of one variable's relationship with a second. The formula is based on the sums of squares model, described above, but is extended to a second variable. Note too that the denominator differs from that for covariance:

$$r = \frac{\sum (X_i - \overline{X}) (Y_i - \overline{Y})}{\sqrt{\sum (X_i - \overline{X})^2 \sum (Y_i - \overline{Y})^2}},$$

where r is the correlation coefficient. Values for r range from -1.0 (perfect negative correlation; in other words, when one variable goes up, the other goes down) and $+1.0$ (perfect positive correlation; the two variables vary together). A value of 0.0 means that there is no relationship between the two variables. There are several forms of correlations, and researchers often consult statistics books to determine if one algorithm is more suited to their datasets than others.

One bivariate analysis that builds on correlation is *regression*, which is also used for prediction. There are several forms of regression. The most popular is the linear regression model, which determines whether y (called the response variable or independent variable) varies with x (called the scalar or dependent variable) in a predictable way. Linear regression calculates the parameters of a line (its slope, b, and its y-intercept, a) that minimizes the sums of squares:

$$y = a + bx,$$

where

$$b = \frac{\sum_{i=1}^{N} X_i Y_i - N\overline{X}\,\overline{Y}}{\sum_{i=1}^{N} X_i^2 - N\overline{X}^2},$$

and

$$a = \overline{Y} - b\overline{X}.$$

The linear regression must meet two criteria to be valid: each of the variables must be *normally distributed* (or close to it), and variances cannot be *heteroskedastic*, a condition in which the variances are not equal in magnitude across variables.

10.2.2. Visualizing Trends

A common visualization technique for examining the values within each variable is the *box plot* (Fig. 10.1C), also called a "box-and-whisker plot." The standard box plot displays several elements:

Median: The value in the middle of the distribution of all values of the variable.

First quartile (Q_1): The first 25%, or first quarter, of the values as they are sorted from lowest to highest. This term also refers to the median value between the lowest value and the overall median value of the variable.

Third quartile (Q_3): The first 75%, or third quarter, of the values as they are sorted from lowest to highest. This term also refers to the median value between the overall median value and the largest value of the variable.

Interquartile range (IQR): The difference between the first and third quartile values.

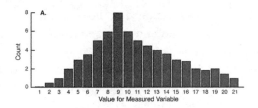

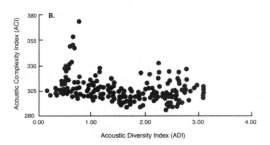

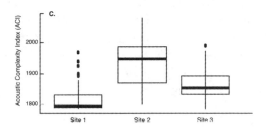

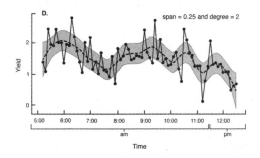

Figure 10.1 Common approaches to visualizing data. (A) Histogram showing the distribution of values for one variable that has values ranging between 1 and 21. (B) Scatterplot for two variables (ACI and ADI). (C) Box plots for one variable, ACI, at three sites. (D) LOESS curve for ADI for recordings made every 5 minutes.

A box plot uses these values in the following way. First, a box is constructed such that the bottom and the top represent the first and third quartiles, respectively. A horizontal line displays the median value. Two "whiskers" extend to a distance of 1.5 times the IQR above and below the median value. Finally, any values that fall outside the 1.5 IQR range of the whiskers are plotted as separate values (i.e., as dots or asterisks) and are noted as outliers.

Data scientists use several forms of data visualization based on linear regression techniques. For example, *LOESS curves* (Cleveland 1979; Cleveland and Devlin 1988) are commonly used to visualize data when the *x*-axis represents time, in hours, days, weeks, months, or years, and there are sites that the scientists wish to compare (e.g., inside and outside a marine protected area). LOESS (*Locally Estimated Scatterplot Smoothing*) is a method of visualizing means and some measure of variances for small sections of the *x*-axis that are fit to a linear regression model; these are then summed along that axis. Typically, a LOESS curve presents data as a scatterplot (all *x*,*y* values are shown as dots), a curve that represents a local mean (i.e., the mean for a subset of values in a defined range along the *x*-axis), and a shaded ribbon around the curve of the local mean that shows some measure of variance (e.g., standard error). Another common version of LOESS is the *LOWESS curve*, which uses a weighted value approach to plotting data. A LOESS curve for a hypothetical variable measured over a 7-hour period with recordings made every 5 minutes is shown in Figure 10.1D. A similar approach is to use polynomial equations, rather than linear regression models, to fit data across the *x*-axis. One model for doing this, called the *adaptive Savitzky-Golay filter*, has been used by remote-sensing experts for decades (Chen et al. 2004).

10.3. Reduction of Dimensionality

When researchers are working with big data, each record could contain dozens to even thousands of fields (i.e., variables). Many of these variables can strongly covary, and some might even be similarly constructed. For example, in landscape ecology, consider two landscape metrics: number of patches and patch density. These two metrics are likely to be highly correlated, as they are measuring the same property but with slightly different equations. In this case, a strong correlation is simply an artifact and does not provide independent information. Researchers in soundscape ecology and landscape ecology have discovered that many of the metrics they have developed to measure spatial patterns and acoustic patterns, respectively, rely on similar assumptions, or their derivations use similar algorithms. For example, ADI and NDSI might be expected to provide similar values in areas where there is little to no highway traffic. An analysis that contains many variables that are correlated could potentially be biased. In addition, mining such data would be very time consuming, as enormous numbers of pairwise combinations would occur if there were dozens to thousands of variables in the database (e.g., n-dimensional data would generate $n!/(n - 2)! \times 2$ different combinations). Thus, a common practice of data scientists is to use specially designed algorithms to reduce the dimensionality of their data, using approaches that include finding mathematical solutions to all the covariance matrices (which are similar in structure to error matrices), which are then used to combine similar variables into one, albeit abstract, variable. The dimensionality of a dataset can be reduced by means of linear or nonlinear dimensionality reduction models. Understanding both types of models requires a knowledge of linear algebra and how vectors, scalars, and matrices are multiplied, subtracted, and added. However, non-mathematical descriptions of each type follow.

10.3.1 Linear Dimensionality Reduction Models

Numerous linear models have been developed in traditional statistics to reduce dimensionality in data. The best known is arguably *principal components analysis* (PCA; Hotelling 1933). The aim of PCA is to generate a set of new dimensions using the linear transformation of data that explains the most variance in the database across all variables. The result is several (generally three or four) summary variables (i.e., components), which are easier to explore than, say, dozens to thousands. Typically, researchers infer what each of the principal components means as they examine how each variable "loads" to each component. Components are then examined as scatterplots in two-dimensional or three-dimensional space. The first, second, and third principal components often explain most of the variance within the data, although there are many rules for determining how many components to use (see Sokal and Rohlf 1987). For example, one such rule is to select the number of principal components that explain, in sum, at least 65% of the total variance. Axelsson et al. (2010) used PCA to reduce over 120 human perception responses to urban soundscapes so that three key components emerged (see Section 13.3). Approaches similar to PCA include canonical components analysis (CCA), linear discriminant analysis (LDA), and independent components analysis (ICA).

10.3.2 Nonlinear Dimensionality Reduction Models

Because many linear dimensionality reduction models lack the ability to detect nonlinear trends in data, many researchers have developed nonlinear dimensionality reduction models (NDRMs) for use in data mining (see Lee and Verleysen 2007 for an excellent review of issues and models). A method used by many soundscape ecologists is *non-metric dimensionality scaling* (NDMS), in which ordination is based on a user-specified dissimilarity measure. One of the most widely used NDRMs in data mining is *stochastic neighbor embedding* (SNE; Hinton and Roweis 2002) and the associated visualization approach, called *t-SNE* (van der Maaten and Hinton 2008). SNE and t-SNE use a variety of algorithms commonly used in data science (e.g., transformation of data using Gaussian models, fitting data so that a minimization of Kullback-Leibler divergences [see Section 10.6 below] across all data points is achieved) to ensure that all data points in the reduced dimensionality are projected in two-dimensional and three-dimensional space so that they are not spatially aggregated. In other words, transformation of the data attempts to use as much of the two-dimensional and three-dimensional space as possible to plot out values as scatterplots. Many NDRM techniques, such as SNE, also employ a generalized technique used in data mining, called *manifolds*. These techniques represent data as a function of a unique, specific shape (a popular one is the three-dimensional "Swiss roll") projected into three-dimensional space (some use two-dimensional space). Scikit-learn contains several NDRMs, including t-SNE, Isomap (an extension of MDS; Tenebauam et al. 2000), spectral embedding (Luo et al. 2003), and local tangent space alignment (LTSA; Zhang et al. 2007).

10.4. Data Sampling

There are many reasons why researchers may need to create a subsample of data from a larger, complete database. One of the most common is the need to use sensory evaluation methods (SEMs; see Section 9.5) to label data and to build tools to classify sound sources using machine learning. It is common practice for researchers to use a subsample of their data to develop a model, and then to test the accuracy of the model using another subsample. Often at issue is what the size of the subsample should be, in terms of the number of records, or whether it should represent a certain percentage of the total database. Having too small a subsample may mean that the distribution of the entire database is not well represented, and if the sampling is not done properly, the subsample may be biased and thus may not represent the variability that exists in the complete database. This section considers five major types of data sampling (often called *data partitioning*).

10.4.1. Simple Random Sampling

In simple random sampling, records from a database are sampled at random in such a way that every record has an equal chance of being selected. Random sampling can occur with or without *replacement*. With replacement, records can be selected more than once. Without replacement, records can be selected only once (i.e., their candidacy for selection ends when they have been selected once). Selection without replacement is used if there are no obvious subcategories of data (i.e., if there is variability across space and time, this method is not preferred).

10.4.2. Stratified Random Sampling

When the entire database can be divided into groups, or strata, because values in the database are obviously not homogeneous, then selection within and across those strata is performed using random sampling. For example, if a soundscape ecologist were to conduct a study across five ecosystems in which six PARs in each ecosystem had recorded for two months on the same days, then the researcher might consider stratifying the sampling with even distributions across the five ecosystems, across the six PARs, and then across diel periods (morning, afternoon, evening, night), so that each of the $5 \times 6 \times 4$ strata would be represented by an equal number of samples. If there was more than one season within the two-month study period, then it might even be appropriate to stratify across seasons as well. As more strata are added, however, the number of possible replicates available to sample within each stratum becomes very small, as these data may also be needed for validating any model that is developed. Stratified random sampling can occur in two ways: the same number of samples can be randomly selected from each of the n strata, or samples can be randomly selected from each stratum so that the proportion of each stratum selected is the same.

10.4.3. Cluster Sampling

If data group together into clusters—that is, if data points are closely aligned when plotted in two- or three-dimensional space—data scientists can identify these clusters and then randomly select a cluster, so that all the data within that cluster are used or become candidates for further data collection or labeling.

10.4.4. Systematic Sampling

Systematic sampling is accomplished by randomly selecting a "starting point" (e.g., time, record number in the database) and then selecting from the remaining records at fixed intervals from this point. Soundscape ecologists can use this approach for sampling across recordings; for example, a random starting point along a 10-minute recording could be chosen using a random number generator, and this point could be used for all 10-minute recordings to select a 1-minute portion to label.

10.4.5. *k*-Fold Sampling

The *k*-fold sampling method is commonly implemented by data scientists. This technique uses a value of *k* that is greater than 2 and less than $n - 1$, where *k* is the number of sampling groups and *n* is the size of the data. A set of *k* samples is used to train a model, and $k - 1$ samples are used for testing. A variety of *k*-fold sampling algorithms exist (see Wong 2015 for a review). A special case of *k*-fold sampling is the leave one out (L-O-O) method, also known as jackknife sampling.

10.5. Data Mining

The four core approaches to data mining that should be considered in data exploration and modeling of soundscapes—classification, clustering, association, and time series analysis—are summarized in this section. Let us consider each, as well as a few algorithms that are commonly used in these data mining tasks.

10.5.1. Classification

The classification approach to data mining uses a variety of models—which can be statistical, machine-learning based, or rule based—to place each value or set of values in one of several classes. This process is called *classification*, and the tools developed for it are called *classifiers*. Some classification models are considered *supervised*; in these models, a set of values is labeled (e.g., a signal in a recording is assigned to species A), and this relationship is then used to assign that label to other, unlabeled values using a positive (i.e., it is species A) or negative (i.e., it is not species A) labeling rule. In most cases, the model provides a predictive output in which the target variable is assigned a number between 0.0 and 1.0. Values greater than 0.5 are then assigned a value of 1 (e.g., the model predicts them to be species A), and values less than 0.5 are assigned a value of 0 (e.g., the model predicts them not to be species A). Supervised models are considered predictive models, and they use a statistical, machine-learning or rule-based algorithm to assign a label to an unlabeled value (Tan et al. 2016).

Unsupervised classification is the process of creating a classification model that sorts values into classes that are then described by the researcher. Unsupervised models are considered descriptive models. They are useful to help researchers understand relationships across all variables in a database and how they might relate to one key target variable. For example, an unsupervised classification model could take all acoustic index values from recordings and attempt to associate ranges of values with locations inside and outside a protected area. Or, a group of acoustic features could be used to identify a class of sonic species, such as insects, and illustrate how its members might differ from non-insect sound sources.

Many kinds of classification models are used in data mining (Chen et al. 2004). They include decision trees, the *k*-nearest neighbor (*k*-NN) algorithm, neural networks, and support vector machines (SVMs). Each of these classification models comprises several to dozens of specific algorithms that have been developed by data scientists. A few will be covered in some depth here, and others will be mentioned with key references provided.

Decision Trees

A decision tree takes data values for each variable in the dataset and attempts to find threshold values in a database field to create two or more subgroups of other fields, ultimately placing values from the target field into desired groups. The model is set up so that there is one independent variable and a host of dependent variables. The tree is composed of a hierarchical arrangement of these thresholds that are designated as nodes, which can be top, internal, or terminal (or "leaf") nodes in the decision tree. The most common classification models include classification and regression trees (CART; Breiman 2017), which use what are called ID3 and C4.5 algorithms (Quinlan 1986). Other classification models include random forests, boosted regression trees, and multi-adaptive regression splines (MARS). Figure 10.2A is an example of a regression tree in which four land use spatial indices are used (i.e., amount of urban land use in a buffer around a PAR, amount of agriculture in the same buffer, an index of fragmentation, and an index of the length of roads). Many CART software packages provide visuals such as Figure 10.2A that describe the splits in each independent variable as well as the independent variable summary statistics (mean, standard error, and sample size). These trees commonly have several terminal nodes (labeled TNODE), as shown in the figure. The terminal nodes provide summary statistics for the dependent variables—in this illustration, the value of ADI. The user is left to interpret the outcomes at each terminal node. Here, the researcher has decided, on the basis of ADI values, that the 175 PAR locations include 42 that represent "good" soundscapes (TNode 1), 36 + 29 = 65 soundscapes that are marginal (TNodes 2 & 3), 30 + 18 = 48 soundscapes that are impaired (TNodes 4 & 5) and 20 (TNode 6) that are very impaired (i.e., probably representing ADI values where one frequency band from road noise is prominent and few other sounds are present). One benefit of decision trees is that they can be used on a database composed of just about any type of data: continuous, binary, categorical, or nominal. Classification models can work with data that have labels, too.

The k-Nearest Neighbor Algorithm

The k-nearest neighbor (k-NN) algorithm is another classification routine that is based on very simple rules that calculate distances in variable space between known (i.e., those that have been placed in a group already) and unknown values. To understand how k-NN works, let us consider a plot of two acoustic feature values (along the x- and y-axes, respectively) and the calls of two bird species, which are labeled A and B (Fig. 10.2B). Let us consider a new call, which is unknown, labeled 1 in the key. The k-NN algorithm calculates the distance between 1 and the k nearest neighbors. If k = 3, then the three nearest neighbors of unknown sound 1 on the plot are all members of B, so it classifies sound 1 as coming from sound source B. Using the same rules and nearest neighbors, unknown sound 2 is considered next. Its three nearest neighbors are 1 A and 2 Bs. Based on a majority rule, this sound, too, is classified as B. However, if a nearest neighbor rule of 4 is used, sound 2 would have 2 As and 2 Bs as its nearest neighbors, and the algorithm would then apply other rules (e.g., mean distance) to break the tie and assign the sound to A.

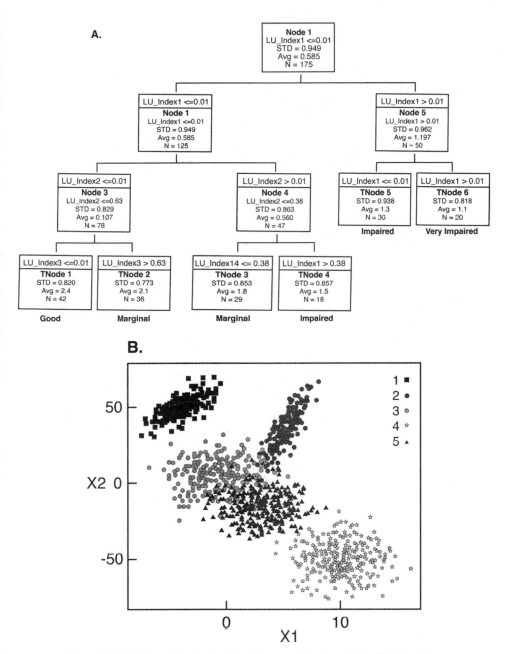

Figure 10.2 Models used in the classification approach to data mining. (A) A decision tree. (B) The *k*-NN algorithm. (C) Architecture of an artificial neural network showing how acoustic features can be used as inputs to a predictive classifier model for a call of bird species 1. This diagram also illustrates the feed-forward nature of weights and the adjustments made after iterative errors are compared. (D) A support vector machine.

C.

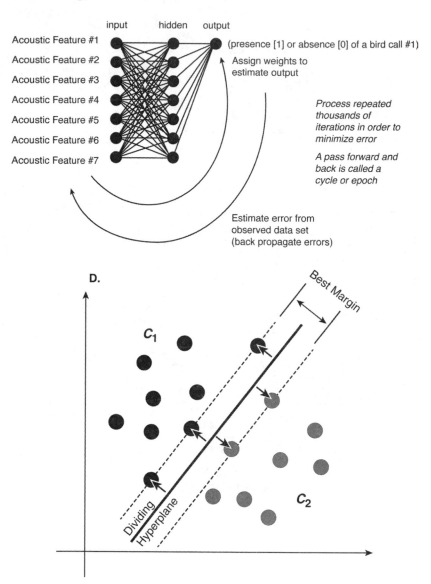

input   hidden   output

Acoustic Feature #1 ● ● ● (presence [1] or absence [0] of a bird call #1)

Acoustic Feature #2 ●

Assign weights to
estimate output

Acoustic Feature #3 ●

Acoustic Feature #4 ●

*Process repeated
thousands of
iterations in order to
minimize error*

Acoustic Feature #5 ●

Acoustic Feature #6 ●

*A pass forward and
back is called a
cycle or epoch*

Acoustic Feature #7 ●

Estimate error from
observed data set
(back propagate errors)

D.

Best Margin

C_1

C_2

Dividing
Hyperplane

Figure 10.2 (*continued*).

Neural Networks

One of the oldest types of machine-learning algorithms is the artificial neural network. This type of algorithm is based on how neurons in the mammalian brain learn about patterns in sensory inputs. The simplest form of a neural network is called a *perceptron*. A perceptron connects inputs to an output using a set of connections that are described as mathematical equations that are referred to as an *activation function* and a *bias* (Fig. 10.2C). The perceptron learns by setting the activation function and bias to random values, then uses these values to try to predict the output value. It uses the difference between the predicted and actual values to calculate an error between them. The algorithm then uses a *delta function* (i.e., an adjustment algorithm) to adjust the weights on another pass. It then calculates the error between the known value and the next predicted value on the subsequent pass. Each pass is called a *cycle*. The perceptron continues to adjust weights, and after many cycles, sometimes hundreds to millions, the learning is halted either by the user, who is following the error trends through the cycles, or by the tool, after it reaches a set minimum error. Most neural networks, however, use an additional layer, called a *hidden layer*, where weights across all possible connections of input, hidden, and output layer are adjusted with each cycle. This process of providing new weights and then calculating errors using the actual and predicted values is referred as a *feed-forward, back propagation* neural network. It has been determined that neural networks with hidden layers perform far better than simple perceptrons. Data scientists have also developed other kinds of neural networks, many for specific kinds of data. For example, convolutional neural networks (CNNs) have been designed to recognize images in photographs. CNNs (e.g., Sethi et al. 2020) have also been successfully used to identify sound sources in complex audio recordings. Neural networks offer several benefits: (1) there are no assumptions about how data are distributed, as there are with statistical models; (2) they easily generalize from large datasets; and (3) they are known to perform well with "noisy" data.

Support Vector Machines

The support vector machine (SVM) is a classifier algorithm that uses the concept of a *hyperplane*, which maximizes the separation of two or more classes across multivariate space. A hyperplane is defined as a vector (line) that separates data in two-, three-, or *n*-dimensional space. To illustrate how an SVM works, notice how the two classes of sound sources in Figure 10.2D (C_1 and C_2, which might be, for example, measures of two acoustic features of a signal) are arranged in two-dimensional variable space. A vector (the heavy line) can be used to separate these two groups, but notice that there is space that can be extended out from this line to the nearest members of the two groups. This space is called the margin width. The SVM will search for vectors that ideally separate all members into their groups and maximize the margin width. Most SVMs work within three or more variables and can use nonlinear hyperplanes to separate groups and maximize the margin space. SVMs are considered good for datasets that are highly dimensional (e.g., have many fields) and in which the data are composed of many records. SVMs have been successfully employed to identify unknown birdcalls (S. H. Zhang et al. 2018) from a large soundscape recording database for which a training dataset was used to create the class member spaces.

10.5.2. Clustering

Cluster analytics, or *clustering*, is the process of dividing data into groups (clusters) that share common characteristics and have meaning (Tan et al. 2016). Clustering is often used to explore complex data that have high dimensionality. Clusters can be mutually exclusive (i.e., not have overlap), have overlap, be arranged hierarchically, or be configured to have many, or just a few, objects or items per cluster (Tan et al. 2016). Clustering does not require data to be labeled. Indeed, one of the main objectives of clustering is to determine how data are grouped together so that a label can be developed for each cluster. Let us consider two common forms of clustering: partitive clustering (i.e., partitioning) and hierarchical clustering.

Partitive Clustering

Partitive clustering groups data points into clusters that may or may not have overlap in variable space. One of the most common partitive clustering algorithms is the k-means algorithm, and another popular one is the k-medoids algorithm. The k-means algorithm (Jain 2010) attempts to minimize the within-cluster square error for k (the number of clusters). It is simple and can be used with a large dataset. The steps are iterative:

> *Step 1.* The researcher selects the number of clusters (k) that they expect to find in the dataset.

Step 2. The *k*-means algorithm then randomly selects mean values for the *k* clusters.

Step 3. The distance between each data point in a cluster and the mean value for that cluster is calculated, and the shortest distance is saved.

Step 4. A total squared error is then calculated for each of these shortest distances, and the errors are summed across the *k*-means.

Step 5. If this is the first iteration, the algorithm returns to Step 2. If not, the algorithm compares the error of this iteration with the error of the previous iteration. If there is a change in error, a rule is used to select a new set of means for the *k* clusters, and the algorithm returns to Step 3. If there is no change in error, the algorithm stops, and the chosen means are used to define the center (often called the *centroid*) of each cluster.

The *k*-means algorithm is easy to understand and interpret, is efficient, and is best used with data that do not have many outliers. It is a popular data exploratory approach (Wu et al. 2008). The deficiencies of the algorithm are that is requires a specification of the number of clusters by the researcher, it is sensitive to noisy data, and the values selected at the start of the iterations could bias the result. If bias is a concern, additional rules can be used to address start conditions (i.e., rerun the algorithm many times). Some of these rules (e.g., see silhouette calculations in Rousseeuw 1987) are also used to test how one can vary *k* across a limited range and compare how well different values of *k* fit the data.

Hierarchical Clustering

Hierarchical clustering involves placing data points in groups that are organized hierarchically, from single data points to one large group that represents all the values in the data. This can be accomplished by using either a top-down or a bottom-up approach. In a top-down, subdividing approach, values for all data points are compared, and those data points with the most similar values are placed together in their own group; this is done iteratively by an algorithm, which sets successively *smaller* dissimilarity values as it continues to split the data into many smaller clusters. This top-down approach is called *divisive* hierarchical clustering. Bottom-up clustering algorithms start with each data point in its own cluster, and each iteration of clustering creates larger clusters until only one cluster exists. This approach is called *agglomerative* clustering.

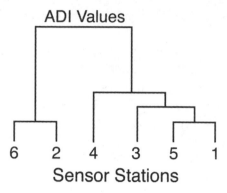

Figure 10.3 A dendrogram, showing how sensor locations can be compared for ADI values.

Clusters can be visualized using several different kinds of techniques. A *dendrogram*, which illustrates clusters in a bifurcated, tree-like structure, can be used for this purpose. For example, if a researcher wants to know the relative similarity of acoustic index values based on sensor location, a dendrogram can illustrate how similar values are for all sensors in a study. The dendrogram in Fig. 10.3 shows that sensors 1 and 5 and sensors 2 and 6 have the most similar ADI values, and that sensor 3 is most similar to sensors 1 and 5 and most dissimilar to sensors 2 and 6.

10.5.3. Association Analytics

The classic association analysis is the discovery of patterns in what are referred to as "basket data" (Agrawal et al. 1993). Association analytics was first described using a case that involved shopping cart (a.k.a. basket) items, in which the analysts asked the question, What items do shoppers buy together? For example, let us consider what a shopper might buy from a grocery store—milk, bread, pizza, and so on—with the list of all items sold in the store comprising all possibilities. An association algorithm would seek to determine what kinds of items are most often purchased together. Such groupings are called *itemsets* (a.k.a. basket contents). Association rules can also be used to explore how items co-occur in a landscape or soundscape. For example, it might be helpful to know what other species are present in a soundscape, given the presence of an endangered species. Likewise, we might want to know what species are most commonly present when another species is absent. Finally, a soundscape ecologist might consider features in landscapes or seascapes, atmospheric conditions (rain, wind), lunar cycles, and other factors that could create common associations across a soundscape, landscape, and a variety of environmental conditions.

Measures of association are made using simple statistics. Consider the soundscape data from Table 10.1, which are scored in a binary form (1 = presence, 0 = absence). Association rules are described thusly: $\{X\} \to \{Y\}$, where $\{X\}$ is one itemset that is examined in relation to another itemset $\{Y\}$. The left-hand part of the statement is called the *antecedent*, and the right-hand part is called the *consequent*. Itemsets are placed within curly brackets (i.e., {}). The rightward-pointing arrow in the statement is a symbol for the word "implies." Consider the statement {greater tinamou, dink frog} → {tropical cicada}. In other words, let us examine the association rule that tests the following condition: the calls of the greater tinamou (X_1) and dink frog (X_2) (the antecedents) imply that tropical cicadas (the consequent) call at the same time. Association rules are assessed by employing one or more of the following three metrics: *support*, *confidence*, and *lift*. Each metric provides useful information about an itemset in the database.

Support measures the how common an itemset is in the database or how commonly an association rule is true in the dataset. It is calculated as the number of itemsets for which $\{X\} \to \{Y\}$ is true divided by the number of all records. Support is measured as a count. Oftentimes, a researcher is interested in the number of cases in which a certain threshold is met. Alternatively, the researcher may be interested in the most common itemsets in the dataset, and these can be calculated and then sorted to discover patterns in the data. Thus, these two forms of support are calculated as

$$\text{support}\,\{X \to Y\} = \text{count of } \frac{\{X \to Y\}}{N},$$

$$\text{support}\,\{X\} = \text{count } \frac{\{X\}}{N}.$$

Confidence is the number of itemsets that occur in the database divided by the number of times that the antecedent occurs. Thus,

$$\text{confidence}\,\{X \to Y\} = \frac{\text{support}\,\{X \to Y\}}{\text{support}\,\{X\}}.$$

Lift measures the support for the association rule $\{X\} \to \{Y\}$ divided by the product of the number of occurrences of $\{X\}$ and the number of occurrences of $\{Y\}$. In other words, lift gives a measure of support for the association rule given how common $\{X\}$ and $\{Y\}$ are in the dataset. It is calculated as the ratio of the number of times the itemset occurs to the number of times it would be expected to occur by chance. A value greater than 1 means that the rule occurs more often than expected by chance, and a value less than 1 is interpreted to mean that the rule is not common.

$$\text{Lift}\,\{X \to Y\} = \frac{\text{support}\,\{X \to Y\}}{\text{support}\,\{X\} \times \text{support}\,\{Y\}}.$$

Thus, using the data from Table 10.1, we see that the support value for our rule is $2/5 = 0.54$, the confidence value is $2/3 = 0.67$, and the lift is $2/(3 \times 3) = 0.22$.

One of the first association algorithms, which is still very popular, is the *Apriori* algorithm (Agrawal and Srikant 1994; Wu et al. 2008). The Apriori algorithm iteratively generates itemsets in the dataset for candidate itemsets of size C. For example, for each iteration C, $C_i = 1$ means that there is only 1 item in each itemset, $C_i = 2$ means there are 2 items in all itemsets, and so on. The Apriori algorithm has two steps at each C_i: (1) candidate generation, which creates a list of itemsets that contain the i number of items, and (2) candidate counting + candidate selection. The researcher specifies the candidate selection criterion, which is the percentage of itemsets that are in the entire database at the start of the algorithm.

To understand how Apriori works, let's consider the soundscape-landscape database from Table 10.1 again. The first step of the algorithm would generate a list of candidate items for $C_i = 1$. The list would be as follows: {howler monkey (HM), greater tinamou (GT), dink frog (DF), tropical cicada (TC), rain (R), old-growth forest (OLF)}. The second step would count the support of each of these, then calculate the proportion of their occurrence in the database as {4/5, 4/5, 4/5, 3/5, 1/5, 2/5}, or {0.8, 0.8,

Table 10.1 A binary representation of hypothetical soundscape-landscape (a.k.a. basket) data

ID	Howler monkey	Greater tinamou	Dink frog	Tropical cicada Fidicina sericans	Rain	Old-growth forest site?
1	1	1	0	0	0	0
2	1	0	1	1	1	0
3	0	1	1	1	0	1
4	1	1	1	1	0	0
5	1	1	1	0	0	1

Source: Modified from Table 6.2 of Tan et al. (2006).

0.8, 0.6, 0.2, 0.4}, and finally to perform the candidate selection based on the criterion specified by the researcher, let us say 25%. Applying this 25% selection criterion would mean that for the $C_i = 1$ itemset, only howler monkey, greater tinamou, dink frog, tropical cicada, and old-growth forest would be retained for the next round, $C_i = 2$ (i.e., rain would drop out).

For the $C_i = 2$ round, the following calculations would be made. First, we would consider all possible pairwise itemsets for those items that remain. This means that the following itemsets would be examined: {HM, GT}, {HM, DF), {HM, TC}, {HM, OGF}, {GT, DF}, {GT, TC}, {GT, OGF}, {DF, TC}, {DF, OGF}, {TC, OLF}, which have proportions of {0.6, 0.4, 0.2, 0.2, 0.4, 0.4, 0.4, 0.6, 0.4, 0.2. After applying the 25% criterion, the following $i = 2$ itemsets would be saved: {HM, GT}, {HM, DF}, {GT, DF}, {GT, TC}, {GT, OGF}, {DF, TC}, {DF, OGF}. These $C_i = 2$ itemsets would then be retained for the $C_i = 3$ round. The algorithm would continue to increment i by 1 and would stop when there were no itemsets left.

Several other association algorithms are used to explore itemsets in databases, including AprioriTid and FP-growth (Sarma and Mishra 2016). Soundscape ecologists could benefit greatly from these approaches, as they are able to integrate a variety of data types (especially after transformation) and can be used especially to integrate soundscape, landscape, and sociocultural variables across space and time.

10.5.4. Time Series

Soundscape recordings are temporally rich, and they capture several temporal characteristics that can be examined by soundscape ecologists. First, animals that participate in the soundscape have activity patterns that are strongly diel. Most notable are the dawn and dusk choruses in terrestrial and aquatic environments, respectively, which are peak times of calling by animals. Diurnal and nocturnal activity periods exist, too; nocturnal patterns are much more diverse, and show greater intensity, than diurnal patterns in most ecosystems. In addition to these diel patterns, seasonal soundscape patterns are characteristic of many biomes (see Section 5.2). Together, these patterns constitute a "rhythm of nature" (Pijanowski et al. 2011b), in which the soundscape has distinct periodicities (i.e., cycles). As described in Chapter 8, these periodicities can be modified by human activities, climate change, and land use change, and the impacts of these disturbances on the soundscape can be assessed through time series modeling. In addition, many soundscapes are composed of a sequence of phases, each dominated by one animal or a group of animals. Measuring the lengths of these phases and assessing how they transition from one to another involves time series analyses. If multiyear data are available, soundscape ecologists might be interested in knowing whether there are long-term trends, such as increasing or decreasing complexity or intensity, in the soundscape. Finally, many triggers of soundscape phases are infrequent, and being able to detect these triggers over time is an important focus of research in soundscape ecology.

Time series analyses are very specialized methods within the fields of statistics and data mining. These analyses also involve advanced mathematics, much of which is beyond the scope of this book; many great resources are available to those seeking to learn more about time series analysis and modeling. Many traditional time series statistical models assume that the process being modeled is stationary. The idea of *stationarity* can be difficult to understand. Fundamentally, it means that the statistical properties of the process remain stable over time (Osborne et al. 2007). For example, a time series is not stationary if mean, variance, or covariance changes over time. A time series is stationary if the data show a regular pattern over time, and no seasonality is present. In short, many traditional time series models, such as ARIMA models (Osborne et al. 2007), assume stationarity, and are thus not likely to be useful to soundscape ecologists, although some of the modeling that is part of these traditional methods (e.g., autoregressive and moving average models that also explore how residuals plot over time) can be applied to soundscape data and thus has tremendous potential for understanding the temporal dynamics of soundscapes. Soundscape ecologists may consider using traditional time series modeling for short-term forecasting (e.g., on the order of hours to days; see Bellisario 2018; Dietze et al. 2018).

The most common times series analyses are summarized here to provide readers with enough information to allow them to understand the principles of each and to explore them in more detail. These analyses also provide a means for examining highly temporal data such as soundscape recordings using traditional tools developed for time series. Examples of applications for soundscape ecologists are also provided.

Aggregated-by-Time Averages

Soundscape ecologists can explore the temporal patterns in data by aggregating time periods and examining their averages and variances. For example, soundscape ecologists may consider how an acoustic index varies through the day by plotting means and standard errors for all recordings made at a particular hour in the diel cycle. The aggregated averages could also be examined across groups or treatments (e.g., inside and outside a marine protected area). Seasonal trends could also be explored by plotting averages of acoustic indices lumped over two-week periods throughout an entire year. The researcher could consider how these trends compare for dawn and dusk chorus periods (i.e., with a window of 1 hour before and after sunrise and sunset) or for midday and nighttime.

Moving Averages

Data scientists often calculate moving averages so that trends can be discerned when data show a lot of variability within a short time period. In essence, moving averages "smooth" the data over time. Another application of moving averages is the use of past values to forecast future values. The technique was first developed in business (Winters 1960) and is commonly used in that field today. Several forms of moving averages (also called running averages, boxcar averages, or rolling averages) exist, including the simple moving average (SMA), exponential weighted average (EWA), and weighted moving average (WMA). Let us consider the simple moving average. Most commonly, a simple moving average is calculated for the n time samples on either side of a central value as one moves along the time axis. This method will result in averages calculated with an odd number of samples (Table 10.2). Soundscape ecologists could apply moving averages to a variety of values across time to determine whether there are trends. For example, a researcher could select data from the same time each day—say, 6:00 a.m.—and then examine whether there are rising or falling trends for a particular acoustic index, such as ADI, over time.

Sampling across Intervals

In the processing of remotely sensed land surface data from MODIS (see Section 9.8), researchers have used the maximum value that occurs over an 8-day and a 16-day composite as a value to represent a window of time. This is necessary because some locations on Earth have cloud cover that makes continuous analysis impossible. A similar approach could be applied to acoustic data when there are missing data so that temporal trends can be assessed. Alternatively, the maximum value of the same acoustic index for each day could be examined as a moving average over time.

Table 10.2 Simple moving average values for acoustic diversity index

Time	ADI	Three-period moving average	Five-period moving average
6:00	2.609		
6:30	2.650	2.470	
7:00	2.153	2.376	2.486
7:30	2.325	2.389	2.442
8:00	2.691	2.470	2.475
8:30	2.395	2.634	2.524
9:00	2.815	2.537	2.500
9:30	2.394	2.430	2.510
10:00	2.531	2.446	
10:30	2.367		

Note: Values are for 10-minute soundscape recordings for ten half-hour periods starting at 6:00 a.m. in Borneo (sensor ID 015088).

Phenology Time Series Modeling Using Polynomial Equations

Ecologists and remote-sensing experts have developed several kinds of temporal models that are based on the fitting of time series data to polynomial equations (e.g., de Beurs and Henebry 2005; Jönsson and Eklundh 2004). Fitting data to polynomial equations has several benefits, of which the most commonly used by ecologists is the ability to use the parameters of the polynomial equation to compare different locations or time periods. These approaches are most often used on daily or weekly greening indices, such as the normalized difference vegetation index (NDVI), that are introduced to the polynomial models, since long-term data (often >4 years) that span decades (e.g., from MODIS) are readily available to ecologists and remote-sensing experts. In some instances, data are smoothed using filters; one that has been used successfully is the Savitzky-Golay filter (Jönsson and Eklundh 2004), which applies a slightly modified version of a moving average to the data as they are ordered over time. These polynomial models are often in the form of a quadratic equation (from de Beurs and Henebry 2005):

$$NDVI = \alpha + \beta AGDD + \gamma AGDD^2,$$

where NDVI is what is being predicted, AGDD is accumulated growing degree-days (i.e., days when it is warm enough for plants to grow), α is the intercept of NDVI at low AGDD, β is a slope parameter, and γ is a curve-fitting parameter that determines shape. AGDD is calculated as the number of days, starting from the first day of the year (i.e., January 1) on which the average, or –(max – min)/2, daily temperature has exceeded a particular threshold (viz., 0°C). In modeling soundscapes, researchers may consider using "growing" degree-day (GDD) thresholds that are known to be effective in modeling the activity patterns of animals. For example, some insects develop only when the air temperate exceeds 9°C (48°F). The AGDD would count the number of days above this threshold and use that number as the input for time in the model, and a measure of ADI would be used as a replacement for AGDD in the polynomial model. For modeling the phenology of the land surface over time, the software package TIMESAT (Jönsson and Eklundh 2004) has been used by many ecologists to characterize the dynamics of landscapes.

Mining Sequential Patterns

A host of algorithms have been developed that allow data scientists to discover patterns in "basket" data ordered over time (Agrawal and Srikant 1995; Srikant and Agrawal 1996; Fradkin and Mörchen 2015; Han et al. 2000). Let us assume that each audio recording can be considered a "basket" and that in each basket is a set of sound sources that are described as an itemset. Building on the basket data examples used in the data mining literature, we can also consider each sensor (i.e., location) to be a different "customer" that is tracked as it visits the "store" over a long period of time. Note, too, that in some instances, researchers may also be interested in subsets of items.

Table 10.3 illustrates the format of the type of data that researchers could consider mining for sequential patterns in labeled soundscape recordings. Mining sequential patterns in such data could allow researchers to examine several kinds of questions. For example, given a selected support, lift, and confidence, does the pattern over time as an ordinal variable change at the same time of day? If it does, then this pattern could represent a set of phases and transitions. Researchers could also ask to what extent these phases and transitions are occurring at all sensor locations. Alternatively, they could ask which items are common across what data scientists call a "span"—the time between events that are measured. They could also ask which set of sequences that have the same itemset (i.e., soundscape composition by sound source) is the longest, and at what times of the day, or seasons, these sequences occur.

Table 10.3 A hypothetical set of sound-source items

Acoustic sensor	Start time of 10-minute recording	Items
A_1	05:00	B_1,B_2
	06:00	B_1,B_2,F_1
	07:00	B_1,B_2,F_1,F_2
	08:00	B_2,B_3,F_3
A_2	05:00	B_1,B_2,F_1
	06:00	B_1
	07:00	B_1
	08:00	B_1,B_2
A_3	05:00	B_1,B_2,F_1
	06:00	
	07:00	
	08:00	
A_4	05:00	
	06:00	
	07:00	
	08:00	

Note: Items occur in 10-minute recordings by four different sensors at times between 05:00 and 08:10. B_1, B_2, and B_3 are three different species of birds, and F_1, F_2, and F_3 are three different species of frogs.

10.6. General Predictive Analytics

Researchers are often interested in developing not only models that describe a pattern, but also models that can be used to predict a value. Predictive models can focus on forecasting the future, predicting what should occur at one place given data from other places, or predicting a value when one is not known. Soundscape ecologists and bioacousticians commonly develop models to predict the source of a sound given a set of acoustic features.

10.6.1. Predictive Modeling Phases

Let us consider a very common pathway of developing, calibrating, and validating predictive models using the data mining approaches that are described in Section 10.5. This pathway is illustrated in Figure 10.4 for a simple model built to use acoustic features and known labels in a sound-scape recording to predict the presence of a particular bird in that recording. Researchers who have a large database should first select a smaller set of re-cords from that database, using the approaches outlined in Section 10.4, and then label those data. Labeling could use, for example, any of the SEMs)/au-ral annotation techniques described in Section 9.5. This labeled database is then partitioned into what are referred to as training, testing, and, in some cases, calibration datasets. With the simpler approach, which uses only test-ing and training datasets, the training dataset is used to build a calibrated model, and the testing dataset is used to validate it. This step is called the *training phase* of modeling. A standard format for the training dataset is given in Figure 10.4, Step A. Note that several acoustic features are stored in the database along with a label indicating its source and a binary label (pres-ence of calls of the bird of interest = 1, presence of other calls = 0). These data are then used in a predictive machine-learning tool (e.g., a neural net-work) with the acoustic features as independent variables and the presence of bird 1 as the dependent variable. The neural network makes many passes through the data, adjusting its own parameters according to user-defined rules, and as it does so, it makes a probability estimate of the dependent variable for each record (Fig. 10.4, Step B). For example, for record 1 in the training dataset, at cycle number i, the neural network estimates that there is a 0.8 probability that the acoustic features indicate the presence of bird 1. During this cycle number i, a *loss function* is calculated as the difference between the truth value (i.e., 0 = not present or 1 = present) and the esti-mated value for each record (see Wang et al. 2020 for an excellent summary of loss functions and their specific uses). For example, if a mean square error (MSE, literally the square of the error between the estimate and the truth value, which is then averaged) is used as the loss function for the five sound sources in the training dataset, the MSE would–be $[(1 - 0.8)^2 + (1 - 0.9)^2 + (1 - 0.2)^2 + (1 - 0.1)^2 + (0 - 0.9)^2]/5 = 1.51$. The researchers building the model will follow loss function values over all training cycles and use one of a few rules to stop the training. They may elect to set, at the beginning, a loss function that is satisfactory, or they may select the ith cycle on the basis of the trends. A loss function typically rises and falls through many training cycles, and so the researcher may opt to select a global minimum across numerous training cycles (see also the discussion about overfitting and brittle models later in this section). Once the researcher has determined which cycle provides the desired model, the parameters from the data min-ing tool—the neural network, in this case—are saved.

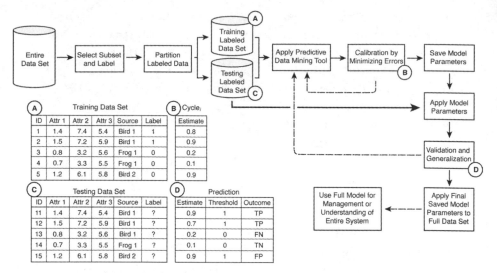

Figure 10.4 Steps in developing and testing a predictive model using machine learning: A = training; B = calibration; C = testing; D = prediction.

The next step of modeling is the *testing phase* (Fig. 10.4, Step C). Note that the labeled testing dataset has a structure identical to that of the training dataset, but is used differently. As the researcher applies the model parameters to the labeled testing dataset, an estimate is made (Fig. 10.4, Step D) as the model uses the acoustic features in this second dataset to predict the presence of bird 1. A threshold rule is then applied to the estimated value to convert it to a predicted value of 0 or 1. To determine whether a model is useful, researchers commonly use *accuracy metrics*, which are summarized in the next section.

All researchers who use data mining tools to develop predictive models know that many of these tools can overfit data. *Overfitting* is the condition in which the data mining tool has learned the patterns in the data so well that it cannot be used with satisfactory results when presented with another set of data—perhaps data collected from places outside the original study area, or during different time periods whose ambient sounds make the original model perform poorly. One approach to avoid overfitting is to use data mining model parameters that fit the training dataset less well so that the model can generalize across data that are different from the original labeled training data. Models that perform well on one dataset but not another are said to be *brittle*. Tests of *generalization* are typically very specific to a modeling activity. For example, a neural network model of land use/cover change between two time periods, developed by Pijanowski et al. (2014), was subjected to three tests of generalization: (1) different land use/cover datasets; (2) different locations; and (3) different time periods. In other words, the model was tested on how well it generalized across different data sources, different time periods, and in different metropolitan areas.

In many instances, researchers apply the validated model to all of their records—most importantly, to records that have not been labeled—to generate an estimate of the presence or absence of the target species' call. This might aid the researcher in developing a more complete assessment of the presence of, say, bird 1 across all study sites and times, which could be useful for natural resource management. If bird 1 is an endangered species or a keynote species, then this information can be used to develop plans for managing the landscape to help support its presence at the locations where it was found.

Modelers have developed several key terms (Table 10.4) that are used during the process of developing, testing, and applying models that are data driven.

10.6.2. Accuracy Metrics for Predictive Models

A host of *accuracy metrics* have been developed by modelers from a variety of fields, most notably the biomedical sciences. In general, there are three kinds of accuracy metrics used by data scientists to measure the accuracy of predictive models: binary metrics, threshold-independent metrics, and outcome distribution measures.

Binary Metrics

Binary accuracy metrics, developed first in the biomedical sciences, use the terms of that discipline to construct their metrics; the most commonly used terms are "positive" (e.g., a drug works or a medical test is accurate) and "negative." Many of these metrics are used in ecology (see Fielding and Bell 1997 for an excellent overview of modeling in conservation biology). They are based on an error matrix called a *confusion matrix* (Table 10.5). A confusion matrix aligns correct outcomes and errors in a 2 × 2 table. The four outcomes are true positive, false positive (the error of predicting that something would occur when in fact it did not), true negative (e.g., correctly predicting the absence of something) and false

Table 10.4 Key terms of predictive model development

Term	Definition
Training	The phase of modeling in which a subset of a data set is presented to a machine-learning tool, which learns the patterns that exist in the data.
Testing	The phase of modeling in which another subset of the data set is presented to a model that has been built, the dependent variable is held back, and metrics are used to measure how well the model predicts the values of the dependent variable.
Calibration	The changing of model parameters, through either an automated or a human-directed process, to ensure that the model performs satisfactorily. Calibration may involve changing model structure (e.g., changing the number of hidden layers in a neural network) or changing the algorithms used in a machine-learning tool.
Sensitivity analysis	A specific modeling exercise that involves determining how sensitive the model is to conditions such as errors in data, range of data inputs, and features used by the machine-learning algorithm. This exercise could involve some "experimenting" in which errors are added to data to determine how the model behaves; or, alternatively, sweeping across parameters (e.g., methodically changing the number of input nodes from a low to high value to determine how this affects the model's accuracy and/or ability to generalize). Sensitivity analysis can be performed as part of model calibration, validation, or both.
Validation	The process of using a testing data set independent from the training data set to ensure that the model performs satisfactorily on any data that might be used with it. A methodical, multi-step series of validations is common; this practice is called cross-validation.
Parsimony	A model is considered parsimonious if it is simple and has good predictive ability.
Generalization	Ability of a model to perform satisfactorily across space, time, and data sets. For example, a call detection tool that can use soundscape recordings from different sensors would be one that can generalize across different kinds of data sets.
Robustness	Ability of a model to generalize well across all model uses.
Brittleness	Inability of a model to perform satisfactorily when new data are presented to it because it has learned the patterns in the data that were used to construct it.
Ensembles	Multiple models (i.e., different data mining algorithms) used to generate predictions, which are then combined in a variety of ways (e.g., voting, averaging, or using decision trees).

Source: Modified from Gardner and Urban 2003.

Table 10.5 A confusion matrix for scoring predictive models

		Actual/Observed	
		+	−
Predicted	+	a (TP)	b (FP)
	−	c (FN)	d (TN)

Note: Data in the matrix are presented as counts. The total number of samples in the model is given as N, which is $N = a + b + c + d$. TP = true positives, TN = true negatives, FP = false positives, and FN = false negatives. + = positives or values of 1, − = negatives or values of 0.

negative (e.g., the error of predicting the absence of something when in fact it was present).

A variety of accuracy metrics have been developed that use the number of occurrences in each cell of the confusion matrix as part of a simple arithmetic expression, such as a ratio of counts (Table 10.6). For example, positive predictive power (PPP) is the measure of how well a model performs for identifying the presence of, for example, a birdcall in a soundscape recording. Some of these accuracy metrics are easy to understand and communicate to non-modelers but might not be comprehensive measures of a model's performance, and in fact, may have a bias based on the nature of the model or the data. For instance, some accuracy metrics are not considered good measures of model performance if there is a large imbalance of 0s and 1s (i.e., if there are significantly more 0s than 1s in the data, or vice versa, then these accuracy metrics are not used). Luque et al. (2019) assessed the effect of such imbalances on the potential bias of common accuracy metrics for testing predictive models. They found that several worked well when there was a large imbalance of 0s and 1s; these metrics are indicated with a (1) in Table 10.6. Those that performed moderately well with imbalanced training and testing data are indicated with a (2), and those that did not perform well and had obvious bias are indicated with a (3). In general, those accuracy metrics that use all four cells in the confusion matrix do better than those that use only two or three.

Threshold-Independent Metrics

Because a majority of machine-learning tools do not create binary outcomes directly from their algorithms, but rather a range of values between 0.0 and 1.0, a few accuracy metrics have been developed that use these values di-

Table 10.6 Accuracy metrics derived from the confusion matrix in Table 10.5

Metric	Calculation
Prevalence	$(a+c)/N$
Overall diagnostic power	$(b+d)/N$
Correct classification rate or total accuracy (ACC) (2)	$(a+d)/N$
Sensitivity (SNS) (1)	$a/(a+c)$
Specificity (SPC) (1)	$d/(b+d)$
False positive rate	$b/(b+d)$
Positive predictive power (PPP) or precision (PRC) (3)	$a/(a+b)$
Negative predictive power (NPP) (3)	$d/(c+d)$
Misclassification rate	$(b+c)/N$
Odds ratio (OR)	$(a \times d)/(c \times b)$
F_1 score (3)	$2 * \dfrac{PPP \times SNS}{PPP + SNS}$
Geometric mean error (GM) (1)	$\sqrt{(SNS \times SPC)}$
Matthews correlation coefficient (MCC)* (2)	$\dfrac{(a \times d)-(b \times c)}{\sqrt{(a+b)(a+c)(d+b)(c+d)}}$
Bookmaker informedness (BM)* or arithmetic mean error (1)	$SNS + SPC - 1$
Kappa	$\dfrac{(a \times d)-\{[(a+c)(a+b)(b+d)(c+d)]/N\}}{N-\{[(a+c)(a+b)(b+d)(c+d)]/N\}}$
NMI	$\dfrac{-a \times \ln(a)-b \times \ln(b)-c \times \ln(c)-d \times \ln(d)+(a+b) \times \ln(a+b) \times (c+d) \times \ln(c+d)}{N \times \ln(N)-[(a+c) \times \ln(a+c)+(b+d) \times \ln(b+d)]}$
Markedness (MK)* (2)	$PPP + NPP - 1$

Source: Modified from Fielding and Bell (1997) and Luque et al. (2019).

Note: All metrics are values between 0 and 1 except those indicated with an asterisk (*), whose values range from −1 to +1. Metrics assessed for imbalance bias by Luque and colleagues (2019) are placed into three categories following their Table 14 (p. 229): (1) null/low bias, (2) medium bias, and (3) high bias. Those metrics not marked with a (1), (2), or (3) were not assessed by those authors.

rectly and thus are not based on the confusion matrix described above. The most common non-binary accuracy metric is the area under the receiver operating characteristic curve (denoted as AROC, AURC, AUROC, or ROC by various disciplines). This metric plots out the range of SNS (sensitivity, meaning the true positive fraction– and 1 – SPC (specificity, meaning the true negative fraction) for all possible threshold values in an x–y plot, as shown in Figure 10.5. Most algorithms use a threshold value of 0.5; values below that threshold are firmly set to 0 and those above it are firmly set to 1. Varying the threshold value from 0.01 to 0.99 (rather than using just 0.5) and plotting the SNS and SPC values from these thresholds produces curves like the one shown in Figure 10.5. A random model would produce a diagonal line (shown as chance performance). Most models should perform better than a random model, and thus their plots should extend into the upper left portion of the AROC plot. The accuracy metric that is used is the area under the curve. The area under the chance performance curve takes up exactly half of the graph, so the area under that curve is 0.5 (sometimes called the no skill condition). A perfect model would produce a curve that extends up to 1.0 on the y-axis when $x = 0.0$ and thus would give an AROC value of 1.0. Figure 10.5 shows a set of training and testing AROC values. Many data scientists prefer to use this metric, as it removes any potential

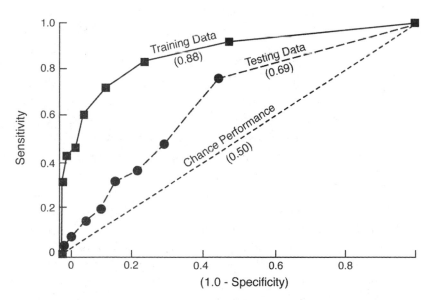

Figure 10.5 How the area under the receiver operating characteristic (AROC) curve is calculated.

bias involved in selecting a threshold value to bin the data into 0s and 1s. Because many machine-learning tools are based on nonlinear algorithms, and because the assumption that 0.5 is the proper threshold may not be valid in all cases, data scientists prefer to use AROC along with those metrics listed in Table 10.6 to report the goodness of fit of models. A similar kind of plot, called the Precision-Recall Curve (PRC; Davis and Goodrich 2006; Sofaer et al. 2019), is often used for class imbalanced training and testing data.

Comparison of Outcome Distributions

The outcomes of model training and testing will produce a distribution before they are binned into 0s and 1s, and methods have been developed that allow data scientists to compare the distributions of different models. One metric used for this purpose is *cross-entropy*. Cross-entropy can be applied to modeling situations where the true distribution is a binary outcome (i.e., 0 or 1) or a probability (i.e., ranges between 0.0 and 1.0). To understand cross-entropy, let us first examine the equation for entropy, which is based on the information theory of Shannon, using the form for species and individuals:

$$H(p) = -\sum_{x=1}^{N} p(x) \log p(x) ,$$

where H is entropy, $p(x)$ is the proportion of individuals in species x, and log is the natural logarithm. Cross-entropy incorporates two distributions, p and q, into the same equation:

$$H(p,q) = -\sum_{x=1}^{N} p(x) \log q(x) .$$

To understand how cross-entropy is calculated, let's consider a case in which the machine-learning algorithm is trying to learn about the presence of labeled event B (the target outcome) in a soundscape recording in which there are three possible outcomes (events A, B, and C). We would compare the distribution of what the machine has learned with that of the truth dataset (i.e., dataset of known values) as shown in Table 10.7.

By calculating cross-entropy from the values for predictive model 1 in Table 10.7, we get $0.0 \times \ln(0.228) + 1.0 \times \ln(0.619) + 0.0 \times \ln(0.153)$, which is −0.479, and, as with all calculations of entropy, we take the negative of this number to get the cross-entropy value, 0.479. Let us compare this result with those from models 2 through 4 in Table 10.7. Model 2 gives us an 80% correct hit and 20% miss rate, which leads to a smaller cross-entropy value of 0.223. An equal distribution of hits and misses across all three pos-

Table 10.7 Machine-learning outcomes and cross-entropy measures of fit to training or testing data

Condition	Class A	Class B	Class C	Cross-entropy
Truth	0	1	0	
Predictive model 1	0.228	0.619	0.153	0.479
Predictive model 2	0.1	0.8	0.1	0.223
Predictive model 3	0.33	0.34	0.33	1.108
Predictive model 4	0.001	0.998	0.001	0.002

Source: Modified from stackoverflow.com/questions/41990250/what-is-cross-entropy.

sible outcomes, as in model 3 (essentially a random model for getting the target outcome) leads to a very high entropy value of 1.1. A nearly perfect predictive model, model 4, gives an entropy value of nearly 0. The type of calculation described in this paragraph is called a *one-hot distribution*, a term often used in accuracy metric calculations, as there is only one possible target outcome, which is given the binary value of 1, and all others are 0.

Cross-entropy can also be used for non-one-hot distribution comparisons. If p and q have distributions that have values between 0.0 and 1.0, then cross-entropy can also be calculated using the same equation.

Finally, another form of entropy that is calculated commonly in ecology is the Kullback-Leibler divergence score (D_{KL}; Kullback and Leibler 1951). It is also known as *relative entropy*. D_{KL} uses the following equation to calculate entropy across two different distributions:

$$D_{KL}(P||Q) = -\sum_{x \in X} P(x) \log \frac{Q_x}{P_x},$$

where P and Q represent the two distributions from two different models, the reference distribution and the sample distribution, respectively. The D_{KL} represents the probability that a model built to predict a bit of information (as 0 or 1) using P can predict Q.

10.7. Narrative Analysis

10.7.1. Key Terms

Soundscape ecologists may venture into as many as three different kinds of data sources that require textual analysis. Textual analysis is designed to analyze (1) *interviews, participant observations*, and other *expressive narrative information* (e.g., songs, stories, poems) collected for research on individuals or communities related to the sounds they hear and perceive, or to how sounds are used for supporting lifeways; (2) *social media*, such as Facebook, Twitter, and SnapChat, which can be used to determine, for example, the sentiment of people toward local conservation efforts; and (3) *scientific literature* in online digital library collections, which can be mined in order to discover patterns in research approaches, concepts, themes, and results across large numbers of academic and nonacademic articles.

Two approaches to textual analysis are described here. The first, *textual data mining* (TDM), is based on tools developed to mine large online literature collections for data related to a research question that is in a particular domain. TDM has intellectual roots in linguistics, medicine, business, and national security, but has been used by only a few disciplines thus far—most notably in the medical field, where research articles on, for example, a particular disease may span several decades and appear in hundreds of journals. It is introduced here at a high level with the hope that soundscape ecologists may begin to use these powerful tools. The second approach, which has been central to the work of social scientists and scholars in the humanities for decades, is *content analysis* guided by *grounded theory* (see Section 9.9).

10.7.2. Textual Data Mining

Textual data mining shares some approaches with the numerical and association machine-learning algorithms described above. Interestingly, many of the tools used for TDM are found in some of the same online toolkits, such as Scikit-learn and R, that are used for other kinds of data. Not surprisingly, however, there are several important differences between TDM and other kinds of data mining that researchers interested in mining large text documents need to understand. An informative hands-on reference on the topic is *The Text Mining Handbook* by Feldman and Sanger (2007), and another, which provides detailed step-by-step instructions for researchers using R, is Feinerer et al. (2008).

The steps for processing documents from online textual data repositories for knowledge discovery in textual databases (KDT) are shown in Figure 10.6. Note that text documents can come from a variety of online resources, from which they are extracted using "fetching tools" (for static documents) and/or "crawling tools" (for documents frequently updated). The text documents are then archived, and a number of automated or user-controlled

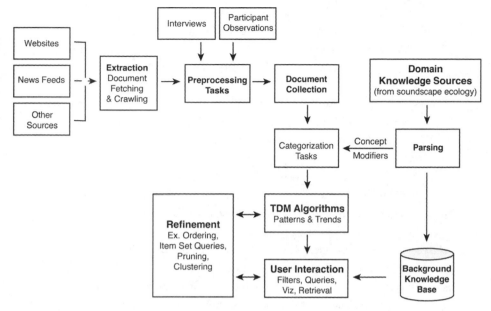

Figure 10.6 Processing steps for textual data mining. (Modified from Figure I.7 in Feldman and Sanger 2007.)

tasks are completed. One key set of tasks needed for domain-specific lexi-cons is the creation of knowledge and domain ontologies, as shown along the right side of Figure 10.6. An *ontology* is an organizational framework of key concepts and terms and how they are related. An ontology for soundscape ecology might include "biophony," "geophony," and "anthrophony," and the fact that they are the three major sound source types in a soundscape. An ontology could also include a list of biological sound sources that are part of the biophony. The following steps are typical in conducting KDT.

Extraction. Automated tools can fetch and crawl through hundreds to mil-lions of web pages, receive real-time news feeds, and extract information from other internet-based sources. In addition, TDM tools can extract databases created from interviews, participant observations, and other textual information. In the latter case, these data would also be collated and stored together.

Preprocessing tasks. A variety of tools have been developed that allow re-searchers to transform documents from original formats into a format necessary for TDM. The first step of such transformation is called text preprocessing. Methods of text preprocessing include the use of what

TDM experts call *natural language processing* (NLP) tools, which extract text to produce domain-independent textual features. This process also involves *tokenization*, which is the parsing of large documents into smaller segments of text, such as sentences, words, or even syllables. These features are often labeled with *parts of speech* (POS) tags, which specify the function that the word plays in a sentence (e.g., noun, verb); however, there are no POS standard libraries to use.

Document collection. All text-based documents, such as those from interviews, researcher's notes on participant observations, emails, web sites, newspaper articles, reports, lyrics from songs, and poems, are assembled into a collection. These documents contain POS tags and are also categorized (annotated with date/time, place, keywords) as part of the document preprocessing step.

Domain knowledge sources. In a set of tasks separate from the document preprocessing tasks, a *domain knowledge* textual database is introduced to constrain categorization or to introduce additional labels that define relationship rules for technical domain terms, or to create a hierarchy of the terms that are commonly used in a field. In the latter case, a user could introduce the term "soundscape" and then its subcomponents "biophony," "geophony," and "anthrophony" as a vocabulary tree. In some fields, such as genetics, background knowledge textual databases have been developed by a large community of scholars so that TDM can be accomplished with little input. Many TDM tools have options to build domain knowledge databases for specific textual data mining purposes.

Parsing. A *parsing* step, which organizes text into grammatical structures that align with one of several grammatical theories; is built into many TDM tools. The last preprocessing step is to use the results of the previous steps to create the complete textual database, which can be organized either for *categorization* (i.e., to discover patterns in text across all documents) or for *information extraction* or text feature retrieval (i.e., to query the database in order to develop summary information across all of the documents).

Categorization tasks. Categorization tasks are associated with classification, the placing of documents or text features into categories (e.g., topics, subjects). Categorization can be organized around a *controlled vocabulary* (in which case it is called text indexing), or it can use *time stamps* (e.g., dates of publication for journal articles) or *text filtering* tools (for binning text into relevant vs. irrelevant groups). In text filtering, as users select text that is relevant, this information is used by an adaptive labeling algorithm to label other text features as relevant or irrelevant.

Parsing. Routines that are designed to reduce the number of terms, concepts, and other text features can be employed to increase efficiency. Text tools that convert words of different forms to a standard form can be part of this step.

Background knowledge base. This textual knowledge base is organized for human user interaction that would allow for simple queries (e.g., searching for words) of the text database, subsetting it for visualizations (e.g., for viewing a map of places associated with keywords), or for counting the number of times a set of words occurs across all documents.

TDM algorithms. Many of the discovery tasks in TDM parallel those used for other types of data mining. Specifically, TDM discovery seeks to determine what patterns exist across the large number of documents that are being considered in the analysis. TDM discovery tasks include (1) *clustering*, which places similar documents together in groups as they are assessed for terms and concepts (e.g., Steinbach et al. 2000); (2) *classification*, which places documents in different categories according to their content; (3) *associative assessment*, which relates documents across itemset rules as described above for labeled data; and (4) *trends analysis*, which determines how terms and concepts change over time. Trends analysis is a very common application of TDM in the sciences because it attempts to examine trends in research questions, concepts, approaches, and conclusions over time—usually decades—using published research. It has been argued by systems biologists (Ananiadou et al. 2006) that some sciences have become data-rich and hypothesis-poor (e.g., functional genomics) and thus require the use of TDM along with data mining to generate new knowledge quickly. Some of the more common core TDM operations include quantifying distributions (e.g., the number or proportion of concepts and terms per document or interviewee), frequent and near frequent concept sets, and associations (Feldman and Sanger 2007).

Refinement. Also called post-processing, refinement consists of operations that reorder, create further subsets of, and use clustering techniques to prune the textual database, and use algorithms to generalize (e.g., combine) across terms, concepts, and other patterns in the data.

User interactions. Users interact with two text databases, the document collection and the background knowledge base, to perform such tasks as browsing, querying, visualizing, and text retrieval.

10.7.3. Content Analysis

Sociocultural scientists, as well as those in education, medicine, and areas of the humanities such as ecoethnomusicology (e.g., Post and Pijanowski 2018), conduct interviews, make participant observations, and analyze other written and expressive forms using content analysis software (i.e., CAQDAs) and approaches guided by grounded theory (see Section 9.9 for these concepts). Tools such as CAQDAs allow researchers to examine human behaviors, beliefs and worldviews, sentiments, emotions, roles, and relationships between individuals. Content analysis of qualitative information is less strict and procedural than analysis of quantitative information. Much like KDT, described above, this related field of qualitative analysis relies on users to constantly search, code, visualize, and quantify textual features across qualitative information. Several key tools and associated activities are involved in content analysis of qualitative information:

> *Projects*. When a user sets up a project, CAQDAs can assist in assembling the documents that are imported into the system. The software keeps a list of these documents for use.
>
> *Nodes*. An important feature of many CAQDAs is their creation of nodes that are used to associate text displayed by the software. A *node* is a "code, theme, concept, or idea about the data in the project" (Wong 2008, 5). Node names are generated by the researcher; for example, a node could be "Attitudes," "Barriers," or "Decision Making." Nodes can be organized as "free nodes" or as "tree nodes." Tree nodes are hierarchically structured; for example, "Attitudes" could be divided into "Attitudes toward Bird Sounds," "Attitudes toward Noise," and so on. Text in any document is then selected and associated with a node. Multiple nodes can be associated with the same text. The activity that associates text in a document with a node is called *coding*.
>
> *Attributes*. Many software tools also allow users to assign important attributes, such as gender, age, or status, to a document or to a set of text (if, for example, a researcher is conducting a focus group interview).

Summary

Because research activities in soundscape ecology can span the range from extracting high volumes of numerical data from audio recordings to examining human perceptions through qualitative analysis to mixed-methods approaches that combine quantitative and qualitative information, the number of possible analytical tools and methodological approaches is truly endless. Within the field of soundscape ecology and the parallel fields of bioacoustics, ecoacoustics, and acoustic ecology are a rich variety of approaches that have matured to a point where many "best practices" have been established. What is also apparent to many who have worked across this methodological and analytical spectrum of approaches is that research will require collaborations of experts across diverse disciplines, probably with the addition of data scientists who have broad knowledge of engineering tools that can be exploited. In addition, much work that involves research into human perceptions should consider the use of knowledge co-production approaches to ensure that outcomes are as unbiased as possible. There are many challenges to completing research that combines qualitative and quantitative approaches and experts from the sciences, humanities, and social sciences. The most significant of these challenges may be developing trust and investing in some level of cross-training that leads to an appreciation of the sophistication that all disciplines bring to addressing a problem.

Discussion Questions

1. Describe the kinds of analyses that would be needed to address the coupled human and natural system of the herding community described in Figure 8.6. What qualitative, quantitative, and mixed-methods analyses could be undertaken if you were addressing the question of what would happen if the herders were to abandon their traditional practices of herding livestock in rural environments and move to a city, where their livestock would be raised in overgrazed landscapes in confined locations surrounded by noise from traffic?

2. Sketch out a set of steps that would allow you to combine sound-source data from a labeled training dataset with text data from interviews with indigenous community members who depend on the natural resources (e.g., wildlife for food, trees to build houses, water for drinking, open areas for livestock grazing) of a national park and areas outside the park.

Further Reading

Feldman, Ronen, and James Sanger. *The Text Mining Handbook: Advanced Approaches in Analyzing Unstructured Data.* Cambridge University Press, 2007.

Tan, Pang-Ning, Michael Steinbach, and Vipin Kumar. "Data Mining Cluster Analysis: Basic Concepts and Algorithms." In *Introduction to Data Mining* (2013): 487–533. Pearson Education India.

Welsh, Elaine. "Dealing with Data: Using NVivo in the Qualitative Data Analysis Process." *Forum qualitative sozialforschung/Forum: Qualitative Social Research*, vol. 3, no. 2 (2002).

APPLICATIONS

The extent to which soundscape ecology has applied its concepts, tools, and methods to address the grand societal challenges that were outlined in Chapter 1 is presented in Part IV. These challenges include the biodiversity crisis, the loss of human connectedness to nature, noise impacts on human and wildlife health, and the loss of traditional ecological knowledge. Part IV comprises four chapters, each focusing on coupled human and natural systems: terrestrial, marine/aquatic, urban, and rural/cultural. Cutting across these chapters is a focus on how soundscapes and their measures, particularly acoustic indices, have been used to study problems in these systems.

11

Terrestrial Applications

OVERVIEW. This chapter summarizes the most recent (since 2005) research in acoustic index (AI) development and assessment, biodiversity assessment using soundscape recordings, quantification of relationships between landscapes and soundscapes across space, time, and disturbance dimensions, and research that has focused on quantifying the extent and impact of noise in protected areas in terrestrial systems (i.e., a focus that is not entirely on the urban/built environment). Historical trends, general conclusions, and gaps in knowledge are described as well.

KEYWORDS: acoustic indices, biodiversity assessments, soundscape index assessments

11.1. Acoustic Indices: Assessments

A large thrust of research in soundscape ecology has been the assessment of the soundscape indices that are referred to in Section 9.4 as acoustic indices (AIs). There has been considerable research on (1) how AIs differ across space and time; (2) how AIs compare with more labor-intensive surveys of animals done using traditional field observation methods or SEMs that generate values such as species richness or numbers of sounds, (3) how AIs compare with simple measures of intensity; (4) how duty cycles affect the efficacy of AIs; and (5) how AIs are affected by the acoustic theater (i.e., the structure of vegetation and major land uses within the landscape).

Some of the early work that used soundscape recordings to estimate the diversity of soniferous animals was focused on determining whether developing a species richness score by listening to the recordings gave the same results as conducting a survey in the field. Several studies (Hobson et al. 2002; Rempel et al. 2005, 2009; Celis-Muillo et al. 2009, 2012; Sedláček et al. 2015) found that in most cases, the results for number of species (i.e., species richness), species composition, and species detection (presence or absence during a time window) were similar. In general, this work suggests that results from field surveys conducted either at the same time as, or just after, a recording was made were very similar to those from the recording. Tests of different microphone systems also revealed that their recordings did not differ greatly in these analyses. Perez-Granados et al. (2018) found that PARs and audio recordings could be used to detect the rare and patchily distributed Dupont's lark. Hingston et al. (2018), however, did not find a 100% match of species composition between PAR recordings and field samples due to two major factors: first, in field surveys, birds that were not vocalizing were detected, but they obviously were not recorded; second, listeners to bird species that were mimics were likely to make mistakes in identification that field observers would not make, since they had visual confirmation of the species that was the vocal mimicker. In summary, the use of audio recordings to detect and score soniferous animals has been shown to be fairly robust, although mostly for birds, and for rare species, it has some advantages compared with field surveys.

Sueur and colleagues (2008) were the first to use acoustic indices to estimate biodiversity from soundscape recordings. They argued that the use of soundscape recordings with acoustic indices can support a cost-effective, noninvasive, rapid survey for animal biodiversity. They presented a set of alpha and beta acoustic diversity indices based on measures of acoustic entropy and applied these indices to recordings made in Tanzanian landscapes. They showed that the alpha acoustic diversity measures correlated strongly with the number of species at a location, and that the beta acoustic diversity measures demonstrated a linear correlation with the dissimilarity of species across these landscapes. Their findings confirmed for the first time that acoustic indices could be used to assess two important dimensions of biodiversity, namely, alpha (i.e., local or within-group measures) and beta diversity (i.e., change in community diversity across space or between-group measures).

One of the first multiple-AI comparative studies was conducted by Fuller et al. (2015). This study examined how ADI, H, AEI, BI, ACI, and NDSI varied over diel cycles (a.k.a. time of day) and across several different kinds of landscapes in eastern Australia. They found that ACI and BI were correlated with avian song intensity but were not sensitive to landscape patterns. ADI, H, AEI, and NDSI were sensitive to nocturnal sound intensities and diversity. ADI, H, AEI, and NDSI were most sensitive to landscape condition. Sueur et al. (2014) also conducted a comparative AI study in several different kinds of landscapes in French Guiana. They examined twenty-one AIs, grouped into alpha diversity (e.g., AEI, ADI, NDSI, ACI, NP) and intensity indices (e.g., LA_{eq}), and found that many of the AIs were sensitive to sounds from humans, which in some cases made them biased when used for biodiversity estimates. Another comparative AI study was conducted by Towsey et al. (2014), who developed fourteen first-order AIs, eight based on measures of an acoustic feature called a wave envelope and six based on spectrograms; they found that even these simple AIs were able to correlate with species richness estimates made at PAR locations. These early studies all came to the same conclusions about AIs. first, AIs differ considerably in the way that they vary across space and time. Second, many correlate with one another, some in predictable ways (e.g., H and ADI use different acoustic features but are both constructed using entropy). Third, some are more sensitive to anthropogenic noise than others.

Several researchers have examined whether AIs correlate with either (1) species richness surveys of all sounds or of calls of specific animal groups or (2) species richness values developed from listening to and labeling recordings. Mammides et al. (2017) examined how seven AIs—H, ACI, AR, AEI, BI, NDSI, and ADI—correlated with point-count surveys of birds at PAR locations across three regions of southern China that differed in their amounts of natural vegetation. They found a strong correlation of H, AEI, and ADI with measures of species richness derived from the in situ surveys. Oddly, AR varied inversely with species richness, which is the opposite of what was expected, but they did suggest that AR was developed primarily for sites with low species richness, and their study area is considered one of the most biodiverse regions of China. BI did not correlate strongly with species richness measures. They found that NDSI did not correlate strongly with species richness values either, which is the opposite of what Fuller et al. (2015) found in their multi-index comparison study. Machado and colleagues (2017) performed a study similar to that of Mammides et al. (2017) in which they made recordings in two landscapes in the Brazilian Cerrado, identified and labeled birdcalls from the known 107 species potentially in the recordings, and then used these data to calculate a species richness value for each recording. They hypothesized that the labeled and scored recordings and the ADI would both show greater species richness in the less disturbed of the two landscapes. Both species richness and ADI were indeed significantly greater in the less disturbed landscape, and ADI and species richness were positively correlated as well. They also showed that NDSI was positively correlated with the distance of the PAR from the road.

An interesting study of the use of AIs was performed by Dodgin et al. (2020), who attempted to use PArs and AIs to assess the species diversity of burying beetles. Although capable of making low-intensity sounds, these beetles are unlikely to be detectable in soundscape recordings. The PARs were set up in tallgrass prairies, where burying beetles were sampled using pitfall traps. Shannon's and Simpson's diversity indices were calculated and compared with values of ADI, ACI, AEI, BI, NDSI, and H. The rationale for the study is unique: The researchers hypothesized that burying beetles, which are present at high densities where there is a high density of dung from animals, would be a proxy for large animals (soniferous and silent), or that the complexity and intensity of the soundscape would reflect high biodiversity across all species in the community. They found that three of the AIs (ACI, BI, and H) did in fact correlate with the abundances of specific species, and that all six AIs varied predictably over diel cycles (dawn, day, dusk, and night); however, none of the AIs were correlated with Shannon's and Simpson's diversity indices or evenness or richness scores developed from the beetles caught in the pitfall traps. In some instances, AIs correlated more strongly with the abundance of a single species of burying beetle than with the abundances of the beetles as a group. The researchers also noted that the AIs were very sensitive to strong winds in the afternoon and to nocturnal insects.

New AIs are often developed using a combination of approaches, but many new AIs are tested by comparing them with the classic AIs (i.e., ADI, ACI, AEI, BI, NDSI, D, and H) and with other data on species composition. Gage et al.'s (2017) presentation of normalized soundscape power (NSP) illustrates this approach. Normalized soundscape power uses an approach similar to the derivation of ADI and AEI in that frequency bands in 1 kHz widths are extracted and metrics are developed from these. Normalized soundscape power for 1 kHz frequency bands between 1 kHz and 7 kHz were calculated for a set of soundscape recordings from a PAR located in an ecological reserve in southern Australia that used a 10-minutes-on, 20-minutes-off duty cycle. The seven AIs listed above were calculated for 1-minute and 10-minute recordings and compared to see how the length of the recording would affect the AIs. A subsample of 180 recordings from morning periods was then selected, and birdcalls (67 species) were used to identify species and number of calls per species in each recording. The researchers found not only that the AIs were positively correlated with species richness and number of calls, but that NSP was positively correlated with both, demonstrating that this new AI has the potential to characterize both species richness and abundance using number of calls as a proxy.

A few studies, however, have shown that many of the classic AIs do not fully correlate with species richness data derived either from in situ surveys or from SEMs. The research of Moreno-Gomez and colleagues (2019) focused on soundscape recordings from three PARs located in a Peruvian tropical rainforest. They calculated the classic AIs for these 1-minute recordings (made once per hour over a 3-month period) and then selected a sample of the recordings to score for the presence and absence of known bird and anuran species. Their analysis had mixed results. They showed, for example, that the species richness of birds, anurans, and both correlated in a positive way only with AEI and H. ADI, ACI, BI, and AR did not correlate with species richness values. The three PAR sites also had different diel patterns of species richness, which often did not match those of the AIs. Although it demonstrated an approach to testing the efficacy of the AIs, the study did have several shortcomings. First, the researchers used a different model of the Wildlife Acoustics Song Meter in each of the three sites. No calibration of the amplitude-sensitive AIs—namely, ADI and AEI—was done. ADI, in particular, plotted at nearly maximum values across time, indicating that the noise floor default value was possibly not proper.

Eldridge et al. (2018) also examined the classic AIs, proposed three new ones (simple acoustic descriptors of root means square, spectral centroid, and zero-crossing rate), and then compared them with species richness and abundance data derived by SEM for birds, anurans, and insects. They found that in both the temperate and the tropical rainforest sites they analyzed, the AIs were better predictors of bird species richness than anuran species richness; and that a combination of the AIs and simple acoustic descriptors could enhance the characterization of soniferous species diversity in these two major types of forest landscapes.

In one of the most comprehensive analyses of AIs conducted to date, Buxton et al. (2018) conducted a literature review of all AIs proposed by soundscape ecologists, ecoacousticians, and bioacousticians and found sixty-nine, including the classic AIs frequently studied by soundscape ecologists. They calculated values of all sixty-nine indices for a set of labeled recordings, most from terrestrial but some from marine environments, that had been used to calculate species richness, Shannon's diversity, and the total number of biological sounds in a recording. They reported several important findings. First, only thirty-six of the indices were sensitive to testing criteria (that the AI must vary over time, and that there must be reported evidence that it was correlated with an independent measure of species richness). Second, they used those sensitive and relatively proven AIs with the species richness scores in a random forest model, which showed that the AIs eventually grouped into three classes based on how well they predicted three independent variables: species richness, Shannon's diversity, and total biological sounds. A few of the classic AIs were good predictors of richness and Shannon's diversity (e.g., ACI, ADI), while some indices satisfactorily predicted only one (e.g., NDSI and AR predicted only total number of sounds). This work shows that there are potentially several dozen AIs that can be used, and that many are useful to predict important ecological measures such as species richness and Shannon's diversity. It also shows that each of these AIs provides different information useful for biodiversity conservation, and that machine-learning tools such as random forest models are powerful tools for analyzing many recordings and AIs.

One of the most important questions facing soundscape ecologists is the best duty cycle to use, given that bias might be introduced if recordings are too short or too long. In other words, could a 1-minute-on, 59-minutes-off duty cycle be as effective as using a full duty (a.k.a. continuous) cycle (59 minutes on, 1 minute off to write the audio file to storage). Francomano et al. (2020) examined a set of continuous recordings made in eight different landscapes/seascapes: a Magellanic subpolar penguin colony, a Neotropical rainforest, Mongolian grasslands, a Caribbean coral reef, a miombo swamp, a North American prairie, a California woodland, and a Magellanic lowland forest. They subsampled the continuous recordings to calculate values for H, ACI, and BI across these sites and across four major time periods (nighttime, daytime, dawn, and dusk) and found that 1-minute recordings were enough to minimize sampling rate variances of AI values. Their results are consistent with those of Pieretti et al. (2015), who also found that 1-minute samples were adequate to characterize temporal variability and spatial differences using ACI.

There are several benefits to using 1-minute recordings. First, many of the current terrestrial and marine PARs can operate for up to a year on a duty cycle of 1 minute on, 59 minutes off. Thus, the AIs developed from these recordings for use in sentinel PARs and long-term studies are likely to be enough. A 1-minute recording can also reduce the computational time required for analysis, reduce costs of storage, and reduce labor costs of servicing PARs in the field. However, it should be mentioned that 1-minute recordings will not allow soundscape ecologists to characterize changes in soundscape phases, and may miss important triggers that would allow them to understand the coupling of triggers and changes in soundscape phases. In addition, rare acoustic events may be missed.

The use of acoustic indices to monitor species richness, abundances, and activity patterns has come under some minor criticism. Darras et al. (2019) conducted a study of the use of PARs and soundscape recordings in several different landscapes that varied in vegetation structure and land use type. They measured understory vegetation, tree structure, and sound transmission (across frequency ranges and intensities) in thirty-eight plots in lowland rainforest and rubber plantation landscapes to assess what they called "acoustic detection spaces" (a.k.a. acoustic theaters). They found that sound transmission varied with land use, measures of vegetation structure, sound-source height, and ambient sound pressure level. They also found that the ability of a PAR to detect a sound is affected by vegetation structure as well as sound-source height. This work emphasizes the need to more fully understand how the acoustic theater may bias species detection in comparative studies.

11.2. Landscape-Soundscape Relationships

How the structure and dynamics of the landscape are reflected in the soundscape are important considerations for the management of landscapes, especially for biological conservation. This section focuses on research that has made efforts to quantify spatial features in landscapes and determine how soundscape information in the form of AIs can be used to understand these important relationships.

Two early soundscape-landscape studies appeared in the literature in 2011: those of Pijanowski et al. (2011a) and Joo et al. (2011). Both studies set up PARs across a land use gradient from an urban environment to a rural landscape to examine how landscape patterns, particularly the major land use/cover type at each site, contributed to soundscape composition over space and time. Both research projects examined soundscape composition by quantifying the sounds in the frequency ranges of the sound sources present in the landscape. Pijanowski and colleagues (2011a) placed seven PARs along an urban to rural to natural gradient from the end of the Purdue University campus to rural research stations located about 10 km west of the university, at (1) a forest patch in an urban location, (2) an abandoned orchard in a rural landscape, (3) a secondary forest stand at a forest research center, (4) a wetlands complex set within a mix of forest and agricultural fields (mostly row crops), (5 and 6) two agricultural fields, one soybeans and one corn, and (7) an old-growth forest at a forest research station more than 1 km from the other forest stands. These PARs (Wildlife Acoustics SM1) were set to record starting in late March and ending in November of 2008 on a 15-minutes-on, 45-minutes-off duty cycle with a 44.1 kHz sampling rate. Several acoustic measures were reported for each site and were plotted over time and across frequency bands related to the sound sources (<1 kHz road noise, 2–3 kHz frogs, 3–7 kHz birds, 1–10 kHz strong wind and rain). Recordings with strong wind and rain sounds were removed from the analysis. Pijanowski et al. (2011a) found that ADI varied predictably over diel cycles, with strong dawn and dusk chorus peaks. ADI was also greater at night than during the day. They also found that as one moved from the more urban areas to areas that were more rural and natural, but not agricultural, ADI increased for all hours and across the entire time of the study. The agricultural sites had very small variation across the diel cycle and throughout the year (i.e., the ADI was "flatlined"). Frequency band activity (percentage of a band that has spectral energy above –50 dBFS) was greatest at 1–2 kHz early in the spring, but by early fall, the 4–5 kHz band was the most active, indicating that sounds from vehicles dominated early and sounds from insects (e.g., katydids) dominated later in the fall. The greatest diversity of sounds also occurred in the fall, as sounds from birds, amphibians, and insects were all present at this time of year. Joo and colleagues' (2011) study was conducted before electronic PARs were available, so they used automated tape recorders set in seventeen locations around the Michigan State University campus. Their sites covered a land use gradient similar to that used by Pijanowski et al. (2011a), including urban (residential, commercial), agricultural (i.e., row crops), forest urban park, and rural sites. This study was the first presentation of their power spectral density (PSD) metric, which was measured for 1 kHz frequency bands. The researchers recorded for 2 days per month (days without rain) from February to December in 2006. In addition, they listened to and scored for the presence of birds in all recordings. They too found that the peak in biological activity occurred in the fall. The soundscape analysis was consistent with their analysis of bird species richness across all land use categories, further providing evidence that the acoustic measures they developed were good indicators of animal diversity. A more recent study that is similar in design to those two studies is that of Buxton (2018), who made recordings at twenty-eight sites across a land use gradient (from natural to managed landscapes including tea, coffee, and carda-mom) in the Western Ghats of India. Over thirty-six AIs were examined in relationship to the land uses characterizing each site. The study found that AIs varied predictably across the land use gradient and were positively cor-related with bird species richness and total number of avian vocalizations.

Tucker et al.'s (2014) study of landscapes and soundscapes is one of the first that quantified landscapes beyond major land use type or proportion of the major land use type within a buffer area around PAR locations. Tucker and colleagues developed the BioCondition landscape assessment tool to generate scores for landscape features (based on patch size and patch con-nectivity metrics for locations where PARs were deployed) and vegetation condition (based on percentage of leaf litter, weed cover, ground cover, and six other metrics). They used Wildlife Acoustic SM2 PARs, recording on a 1-minute-on, 29-minutes-off schedule for a period of 1 month in spring. They also applied a relative power spectral density (rPSD) metric (based on Welch 1967) for each frequency band, which divides the PSDs by a maximum value, and then correlated these results with landscape pattern metrics and vegetation condition. They found that rPSD from soundscape recordings was correlated with positive scores for landscape features (e.g., large patch sizes of natural vegetation) and positive values for vegetation condition.

As extensive work in exploring specific relationships between landscapes and soundscapes has been done over the past fifteen years, a summary of that work is provided in Table 11.1 to illustrate the diversity of the land-scapes, approaches, and findings involved.

Some of the more advanced landscape characterization involves the use of airborne LiDAR, which is used to quantify the three-dimensional structure of vegetation and surface features such as buildings and other objects in the built environment. Two studies have examined how the three-dimensional structural complexity of vegetation relates to the complexity of the sound-scape. Pijanowski (2011a) hypothesized that the complexity of landscapes, and in particular the complexity of vegetation, should be positively corre-lated with animal species richness and thus with measures of acoustic diver-sity. This hypothesis was first tested by Pekin et al. (2012) using LiDAR data for the forests and wetland complexes in the lowland Neotropical rainforest of Costa Rica. They developed eight LiDAR metrics to quantify the structure of vegetation, including canopy height, the number of vegetation strata, the size of gaps in the lower and upper canopy, and the density of foliage near the ground. They then calculated ADI for 6 days of recordings during the dusk chorus at fourteen locations. They found that the canopy structural

Table 11.1 Summary of studies focusing on relationships between landscape patterns and soundscape composition

Reference	Types of landscapes, sites, and how quantified	AIs employed	Other data used	General conclusions
Bormpoudakis et al. 2013	Four forest and two grassland landscapes in Greece (32 sites). Classified based on EU natural resource habitat type system	CENT, SD, KURT, SKEW, ZCR, H, SFM, 1/f	No labeling performed, no quantification of landscapes outside of habitat type class	Sounds in complex landscapes are also complex, supporting notion that soundscape complexity could provide animals with acoustic cues to the complexity of habitat
Ng et al. 2018	Cleared, regrowth, and remnant forests, Australia	AEI, ADI, H	BioCondition metrics of landscapes	AEI and H were positively correlated with BioCondition scores
Burivalova et al. 2018	Paleotropical rainforests that are selectively logged (two forms) and those in protected areas, Indonesia	Acoustic power, soundscape saturation applied as a dissimilarity metric to compare pristine with managed landscapes	Qualitative measures of landscapes characterize the change in forest composition from pristine condition into managed landscapes as reflected in subtractive vs. additive amounts of forest, and homogeneity versus heterogeneity of vegetation	Homogenization of the vegetation resulted in homogenization of the soundscape; in some cases, homogenization of the soundscape at night differed from that during the day
Furumo and Aide 2019	Six habitats studied: Oil palm landscapes, cattle pastures, rice fields, banana plantations, riparian forests, all compared with remnant forests in Colombia	Soundscape frequency/time/activity (SFTA) bins based on threshold amplitude	Used an NDMS ordination of SFTAs for the six locations and examined patterns based on simple vegetation structural features	All landscapes have distinct soundscape characteristics; those that had closed canopies were more similar than those with open canopies, and those with closed canopies were more similar to remnant forests

Reference	Sites	Indices	Methods	Results
Hayashi et al. 2020	Oil palm landscapes and remnant forests in South Sumatra, Indonesia	D_f and D_{cf}	Compared sites and times of day using dissimilarity measures that quantify beta acoustic diversity	There was a significant difference between oil palm locations and remnant forests
Shamon et al. 2021	Grasslands in Montana	BI, AEI, ADI, ACI	Bird species identification and an acoustic richness score (insects, amphibians, wind and rain) from audio recordings; landscape features including land cover, soil type, and elevation (e.g., roughness) characteristics	BI positively correlated with grassland landscape indices; ACI had a negative correlation (opposite to predictions); ADI, AEI did not perform well (probably gain was set too high and so threshold should be greater)
Rajan et al. 2019	Compared an urban park, a sacred grove, and a bird sanctuary in southern coastal area of India	ADI, AEI, ACI, BI, and NDSI	Created a list of known bird species at each location based on a recent survey and used it as a proxy for bird species richness	Sites with the greatest number of bird species (sanctuary, followed by sacred grove and then urban park) had highest AI values; ADI, BI, and NDSI showed the strongest relationship between landscape condition and soundscape diversity
Myers et al. 2019	Compared olive grove landscapes that are organic (11 PARs) versus conventional (11 PARs) in southern Greece	ADI, BI, and ACI	Compared management types, also included vegetation height, elevation, tree species richness, distance from road, open water, and forest	ACI and BI were consistently greater in organic olive groves; ADI did not perform as well as the other two. Other landscape features were not as significant as management type
Campos-Cerqueira et al. 2019	Compared sites with Forest Stewardship Council (FSC) certification (N = 24) with non-FSC managed lands (N = 20) and reference managed forests (N = 23) in Peru	Acoustic space use (ASU) of Aide et al. (2017)	—	Variation in ASU was best explained by management type (mean ASU was greatest in FSC forests; most overlap in ASU was between FSC and reference forests)

(continued)

Table 11.1 (continued)

Reference	Types of landscapes, sites, and how quantified	AIs employed	Other data used	General conclusions
Nascimento et al. 2020	Six kinds of natural landscapes and two human-dominated landscapes in Brazil (143 sites), grouped into three classes: open, flooded forests, and non-flooded forests	Thirteen AIs including the classic AIs	Canopy height, percentage cover, litter depth, DBH of trees, shrub cover using line-point intercept method	Most of the indices were positively correlated with canopy cover and other vegetation variables; some had a negative relationship to vegetation variables (ACI, AEI, KURT and SKEW)
Scarpelli et al. 2021	Deployed PARs across an Atlantic tropical forest landscape that contains a mixture of remnant forests, agriculture, and urban uses	Used Towsey et al. (2018) metrics that are waveform based (3), intensity based (3), and entropy based (4), plus ACI, CLS, SPD, and NDSI	Used a land cover map that was summarized for percentage cover for 100 m, 200 m, 500 m, 1 km, and 2 km buffers	Only the 100 m and 2 km buffer size classes showed any discernible relationship between land cover proportions and AIs examined. There was a complex relationship between the AIs and the two major buffer sizes, suggesting that AIs may be sensitive to spatial scale of habitat
Dooley et al. 2020	Examined how the soundscape varied with urban use intensities in central Florida, USA	Average PSD, AIP, AggEnt Delta Power, Center Frequency	Landscape development intensity (LDI) of Reiss et al. (2010) for each land use category within 100 m, 500 m, and 1,500 m buffers	Soundscape complexity (AggEnt) decreased with increasing intensity of human use as measured by LDI
Depraetere et al. 2012	Introduced a new AI called acoustic richness and compared it with other AIs in a temperate woodland	AR and D_f	Aural identification of bird species from soundscape recordings	AR varied expectedly across space and time and correlated with D_f and with avian species richness developed from the recordings

metrics explained 75% of the variation in ADI and 60% of the variation in acoustic frequency band composition. They were also able to use LiDAR data to develop a spatial model of acoustic diversity and to identify acoustic diversity hot spots in the forest. This, they argued, was one way to create a management tool that would identity high-priority areas for conservation, as these areas would harbor the greatest variety of biological sounds and thus would be likely to support the most animal species. Following on from the work of Pekin et al. (2012), Rappaport et al. (2020) used airborne LiDAR to create an acoustic space occupancy model (ASOM), which represented not only the structural measures of the three-dimensional space around a sensor but also the potential for the vegetation to attenuate the sounds. They used the Aide et al. (2013) ASU values derived from soundscape recordings to develop the ASOM so as to determine how frequent each of the ASU was at each site, and analyzed ASU in relation to vegetation structural indices from LiDAR. They were able to demonstrate potential frequency-based detection biases that may exist due to vegetation structure and vegetation density. Although clearly at early stages of development, this work represents a promising approach to combining AIs and vegetation structure within the acoustic theater by incorporating sound propagation processes into a comprehensive landscape-soundscape-AI integrated model.

Mountains have been particularly useful landscapes in which to study large-scale patterns and processes. Indeed, many of the classic studies in biogeography have been conducted on mountains, which represent an important natural environmental gradient (Lomolino et al. 2015; Lomolino and Pijanowski 2020). Two noteworthy soundscape ecology research projects that were conducted along mountain elevational gradients illustrate the potential for using mountains to understand spatial patterns and hence beta diversity. The study by Campos-Cerqueira and Aide (2017) used fifty-eight PARs located along three elevational transects that spanned lowland tabonuco forest, mid-altitude palm forest, and high-altitude elfin forest located in Puerto Rico. The researchers calculated the number of amplitude peaks for frequency bin sizes of 172 Hz, and they summed all recordings by hour (i.e., generated 24 summaries) for each PAR site. They then examined these metrics across the elevational gradient. They found that elevation had a significant effect on the complexity of the soundscape as well as the frequencies that existed at each location, reflecting different classes of vocalizing vertebrates. They were able to detect wind and river sounds and associate these with distinct elevation zones. In short, they were able to demonstrate that soundscapes vary predictably along elevational gradients for both biological and geophysical sound sources. Gasc et al.'s (2018) study of Sonoran Desert sky islands (isolated mountain ranges) in Arizona, near the location of the classic biogeographic studies of Whittaker in the 1970s, involved the use of PARs aligned along an elevational gradient that contained four major habitats, from lowland grasslands to high-elevation pinyon pines. A massive wildfire burned much of this protected area in 2011. In 2013, twenty-four PARs were placed in burned and unburned locations in the four major habitats, where they recorded on a 10-minutes-on, 50-minutes-off duty cycle for 9 months. The researchers used the BI and SEMs to label a subset of 1-minute recordings for the presence of five geophysical sounds (wind, rain, water flow, waves, thunder), five biological sound types representing the five major animal groups (bird, insect, mammal, amphibian, and reptile), and eight anthropogenic sounds that could occur at the sites (talking, walking, plane, car, boat, other motors, gunshot, and alarm). They found that the burned sites had fewer insects than the unburned sites, that the peak in daytime biological activity (as reflected in the BI values) occurred just prior to the monsoon season (mid-July to early November), and that the peak in biological sounds occurred during the later portion of the monsoons. The peaks in BI values were also determined by elevation, time of year, time of day, and whether the landscape was burned or not. BI values were greatest in unburned areas, although only slightly so for nighttime periods, and unburned areas had significantly greater daytime BI values for all four elevation-based ecological zones. These observations suggest that animals that are active during the day are not recovering from the wildfire in any of the four habitats, but least of all at higher elevations.

A more recent study on the animal community's response to wildfire was conducted by Duarte et al. (2021) in the Brazilian Cerrado (a.k.a. savanna). They used a soundscape ecological approach to examine the effect of a 2012 wildfire using sound recordings from before and after the burn; the burn was accidental and not originally part of the study. The researchers note that burns at the site are frequent, and that a large one covering most of the area occurred in 2011. They used the ACI and PSD, as well as an SEM, to score 1-minute recordings for the presence of birds, anurans, and insects, the latter as acoustic morphospecies. They found that insects and bats were missing soon after the wildfire but were able to determine that most soniferous species returned about a year after the wildfire occurred.

Only a few studies have examined AIs over long periods of time (>1 year). Gage and Axel (2014) describe a four-year study at six PAR locations on an uninhabited island in northern Michigan; most sites were a mix of coniferous forest and wetland complexes. The study was conducted with soundscape recordings made from April to October in the years 2009–2012. Several temporal patterns were evident. First, the soundscape power (i.e., rPSD) for lower frequencies differed in periodicity from that for higher frequencies: for lower frequencies it was greatest in the early evening hours, when traffic was greatest. For higher frequencies, it was greatest at dawn and then at the dusk chorus. Second, seasonal patterns were very evident for all four years. The higher frequencies (4–6 kHz) had the greatest soundscape power in the fall; the medium frequencies (3–4 kHz) had the greatest in the spring. The lowest frequencies had the greatest power in the month of December, when leaves were completely off the trees and sounds from highways penetrated the forest from the nearby mainland. This work demonstrated that there are complex cyclical patterns to the soundscape and that measures of individual frequency bands provide information about the prevalence of the major sound sources.

11.3. Soundscape Research in Support of Biodiversity Assessments

One of the first applications of soundscape ecology was the use of audio recordings to assess animal biodiversity in more than one taxonomic group. Sueur et al. (2008) were among the first to use soundscape recordings to assess biodiversity at the community level (but see Riede 1993, 1998; Hobson et al. 2002; Brandes 2005, 2008; Brandes et al. 2006; Acevedo and Villanueva-Rivera 2006 for similar efforts). They developed two sets of acoustic indices, one that measured acoustic entropy, thus serving as a measure of alpha acoustic diversity, and another that measured beta acoustic diversity. The two sets of indices were tested on more than 240 simulated soundscape recordings, which were made by mixing individual recordings of birds, amphibians, and insects to create choruses with different species richness numbers. The researchers found that both their alpha and beta acoustic diversity indices were significantly correlated with species richness. They also applied these measures to soundscape recordings made at three degraded woodland sites in Tanzania, and found that their acoustic diversity indices were lowest in the most degraded woodland and were similar in the two less degraded ones. They argued that these acoustic indices could serve as a measure of the diversity of all acoustic morphospecies across multiple taxa; that this approach could serve to rapidly assess biodiversity at a location; and that it would be more cost-effective than rapid in situ biodiversity assessment methods and all-taxa biodiversity inventories, which are common ways to assess biodiversity in many locations, particularly in protected areas.

Gasc et al. (2013) extended this early work of Sueur et al. (2008) by examining how acoustic diversity (which they called community acoustic diversity, or CAD) correlated with the phylogenetic and functional diversity of bird communities at French Breeding Bird Survey (FBBS) locations. They used the FBBS to generate species lists, then used those lists to create a community acoustic file. They also compiled phylogenetic data using measures of species taxonomic relationships as distance measures, and calculated five measures of functional diversity from the literature (e.g., behavior, diet, morphological measures such as body size, reproduction measures such as nest type, and body mass). These researchers were able to show that CAD values for their simulated acoustic files correlated with measures of phylogenetic diversity and functional diversity, especially body mass and nest type. Thus, there is some evidence that soundscape recordings and measures of acoustic diversity can correlate with important dimensions of diversity other than species richness.

By employing PARs situated across a tropical rainforest in French Guiana, Gottesman et al. (2018) attempted to determine the spatial-temporal variability of animal sound across vertical and horizontal space. The 43-day study used twenty-four PARs placed in two twelve-sensor grids, with one grid located in the understory and the other in the canopy above it. Recordings were used to calculate the D_f acoustic index, which was then used to compare sites within and between the sensor grids. The researchers found that there were four distinct periods of activity during the day, that the vertical differences existed in "subtle ways," and that acoustic diversity was very heterogeneous across space. These results suggest that soundscape recordings can be used to assess not only species richness but also the activity patterns of animals as they occupy different vertical strata of the forest, that PARs should be deployed in sensor arrays, and that researchers should consider vertical stratification as a main feature of the relationship between vegetation structure and animal biodiversity.

One of the major objectives of biodiversity assessment is to understand how human activities affect species richness. The contributions of soundscape ecologists to this type of work are varied and extensive. For example, they have demonstrated that PArs and AIs are useful for understanding how biodiversity is affected by invasive species (Gasc et al. 2018), how animal communities recover from human-caused wildfires (Gasc et al. 2018), how biodiversity changes with the age of forest stands (Pekin et al. 2012;

Burivalova et al. 2017), whether wind turbines affect avian biodiversity in grassland landscapes (Raynor et al. 2017), how natural gas exploration affects biodiversity in a tropical rainforest (Deichmann et al. 2017), how the degradation of tropical rainforests can be measured in Borneo (Mitchell et al. 2020), Malaysia (Sethi et al. 2020), and Panama (Bradfer-Lawrence et al. 2020), and whether biological sounds are more common near or far from highways (Ghadiri Khanaposhtani et al. 2019).

11.4. Soundscapes, Wildlife, People, and Noise in Protected Areas

Acousticians have long studied the occurrence of noise, especially in urban areas (e.g., Kryter 1972; Beranek 1966; Rosen 1974; Purkis 1964; Anthrop 1970; Bonvallet 1951; Ling 1947). As has been mentioned before (see Section 3.3), "noise" is often used as a term to denote the presence of sounds produced by humans, most often those created by machines (e.g., cars, planes, radios). Despite the fact that soundscape ecologists who study sounds in the urban environment have resorted to avoiding the term "noise," as it has negative connotations for some (though not all) people who experience these sounds, the study of noise in natural areas has remained an important area of research for conservation biologists, who are concerned with how these sounds affect animal communication and animal well-being. As such, it is an important focus for soundscape ecologists who study protected areas as well.

Soundscape ecologists and acousticians have focused on several aspects of anthropogenic sounds in natural areas, protected areas, and national parks. Anthropogenic sounds are often subdivided into sound-source categories. Shannon et al. (2016) conducted a broad literature review on the documented effects of noise on wildlife in which they used the following broad sound-source categories: environmental (if a source is not known), transportation (planes, road traffic, ships, rail, etc.), industrial (e.g., construction, energy extraction), military (e.g., gunfire, explosions, training exercises), recreational (e.g., transportation vehicles for tourism) and other. They examined the effects of anthropogenic sounds on wildlife physiology, survival and fecundity, mating behavior, feeding behavior, vigilance (e.g., predator avoidance behavior), vocalizations, population metrics (e.g., numbers of individuals in areas with and without anthropogenic sounds), and community metrics (e.g., number and abundances of species in "noisy" vs. non-noisy areas). They found considerable evidence that nearly all anthropogenic sound sources affect nearly all measures of wildlife well-being, that these effects are more pronounced at 50 dBA and above, and that these effects are measurable for nearly all taxonomic groups, terrestrial and aquatic. They pointed to the facts that most studies on the effects of noise on wildlife focus on North America and Europe, and that many have examined a single species or group of species. They argued that noise must be managed in areas that are designed to support conservation efforts for large numbers of species or for endangered and threatened species.

One of the first studies to examine noise impacts on wildlife behavior in landscapes was that of Francis and colleagues (2011), who studied the effects of gas well compressors placed in a rural pinyon pine landscape in northern New Mexico. They examined the effect of compressor sounds on bird nesting locations and the potential for those sounds to mask bird vocalizations. Several locations in the landscape where compressors were not present served as controls. They found that bird species whose vocalizations would be masked by the low-frequency compressor sounds avoided nesting near the compressors, while those that had high-frequency calls remained near the compressors. They argued that low-frequency sounds produced by humans everywhere could act as an evolutionary selective force favoring some bird species over others depending on the frequency of their calls. Similar work (Ghadiri Khanaposhtani et al. 2018) conducted at the Aldo Leopold camp in Baraboo, Wisconsin, also found that highway noise "filtered" grassland and temperate forest birds, affecting both nest placement and singing perch locations in this historically important landscape (where Leopold wrote most of his *Sand County Almanac*).

Anthropogenic sounds in protected areas have become a focus of research by the Natural Sounds and Night Skies (NSNS) Division of the US National Park Service. The Natural Sounds program was established in 2000 by Director's Order #47 on "soundscape preservation and noise management" (NPS 2000). This order declared the soundscapes of national parks to be a resource that is important not only to visitor experiences but also to the wildlife that exists in these landscapes. The order defined and described a soundscape as follows: "Soundscape refers to the total ambient acoustic environment associated with a given environment (sonic environment) in an area such as a national park. It also refers to the total ambient sound level for the park. In a national park setting, this soundscape is usually composed of both natural ambient sounds and a variety of human-made sounds." The order considered "noise" to be sounds that are "unwanted, often unpleasant in quality, intensity or repetition." A later initiative added "light pollution" to the program's mission. The mission of the NSNS is to study the presence of natural and cultural sounds in US national parks, to determine the impacts these sounds have on visitor experiences, and to understand how management and planning decisions affect the occurrence of these sounds. Soon after the opening of its office in Fort Collins, Colorado, the NSNS established an extensive monitoring program for all parks. Using some of the same monitoring data that were used to create noise models for each of the national parks, Buxton et al. (2017) determined that anthropogenic sounds doubled the level of background (i.e., ambient) sounds in 63% of the protected-area units of the park system, and that for 21% of the units, there was a tenfold increase in ambient sound levels. Most alarming was the fact that this tenfold increase occurred in 14% of critical units—those that harbor endangered species. Major sources of anthropogenic sounds were transportation, construction, and energy extraction (e.g., gas well compressors) near and within the parks.

Francis and Barber (2013) have presented a comprehensive framework (Fig. 11.1) that describes the effects of two major forms of noise—*acute* (abrupt, sudden, or erratic sounds) and *chronic* (frequently occurring or continuous sounds)—on behavioral, physiological, and survival aspects of wildlife well-being. With this framework, they suggest that many of the aspects of behavior modified by anthropogenic sounds are interdependent, and that animal physiology and fitness are affected by these behavioral changes. They also suggest that anthropogenic sound sources take three major forms: (1) single sources, from which sounds radiate outward in space such that their intensity decreases 6 dB as a function of distance; (2) linear sources, such as roads, which produce sound patterns that are cylindrical, such that their intensity decreases 3 dB as a function of distance; and (3) volumetric sound sources, such as helicopters or planes, which produce sound patterns whose intensities decrease 6 dB with each doubling of distance through the air.

Anthropogenic sounds, as well as natural sounds, in protected areas have important effects on the *quality of visitor experiences*. Some of the earliest research on this topic was conducted by Kariel (1977, 1980, 1990), who studied two kinds of campers (a.k.a. recreationalists) in Canadian national parks. Campers of the first kind, who camped in remote areas, were referred to as "mountaineers," and those of the second kind, who camped at developed campgrounds, were referred to as "development" campers. Kariel (1980) discovered that both groups found the sounds of wind, water, birds, insects, and campfires to be pleasing and the sounds of chainsaws, motor bikes, and cars to be annoying. "Development" campers were more accept-

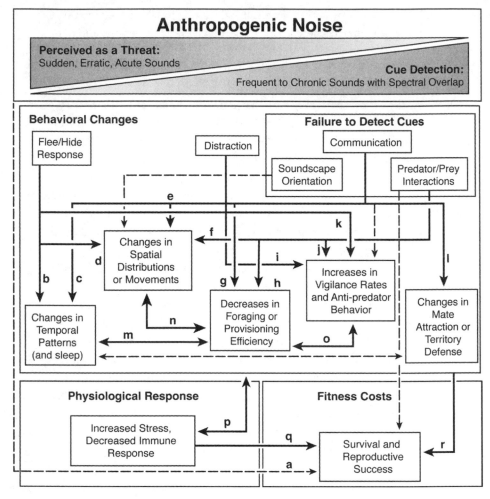

Figure 11.1 The impacts of noise on animal behavior and physiology. Lowercase letters refer to references used that illustrate the relationships. Readers are encouraged to consult the original diagram for these references. (Modified from Francis and Barber 2013.)

ing of anthropogenic sounds than were mountaineers. A few years later, Anderson et al. (1983) studied the aesthetic qualities of various sounds as judged by undergraduate students and found that they rated natural sounds as positive, engine noises as negative, and the sounds of people taking, yelling, and laughing as neutral. Numerous subsequent studies have also demonstrated that anthropogenic sounds in protected and recreational areas decrease the quality of visitor experiences. One review (Gramann 1999), conducted by the US National Park Service, introduced the new terms "natural quiet" and "soundscape" and contrasted them with the general term "noise"; these new terms were introduced to ensure that research on visitor experiences considers all sounds and the variability of how people perceive individual sounds and their sources. This review also suggested the need to study visitor experiences using a *psychoacoustics* approach, rather than a psychological approach (e.g., measuring sentiment), and urged that the measures used be correlated with environmental acoustic measures (such as dB). By using psychoacoustics approaches, researchers and park managers can assess the role of sound in all aspects of perception, including whether people find each sound annoying or pleasing, but can also focus on a dose-response relationship (i.e., how intense, common, repetitive, or random sounds or sets of sounds are and how they are then perceived). This recommendation was made because many studies up to that point had shown that some sounds are perceived the same no matter how intense or common they are, but others are perceived differently and scale with intensity and/or frequency. This summary work also helped to initiate the emphasis on understanding soundscapes as visitor experiences and as natural resources, and it set a standard for the evaluation of soundscapes in natural recreational areas (Anderson 1983; Sutton 1998; Miller 1999). Similar work has occurred outside the United States, most notably in New Zealand's national parks (Cessford 1999; Harbrow et al. 2011), Chilean Patagonia (Ednie et al. 2020), and provincial parks in Canada (Lemieux et al. 2016).

Park visitor experience research has also focused on relationships between sound sources and behavior. Pilcher et al. (2009) conducted what is known as a classic *dose-response psychoacoustics* study at the Muir Woods National Monument north of San Francisco, California, to understand how different sounds affect perceptions of the soundscape. The researchers created five different recordings spanning a scale from completely natural to mostly anthropogenic, with the first recording containing only natural sounds of the park and each subsequent recording adding more anthropogenic sounds (talking), so that the fifth recording contained very intense crowd sounds that masked the natural sounds. The recordings also varied in intensity, ranging from a low of 31 dBA for the natural sounds to 47 dBA for the crowd sounds. Visitors used headphones to listen to the five recordings and then answered a short questionnaire. The researchers found that natural sounds were most "pleasing" to people who were visiting national parks and that anthropogenic sounds were most annoying. However, they also found that a threshold existed at which the anthropogenic sounds began to elicit an "annoying" response. Anthropogenic sounds that were intense were highly annoying, suggesting that for some sound sources, intensity can also affect visitor experiences. This work suggests that thresholds in sound sources and intensities could be used as indicators to characterize the potential quality of visitor experiences in natural areas.

Summary

Terrestrial soundscape ecologists have developed a variety of tools, especially acoustic indices, and have tested them in a variety of landscapes around the world. Although much more work is needed to fully demonstrate their efficacy in all landscapes, in all situations, and at all times, most of the findings have been positive with regard to their utility. As many soundscape ecologists have concluded, AIs calculated from PAR data are relatively useful and powerful tools for conducting rapid assessments of biodiversity. The work that has focused on quantifying soundscape-landscape relationships has also been relatively positive as well, as land use/cover patterns, networks of roads, rivers, and topography have been shown to correlate with soundscape measures, mostly AIs. Finally, work that has been conducted in protected areas has shown that most of these landscapes experience anthropogenic sounds on a regular basis and that in some cases, these sounds detract from visitor experience and/or alter the behavior of animals in areas that are important for conservation.

Discussion Questions

1. Design a study that would test four AIs, as well as a new one that you developed, using PARs deployed in three forested landscapes: old-growth forest, secondary forest, and an urban forest park. Describe the study by specifying a duty cycle, the four AIs you would select to compare with your new AI, and other methods that might be necessary to test the usefulness of this new AI to monitor bird, amphibian, and insect diversity.
2. Design a study that would determine whether a planned highway expansion would have any effect on the experiences of visitors to a small park. Consider the kinds of questions you might ask, the theoretical concepts that might be taken into consideration, and how you might do this given that the highway has not been built yet. Also consider the types of collaborations that you might need and any special paperwork that might need to be filed with your university to do this work.
3. The Green Soundscape Index and the CUQI have been proposed to compare

sounds across different landscapes where natural and urban landscapes are integrated over a large region. Design a study in which you would place PARs across the region, calculate GSI and CUQI for each PAR location, and use spatial data from a GIS to analyze spatial patterns that you would then relate back to your GSI and CUQI measures. In your design, state at least three research questions that you would like to have answered as part of this study.

Further Reading

Buxton, Rachel T., Megan F. McKenna, Mary Clapp, Erik Meyer, Erik Stabenau, Lisa M. Angeloni, Kevin Crooks, and George Wittemyer. 2018b. "Efficacy of Extracting Indices from Large-Scale Acoustic Recordings to Monitor Biodiversity." *Conservation Biology* 32 (5): 1174–84.

Fuller, Susan, Anne C. Axel, David Tucker, and Stuart H. Gage. "Connecting Soundscape to Landscape: Which Acoustic Index Best Describes Landscape Configuration?" *Ecological Indicators* 58 (2015): 207–15.

Sueur, Jérôme, Almo Farina, Amandine Gasc, Nadia Pieretti, and Sandrine Pavoine. "Acoustic Indices for Biodiversity Assessment and Landscape Investigation." *Acta Acustica united with Acustica* 100, no. 4 (2014): 772–81.

12

Aquatic Soundscape Ecology

OVERVIEW. This chapter provides an overview of the use of passive acoustic recording technologies, acoustic indices, and in situ methods such as biological surveys to assess the status and trends of biodiversity in aquatic systems. The variety of marine and freshwater systems that have been studied to date from a soundscape ecology perspective is impressive; not only have temperate and tropical seas been studied, but systems such as kelp forests, seagrass ecosystems, coral reefs, mangroves, marine canyons, and ocean glacial systems have also been the focus of research. Freshwater systems that have been studied using the soundscape ecological approach include freshwater ponds, lakes, streams, and rivers. The effects of overfishing and invasive species, the importance of marine protected areas, the responses of sensitive marine systems to hurricanes and typhoons, the masking of biological sounds by anthropogenic noise, and the effects of artificial objects such as shipwrecks on local marine biodiversity have been the focus of marine and freshwater soundscape ecological research.

KEYWORDS: fish, hydrophone, riverscape, snapping shrimp, soundpeaking, soundscape orientation

---

12.1. Brief History of Acoustic Monitoring of Oceans

The use of PARs has a long history in research and monitoring in oceans. Much of the pioneering work was conducted in the 1960s, when marine biologists began to study the sounds of marine mammals such as whales (Schevill 1964, as reported in Howe et al. 2019). Recording technologies were not yet well developed, however, so recordings were often made using non-digital media (i.e., analog tape). In addition, many of the sound sources, especially fish and marine crustaceans, were poorly known, so interpretation of these recordings beyond identifying one or a handful of species was not practical (see the excellent historical summary in Mann 2012). Recent advances in underwater technology and a larger catalog of aquatic animal sounds have led to advances in our understanding of the spatial-temporal dynamics of ocean soundscapes as they relate to natural and anthropogenic changes. Howe et al. (2019) point to the development and deployment of PARs in the ten-year (1996–2006) Acoustic Thermometry of Ocean Climate (ATOC) pilot project, which attempted to link acoustic information with climate change patterns, as a watershed event that facilitated the ability of researchers to deploy automated PARs for long periods and to relate sounds to biological, geophysical, and anthropogenic sound sources. The ATOC pilot project focused on monitoring not only biological sound sources, which were known to be numerous and complex, but also sounds related to geophysical dynamics, such as wind, rain, earthquakes, volcanoes, and ice, and sounds produced by humans, such as ship noise and underwater construction. As the cost and ease of use of underwater PARs is now making this technology a very practical means of monitoring large areas of the world's oceans, embayments, and other marine systems, there have been numerous calls for the creation of acoustic sensors and application of acoustic indices in the existing *Global Ocean Observing System*, which would introduce passive monitoring of sound as one of the Essential Ocean Variables (EOVs; see Ocean Sound EOV 2018).

Much of the early work on developing acoustic indices for assessing the status and dynamics of aquatic systems focused on simple measures of sound pressure levels (full and quantified within biologically active frequency bands), weighted sound exposure levels (SELs), and metrics such as roughness and kurtosis (Miksis-Olds et al. 2018). As advances in the development and use of acoustic indices in the terrestrial environment progressed between 2008 and 2014, many of the AIs used to assess status and trends in those ecosystems began to be tested in marine, freshwater, and estuarine environments (Pieretti and Danovaro 2020).

12.2. Importance of Sound to Aquatic Organisms and Other Considerations

In the early 2000s, many researchers argued that all ocean sounds are useful cues for animals. Slabbekoorn and Bouton (2008), for example, posited that biological sounds serve several important functions for animals underwater. These functions include the obvious purpose of communication between conspecifics, but underwater sounds, because they can travel nearly 5 times farther in water than in air, can also serve as long-distance cues for orientation and navigation. Slabbekoorn and Bouton referred to this use of sounds as *soundscape orientation*.

Much of the work in ocean systems is based on the principles of acoustic properties described by Wenz (1962) (see Section 3.2.7). This research also focuses on characterizing soundscapes in the same way that terrestrial soundscape ecologists have for nearly twenty years; that is, by determining the sources of specific biological sounds and then characterizing how these sounds change over space, time, and disturbance state; how sounds from the geophysical environment change depending on the time of day, season, and geographic position on Earth; and the role that anthropogenic sounds play in the ability of aquatic animals to communicate and whether those sounds affect their behavior, physiology, and ability to survive. In marine systems, it is important to note that there are two major geophysical sources of sounds: wave turbulence at the surface, which creates low-frequency sounds, and wind-dependent processes, which produce high-frequency sounds. Most soundscape ecological research assesses biological sounds at low Beaufort scale numbers (often at, or below, 1).

Prior to the development of AIs for terrestrial soundscapes (including the classic AIs summarized in Section 9.4), many marine scientists focused on developing intensity-based AIs. The AIs most commonly used in marine studies focus on measures of sound pressure level (SPL) for the frequency bands of common sound sources; these frequencies include those of turbulence- and wind-dependent sounds (which vary with surface sea states), sounds from fish, sounds from snapping shrimp, sounds from other animals (e.g., sea urchins, whales), and sounds from ships and underwater construction.

The next section provides a summary of the advances in the early years of the transition from the sole use of intensity-based marine AIs to the exploration of classic AIs (ca. 2012–2015). During this time, soundscape ecologists examined how the classic AIs covary over space, time, and disturbance state, and how they correlate with other in situ and environmental variables.

12.3. Acoustic Indices: Assessments in Marine Systems

In aquatic systems, there has been considerable research on (1) how the classic AIs first developed in terrestrial systems compare with the acoustic measures more traditionally used in marine systems; (2) how the classic AIs differ across space and time; and (3) how AIs compare with more labor-intensive surveys done using traditional field observation methods or SEMs as a means of generating species richness, number of sounds, or other metrics.

One of the first studies to apply terrestrial AIs to marine soundscape recordings was that of McWilliams and Hawkins (2013). They applied modified versions of ACI (Pieretti et al. 2011) and ADI (Villanueva-Rivera et al. 2011) to marine recordings from forty-five locations off the coast of Ireland. The PARs were placed in three patches of different sizes in each of three benthic habitats (with five PARs per habitat patch). The researchers modified ACI to calculate a peak ACI for 1 kHz frequency bands, and they applied ADI to biologically active frequency bands; this method differs from the terrestrial use of AIs, which considers frequencies within a range from 1 kHz to the Nyquist frequency. The researchers also created two other AIs, one configured to identify pulse trains within a frequency range that was then used to calculate snapping shrimp snap counts, and another that quantified snap amplitude. Their modified versions of ACI and ADI correlated with snap count and snap intensity, which also covaried, and they found that that ACI and ADI differed among habitat types and patch sizes. Peak ACI values were in the 2–4 kHz bands. They also reported that ACI and ADI were sensitive to anthropogenic and geophysical sounds.

To determine whether there was a difference in the soundscapes of two very different reef systems, Staaterman et al. (2013) used power spectral density (PSD) measures (in dB re 1 μPa^2 Hz^{-1} for 100 Hz bins) across the 0–10 kHz frequency range. Recordings were made for 12 seconds every 5 minutes over 2 days. PSD values were then compared between sites and over time within the PSD bins. It was clear that the relatively healthy Panama reef systems had a much greater PSD across all PSD bins, and that the PSD peaks occurred during the nighttime for the Panama site, but there was no discernible PSD temporal pattern in the degraded Florida site, which was also much quieter (i.e., PSD values were always low across all PSD bins).

Another early marine soundscape study was conducted by Parks et al. (2014), who examined H_t, H_f, and H across three ocean basins (Pacific, Indian, and Atlantic) and compared the H AI with that of the content of a database developed from the same recordings, in which sounds from biological, geophysical, and anthropogenic sources had been labeled. A unique H_n (i.e., H adjusted for noise) was also calculated and compared with H for biological sounds in the database. The researchers found no relationship between H for the ocean basins and for the biological sounds in the database, but H_n did show positive correlations between the two. They suggested that there was considerable promise in the future for developing AIs that have efficacy in marine systems.

Staaterman et al. (2014) applied several traditional marine AIs, such as RMS amplitude, to the full spectrum and also across the 1 Hz to 10 kHz bands, divided into low-frequency geophysical sounds (<10 Hz), which come from turbulence and fluctuations of waves; low-frequency biological (a.k.a. "fish") sounds (10–1,000 Hz); and high-frequency sounds (to 10 kHz), which come from wind. They also calculated ACI for the low-frequency fish bands. The study used data from long-term marine soundscape recordings taken for 12 seconds every 5 minutes in two reef systems off the coast of Florida to determine how variable the soundscape was at night through-out the year. Wind speed, water and air temperature, atmospheric pressure, and water levels were used as variables to understand how these geophysi-cal dynamics influence the soundscape. They found that high frequencies varied over a diel cycle, but low frequencies varied over a lunar cycle, with peaks during the new moon. The two reefs did vary in the strength of these two temporal cycles. In all instances, ACI varied predictably across diel, sea-sonal, and lunar cycles, and the differences between sites were also signifi-cant. The researchers concluded that ACI had great promise as an indicator of biological activity and as an assessment tool for determining the health of marine ecosystems.

One of the next studies on marine soundscapes attempted to determine if any of the classic AIs or traditional marine AIs were positively correlated with in situ surveys of fish diversity and coral cover. Kaplan and colleagues (2015) studied three reef systems in the US Virgin Islands National Park. They deployed PARs and recorded for over 100 days using a continuous re-cording schedule. SPL were calculated for each recording for the 100–1,000 Hz and the 2–20 kHz frequency bands, which are the active frequency bands for fish and snapping shrimp, respectively. Octave band SPL (dB re 1 μPa) were also calculated for five frequency centroids. Relative power spectral densities, using the Welch (1967) method, and two of the classic AIs—H and ACI—were also calculated for each recording. Standard fish and benthic biodiversity surveys were conducted at each of the three reefs at the start and end of the study. The study found that diel trends in the low-frequency bands of the recordings were correlated with fish density and coral cover, but that diel trends in the high-frequency bands (where snapping shrimp create sounds) were not. H was also strongly correlated with SPL for low and broadband frequencies, but H calculated for all frequencies was greatly influenced by snapping shrimp, and the densities of these animals did not reflect biodiversity; thus calculations for H need to be done for frequencies of fish and other non–snapping shrimp sound sources. The researchers did, interestingly, find that ACI values did not correlate with fish density or coral cover, nor did ACI have a positive correlation with SPL for low "fish" frequency values. They concluded that the use of AIs in the marine environment will require modifications, as certain sound sources can unduly influence values across space and time, but with those modifications, AIs can correlate well with fish and coral cover measures.

This early work in applying classic AIs to the study of soundscapes encouraged further exploration of the use of AIs to examine a variety of problems in marine environments. Much of the work done since that time is summarized in Table 12.1. Most of this work, not surprisingly, has focused on coral reefs, as they are biodiversity hot spots for marine systems and are considered the most threatened of all ecosystems due to climate change (e.g., warming and acidification), noise, and overfishing pressures.

12.4. Assessments of Freshwater Soundscapes

One of the last ecosystems that soundscape ecology research has ventured into is freshwater systems. Compared with marine, terrestrial, and urban systems, freshwater systems, including lakes, ponds, rivers, and streams, have only recently become a major focus for researchers interested in studying biodiversity in these ecosystems. The sound-source catalog for freshwater fish is very poor (Rountree et al. 2002), as are those for freshwater turtles (Giles et al. 2009) and macroinvertebrates (O'Shea and Poché 2006). The freshwater fish sound libraries that do exist are biased toward North American and European systems and are based on species-specific fish bioacoustics work dating back to the 1980s. This section presents an overview of the work done in both lentic (e.g., still-water bodies such as lakes) and lotic (e.g., moving water, such as streams) systems.

Identifying sources of sounds in recordings of freshwater systems is challenging (Rountree et al. 2018). The biological sounds in these systems include snorts, grunts, fast repetitive ticks (FRTs), and surface acoustic events. Fast repetitive ticks are known to be common, but the source of these sounds is still unknown; some researchers speculate that they are created at the anus of a fish. Another common sound made by freshwater fish is produced when an individual breaks the surface of the water to gulp air or

Table 12.1 Summary of approaches used in recent soundscape ecological research in marine ecosystems

Reference	Objective (specific ecosystem studied in boldface)	AIs used	Supplementary data used	Findings
Nedelec et al. 2015	Assess the spatial-temporal variability of soundscapes in **coral reefs** and how it covaries with survey data	SPL in specific frequency bands, one-third octave SPL, and number of snaps by snapping shrimp per recording	Fish surveys at each site (daytime and nighttime) and habitat surveys, sea state	All measures correlated with in situ data. Positive correlations of AIs with fish and habitat surveys and negative correlations with snap number (expected)
Erbe et al. 2015	Determine how environmental variables like wind speed, rain, and ocean currents, and human activities like ship traffic, affect soundscapes of a **submarine canyon** that harbors high biodiversity	SPL in specific frequency bands, PSD for one-third octave bands, SNR for four whale species' frequency ranges, fish chorus frequency ranges	Wind speeds, precipitation amounts, ship traffic data (e.g., speed, size, position)	Many of the biological, geophysical, and anthropogenic activities correlated with SPL in their frequency ranges, thus demonstrating the full efficacy of passive acoustic monitoring as a means to characterize dynamics in marine systems
Rossi et al. 2016	Determine if soundscape studies can measure effects of ocean acidification on sounds produced by snapping shrimp, sea urchins, and fish in **temperate ocean CO_2 vent** areas, which are proxies for future CO_2 concentrations due to climate change (two CO_2 sites in Italy, control in Australia)	Full-bandwidth SPL (as RMS), PSD between 300 Hz and 10 kHz for three CO_2 vent conditions (extreme, high, and control) at seven sites for 3 hr after dark (i.e., dusk chorus); number of snaps; and a parallel study that examined impact of snapping shrimp	Previous survey data on sea urchin densities at sites, pCO_2 at sites	Elevated CO_2 reduced sounds from snapping shrimp and sea urchins as measured across the 3 hr period. Lab experiments with snapping shrimp demonstrated reduced snap numbers and intensities. As these snaps are important orientation cues for fish and other marine mammals, reductions could have serious consequences during future climate change

(continued)

Table 12.1 [*continued*]

Reference	Objective (specific ecosystem studied in boldface)	AIs used	Supplementary data used	Findings
Coquereau et al. 2016	Compare the soundscapes of two relatively similar **maerl beds** as assessed in 2004, when one had been opened to fishing and the other protected from fishing	SPL and peak frequencies for 10 s temporal bins of 10 min recordings made in 2014 and 2015	Fish and maerl cover surveys conducted three times per year, with comparisons of species richness for years 1992–2013	SPL and diversity of peak frequencies were greater in the unfished bed than in the fished bed; species richness in unfished bed did not change, but in fished bed, species richness declined by 30%, maerl cover by nearly 50%, and thickness by 90%
Harris et al. 2016	Examine multiple classic AIs as proxies for biodiversity in **temperate reefs** of New Zealand	Studied nine reef systems, used ACI, H, and AR and compared with sites where visual fish surveys were conducted	Fish species evenness and diversity, wind speeds, and ship traffic	ACI was positively correlated with fish diversity and evenness. AR and ACI were not affected by changes in spectral resolution, but H was. H was positively correlated with fish species richness above 140 Hz. Wind and ship sounds did not unduly influence ACI and H. ACI met four criteria for robust AI.
Freeman and Freeman 2016	Examine the use of soundscape recordings to assess health of **cool tropical coral reefs** in Hawaii at 23 sites that vary from healthy (e.g., protected) to very degraded.	Used PSD calculated for narrow frequency bands associated with the known sounds produced by urchins, several common fish, and crabs. Examined these at different times of day (morning, afternoon, evening, pre-sunrise).	Reef health quantified by comprehensive ecological assessment of urchins, crabs, fish, shrimp, and brittle stars, bathymetric complexity, and crustose algae (indicator of degraded reef) derived from time-lapse photographs	There was a positive correlation between biological sounds below 2 kHz and ecological assessment variables indicating healthy reefs, and a positive correlation between daytime sounds in the 2–20 kHz range (possibly from snapping shrimp) and variables indicating unhealthy reefs

Study	Objective	AIs/Methods	Validation	Findings
Pieretti et al. 2017	Compare variation over time in a traditional marine AI (PSD) and in ACI for a **shallow hard bottom** marine ecosystem in the Mediterranean Sea during a 2-month recording study	A specially configured set of ACI metrics based on frequency and fast Fourier transform (FFT); amplitude filters used; PSD calculated on 1 min recordings across 1–40 kHz frequency range using 312 Hz frequency bins	Checklist of sound-producing organisms developed by scuba-diving survey experts	ACI and PSD, when examined >600 Hz, varied together over time and at both of the sites. Nighttime ACI and PSD were greater than daytime. Diel rhythms for both AIs were very distinct.
Rice et al. 2017	Compare recordings over a 5-month period at four sites off **sandy bottom coasts** of Georgia and Florida as part of project for monitoring whales	Compared H, ADI (calculated with 200 Hz frequency bands), and ACI with L_{eq} at 5th, 25th, 50th, 75th, and 95th percentiles	—	H, ACI, and ADI varied over time and also among the four sites. Each was sensitive to different kinds of sound sources: H, ADI, and ACI were sensitive to ship sounds; black drum (fish) chorus influenced H and ACI but not ADI.
Elise et al. 2019	Determine the most efficient sampling settings across days, time of day, and moon phases for two **coral reef** systems in different states (healthy and degraded), one off Réunion and another off New Caledonia	Used ACI and SPL calculated on four bandwidths: 0.1–0.5 kHz, 0.5–1 kHz, 1–2 kHz, and 2–7 kHz; also examined four sample durations: 5 s, 1 min, 5 min, and 30 min	Ecological surveys of fish and benthic communities that confirm status of coral reef systems	ACI and SPL were able to determine coral reef state for all sample durations and across all frequency bands; short recordings and higher-frequency bands (<2 kHz) were adequate at night, but longer recordings (5 min) were most reliable for low-frequency bands (<1 kHz). Study emphasizes that recording duration and frequency bands are likely to be important to assess coral reef health.
Elise et al. 2019	Compare 2-hr audio recordings from two relatively pristine tropical **coral reefs** with fish assemblage characteristics from 2-hr video recordings	Used six AIs: H_s, H_t, H, ACI, BI, and SPL	Six ecological functions were measured using 12 variables (3 for coral cover, 9 for herbivore, planktivore, and tertiary consumers, described as abundance, diversity, and biomass)	This short snapshot of audio and video demonstrated that the AIs used had a positive correlation with all ecological variables for the two contrasting coral reef sites

(continued)

Table 12.1 (continued)

Reference	Objective (specific ecosystem studied in boldface)	AIs used	Supplementary data used	Findings
Heenehan et al. 2019	Examine the spatial and temporal occurrence of sounds from major sound sources recorded over 14 days in nine **Caribbean Sea** locations	Created long-term spectrograms that were used to score the presence or absence of five biological sound sources (humpback whales, minke whales, sperm whales, snapping shrimp, dolphins), an earthquake, and ship sounds	—	This study used no AIs but created a visualization from which data on the presence or absence of each of the major sound sources were keyed into a database for further analysis. An assessment of the diel patterns of whale calls and vessel noise showed that there were nearly continuous vessel sounds in two of the nine sites and that the most overlap of whale songs and vessel noise was during the day
Chary et al. 2020	Use soundscape recordings to characterize differences in spatial-temporal changes in biodiversity on the tropical seabed at three high-biodiversity sites, a **shipwreck site** and two nearby **natural reef sites**	Used SPL within frequency bands of known sound sources; compared three different kinds of microphone installations (seabed mount, moored, and hanging hydrophone); study focused on 7 days of recordings	Photo and video surveys conducted for fish (reported as percentage abundance for each species) and used to compare with SPL for specific frequency bands	All three types of PAR installations were able to characterize the temporal and spatial variability of the sites
Monczak et al. 2020	Characterize phenology of soundscapes in **estuarine** systems to assess biodiversity shifts due to climate change using 6 years of acoustic recordings	Plotted time series of SPL for high and low frequencies	Related these long-term acoustic trends to trends in mean spring water temperature, and other climate trends, and species richness	SPL varied predictably with trends in climate shifts in the spring, and SPL and species richness were positively correlated

to feed off the surface; these sounds are known as surface acoustic events (Rountree et al. 2018). To determine how sounds are made by individual species of freshwater fish, Rountree and colleagues (2018) used a hydrophone and an underwater video camera to record the behavior of six widespread species of freshwater fish in the New England region of the United States.

One of the first studies to characterize the diversity of freshwater soundscapes was that of Desjonquères et al. (2015). These researchers were interested in determining whether there were between-site and within-site temporal differences in the diversity of sounds in three different ponds in France. The three ponds differed in the type of vegetation cover in the surrounding landscape (forest, semi-closed, and open for ponds 1, 2, and 3, respectively). The researchers used three methods to quantify sounds recorded in the ponds. First, they used SEM to label sound types characterized by similar amplitudes and dominant frequency contours in a subset of their recordings. For each pond recording, sound-type richness and abundance were calculated. Second, they computed signal-to-noise ratios (SNR) by comparing small samples of the pond recordings that lacked any biological sounds (i.e., had only noise) with samples that contained biological sounds, and expressed these values as the ratio of the dB of the noise samples to that of the biological sound samples. A third assessment was made using six AIs: H_f, H_t, M, AR, NP, and ACI. They identified 48 sound types in the forested pond, 22 in the semi-closed-canopy pond, and 9 in the open-canopy pond. The ponds shared some sound types, but many were unique to each pond. The peak number of sound types varied across the ponds as well. The SNR differed between ponds 1 and 2, and between ponds 2 and 3, but not between ponds 1 and 3. Correlations of the AIs with the sound-type analysis and SNR were mixed. M, ACI, and NP were positively correlated with SNR and with sound-type richness and abundance. H_f and H_t were negatively correlated with SNR and with sound-type richness and abundance, and AR was negatively correlated with SNR but not correlated with sound-type richness and abundance. Although these results were not consistent, they did suggest that the use of acoustic indices and sound-type analysis could provide useful information for management of ponds.

Linke and colleagues' (2018) freshwater soundscape study of two water holes in a river in Northern Australia attempted to characterize the diel patterns of the soundscape. The researchers recorded continuously for 6 days and then created 5-second samples of the recordings at 10-minute intervals, which were then aurally labeled for sound types. This analysis resulted in

22 different biological sound types produced by fish and aquatic insects, along with another 22 sound types that were not biological or were considered to be from an unknown source. Several AIs (H, M, ACI, $ACI_{500-1,000}$ [fish], and ACI_{5-20} [aquatic insects]) were also calculated. Both measures of acoustic diversity were examined as 4-hour summaries. The researchers concluded that aquatic insects produce a peak of sound at night with intense choruses, that fish chorus at dawn, and that the AIs were sensitive to patterns of flow from a connected stream.

Few studies have characterized all sounds in freshwater lake ecosystems. One that has is a study of Minnesota lakes by Putland and Mensinger et al. (2020). These researchers collected 1-hour soundscape recordings from fifteen lakes, one each during the summer, when the water was open, and during the winter, when the lakes were covered in ice. They examined 1-minute samples of these recordings and scored each for the presence or absence of biological, geophysical, and anthropogenic sounds. Geophysical sounds included surface water sounds, sounds of ice breaking, and wind. Anthropogenic sounds that they identified included those from motorboats, canoes, and ice augers. PSD was calculated for 1 kHz frequency bands, and SPL was calculated for each recording. These researchers found that SPL was greater in summer than in winter for all the lakes except one, and that PSD for all frequency bands was also greater in summer than in winter. They suggested that the lack of wind during the winter, when the lakes were frozen, was the primary reason for the lower PSD and SPL in that season. They also found that the sounds produced by humans were in frequency bands that overlap with those of the sounds produced by fish, amphibians, and aquatic insects, suggesting that masking is occurring in these freshwater lakes. Given that over half of the world's 117 million freshwater lakes periodically freeze during the year, the study of soundscapes during winter is crucial to understanding how organisms in these systems survive and are potentially threatened by human activity, such as recreational activity in or at the edges of lake ecosystems.

There have been several noteworthy studies of the soundscapes of rivers, also called *fluvial aquatic systems* or *riverscapes* (Ward 1998). Desjonquères et al. (2018) compared the soundscapes of six channels on the Rhone River in France that varied in their hydrological connectivity to the main river stem. They deployed hydrophones attached to PARs in each of the six river channels. They also made traditional macroinvertebrate inventories of these locations, which recorded 142 animal taxa across all sites, and they recorded water temperatures. Labeling of the recordings revealed 128 distinct sound types, which were then divided into seven classes on the basis of acoustic similarities, and each recording was scored for the presence or absence of each sound class. The researchers then examined how sound types, water temperature, and macroinvertebrate inventories covaried. They found that amount of connectivity and water temperature each had a strong effect on acoustic community composition. Acoustic community diversity was positively correlated with macroinvertebrate diversity as well. These findings demonstrate that acoustic community composition correlates with in situ species inventory data as well as with the structural features of rivers and streams, and that a soundscape-riverscape relationship thus exists for rivers and streams.

Just as researchers have found that diel patterns exist in terrestrial and marine systems, long-term soundscape studies of rivers have shown that these patterns also exist in lotic systems. Decker and colleagues (2020) studied twelve streams across a large region of southern Australia by recording continuously for 3 days in each stream. They calculated AIs (M, H, and ACI) for each 1-minute segment of the recordings and used PCA to determine how the AIs grouped according to time of day. They identified five distinct statistical groupings of soundscape compositions (which they called silent, faint, day/night, dailyday, and flow). Dailyday had the greatest acoustic diversity of the five phases. When they examined temporal variability with hour of the day as an alternative approach to grouping using PCA, they found three distinct soundscape phases: night, twilight, and day.

A study of the soundscape of a tropical pond produced some interesting results. Gottesman et al. (2020b) deployed a hydrophone in a pond located at the La Selva Biological Station in Costa Rica and analyzed the recordings using aural annotation methods that resulted in a list of distinct biological sound types (probably aquatic insects and amphibians). The list was then used to calculate a sound-type diversity metric based on Shannon's entropy. The diel pattern of this metric demonstrated that the greatest diversity of sound types occurred late in the evening as a dusk chorus, and that nighttime soundscapes were more diverse than those during the day.

One of the unique aspects of freshwater systems is that they are commonly regulated by dams, which create reservoirs and unnatural fluctuations in streamflow. A major source of sounds in rivers is the movement of streambed sediment particles and their collisions with objects (e.g., rocks) in the stream. Another is the turbulence created by water flowing over structures in streams (e.g., tree limbs, boulders, stones, human-placed structures such as stream bank impediments). Fluctuations caused by the rapid discharge of water and sediment from hydroelectric dams can occur several times a day. These fluctuations in flow constitute a phenomenon called hydropeaking. The sounds of these discharges produce fluctuation in the acoustic environment of rivers and streams that has been called *soundpeaking* (Lumsdon et al. 2018).

To determine the effects of stream regulation by hydroelectric dams and the subsequent changes in sediment movement and flow rates, Lumsdon and colleagues (2018) measured the soundscapes of streams affected by these dams and compared them with those of undammed streams. They examined two different channel configurations—riffles and pools—in each stream. They installed hydrophones and acoustic particle velocity sensors in riffle and pool locations in both affected and unaffected streams. In the dammed streams, data were collected on dam turbine capacity, and discharge was measured at 15-minute intervals. Acoustic analysis focused on peak flow and base flow periods. The researchers calculated SPL for ten octave bands in 1-minute samples of the recordings. Some manual "filtering" was performed first, as there were several sources of artifacts in the data. The researchers found that the soundscapes were greatly influenced by dam discharge rates, and that the sounds most affected by increasing flow rates in both riffles and pools were those in the low-frequency range, which is the hearing range of most species of freshwater fish. This study points to important effects that dams built by humans may have on freshwater systems via hydropeaking (see Moreira et al. 2019 for review), as not only are most major rivers in the world already dammed, but society's need to increase renewable energy supplies by increasing the use of hydroelectric power is likely to increase dam construction in the near and distant future. Dams are also known to change how aquatic organisms move within stream and river systems, as they often create impediments to movement upstream and downstream. Thus dams can alter how aquatic communities are structured, and their biological sound sources can be altered as well.

How sounds are created, and how they can be perceived, by aquatic animals is often regulated at relatively fine spatial scales. As water flows through a stream (shown moving left to right in Fig. 12.1), the configuration of the channel can create *aquatic acoustic theaters* that vary depending on whether a location is a pool or a riffle. A riffle will create considerable turbulence that, depending on the three-dimensional streambed profile, may not penetrate pools (shown as masking by riffle noises). The channel configuration can also create quiet zones where bubbles dissipate (bubble entrainment). Stream depth also acts to filter certain frequencies of sound: wavelengths greater

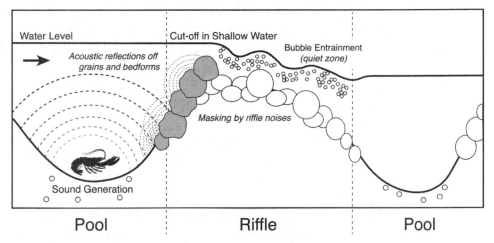

Figure 12.1 Effects of channel configuration on sound production
in streams. (Modified from Johnson and Rice 2014.)

than 4 times the stream depth are known to attenuate greatly (Tonolla 2009). Due to this *cut-off phenomenon* (Officer 1958), shallow water reduces the propagation of low frequencies; so the signals of aquatic animals that produce low-frequency sounds will not travel far. It has also been recognized that in river systems, "noise windows" (Lugli and Fine 2003), which are quiet acoustic spaces (generally 100–300 Hz), exist and are used by many aquatic animals, especially fish. John and Rice (2014) suggest that stream channel configuration and habitat, by facilitating certain "noise windows," are likely to influence aquatic animals' selection of sites to live and produce sounds in.

Hearing is only one of several senses used by aquatic organisms, and the characteristics of its *nonacoustic sensory capabilities* may influence the importance of sound to an aquatic organism. In addition to sounds, aquatic animals may use visual (e.g., light and turbidity) and hydrodynamic (e.g., flow rates and particle densities detected by hairs and air bladders) properties of their environment, as well as vibrations (e.g., how water movement affects substrate movement) and chemical cues. Acoustic cues can become more important, it has been argued (Johnson and Rice 2014), to aquatic organisms in deep pools where light does not penetrate due to depth, turbidity, the presence of woody debris, or the three-dimensional contours of the river bottom. There is also evidence (e.g., Popper et al. 2003) that aquatic animals can detect modifications of sound by the three-dimensional configuration of an aquatic area to determine the shape and configuration of the spaces around it.

Because freshwater systems are located within terrestrial systems, human activities on land affect the dynamics of aquatic systems—thus land and water function as a coupled human and natural system. Studies that attempted to use soundscapes to quantify terrestrial dynamics and patterns that are also related to aquatic dynamics would constitute comprehensive studies of soundscapes at the landscape scale. Proulx et al.'s (2019) review of the current knowledge of anthropogenic sound production in terrestrial and aquatic systems, which employed many of the basic concepts of landscape ecology (see Section 5.1), determined that anthropogenic sounds are present in both systems, that these sounds affect the communication of animals in both systems, that the habitat and its modification by humans alter sound propagation, and that some species appear to adapt to low-frequency sounds produced by vehicular traffic. They concluded that anthropogenic sounds in both terrestrial and aquatic systems are likely to mask the ability of many animals to use acoustic communication, that the modification of both terrestrial and aquatic systems will have confounding effects on the propagation of sounds produced by animals, and that changes in the landscape will also alter the physical characteristics of aquatic systems (e.g., by introducing more sediments to lakes and rivers, which will change temperatures and the fine-scale configurations of stream channels), resulting in altered animal community composition and thus changes in the acoustic community. Fish in streams differ in their acoustic sensitivities and thus may be affected differently by changes in those streams, whether caused directly by landscape changes or by anthropogenic structures such as dams that ultimately affect predator-prey relationships (Smith et al. 2018). There are some species, such as salmonids (e.g., trout), that have a high auditory threshold for sound perception and are thus able to live in river segments where a lot of sounds are created by turbulence (Kacem et al. 2020); these species could adapt better to areas where low frequencies are introduced, either directly by changes in streamflow or by sounds produced by motorboats.

12.5. Noise in Aquatic Systems and in Marine Protected Areas

The spatial-temporal pattern of noise (i.e., anthropogenic sounds) in aquatic systems is relatively well known, and documentation of the effects of noise on specific animal species—on their behaviors, their ability to communicate, their physiology—is rather extensive, particularly in the marine environment.

Anthropogenic sounds in the oceans have increased considerably over the past fifty to one hundred years (Hildebrand 2009), and the spatial extent of these sounds now encompasses most of the oceans (Duarte et al. 2021). Major sources of anthropogenic sounds include ships, with major shipping channels across the Atlantic, Pacific, and Indian Oceans producing the greatest intensities; fishing vessels; shoreline and offshore construction (especially blasts), oil and gas extraction sites; and recreational vessels. The frequency ranges of these anthropogenic sounds overlap, and thus potentially mask, acoustic communication signals from a broad range of animals spanning crustaceans, fish, marine mammals, and cephalopods (Fig. 12.2). The natural sounds of the oceans have also been modified, as many species whose signals propagate long distances (e.g., whales) have been hunted to very low population levels or even to extinction (e.g., the Atlantic guild of the gray whale). As human-induced climate change is leading to more intense and frequent hurricanes, the sounds of these storms may also be con-

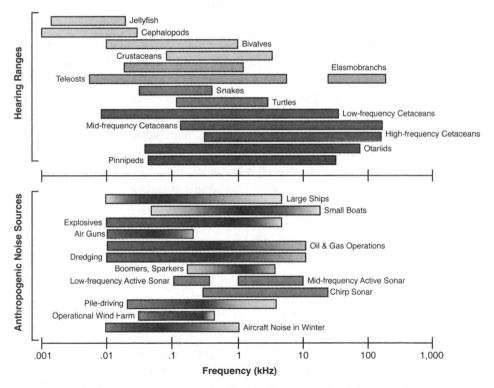

Figure 12.2 Frequencies of animal hearing ranges and anthropogenic sounds in the modern ocean. (Modified from Duarte et al. 2021.)

sidered a form of anthropogenically driven geophony that can mask acoustic communication by marine mammals. There is probably a greater proportion of anthropogenic sounds in marine systems than there is on land.

A few studies have examined the composition of soundscapes in marine protected areas (MPA). For example, Kaplan and colleagues (2015) studied the soundscapes of three reefs in the US Virgin Islands National Park using submerged PARs. The recordings were analyzed by (1) calculating a mean peak frequency; (2) visually and aurally inspecting the recordings to produce an LTSA plot, used to aid the listener in identifying boat sounds; and (3) calculating mean SPL for three frequency bands: 100–1,000 Hz for fish sounds, 2–20 kHz, in the range of snapping shrimp, and 100 Hz–20 kHz for all sounds. The researchers found that anthrophony was common in all three reefs, occurring in 6%–12% of the recordings; that boat noise occurred in the 100–1,000 Hz range, which overlaps the range of fish; and that the sound level in boat recordings was over 7 dB above ambient levels (i.e., in recordings with no boat noise).

Kelp forests are unique marine systems that support a diversity of plant and animal communities. A study of the soundscapes of Channel Islands National Park (Gottesman et al. 2020b) focused on comparing soundscapes within the marine protected area with those outside of the MPA. A 6-week set of recordings was analyzed using acoustic indices, and aural annotation was used to score a subset of the recordings. Prior surveys of these areas had shown that the MPA had a high density of kelp and a low number of sea urchins, which produce sounds by scraping their mandibles against rocks on the bottom. Outside of the MPA, kelp densities are low and urchin densities are high. The AIs showed low-frequency sounds, produced mostly by fish, to be greater inside the MPA than outside, and that sounds from sea urchins were dominant outside the MPA.

The presence of anthropogenic sounds in freshwater soundscapes is less well documented. In a study of the temperate freshwater lentic and lotic systems of New England (Rountree et al. 2020), 19 lakes, 17 ponds, 20 rivers, and 20 streams were recorded using a hydrophone and acoustic data logger. The composition of the soundscape was analyzed using acoustic indices developed from aural annotation of the recordings, and PSD was calculated for each of the sound types across its frequency bandwidth. An average PSD for ambient sound (no sound types) at each location was subtracted from the PSD of the sound types that occurred there. SPL for entire recordings were also used as an acoustic index. The acoustic indices developed from the aural annotation included sound-type rate (i.e., number of sounds per minute by type), percentage of recording in which a sound type occurred, and the amount of temporal overlap of biological and anthropogenic sounds. Sound types (Fig. 12.3) that occurred across all sites showed that many anthropogenic sounds (e.g., running boat, fishing) have frequencies that overlap with common sounds from fish (e.g., surface sounds, air movement, and FRTs). Anthropogenic sounds were the most common (occurring in 63% of all recordings and 92% of the time across all recordings), followed by biological sounds (occurring in 57% of all recordings). Rountree et al. (2020) also describe a phenomenon they call the *holo-soundscape*, which is the occurrence of terrestrial sounds that penetrate the water and are used, for example, by fish to avoid predators attacking from the air (e.g., gulls). The penetration of water bodies by sounds from the air is common; sounds of planes and birds can be heard in underwater recordings. Sounds from traffic, however, can vary depending on the topography around a water body; small hills can attenuate traffic sounds greatly (Pijanowski, pers. obs.). This study underscores the fact that most freshwater systems have a considerable amount of anthropogenic sounds, which have the potential to mask acoustic communication signals of freshwater animals.

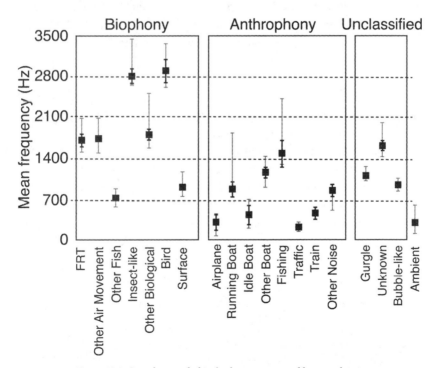

Figure 12.3 Sounds recorded in freshwater, grouped by sound type and mean frequency. (Modified from Roundtree et al. 2020.)

Summary

Many research efforts that parallel terrestrial soundscape work are evident in aquatic soundscape research. Comparing AIs with traditional marine metrics has been a particular focus of researchers for the past five to ten years. The use of AIs in aquatic systems is still relatively young, so more work is needed to test existing AIs developed for terrestrial landscapes, and there is probably a need for new AIs that are less sensitive to the different kinds of sounds in the aquatic environment. It is also clear that the marine system is far "noisier" than the terrestrial system, as sound travels farther in water than in air. Communicating this fact to the public may be challenging, as people rarely experience underwater sounds and thus might not be able to connect to the problem as readily as they could on land. Finally, it is clear that in comparison with terrestrial and marine systems, our knowledge of the soundscapes of freshwater systems is less. Given that most animals that are threatened on Earth either live in freshwater systems (e.g., fish and invertebrates) or depend on them for a portion of their life cycles (e.g., amphibians), more work in this area should be a high priority. As noted above, some freshwater systems, especially lotic systems, have been altered significantly by humans over the past several hundred years, and the effects of these alterations on the soundscape are poorly understood.

Discussion Questions

1. Compare and contrast the challenges and opportunities that soundscape ecologists face in working in marine versus freshwater environments.
2. Design a study that might compare several AIs in tropical rainforest ponds where the knowledge of species is poor. Consider the use of sound-type indices paired with aural annotation methods and how these measures would vary over space and time.
3. How might you make people aware of the level of noise that is occurring in today's oceans? Develop a set of public relations activities that would help to focus public attention on this major problem in the oceans.

Further Reading

Miksis-Olds, Jennifer L., Bruce Martin, and Peter L. Tyack. "Exploring the Ocean through Soundscapes." *Acoustics Today* 14, no. 1 (2018): 26–34.
Rountree, Rodney A., Marta Bolgan, and Francis Juanes. "How Can We Understand

Freshwater Soundscapes without Fish Sound Descriptions?" *Fisheries* 44, no. 3 (2019): 137–43.

Linke, Simon, Toby Gifford, Camille Desjonquères, Diego Tonolla, Thierry Aubin, Leah Barclay, Chris Karaconstantis, Mark J. Kennard, Fanny Rybak, and Jérôme Sueur. "Freshwater Ecoacoustics as a Tool for Continuous Ecosystem Monitoring." *Frontiers in Ecology and the Environment* 16, no. 4 (2018): 231–38.

13

Urban Soundscapes

OVERVIEW. This chapter focuses on the rich history of research and applications of urban soundscapes. It is organized around the following topics: (1) the major sources of sound in the urban environment and how they vary over space, time, and intensity; (2) how residents of the urban environment perceive these sounds and what approaches are used to assess those perceptions; and (3) urban planning approaches used to make urban spaces more sonically pleasant and healthy for humans.

KEYWORDS: sound perception, traffic noise, urban noise, urban park

---

13.1. Major Sources of Sound in the Urban Soundscape: Working Taxonomies

Research in urban soundscapes has been relatively isolated from that occurring in natural terrestrial and aquatic systems. There are probably several reasons for this, including the facts that most urban soundscape research is conducted by social scientists interested in planning, and that the focus of that research is on human perception of city acoustic environments. Furthermore, much of that work does not focus solely on the concept of "noise," in recognition that any sound can be perceived differently by different people; the term "noise" in this research community has a limited use, as it carries the connotation of sounds that are unwanted (by everyone), which may not be the case. The work of the urban soundscape community also has closer ties to that of Schafer and his colleagues, and it is also the community that led the effort to create an official international standard definition of the term "soundscape." Finally, one of the objectives of this area of research is the improvement of urban planning, which is, interestingly, why the term "soundscape" was originally coined by Southworth in 1967. As there are now more people living in urban environments than in rural environments, urban soundscape work has the potential to affect more people directly than the other areas of research and applications covered in this book.

A lot of the early work in urban soundscapes evolved from the need to understand the sounds of the urban environment for the purposes of improving the acoustic experiences of its residents. A clear merging of the research on urban noise with the Schafer soundscape concept began in the late 1990s and early 2000s. Several major advances occurred during this time. The first was to distinguish urban soundscape research from the approach that had traditionally focused on environmental noise management (Table 13.1). The urban soundscape approach was aimed at understanding how people perceive and react to all the sounds that occur in the urban environment and using this information for planning and management purposes. It valued sounds as a resource, assumed that all sounds are subject to personal preference (i.e., one either likes, dislikes, or has no opinion of each specific sound), and that sounds can be managed once sound sources and their patterns are measured and understood.

One of the other critical elements of this work was the development of sound-source taxonomies so that researchers could better understand the spatial, temporal, and intensity variability within the urban landscape.

Table 13.1 A comparison of the environmental noise management and soundscape approaches

Approach	Environmental noise management	Urban soundscape approach
Assessments	• Sound managed as a waste • Sounds as a source of discomfort • Measured as all sounds together (i.e., dBA)	• Sound perceived as a resource • Sounds as a source of preference • Sounds as differentiated by sources in space and time
Outcomes	• Human response to level of sound • Management by reducing or removing	• Quiet is not the object, but understanding sound source is • Management by focusing on planning methods that allow for promoting wanted sounds and reducing or masking unwanted sounds

Source: Modified from Brown (2010, 2012).

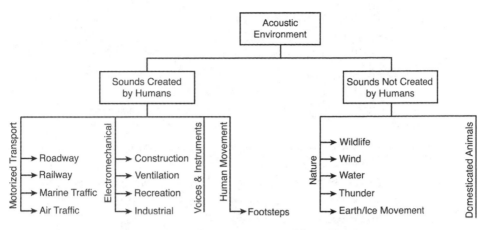

Figure 13.1 A sound-source taxonomy widely used in urban
soundscape research. (From Brown 2010, 2011.)

The urban sound-source taxonomy of Brown (2010), shown in Figure 13.1 organizes sound sources into two major categories: (1) sounds produced by human activity (both sounds from machines and non-machine sounds such as voices, music, footsteps), and (2) sounds not produced by human activity, which include natural sounds from sources such as wildlife, wind, and domestic animals. The more detailed portions of the taxonomies organize these sound-source types into a hierarchy of subcategories, as Figure 13.1 shows.

13.2. Patterns of Sounds in the Urban Landscape

The early work in urban soundscapes focused not only on how to categorize sounds by source, but also on how sound sources vary across space and time. For example, Botteldooren et al. (2006) developed a unique measure of sound that compared the temporal structure of all sounds (i.e., the soundscape) of a city with that of music. The motivation for this idea was to move beyond a classification of sound sources by intensity to a classification based on temporal structure. In addition, because the researchers were interested in how people perceive sounds of all types, they wanted a "baseline" metric that would be measurable in terms of the temporal structure of a sound rather than whether is it unwanted or "loud." Botteldooren and colleagues developed a measure of soundscapes, called *music-likeness*, that compared their temporal structure with that of several classical music compositions. They found that many urban and rural soundscapes had music-likeness, and that those that did not often had a single sound source. They also found evidence that highway and road noise could, in many instances, possess a music-like quality if the cyclical patterns of the sounds were not random. The researchers argued that music-likeness could be a more objective measure of sounds in the urban environment than intensity, as it removes the subjective judgment that is often made with some sound sources, particularly those in the urban environment.

Another approach that has been used by urban soundscape researchers has been to model sound sources as they vary across space and time. An example is the work of De Coensel et al. (2005), who developed a GIS-based traffic microsimulator that created maps of road sounds as a function of the source amplitude, distance from the source (i.e., propagation), and building profiles (height, depth, and width) using simulations of individual vehicles. The model was eventually compared with recordings of sounds in a Belgian city. As the researchers argued, developing a spatial-temporal model of a common sound source in the urban landscape is the first step to understanding the dynamics of the sounds it produces, how those sounds are experienced at all places in the landscape by people, and how those sounds may be one component of the soundscape of a place. They argued that modeling of sound sources could also be used to address what-if questions that might be asked during an urban planning exploration exercise.

One of the first landscape-soundscape assessments made was that of an urban area, Rostock, located in Germany. Liu et al. (2013) developed GIS maps of the city using Landsat TM–based land use/cover imagery, road and river networks, and satellite-based measures of vegetation density (e.g., NDVI) to create metrics of urban density, distance from roads and other spatial objects (e.g., parks), and land use fragmentation. Surveys were conducted at locations throughout the city on the presence or absence of twenty-eight sound sources in the three major categories (anthropogenic, biological, and geophysical) during eight times of the day (2-hour time blocks between 06:00 and 22:00), and each sound was rated for loudness on a Likert scale (1 = very quiet to 5 = very loud). They then used the GIS to compare the sounds present at each location, how variable those sounds were over time, and how they varied in loudness, with the structure of the urban landscape at that location. Their work showed that the urban soundscape was complex, driven spatially by the structure of land use/cover, and that its temporal patterns were driven by human activity and the activity patterns of wildlife. They also concluded that birds can be one of the few loud sound sources that are important components of the urban soundscape. Similar sound-source mapping and propagation modeling was also done by Aletta and Kang (2015), who showed that altering physical barriers around highways and introducing water sounds in nearby parks could improve the quality of the park soundscapes and reduce the spatial extent of traffic noise throughout the landscape.

One common approach used by planners to increase the amount of natural area–like space in cities is to provide what is sometimes called *green infrastructure*. They may create parks with lots of vegetation, vegetated rooftops, and tree corridors along streets and walkways, and may encourage residents to plant trees on their properties. Margaritis and Kang (2017) examined whether the amount of vegetation in a city—a measure of its green infrastructure—corresponded to the intensity of sound in twenty-five cities in Europe. They found strong evidence of a negative correlation between the amount of green space and the intensity of all sounds in those cities (i.e., the more green infrastructure, the quieter the city).

A study by Fairbrass et al. (2017) was one of the first to apply classic AIs in the urban environment. The objective of the study was to determine whether any of four classic AIs—ACI, NDSI, ADI, and BI—could be used to assess biodiversity in urban landscapes, especially in areas with a large amount of green infrastructure. They placed PARs at fifteen sites within the London metropolitan area that differed in their amount of green infrastructure, and they rated each site with a green infrastructure score based on spatial features from Google Earth (size of the vegetation patch and density of urban cover within a 500 m radius of the PAR). The PARs were used to collect soundscape recordings with a nearly continuous duty cycle (29 minutes on, 1 minute off), and AIs were calculated for 1-minute samples of these recordings. Samples were also labeled for the presence or absence of thirty-three common sounds. The annotated sounds were then summarized to soundscape class (i.e., biological, geophysical, or anthropogenic), with measures of the amount of activity (i.e., number of spectrogram pixels for a labeled sound source) and diversity (i.e., the number of unique sound sources, many identified to species) for each recording. The researchers found that ACI, BI, and NDSI were positively correlated with biological sound activity and diversity in the labeled database; that these same three indices were also positively correlated with anthropogenic sound activity; and that BI and NDSI were positively correlated with anthropogenic sound diversity. This study emphasized that the classic AIs provide different kinds of information about the composition of sounds in the urban environment. Some are more sensitive to biological sounds than others, although most are sensitive to anthropogenic sounds.

Several noteworthy examples of "what-if" scenarios for urban sound sources have helped researchers to understand urban soundscape dynamics. One was reported by Can et al. (2020), who used an expert focus group discussion and surveys to collect data on how residents would predict future sound intensity levels in urban landscapes given current "mobility" trends: that by 2050, the number of passenger miles driven is predicted to increase by 200%–300% and freight activity by as much as 250%, and that a doubling of vehicle flow would result in an increase of 3 dB in traffic noise levels. The researchers asked what kinds of changes in mobility would be required to reduce the sounds of traffic in cities under these circumstances. They concluded that, given urban expansion, increased ground and air vehicular traffic, and increased globalization of supply chains (e.g., goods being moved over longer distances), urban soundscapes are likely to increase in extent and intensity. They also suggested that shifts toward non-vehicular mobility, the introduction of electric vehicles, and shortening of the supply chain could compensate for the increases anticipated for the future. The second noteworthy study of urban soundscapes using a "what-if" scenario is that of Ulloa et al. (2021), who examined how the May 2020 COVID-19 lockdown in Colombia affected human activities and urban soundscapes. They used a citizen science approach in which participants were asked to download an app that would allow them to record their soundscapes and answer questions about sound sources and how they perceived the sounds. The variables the researchers collected (e.g., urban density, sound sources, perception, time of day, weather) were examined using three separate soundscape measures: (1) SPL, as RMS amplitude, for sound intensity; (2) a soundscape perception index, encompassing a host of variables, collected from participants, and (3) a multidimensional score that was reduced to two main components using the t-SNE algorithm (see Section 10.3). They found that after the full lockdown, there was an increase in sound intensities, an increase in low-frequency traffic sounds, and a decrease in birdcalls, leading to a decreased SPI; and that the strongest factor contributing to increased low-frequency sounds was urban density.

13.3. Measuring and Understanding Human Perceptions of the Urban Landscape

Studies on the perception of sounds by people in cities go back to the origination of the term "soundscape." Southworth (1967, 1969) conducted an exploratory study with three groups of subjects who were given tours of the downtown area of Boston. Members of one group were blindfolded and so could rely only on their sense of hearing. Members of a second group were given earplugs and earmuffs and thus could see, but not hear. The third group, which served as a control group, could see and hear. The three groups were given tours along the same route, with stops at thirty-three designated locations that were known to vary in their visual and sonic properties. Southworth (1969) found that all but five locations had enough unique auditory features to characterize the type of place. These locations included a fish wharf, a central area where church bells were common, and quiet residential neighborhoods (e.g., Beacon Hill) that contained sounds of people and various activities but were sheltered from foreground highway sounds. The blindfolded and control subjects paid close attention to unique and unexpected sounds, such as a police officer's whistle, sounds from a creaking boat, and the voices of old men, which, they stated, provided them with information about the location. The tour guide also asked blindfolded and control subjects to describe their level of "sonic delight" at each location. These subjects stated that low- and medium-frequency sounds were the most pleasing, that unique sounds were also pleasing because they provided information and because many were novel and culturally approved (e.g., sounds of birds, fountains, church bells), and that some of the pleasing sounds were too "soft." Sounds of cars were not always deemed unpleasant; those of idling cars were considered pleasant. The research also found positive correlations between visual and sonic scores of informativeness, uniqueness, responsiveness, and continuity for those locations that had a special "sonic identity" (Southworth 1967, 60). There were also interesting parallels between subjects who could only hear and those who could only see. Those who focused on the visual found that objects that did not move or had a common visual appearance were not that informative; objects that moved were also more pleasant than stationary objects. Areas that had unpleasant sounds, however, did not necessarily have unpleasant visual associations with them (e.g., locations with "loud" traffic noise did not have unpleasant visual scores).

One of the early arguments made for the use of the term "soundscape," as opposed to using terms like "noise" to denote the physical and psychological effects of all ambient city sounds on people, was made by Raimbault and Dubois (2005). Their argument for the use of terms like "soundscape" has several important points. First, they argue that a semantic analysis of sounds requires the use of nontechnical terms, consisting of a sound source and a qualifying verb, that ultimately provide the necessary meaning to the subject and the researcher. For example, "brakes squeaking" gives a more meaningful description of the sound for use in understanding the subject's perception and for practical applications, such as land use planning. Second, they argue that studies from the 1990s clearly show that people do not respond directly to the combined sources of sounds, but rather to component sources in ways that are complex and often are integrated with visual experiences as well. Thus, studying noise as measured by a simple "loudness" index (e.g., A-weighted sound pressure level) is not useful for understanding how people perceive their sonic spaces, nor is it useful for urban planning or for assessing potential effects on human health. Finally, they argue that sounds in urban environments should be studied from the perspective of the residents of cities, which emphasizes their functional characteristics, as well as for physical-technical features that are measurable and can be monitored over space and time (Fig. 13.2). They urge that city planners use the kinds of nontechnical terms used by city residents, as described above, as this helps to align planning "levers" for functional characteristics with residents' experiences. Following up on their work, they emphasized that the goal of "psychological investigations [into the sonic urban environment] is to figure out how the physical world affects people and how people elaborate a representation of the world on the basis of their sensory experiences"

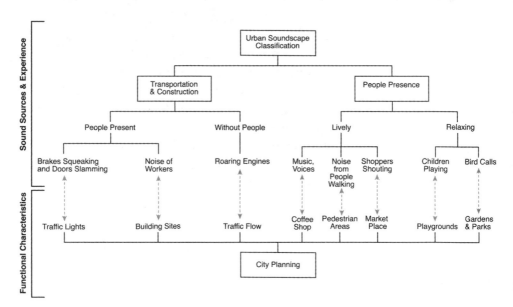

Figure 13.2 Urban sounds and planning spaces illustrating proposed alignment between the two. (Modified from Raimbault and Dubois 2005.)

(Dubois et al. 2006, 866). They distinguish two important aspects of this kind of research: (1) how people give meaning to sounds in the urban environment on the basis of their everyday experiences (the *psychological* dimension) and (2) how people convey that meaning through language that is part of a collective expression (the *linguistic* dimension). The latter dimension requires the development, using participatory methods with subjects, of a semantic sound classification system that can be used to compare the perceptions of different groups of subjects (e.g., age, income).

Research on urban soundscapes has also shown that *sound preference* in city landscapes is influenced by a variety of sociocultural, demographic, physical, behavioral, and psychological characteristics of people living in these environments. Yu and Kang (2010) studied the sounds in nineteen cities in Europe and China and assessed the effects of individual demographic characteristics on preferences for sounds of four types: natural, mechanical, human, and instrumental. They found positive correlations between age and the preference for natural sounds as well as the perception of certain sounds as annoying. Education had an influence on the preference for some sounds, but gender, occupation, and residential location did not greatly influence sound preferences except for bird songs, which was gender specific. They also suggested that sociocultural influences (e.g., when Europe was compared with China) are likely, but are not as great a contributor to sound preferences as the other factors.

A simple three-component model of soundscape perception was developed by Axelsson et al. (2010) using a set of 116 soundscape attributes, generated from a variety of environmental psychological lists, of perception (e.g., joyful, tranquil, messy), eventfulness (e.g., common, rare, lively; see Warren and Verbrugge 1984 for a more detailed description of acoustic events), and place (e.g., rural, city park). Fifty 30-second excerpts from binaural recordings of urban soundscapes (i.e., recorded so as to simulate the way sounds enter both ears) were then played to 100 listeners fitted with headphones. A list of the soundscape attributes was presented to the listeners on a computer, and each was asked to score five soundscape excerpts for those attributes on a scale of 0% to 100%. Researchers also listened to all fifty soundscape excerpts and labeled each with the types of sounds that were dominant in the foreground: natural, technological, and/or human-made. The recordings were also scored for intensity using several L_{eq} acoustic measures. The dimensionality of the data from the survey and the labeled recordings was reduced using principal components analysis (see Section 10.3) to determine which of the attributes grouped together. The resultant model (Fig. 13.3) produced three main components: pleasantness, eventfulness, and familiarity. Major sound-source types were strongly correlated with pleasantness and eventfulness and were weakly correlated with familiarity (Table 13.2). It is apparent from this table, and from Figure 13.3, that urban soundscapes can have positive perceptual attributes, that natural and human sounds have the most positive affect for urban residents, and that technological sounds have the most negative affect, but the features of these sounds are also important to a person's perception of soundscapes.

Several studies have examined how the backgrounds of urban residents and their demographic characteristics influence their perceptions of soundscapes. Liu and Kang (2016) conducted interviews with fifty-three residents of the city of Sheffield, England, and used grounded theory to

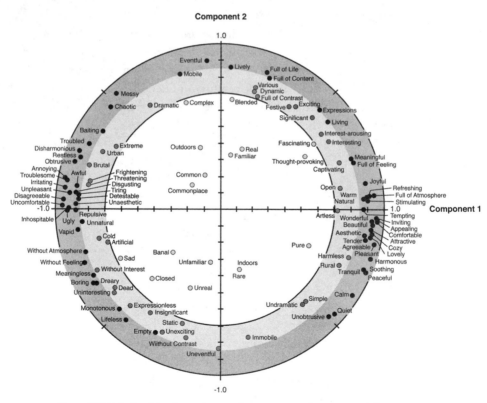

Figure 13.3 Relationship of sound perception terms to major perceptual dimensions in urban landscapes. Three zones (distinguished by shading) divide the component space into distances based on variance measures. (Modified from Axelsson et al. 2010.)

Table 13.2 Relationships between major sound-source types and a three-component soundscape perception model

Major sound sources	Pleasantness	Eventfulness	Familiarity
Technological	Strong negative	Weak positive	Weak positive
Human	Weak positive	Strong positive	Weak positive
Natural	Strong positive	Weak negative	Weak negative

Source: Modified from Table II of Axelsson et al. (2010, 2842).

develop a subjective understanding of soundscapes in urban landscapes. Three different kinds of interviews were conducted depending on the location of the interview: street, office, or home. Home interviews were conducted with elders to understand historical trends. Using the grounded theory approach of Glaser (1992), the researchers analyzed these interviews and found that subjects had five different perspectives on soundscapes in urban settings: (1) how the soundscape was *defined*; (2) *soundscape memory*; (3) a *soundscape sentiment* assessment; (4) *expectations* of soundscapes in the future, and (5) the *aesthetics* of soundscapes. They found that soundscape perception was not associated with the sound itself, but rather with the activity that produces the sound. Soundscape sentiment was often described using favorite sounds and annoying sounds as comparative descriptors of any sound. Emotion was a strong descriptor of any sound. The aesthetics of a soundscape increased with the number and intensity of natural sounds.

More recent work on perception of urban soundscapes has examined how positive and negative attributes of soundscapes influence the physical (i.e., physiological) and mental (i.e., psychological) health of urban residents. Much of this work has been based on earlier work by environmental psychologists who studied the health benefits of natural sounds, including research in attention restoration theory (see Section 6.8) (e.g., Kaplan and Kaplan 1982; Kaplan 1995), and research in the 1980s and 1990s on the harmful effects of urban "noise" on human health (see Westman and Walters 1981; Brown and Lam 1987; Lercher 1996; Schulte-Fortkamp 1999; Job and Hatfield 2001; Van Kempen et al. 2002). Aletta et al. (2018) reviewed the published literature that focuses on urban soundscapes and human health, and found two general trends. First, for large-scale studies that involved multiple cities and large numbers of subjects, a positive perception of soundscapes (e.g., reduced annoyance from less noise) was correlated with positive self-reports of health, and for small-scale studies that focused on small groups in one location, positive perceptions of soundscapes (e.g., pleasant sounds, calm soundscapes) reduced stress of residents. Specialized attributes of urban soundscapes, such as *quietness* (Tsaligopoulos et al. 2021), have also been explored for effects on human health. As many authors have argued, "quietness" is a difficult attribute to quantify. It can refer to low sound pressure levels, or to conditions where noise is neither absent nor dominant, but it also refers to specific spaces in the urban landscape that are visually pleasant and have soundscapes with few mechanical sounds and high amounts of eventful sounds like birdcalls and people talking. To examine how "quietness" may impact human health, Tsaligopoulos et al. (2021) developed a quietness index that examined three AIs—ACI, NDSI, and L_{eq}—both separately and combined as a composite urban quietness index (CUQI). When they applied it to soundscape recordings from parks and other natural areas (e.g., grounds of an ancient castle) across a city in Greece, they found that CUQI could serve as a proxy for quietness, as it had the largest value in the core areas of parks and a lower value at park edges.

13.4. Urban Planning and Soundscapes

Several national and continental initiatives have resulted in a set of policies that regulate sound (a.k.a. "noise") in urban environments. Although far from being comprehensive, this section's summary of the best-known noise regulations that have been enacted by national, international, and local governments is reflective of the currently poor state of governance of sound quality in urban landscapes.

13.4.1. Global Guidelines

A very broad set of noise policy guidelines have been developed by the World Bank's International Finance Corporation (IFC) and are often applied if no local, state/province, national, or intergovernmental/continental policy exists. Known as the Noise Management Environmental, Health, and Safety (EHS) guidelines, they suggest that controlling noise from stationary sources should involve measures at the source with the objective of reducing the noise below 55 dBA during the daytime and 45 dBA during the nighttime for residential areas. They also recommend that a noise-monitoring program be established to verify the presence of noise at levels that are considered unhealthy for residents.

13.4.2. European Union's Environmental Noise Directive

In 2002, the European Union enacted the Environmental Noise Directive (called Directive 2002/49/EC), which comprises sixteen articles that suggest how communities in the member states should control noise in cities and rural areas. The principal objective of Directive 2002/49/EC is to "define a common approach intended to avoid, prevent or reduce on a prioritized basis the harmful effects, including annoyance, due to the exposure to environmental noise." The directive seeks to establish definitions and measures of noise as well as outcomes from noise assessments (e.g., noise maps, action plans, acoustic land use planning measures, data sharing, public outreach, reporting and reviewing, and establishment of governing rules across local, national, and intergovernmental organizations). A detailed set of guidelines are also provided to local governments on standard measures of noise, procedures for developing noise maps, and guidance on coordinating planning across local and national units of government. As argued by Aletta and Kang (2018), the directive has led to the concept of "urban sound planning."

13.4.3. US Noise Control and Abatement Act of 1972

In 1972, the US Congress enacted broad environmental legislation that focused on protecting air and water quality. One of these measures was the Noise Control and Abatement Act of 1972, contained in the Clean Air Act (as Title IV), which instructed the US Environmental Protection Agency to promote an environment free of noise for all US citizens. As a result, the EPA established an Office of Noise Abatement and Control, whose purpose was to determine how to control the noise that was considered a "growing danger to the health and welfare of the Nation's population particularly in urban areas" (Sec. 2, 42 U.S.C. § 4901) and to coordinate all noise control policies with "state and local governments." Congress allocated funds to the office to help support its mission. The act focuses on three main goals: (1) coordinate and fund research on noise; (2) establish a set of standards for monitoring and measuring noise produced by products distributed in commerce; and (3) create a public education program on noise and its effect on public health. A few years later, in 1978, the US Congress enacted the Quiet Communities Act (S. 3083), whose purpose was to "promote the development of effective State and local noise control programs." This law, which was an amendment to the 1972 Noise Control and Abatement Act, focused on funding research on noise and its impact on residents in urban areas, noise monitoring, and public education programs. In 1981, EPA defunded the Office of Noise Abatement and Control, as it decided that noise abatement and control measures were best employed at the state and local levels. Despite this shift in funding, the 1972 and 1978 acts are still on the books today.

A not-for-profit group called the Noise Pollution Clearinghouse (www .nonoise.org; accessed April 3, 2021) conducted a study of 491 noise ordinances (local policy laws) of the nation's largest communities. They found that most communities had noise ordinances based on a nuisance standard (85% of communities), but that some had restrictions based on zoning (65%), decibel levels (55%), or time of day (e.g., quieter at night, 47%), and that some defined specific quiet areas in land use plans. In terms of dBA standards, communities varied widely in their allowable sound intensities (which ranged from 50 dBA to 85 dBA). As pointed out by the author of the report, coordination across all units of local government, of which there are well over 40,000 at the municipal, township, and county levels, represents a complex and far from standard approach to monitoring and controlling noise in urban environments. According to the Noise Pollution Clearninghouse, there is also little guidance on how to enforce any local noise ordinance.

13.4.4. Other Countries

Several notable national soundscape initiatives in areas outside the United States and Europe are assessing soundscapes in urban landscapes for the purposes of land use planning. Nearly all these national initiatives were inspired by Schafer and colleagues' World Soundscape Project (see Section 14.1). In the 1990s, the Japanese Environmental Protection Agency initiated a national program called "100 Soundscapes of Japan: Preserving Our Heritage," in which the agency requested submissions of important natural and cultural soundscapes that should be protected. One hundred soundscapes were selected for further research on their soundscape composition, how those sounds are changing, and how Japanese people perceive these soundscapes and how they are changing (Ge and Hokao 2004). The goal of the program is to preserve Japan's unique natural and cultural soundscapes for future generations through better planning and noise control. The Commonwealth of Australia (Australia Department of Health 2018) recently reviewed the current literature on the human health effects of noise and the benefits of quiet spaces and natural sounds and produced a report to be used for the purposes of planning. Citing the European Union's Environmental Noise Directive, the report concluded that there is considerable evidence that noise affects human health; that increased efforts to reduce noise, especially in urban settings, are needed; that more research is needed to increase the understanding of noise, noise control and abatement, and how these measures can improve human health in the urban environment; and that environmental noise needs to be addressed in land use planning (EN 2018, 64). Colombia's noise control laws have origins similar to those of the United States. In 1974, when Colombia's government passed laws that established guidelines for air and water quality, one of the air quality measures addressed noise. This original law was then expanded in the 1980s and 1990s to include monitoring and control of noise for public health (Blackman et al. 2006, 60).

13.4.5. Research on Soundscape Planning in Urban Landscapes

Although land use planning has traditionally taken a noise-based approach to managing sounds in the urban environment, many researchers in urban soundscape planning have stressed that there is a difference between managing the acoustic environment for noise and managing all sounds as a soundscape. Brown (2010, 2012) has argued that "soundscapes and environmental noise management are complementary, not competing, approaches with respect to the study and management of the outdoor acoustic environment." (2010, 493). As such, sound management needs to focus on both the composition of the soundscape and listener perception of the soundscape. Much of the work in urban soundscapes emphasizes the ISO definition that acknowledges the duality of the soundscape—the physical presence of all sounds but also how sounds are perceived by those who experience them. Urban soundscape planning has also evolved to a point where it can focus not only on the visual identity of a city or place within the urban landscape, but also on its "sonic identity" (*sensu* Schafer 1994), which should match its visual identity. This section reviews the work that has focused on urban soundscape planning.

Some of the early work in urban soundscape planning focused, as it continues to do today, on participatory activities involving soundscape researchers, land use planners, architects, and the public. De Coensel et al. (2010) reported on the redevelopment of a former industrial site in Antwerp, Belgium. The planning process involved a host of public information meetings, planning scenario focus groups, and input from the public over several stages (i.e., from early design ideas for buildings and landscapes to simulations of three-dimensional spaces to more detailed plans along with aural simulations of the spaces). Researchers analyzed the existing soundscape of the area proposed for redevelopment and reported their findings to the planners and architects. Interviews were conducted with nearby residents relating to their quality of life and preferred environmental acoustics for the new space. Researchers also used land use planning scenarios that included architectural designs of buildings to develop soundscape maps of each redevelopment option. Residents and land use planners were able to select their most preferred option from these planning scenarios.

A considerable amount of effort in urban soundscape planning has centered on the design of urban parks, an activity sometimes called *urban green space planning*. For planners, urban parks serve a variety of functions in the urban environment. They are places for recreation, or "escape" from the stress of the city, where natural features are preserved so that residents can experience the visual and auditory features of dense vegetation and natural features such as streams, ponds, and wetlands. Urban soundscape ecologists have sought to understand how people who visit urban parks perceive the soundscapes of these types of urban landscapes. Some research on urban park soundscapes has suggested that for residents to have a fully positive experience in an urban park, its visual qualities need to align with its sonic qualities. In other words, if sounds from traffic and construction permeate an urban park, the full sensory experience is not a good one for visitors (Brambilla and Maffei 2006; Jeon and Hong 2015; Wilson et al. 2016; Jaszczak et al. 2021).

Many cities around the world have historic sites where certain attributes of soundscapes are culturally important. For planning purposes, these soundscapes should be preserved, as they are part of the historical context of the place. Jia and colleagues (2020) examined how various locations in the ancient city of Tianjin, China, should be characterized and preserved using surveys of its residents. They issued a survey to residents, tourists, and planners in Tianjin, which was used to identify locations in the city that should have their soundscapes preserved. In the second phase of the study, twenty-three residents (architectural students from the local university) participated in a soundwalk at five locations and then answered a questionnaire. The respondents expressed the beliefs that the need to preserve the natural and cultural soundscapes of urban parks and historic sites is urgent;

that social/communal sounds, animal sounds, and the sounds of water should be preserved; that the experience sought for these spaces is relaxation; and that vibrancy and strength should be a goal for planning as well.

Kogan et al. (2018) argue that their Green Soundscape Index (GSI, described in Section 9.4) is useful for urban soundscape planning. They developed the index using survey information from subjects listening to sounds at eight locations in the city of Cordoba, Argentina, who were asked to score the presence of natural sounds and sounds from traffic on a five-point scale. They also collected additional information from the subjects to quantify their visual and acoustic experiences, using approaches like that of Axelsson et al. (2010). The researchers concluded that their simple GSI could be used to create maps of the experiences of residents in urban landscapes, which could then be used for planning purposes.

One of the important goals of land use planning is to address inequities in services, exposure to hazards, and other socioeconomic disparities that commonly occur in urban areas. Brainard et al. (2004) conducted one of the first noise exposure studies that examined levels of urban noise occurring in neighborhoods of people of color and people with low incomes. They conducted a spatial analysis that used census data and a noise map created using simulation models. They concluded that there was some evidence, although it was not strong, of a relationship between noise exposure (intensity and duration) and ethnicity and socioeconomic deprivation. A similar study was conducted by Casey et al. (2017), who examined these relationships at a national scale in the United States. By comparing US Census data on race/ethnicity, socioeconomic status, and land use types (e.g., industrial, residential) with measures of noise exposure (intensity and duration), they showed that noise levels, especially at night, were greater for those areas with a high proportion of nonwhite and low-income residents.

As there has been an emergence of research that leverages social media platforms and tools, some urban soundscape researchers have begun to use mobile phone apps to engage the public as "active sensors" in their research. Radicchi et al. (2018) developed a mobile app, called Hush City, that allows residents of a city to key in information about the sounds of a place and planning options that might exist for places such as urban parks and other quiet areas. They argue that apps such as the one they have developed allow residents, researchers, planners, and architects to communicate about the needs and desires of residents in near real-time and thus transform urban spaces into "smart cities" that use technology to integrate a variety of activities of residents and service providers.

Summary

There have been considerable advances in research in, and planning for, urban soundscapes. Substantial effort has been invested in trying to understand human perceptions of sounds in urban landscapes and how they differ from perceptions of sounds in natural or rural areas. This community of scholars has attempted to develop several standards that could be applied to many areas of the world, although most of the motivation for their work has been to focus on improving the soundscapes of urban areas in the European Union. Standards that have been developed include an ISO definition for the term "soundscape," the use of a series of descriptors that relate to how humans perceive soundscapes, and the simplification of these descriptors into a set of scales usable by land use planners. There have also been attempts to study multiple urban systems to determine generalizations that may apply to either cultural dimensions of soundscape perceptions or use of research results in urban planning. Social science techniques are applied often, and the classic AIs have also appeared as part of these research studies. The work in urban soundscape ecology is also designed to apply to several multinational efforts to improve the quality of sounds in cities.

Discussion Questions

1. Design a study in which a large university class would be asked to perform soundwalks to various places on campus. First, develop a map of your campus that includes the known soundscapes and the properties that exist there. Consider including, for example, places such as fountains, reflection spaces, sporting event venues, and locations of high and low traffic. Next, design a semi-structured survey, of no fewer than ten closed-ended questions and one open-ended question, that would assess how students perceive these spaces. Think creatively, as university/college campuses provide unique experiences for students. These experiences are cultural and have important sense of place characteristics. Finally, propose how you would analyze these data.

2. A land use planner has requested your help in determining how to improve the soundscapes of their city. The city is currently considering transforming several key spaces, perhaps into parks, walkways, and other green spaces. How would you go about helping them determine what should be done to improve the soundscapes of these spaces? In your plan, consider the use of

local media, town hall forums, online surveys, street intercept surveys, and the use of recording devices such as PARs or just handheld phones.

3. Discuss what the future might be like given the technological changes predicted to occur in the next twenty to fifty years that would affect the soundscapes of urban spaces. Consider, for example, the complete conversion of all vehicles from combustion to electric engines, the placement of major highways that run through cities underground using boring technologies, and the use of building materials that could absorb sounds of the city rather than reflecting them back to spaces that people occupy.

Further Reading

Kang, Jian. *Urban Sound Environment*. CRC Press, 2006.

Aletta, Francesco, Tin Oberman, and Jian Kang. "Associations between Positive Health-Related Effects and Soundscapes Perceptual Constructs: A Systematic Review." *International Journal of Environmental Research and Public Health* 15, no. 11 (2018): 2392.

Brown, A. Lex. "A Review of Progress in Soundscapes and an Approach to Soundscape Planning." *International Journal of Acoustics and Vibrations* 17, no. 2 (2012): 73–81.

14

Sociocultural Soundscapes

OVERVIEW. Several areas of scholarship have focused on the sociocultural dimensions of soundscapes. As such, this area of soundscape applications has been dominated by scholars who consider themselves acoustic ecologists, ecomusicologists, ethnomusicologists, or ecoethnomusicologists (see Sections 6.2–6.4 for more detail). The sociocultural soundscape scholarship areas summarized in this chapter include soundscape compositions, sonification of scientific data, ecoethnomusicology, and environmental psychology. Much of the work presented in this section was inspired by R. Murray Schafer's (1977, 1994) seminal work but also draws from the anthropological work of Feld (1986) and Seeger (1987), the perceptions work of Ingold (2000), Gibson's articulation of perception in animals and people (1950), the sense of place conceptualizations of geographer Tuan (1977), the ecomusicology work of Titon (2013) and colleagues (Allen and Dawe 2016), and the sensory ethnography of Pink (2015).

KEYWORDS: acousmatic music, attention restoration theory, nature connectedness, sonification, soundscape compositions, World Soundscape Project

14.1. Acoustic Ecology and the World Soundscape Project

R. Murray Schafer emphasized the need to move beyond the previous work in studying acoustic environments, which focused on the "physical specimen to be dissected" to the exclusion of "what may be called the perceptual and cognitive primitives and procedures or concepts derived by them" (Truax 1996, 37). In other words, the soundscape needs to be understood holistically through cognitive and perceptual means rather than solely as measured by scientific instruments.

To address this deficiency in scholarship on environmental acoustics, Schafer launched the *World Soundscape Project* (WSP; Truax 1974, 1996), at Simon Fraser University in Vancouver, Canada. Founding members included Schafer, Barry Truax, Howard Broomfield, Bruce Davis, Hildegard Westerkamp (a student of Truax), and Peter Huse. The main activities of the WSP (Truax 1974) included making field recordings, employing electronic tools to develop soundscape compositions, hosting radio shows containing natural sound recordings and narrated stories, developing and hosting a variety of pedagogical methods, such as soundwalks, the development of instructional materials for the classroom and for the public, a comprehensive study of the local Vancouver soundscape, and a survey of community noise laws and ordinances in Canada. The seminal books produced by the WSP included Schafer's *The New Soundscape* (1969), *The Book of Noise* (1972), *Tuning of the World* (1977), *Music of the Environment*, and *Five Village Soundscapes* (1977; see also Järviluoma 2009) and Truax's *Acoustic Communication* (1984) and *Handbook for Acoustic Ecology* (1978). Many scholars from around the world came to Simon Fraser University to learn about the WSP's approach to studying the soundscape. The WSP developed the World Forum for Acoustic Ecology (WFAE) in 1993. The WFAE has supported *Soundscape: The Journal of Acoustic Ecology*, started in 2000; the forum also endorses and promotes conferences. Regional WFAE chapters also exist in Australia, Japan, Greece, Finland, the United States (as the Midwest Forum for Acoustic Ecology), and the UK/Ireland. Most WFAE members consider themselves artists interested in collaborating with scientists and engineers focusing on composing, public education, or multidisciplinary analysis of specific soundscapes.

14.2. Soundscape Composers

Many artists have created compositions that either incorporate soundscape recordings or create a unique experience that moves beyond simply using instruments or voices. A selection of key soundscape composers, along with the rationale for their soundscape compositions is provided here, with references to specific pieces.

Barry Truax

Barry Truax (1996) argues that simply inserting or integrating environmental sounds into musical compositions is problematic for several reasons. He maintains that environmental sounds are more complex than sounds in music, and that there is a lack of tools and methods that can be used to "parameterize" (Truax 1996, 52) environmental sounds with ordering schemes that exist in music composition. In addition, environmental sounds do not have a syntactic organization; in other words, there is no process of encoding and decoding like that in music or speech. Environmental sounds are also context based, and thus their meaning is grounded in place and time; in this way, they differ entirely from traditional musical compositions, for which place is not part of the context. Music often draws focused attention, whereas environmental sounds are perceived with less conscious effort by the listener. Finally, the normal process of listening to environmental sounds differs from that of listening to music: environmental sounds surround the listener and are constant; traditional musical compositions are typically hosted in concert halls with stationary speakers and stages for musicians.

To address these challenges, Truax (1996) proposed a different framework for creating soundscape compositions. He argues that they can be created by integrating knowledge, generated from soundscape studies, of how people perceive environmental sounds; approaching composing and listening as a form of acoustic communication (a multi-level, information-exchange form of listening to all sounds, environmental and non-environmental); and then presenting a composition to listeners that "evokes" the knowledge of the environment and psychological context so that the composition can be experienced with meaning. A soundscape composition is part of an *acousmatic music* experience—that is, one that is composed with the intent of its being presented with digital devices (e.g., computers) through speakers.

Truax developed a technique he calls "real-time granulation" or "time stretching" of sample sounds (a.k.a. environmental sounds). This technique involves taking a small slice (~50 milliseconds) of sampled sound (e.g., waves from an ocean, the bells from a church) and reproducing it at hundreds to thousands of repetitions, which ends up creating a composition without sharp transitions, as there is a level of randomness of sound patterns introduced by the technique. This produces an acoustic "volume" in the sampled sounds, which become "temporally rich" and have a "larger than life vocal characteristic" (Truax 1992). His works *Pacific* (1990), *Dominion* (1991), *Basilica* (1992), and *Song of Songs* (1992) employ this technique (which the composer explains at https://vimeo.com/12474371). *Pacific* is a four-movement piece that uses recordings from the local ocean environments to create four distinct aural spaces (Truax 1992). The first movement,

Ocean, uses the sounds of waves breaking along a shore that are then temporally expanded using his "time stretching" technique. Certain sounds
(high frequencies) are filtered out to create sonic symbolisms of water and
life. The second movement, *Fog*, is based on recordings of ferry-horn sounds
from boats in the Vancouver harbor. *Harbor*, the third movement, uses the
calls of gulls that are fighting among themselves for food with background
sounds of ocean waves. The sounds of the Chinese New Year, created from
drums, cymbals, and firecrackers, are part of the last movement, called
Dragon. The four movements together present a location with contrasting
energy and mood, communicating the life of this place through the geophysical, biological, and human dimensions that exist there. The use of
the real-time granulation also adds to the expansiveness and power of the
ocean (Truax 1996).

Aside from being a soundscape composer, Truax has written extensively
on the psychology and ecology of sound (Truax 2001), on "acoustic sustainability" (Truax 2012b), and on "aesthetics" as it relates to natural soundscapes and soundscape composition (Truax 2012c).

Hildegard Westerkamp

Another original member of the WSP, Hildegard Westerkamp, has produced
several well-known soundscape compositions that are often cited as innovative pieces by the acoustic ecology community. Westerkamp (2002, 3)
defines a soundscape composition as "the artistic, sonic transmission of
meanings about place, time and environment and listening perception."
Westerkamp (1999, 3) has argued that the "schizophonic" listening experience that can occur when natural sounds are mixed with, and even masked
by, sounds produced by machines, thus creating a situation of sounds that
are placed-based and those that are not, can be a useful tension ("a medium
to awaken our curiosity"). Westerkamp (1999) has also argued that a soundscape composition can be a meaningful place for listener and composer to
"explore together." Westerkamp (2006, 2) has also promoted the use of the
soundwalk (see Section 9.9), which can be introduced alongside any soundscape composition or used by itself, as listeners and the environment can
create a unique "piece together."

The most noteworthy works of Westerkamp include *Kits Beach Soundwalk*
(1989), *Cricket Voice* (1987), and *Beneath the Forest Floor* (1992). *Kits Beach
Soundwalk* is a soundscape composition made from recordings at Kitsilano
Beach in Vancouver. Sounds from the ocean on a calm, early summer morn-

ing are mixed with a soft voice narrating the visual elements of the scene of the natural environment. The piece weaves city sounds in and out after a sequence of natural sounds, mostly of water, presented at the beginning. The latter half of the piece transitions to dream sequences that contain natural sounds along with stories about visits to rural homes that had been lived in for generations. The composition introduces the listener to the calmness of the beach, but also the stresses of modern society, its sounds, and the tragic events that are common in urban environments. Westerkamp's *Cricket Voice* (1987) focuses on the sounds of one cricket in a Mexican desert at night. A recording of the cricket is used to create a series of rhythmic sounds that are played on top of low-frequency electronic pulses. Another set of recordings made that same night, of people using objects to play on the spikes of various cacti and other desert plants, are integrated with the sounds of the cricket. This composition is meant to describe the rhythm of the desert during its most quiet period. In *Beneath the Forest Floor*, low thumping sounds, mixed with the sounds of birds, insects, small mammals, and water, are the focus. The sounds are mixed in a way that is unrealistic; they are sounds that can co-occur, but their sequencing and quick fade-ins and fade-outs make this a surreal natural soundscape. The low thumping sounds, another surreal feature of this composition, create a sense that underground sounds dominate this forest soundscape.

Westerkamp's work has, over the past few decades, become almost "iconic" to the WSP movement, not only characterizing sounds of all places and their effects on human perceptions of place, but also demonstrating how compositions can transcend that purpose to promote the preservation of the natural sounds of any place.

David Monacchi

David Monacchi's *Fragments of Extinction* project highlights the biodiversity crisis using natural soundscape recordings from tropical rainforests, such as those from Borneo, made with advanced microphone arrays (Monacchi 2013). These eight-channel recordings are played back in a bioacoustic theater: a dome that allows listeners to hear and view the spectrograms of soundscapes as manipulated by a conductor. Three different kinds of soundscape compositions can be experienced in the theater: a sound documentary, acoustic transformations, and ecoacoustic compositions. The piece *Integrated Ecosystem* is played as an interactive ecoacoustic composition that allows for a performer to make alterations in temporal and spectral *acoustic niches* that would occur if a species were removed from the soundscape (i.e., went extinct). Like the works of many other acoustic ecologists, Monacchi's *Fragments of Extinction* and his *Integrated Ecosystem* piece are motivated by the desire to increase awareness of the biodiversity crisis (Norman 2012); their use of visuals such as spectrograms, a standard visualization tool used by scientists and audio engineers, integrates art and science for their presentation.

John Luther Adams

Although arguably not exclusively a soundscape artist, the Pulitzer Prize–winning classical composer John Luther Adams has on occasion used environmental sounds as part of his compositions, which urge listeners to consider the dangers of climate change to society and nature. A few of his works are worthy of consideration here as further examples of soundscape compositions that are relatively current. His album *The Place We Began* (2009) contains four music sketches, all based on audio recordings he made in the 1970s. "In the Rain," for example, uses a recording of rain striking the ground and several objects, some of which sound like they are made of metal. Electronic sounds, sounds from a piano, and faint calls of birds introduced throughout the piece contrast with the ambient sounds of the rain shower. Adams writes in the album liner notes: "Last summer in my studio, I discovered several boxes of reel-to-reel tapes I'd recorded in the early 1970s. Using those found objects, I sculpted these new soundscapes from fragments of my past." *The Place Where You Go to Listen* is a piece that uses geological sounds from real-time sensors that are translated into music and sounds presented as a museum exhibition.

Adams's *Inuksuit* is a unique outdoor piece that uses the loud drumming of percussionists, natural soundscape recordings played through speakers, and a group of musicians playing a piccolo, a trumpet, handbells, triangles, and cymbals (Hanslowe 2015). Listeners are asked to move around the musicians and speakers, and may come and go any time, thus creating a customized listening experience; they are even encouraged to walk between the musicians. As listeners walk around the performance, the experience is much like that of Westerkamp's soundwalk, as sounds and experiences change as one moves about and around the three performance circles. The piece is also designed to convey special meaning to the listeners as the timeline unfolds: that of an environmental collapse, or more specifically, of the melting of the polar ice caps due to climate change. The end of the piece is dominated by soft sounds of birdcalls and a piccolo at intensities so low that it is difficult to know when the piece has ended. Hanslowe (2015) suggests that the piece reflects Adams's fear that the future of nature is not good, perhaps even apocalyptic, unless society changes its behavior, a sentiment that other acoustic ecologists share.

Francisco Lopez

A Spanish composer and biologist, Francisco Lopez uses a variety of sound-scape recordings to present natural sounds to listeners, in what Akiyama (2010) calls "densely layered soundscapes" that are "overwhelming" but, as Lopez claims, are not presented with a political motive to change human behaviors. In an essay called *Sonic Creatures* (2019), Lopez describes his philosophy of soundscape composition, contrasting it strongly with that of Schafer and colleagues, who argue that soundscapes are composed of sounds from sound sources, some of which are not worthy of saving and others that need to be removed. Instead, Lopez argues that sounds are "things-by-themselves" and are already detached from the objects that produced them. Lopez (2019, 3) also contends that "[so] long as sounds remain properties of other things and ontologically subservient to them, we will never be able to truly hear *them*" (emphasis in original). Many sounds that are listened to have unknown sources, and thus if we contextualize every sound with a source, we may not hear all the sonic creatures. The essay describes the techniques Lopez uses for *La Selva* (1977), a piece that uses sounds from the Neotropical rainforests of Costa Rica, as attempts to "deepen the connection" to the sound as its own entity without associating it with an object (such as a frog). To develop a sense of listening in complex soundscapes like that found in the rainforest, Lopez employs transcendental field listening exercises to understand sounds as things as themselves.

14.3. Soundscape Composition and Ethnography

Several scholars (e.g., Drever 2002; Feld 1994; Martin 2018) have argued that soundscape composition can be driven by ethnographic research. Or, more precisely, as Feld (1994, 328) has described, "soundscape research really should be presented in the form of a musical composition. This is the one way to bend the loop back so that research and the artistry come together and we can auditorally cross those rivers . . . without the academic literalisms, the print mediation." Martin's *A Bit Closer to Home* (2014) merges the use of the Schafer concept of "earwitnesses" as an ethnographic research approach, which explores life histories through interviews, with sound-scape composition. The piece merges audio recordings of an interview about sounds of the past with environmental sound recordings and electronic sounds to tell a story about how the acoustic surroundings have changed. Stollery (2013) also uses this technique of integrating the spoken narrative from an interview with current and past sounds that are discussed.

14.4. Sonification of Data

Engineers and artists have developed several techniques for sonifying data that are traditionally "silent" (Hermann and Ritter 1999), such as measures of soil moisture, human population density, or air temperature. Sonification has been used to provide scientific researchers with a means to understand complex data structures through the human ability to perceptualize this complexity aurally by translating numerical data to sound.

One of the first attempts at *digital sonification* was made by Fitch and Kramer (1994), who argued that perceptualizing complex data could best be accomplished through auditory, rather than visual, means. An interesting early approach of Hermann and Ritter (1999) used statistical and machine-learning methods (see Section 7.3) to convert data to audio such that the mixing of data values created groups of sounds that represented forms of either clustering or classification of data. Silent data have also been converted by some researchers into spectrograms that can be visualized and listened to at the same time.

More recently, a taxonomy and a set of formal definitions for use in data sonification have been presented (Hermann 2008), and a formal organization, the International Conference on Auditory Display (ICAD; www.icad .org), now convenes annually to present new developments in this field and publishes those advances as a set of annual conference proceedings. ICAD has developed a formal definition of *sonification*: "Sonification is the trans-formation of data relations into received relations in an acoustic signal for the purposes of facilitating communication or interpretation" (Hermann 2008, ICAD08–1).

ICAD (Hermann 2008) also describes several key features that need to occur if a transformation is to be considered sonification: (1) that sound reflects objective properties or relations in the input data; (2) that the transformation is systematic; (3) that the sonification is reproducible; and (4) that the system developed can be used with different data and thus be transferrable.

Instrumental sonification is another form of data sonification in which various musical instruments, either played solo or as a group, relate pitch to a data value. Several examples of sonification are provided in this section.

14.4.1. Digital Sonification

Several acoustic ecologists have ventured into the realm of digital sonification of data in a form that differs in small ways from that presented by ICAD (Hermann 2008). One such acoustic ecologist, Polli (2005, 2012), argued, as did Schafer, Truax, and Westerkamp, for the need to incorporate an understanding of how people perceive their environment through the soundscape with what the data tell us about a social or environmental problem. Polli argued that sonification is very similar to listening to either music or soundscapes, which are both "inherently narrative" (Polli 2005, 31) and informative. Polli (2005) sonified data on wind speed, barometric pressure, wind direction, moisture, and air temperature from fifteen weather stations from a major winter storm in 1979 (President's Day Winter Storm) and a 1991 summer hurricane (Hurricane Bob) that occurred along the East Coast of the United States. Speakers were aligned to the geographic locations of the weather stations and placed at different heights to represent how high in the atmosphere these events occurred to allow listeners to understand the spatial context of the storm. Artists (e.g., Polli) and atmospheric scientists collaborated on the sound properties. They presented the result as an exhibit, called *Atmospheres/Weather Works*, at a science center in New York City. Polli (2012) has argued that sonification, if placed in a proper context, can be used not only to understand complex structures in data but also to undergird a "sound activism" as espoused by Truax (1994).

An example of sonification of data that focuses on landscape change is the work of Angeler et al. (2018), who used a 71-year time series of landscape and water level data on wetland inundation and change in the surrounding landscape (from natural to agricultural) to sonify the "profound socioecological transformation" of Spanish wetlands. Two dominant electronic voices were produced, a low-frequency bass for inundation (water levels) and a second, higher-frequency voice—a soprano—for precipitation. Angeler and colleagues used a minor key (A minor) and translated data to specific pitches in this musical scale. One can listen to their nearly five-minute-long "song," called *The Lament of Las Tablas de Daimiel*, to understand the dynamics of these changes, especially the rapid shift in inundation patterns that happened in the late 1970s when agricultural transformations occurred in the surrounding landscapes.

14.4.2. Instrumental Sonification

An example of *instrumental sonification* is found in the work of Dan Crawford, who has taken 133 years of global meteorological data and used it to sonify climate change in several pieces. An example of his work is *Planetary Bands, Warming World* (https://www.youtube.com/watch?v=-V2Uc8Kax _g), a string quartet in which each instrument plays the temperature change for a different region of the world (from the equator to the North Pole). The four instruments play the changes synchronously through time, starting in 1880 and progressing to the present. Each instrument plays notes from low frequency (low temperature) to high frequency (high temperature) to represent the annual temperatures and how they change over this time.

14.5. Sonic Ecologies: A Multidisciplinary Framework for Public Engagement

Several acoustic ecologists, community policy experts, environmental educators, and soundscape ecologists have collaborated to create listening spaces and activities that transcend disciplinary boundaries to enhance science communication (Barclay et al. 2020). Inspired by the work of Westerkamp and her soundwalk activities, Leah Barclay (2013), an acoustic ecologist from Australia who has collaborated with ecologists and community leaders, developed a *Sonic Ecologies framework* that represents a comprehensive approach to using soundscapes to represent the status of place and how people are affected by it. The framework has five essential elements: (1) site-specific subject matter; (2) multi-platform dissemination; (3) community education and engagement tools; (4) interdisciplinary partnerships and collaborations, and (5) long-term strategic vision. The framework involves using site-specific recordings, artistic pieces that use those recordings, and electroacoustic sounds played by mixing musical instruments and acousmatic tools (e.g., speakers, mobile devices), integrated with interactive activities for the public. Scientists provide the context of the nature of the sounds and the meaning behind their occurrence. Educational activities are designed to engage youth in K–12 schools with content that aligns with the formal classroom content and the education standards that teachers must meet for their curriculum (see also Ghadiri Khanaposhtani et al. 2018). Many of these activities have been configured as permanent installations in public spaces such as science centers and museums. The collaborative *Listening to the Thames* exhibit (Gifford et al. 2014; Barclay et al. 2014) uses recordings from the river obtained from the freshwater bioacoustics monitoring project of ecologist Simon Linke and colleagues (2013), who used the data to assess biodiversity trends in the river.

14.6. Ecoethnomusicology

14.6.1. Ecomusicology and Ecoethnomusicology

The current focus of scholarship known as *ecomusicology* (Titon 2009; Allen 2011; Allen and Dawe 2016) emerged from the work of acoustic ecology (Schafer 1994; Truax 2019), ecocriticism (Ingram 2010; Rehding 2002), ethnomusicology (Guy 2009), and musicology (Toliver 2004). A major focus of ecomusicology has been the development of new concepts that attempt to provide a better understanding of how music, people, and the environment interact. *Ecoethnomusicology* (see Guyette and Post 2015) emphasizes the importance of ethnography and the study of people and their perception of the environment to sonic practices. This section reviews some of the current developments in ecoethnomusicology and presents a set of case studies that help to provide more details on these intellectual developments. Both ecomusicology and ecoethnomusicology seek to understand the role of sound in coupled human and natural systems, especially in the areas of sound production and perception. Both are focused on solving problems for the benefit of society and the natural world, and both require multidisciplinary approaches. Furthermore, their development is likely to take some time, but their practitioners recognize that knowledge co-production is necessary to understand the complexity of these coupled human and natural systems.

14.6.2. Case Studies in Ecoethnomusicology

Research in ecoethnomusicology has focused on many communities that use ecological songs, sonic practices (e.g., communication with livestock), nature mimesis, and acoustic cues about changes in the environment, and on how the people in those communities perceive and live within their environment. Ecological songs are understood as an expressive form of a community's worldview. Ecomusicologists and ecoethnomusicologists have focused on many music genres and forms of sonic practices. Several collections of their work have been published as special issues in journals (e.g., see Feisst's [2016] overview of the *Contemporary Music Review* special issue on this topic) or books (e.g., Allen and Dawe 2016). The several case studies provided here, mostly from non-Western cultures, are meant to illustrate the role that acoustics play in the lived experiences of people and animals.

Mongolian Herders

The largest intact grasslands in the world exist in Mongolia. Here, herding is a way of life that has existed for over 5,000 years (Fijn 2011). These land-scapes are primarily grassland-steppe ecosystems (Köppen climate classification: Bwk). In their pastoralist-livestock, nomadic way of life, Mongolian herders have developed hundreds of ways to communicate to their herds using a variety of sounds. They also sing ecological songs and frequently mimic natural sounds, such as those of birds, mammals (wildlife and livestock), wind, thunder, and rivers, as they go about their work. Ethnographers and ecomusicologists have studied the use of sounds as herding practices in Mongolian society for over sixty years. The worldview of Mongolian herders is one that considers the universe as consisting of the sky (*tenger*), the earth (*gaxar*), and the natural environment that holds the animals (*am'tan*) and human beings (*khün*). Humans and nonhumans are considered equal occupants of the natural environment. Herders raise sheep, goats, cattle, horses, and camels and live with the herd in encampments (*khot ail*). Herders also know specific characteristics of every member of their herd (Fijn 2011). Sonic practices have been developed to communicate to multispecies herds, to specific species, and to individual animals who may have names.

The long song or *urtyn duu* is a form used by herders to define the human co-presence (*sensu* Titon 2013) with livestock (Yoon 2018). The long song has elongated lyrics with interspersed consonants that are used as "ornaments as part of the improvisatory expression" (Yoon 2018, 93). Using participant observations and interviews, Yoon (2018) found that herders have developed an "animal's language to speak to them." The herd communicates back to the herder as well; in fact, Yoon states that herders believe that animals have better perceptive (e.g., seeing and hearing) abilities than humans. Hutchins (2019) described the sonic interactions of the Mongolian herder culture as building three kinds of multispecies communication relationships: animal-human, animal-animal, and human-human.

Some well-documented sonic practices include those that occur during livestock birthing. To calm a mother during birthing and to reduce the chance of her dying, a herder gently and softly sings songs to her. These songs are specific to the species of livestock. There are also milking songs that are known to increase the flow of milk. When a newborn is near the mother, the herder will produce sucking sounds that mimic the sounds of her suckling offspring. Another form of long song that communicates to livestock is the *giigoo*, which is sung before a horse race. This song contains forceful sounds to signify the strength of the rider. The herder also listens to the sounds used by members of the herd and uses them to determine the status of the herd at any given moment (Hutchins 2019).

A well-known form of singing in western Mongolia is *khoomei*, also known as throat singing, overtone singing, harmonic singing, or harmonic chanting (Levin and Edgerton 1999). To duplicate the natural sounds of rivers, the singers "tune" their harmonic patterns to the same fundamental formant frequencies as the rushing river by lifting the tongue up against the alveolar ridge (upper front portion of the mouth) to push air through the nasal passages.

In nearby eastern Kazakhstan, Kazakh herders are reported to use a variety of sensory modalities to evaluate the status of the ecosystem (Post 2019), including sound as well as sight, touch and smell. The scents of "mint, wild onion and wormwood," for example, are indicators of healthy landscapes, as is the sight of "rocky slopes with honeysuckle and juniper" (Post 2019, 379). This observation emphasizes, as Pink (2015) argues, that scholars such as ecomusicologists and ethnographers need to understand all aspects of sensory perception of the environment.

Landscape and topographic features are also associated with sounds. For example, the sounds of grasses are listened to by herders to determine if the grasses are too dry for the herd to graze, and the sounds of ice breaking are used to determine if it is time for the camp to move from the wintering grounds to the spring grazing grounds (Post and Pijanowski 2018). Many places in the landscape are given names that are associated with sounds. Yoon (2018) describes an instance in which herders describe a location as *ulikh*, which means "to howl." The herder and the place are deeply connected through sound, as herders have been recognized as "deep listeners and deep learners" (Becker 2004, 2). Songs and their content are considered a form of traditional ecological knowledge (TEK; *sensu* Berkes 2017) and as such contain a wealth of information about how people need to exist sustainably in this coupled human and natural system.

Yukon's Tr'ondëk Hwëch'in

The people of the Tr'ondëk Hwëch'in First Nation live in the circumpolar (i.e., near the poles) area of Canada, which is a subarctic boreal ecosystem (Köppen climate classification: Dsc). The Tr'ondëk Hwëch'in use ecological songs as "an active means to maintain respectful, reciprocal social relation-ships with the powerful natural world and its inhabitants" (Ranspot 2019, 478). Using a collaborative ethnographic approach (participant-driven in-terviews and participant observations) that builds on the ecomusicology work of Allen and Dawe (2016) and on the livelihood, dwelling, and skill components of environmental perception described by Ingold (2000), while also exploring parallels to other indigenous communities that use song to describe human-nature relationships, Ranspot (2019) was able to determine how the songs of the Tr'ondëk Hwëch'in reinforce the human-animal rela-tionships that have emotional and reciprocal identity as "kinship" (Ranspot 2019, 487), with the moose, caribou, and salmon as focal animals. Ranspot examined the structure (musical and lyrical, a mimetic form), use (*sensu* Ingold 2000), and ecological and social context (*sensu* Allen and Dawe 2016 and Titon 2015) of seven songs. Several examples were discovered that il-lustrate how ecological songs "promote, reinforce, and embody" (Ranspot 2019, 488) a certain set of relationship values in the biocultural lifeways of the Tr'ondëk Hwëch'in. For example, a "Happy Day Song," which is learned by most members of this community, was sporadically sung at the taking of a moose. It was also sung during the funeral of a family member as a way to "say goodbye," reinforcing that human-human and human-animal relations are equally recognized.

Tsimane' of Bolivia

The indigenous Tsimane' people of Bolivia exemplify a culture that uses eco-logical songs, oral expressions, stories, and speech to transmit traditional ecological knowledge about the behaviors and traits of animals that exist in their landscapes (Reyes-Garcia and Fernandez-Llamazares 2019). Songs also transmit, through their content, social norms about how people should re-spectfully relate to wildlife and the rest of the world and how people should properly hunt (e.g., hunting practices, hunting beliefs, and where and when to hunt). These ecological songs are often presented at community social events that also contain other rituals and provide an important context in which TEK is transmitted. Reyes-Garcia and Fernandez-Llamazares (2019) argue that transmission of TEK in the Tsimane' community does not occur solely through music and other sonic practices, but rather is part of the ho-listic transmission of knowledge that occurs in multiple stages through mul-tiple transmission mechanisms. Reyes-Garcia and Fernandez-Llamazares's (2019) research also suggests that the current sonic TEK practices are being threatened by the introduction of other worldviews (e.g., Christianity), other forms of knowledge transmission (e.g., TV, radio, the internet) and other forms of music and sonic practices that are global (e.g., pop) and regional (e.g., Andean music). They argue that Tsimane' traditions need to be sustained because they contain important knowledge about the natural system and how humans sustainably live in this ecosystem. One means to do that, they suggest, is to document the knowledge in other ways and to use this documentation to help highlight the traditional ways of supporting TEK. Examples that they cite, and ones that have been used, are the development of a book, an exhibit, and a radio show that highlight aspects of the local culture and how it sustains livelihoods.

Omuhipiti of the Island of Mozambique

The small (3 km long and 200–500 m wide) Island of Mozambique is located about 5 km off the mainland of Mozambique, Africa (Köppen climate classification: Am, tropical monsoon). This small island has a rich history of use by indigenous people as a place to hide from danger (from which the community name of Omuhipiti is derived) and a more recent history that involves Islam, introduced to the island in the late eighteenth century (Maculuve 2020). A special set of songs are sung about religious holidays; one, sung on the holy day of Maulide (birth of Muhammad, the Prophet of Islam), includes instruments (e.g., drums, a xylophone-like instrument called an ambira, and bells) and lyrics that describe the history of the song as it relates to the introduction of the religion to the island. Maculuve (2020) describes the soundscape of this island as dominated by the ocean and the patterns of ocean dynamics; at times, the powerful waves along the eastern side of the island can be "felt" as vibrations through the ground. A "fluidity" of the soundscape over time is also evident as tidal patterns shift the sounds of breaking waves throughout the day. The sounds of people's voices, especially the calls of the muezzin (an official that initiates the call to prayer on Fridays), are focused around clustered dwellings and times of religious events. Maculuve (2020) also describes how the soundscape of the island has changed over time. Interviews with elders revealed that ecological songs of past livelihoods of fishing, no longer part of the industry of the island, are rarely sung and are not known by the younger generation; that the calls from crows during the dusk chorus have also been lost, as many of the crows are now gone; and that the past sounds of bicycles are gone, too, as they were recently replaced with motorcycles. Maculuve (2020) concluded that the mobility of the people of this small island, which has been driven by its geography, has shaped the cultural and natural soundscapes of this place in space and time.

Yanyuwa of Australia

The Yanyuwa community lives along the north-central coastal area of Australia in what is considered a tropical savanna ecosystem (Köppen climate classification: Aw). The Yanyuwa identify themselves as a "people whose spirit belongs to the sea" (Bradley and Mackinlay 2018, 77). They use ecological songs to describe the complexities of place and how one perceives place through all the senses; songs also contain beliefs about the "spirituality of place," which is tied to nonhuman kin and the "everyday lived reality . . . immersed in seemingly mundane pragmatic activities" (Bradley and Mackinlay 2018, 76). Within a relatively small area, over a thousand places have been given names based on multisensory experiences (textures, scents, sounds, and views). Place names embody the places both "spiritually" and "historically." Bradley and Mackinlay refer to this human-environment relationship as *interanimation* (*sensu* Basso 1996), which is a "process by which people animate places through their experience, emotions and knowledge" (Bradley and Mackinlay 2018, 79). Songs of the Yanyuwa that reflect this interanimation are also dynamic, because experiences change. A *walaba* ecological song is commonly sung; it describes a place on an island, where people and place are linked through sounds of waves from the ocean, a dreaming narrative that includes individuals from the past, and an activity cast surreally—the cooking of flying foxes. The song invites others to go to this place, to listen to the "crashing waves" and to "embody the sounds through the senses and the mind" (Bradley and Mackinlay 2018, 79). This ecological song thus ties together the landscape, the soundscape, and people (present and past) in an activity that is dreamlike (flying foxes) and also includes daily mundane activities (cooking, sleeping).

The songs of the Yanyuwa are also considered community-owned knowledge of place. Some songs are restricted (can be sung only by those of a specific age, gender, or membership in a group or family) whereas others are unrestricted (can be sung by anyone). These differences emphasize other important components of culture, such as power and status, in the experiential embodiment of sounds and the holding and transmission of the traditional sonic practices. Songs of the Yanyuwa are thus constructed of the details of the place that describe a person's sensory relationship to space *sensu* Tuan (1977) (see Section 6.2) but also tie the place to lifeways (*sensu* Ingold 2000) with metaphysical symbolism (*sensu* Feld 1984).

14.7. Environmental Psychology and Soundscapes

Work in environmental psychology of natural soundscapes has focused on two main areas of inquiry. The first is how the sounds of natural areas, such as those in protected areas, affect the health and well-being of people. Most of these studies focus on subjects who live in urban environments but visit natural areas for recreation. The second is the importance of sounds for nature connectedness. Most of this work is based on Kaplan and Kaplan's (1989) attention restoration theory (ART), Ulrich's (1983) stress recovery theory (SRT), Wilson's (1986) concept of biophilia, Howell et al.'s (2011) concept of nature connectedness, and Nisbet's (2005) nature relatedness scales (all summarized in Chapter 6).

14.7.1. Attention Restoration Theory and Soundscapes

Payne (2013) conducted two studies of participants who were asked to use a perceived restorativeness soundscape scale (PRSS) to rate their experiences of different types of soundscapes. The PRSS was developed using the four main components of attention restoration (Kaplan and Kaplan 1989) as its basis; namely, fascination, being-away, compatibility, and extent. Many researchers have also recognized that being-away can be split into being-away-to and being-away-from, and that extent can also be split into coherence and scope. Table 14.1 summarizes the PRSS used by Payne, with the components grouped into these six categories.

Payne's first study used a convenient sample of university students who were presented with recordings and visuals of three different kinds of places: an urban city center (with sounds from engines, horns, braking, construction, footsteps, crowd sounds), an urban park (with sounds of birds and a waterfall, but with some construction sounds) and a rural landscape located in a national park (with sounds from birds, a river, a waterfall, and footsteps on soggy ground). Participants listened to and viewed each of the landscape-soundscape scenes, and then, for each one, ranked each of the statements in Table 14.1 on an agreement-based scale from 0 (do not agree at all) to 6 (completely agree). The researchers found that the rural soundscape had the greatest perceived restorativeness, followed by the urban park and then the urban city center. The second study involved survey-

Table 14.1 PRSS as used to test three distinct places for the perceived restorativeness of their soundscapes

ART component	Perceived restorativeness of soundscape
Fascination	• I find this sonic environment appealing • My attention is drawn to many of the interesting sounds here • These sounds make me want to linger here • These sounds make me wonder about things • I am engrossed by this sonic environment
Being-away-to	• I hear these sounds when I am doing something different to what I usually do • This is a different sonic environment to what I usually hear • I am hearing sounds that I usually hear
Being-away-from	• This sonic environment is a refuge from unwanted distractions • When I hear these sounds I feel free from work, routine and responsibilities • Listening to these sounds gives me a break from my day-to-day listening experience
Compatibility	• These sounds relate to activities I like to do • This sonic environment fits with my personal preferences • I rapidly get used to hearing this type of sonic environment • Hearing these sounds hinders what I would want to do at this place
Extent (Coherence)	• All the sounds I'm hearing belong here (with the place shown) • All the sounds merge to form a coherent sonic environment • The sounds I am hearing seem to fit together quite naturally with this place
Extent (Scope)	• The sonic environment suggests the size of this place is limitless

Source: Modified from Table 1 of Payne (2013, 257).

ing visitors to two different urban parks using the fourteen of the original nineteen PRSS statements that the first study found to be most significant. Both studies suggest that the human auditory senses can have important restorative potential, adding to a large supporting literature of ART that has demonstrated that the visual perception of nature has this psychological effect on people.

In a comprehensive review of the literature on ART and SRT, Buxton et al. (2021) examined the results of eighteen studies that examined the potential health benefits of natural sounds. The analysis examined the methods and conclusions of studies that had examined attention restoration or stress recovery (both of which are psychological) or the positive physiological effects of natural sounds. Their review found evidence for decreased stress and annoyance, improved health, increased cognitive performance, and positive affective outcomes, such as improved mood, related to natural sounds. The researchers also examined the prominence of natural sounds in protected areas. They concluded that natural parks provide the ecosystem services necessary for restorative effects and physiological well-being.

14.7.2. Nature Relatedness and Soundscapes

Francomano and colleagues (2022) conducted a study at a park entrance and in a city center (Ushuaia) in Tierra del Fuego, Argentina. They interviewed park visitors and people on the streets of Ushuaia to determine their valuations of sounds of four types: (1) the sounds they had just heard prior to the start of the survey, (2) sounds recorded in a forested part of a nearby national park, (3) sounds recorded on a nearby island supporting large colonies of penguins, seabirds, and marine mammals, including sounds from waves and wind, and (4) sounds recorded in urban environments, including traffic sounds. After the soundscape recordings were played, the participants were asked to rank the five soundscape valuation statements in Table 14.2 on a five-point agreement-based scale. Participants were also asked about their perceived nature relatedness using a slightly modified version of the NR-6 scale of Nisbet and Zelenski (2013). Finally, each participant was asked to rate, on a five-point scale, their use of five senses (vision, hearing, touch, smell, and taste) when experiencing nature. Francomano et al. (2022) found that most participants used vision most often, followed by hearing. The participants' nature relatedness was positively correlated with their valuation of the natural soundscapes, and their valuation scores were higher for natural sounds than for sounds made by humans (e.g., machines). The study

Table 14.2 Soundscape valuation statements

Psychological measure	Statement
Preference	I liked the sounds I heard
Cognitive	The sounds I heard triggered memories
Context	The sounds I heard provided me with information about the place in which they occurred
Relatedness	The sounds I heard have an effect (either positive or negative) on the animals living where the sounds occurred
Affect	The sounds I heard made me feel emotions

Source: From Francomano et al. (2022), as modified from Dumyahn and Pijanowski (2011).

suggested that soundscapes play an important role in forming a connection with nature and that soundscapes are an important component in the valuation of places, especially those that have many natural sounds.

Summary

It is clear from the discussion in the previous section that soundscapes are an important area of scholarship in environmental psychology. Sociocultural soundscape scholarship also encompasses several expressive forms, including the soundscape compositions of acoustic ecologists, the sonification practices of both scientists and acoustic ecologists, and the sonic practices of indigenous communities around the world. The channels for these expressive forms are highly varied and thus can be configured for a variety of venues: live performance with instruments and/or voice; acousmatic, radio, or interactive performances directed by the composer or engaging the public; soundwalks that direct listening and dialogues by participants; science museum exhibits that combine science and art; and outdoor exhibits that combine existing ambient sounds with recordings.

There are a few points worth mentioning here. First, the sounds that are made as part of the sonic practices of many indigenous communities encompass the meaning of place and a person's livelihood. Second, although they do not occur all together at the same time, like the sounds of dozens of species of animals chorusing against a backdrop of ambient sounds, these sounds, in sum, do represent a soundscape, as they reflect that place and the relationships that people have with their surroundings. Third, songs sung in a human voice with words provide the meaning of those sounds but also of the relationship of the singer to place and thus constitute a soundscape as well—an expressive soundscape. Fourth, it is very clear that the Western worldview of the interactions of people, animals, and the rest of the environment as a binary relationship (e.g., me vs. everything else) is not universal; worldviews of indigenous people are often very different, and any effort to understand how sounds influence perception of the environment needs to be made in relation to those worldviews. Methods developed to enable Western scientists to study these relationships, such as those espoused by Feld, Yuan, Pink, Ingold, and others, need to be carefully followed, as they allow the researcher to discover these relationships. Finally, as Guyette and Post and many others have argued, soundscape ecologists, ethnomusicologists, acoustic ecologists, anthropologists, ecomusicologists, psychologists, and data scientists, among others, should work collaboratively to understand the role that sound plays in everyday life, especially in the context of the sustainability of the coupled human and natural system that exists within the landscape or seascape.

Discussion Questions

1. Design your own sonic ecologies exhibit. Using the five components of the Sonic Ecologies framework described in Section 14.5, sketch out or describe in words an exhibit that would focus on educating the public about the loss of coral reefs. Use any of the methods that are employed by acoustic ecologists and soundscape ecologists that could be used in the exhibit.

2. Describe a "silent" dataset that you might want to sonify. Using the principles of sonification from either Hermann (e.g., ICAD) or Polli, describe how you might process that dataset for the purposes of understanding its complexities and its inherent spatial-temporal patterns.

3. Using the case study of the sonic practices of Mongolian herders in this chapter, or more detailed readings (e.g., Yoon 2018), describe a project that would examine the importance of these sonic practices to the sustainability of herding in these landscapes. Assume that you would have the following disciplinary experts to draw on for your study: ethnomusicologists, grassland ecologists, data scientists, remote-sensing experts for landscape characterization, and soundscape ecologists who would deploy PARs and use acoustic indices for assessment of soundscape and biological diversity. As part of your description of the study, include a list of potential methods you might use (e.g., interviews, machine learning).

Further Reading

Allen, Aaron S., and Kevin Dawe, eds. *Current Directions in Ecomusicology: Music, Culture, Nature*. Routledge, 2016.

Barclay, Leah. "Sonic Ecologies: Exploring the Agency of Soundscapes in Ecological Crisis." *Soundscape: The Journal of Acoustic Ecology* 12, no. 1 (2013): 29–32.

Gautier, Ana María Ochoa. "Acoustic Multinaturalism, the Value of Nature, and the Nature of Music in Ecomusicology." *Boundary 2* 43, no. 1 (2016): 107–41.

Hermann, T., A. Hunt, and J. Neuhoff, eds. 2011. *The Sonification Handbook*. Logos Publishing House. 586 pp. Available at https://sonification.de/handbook/.

Kelman, A. 2010. Rethinking the Soundscape: A Critical Genealogy of a Key Term in Sound Studies. *Senses and Society* 5 (2): 212–34.

Truax, Barry. *Acoustic Communication*. Greenwood Publishing Group, 2001.

References

Aalbers, Scott A., and Mark A. Drawbridge. 2008. "White Seabass Spawning Behavior and Sound Production." *Transactions of the American Fisheries Society* 137 (2): 542–50.

Abadi, Martín, Paul Barham, Jianmin Chen, Zhifeng Chen, Andy Davis, Jeffrey Dean, Matthieu Devin et al. 2016. "Tensorflow: A System for Large-Scale Machine Learning." In *12th {USENIX} Symposium on Operating Systems Design and Implementation ({OSDI} 16)*, 265–83.

Acevedo, Miguel A., and Luis J. Villanueva-Rivera. 2006. "From the Field: Using Automated Digital Recording Systems as Effective Tools for the Monitoring of Birds and Amphibians." *Wildlife Society Bulletin* 34 (1): 211–14.

Adams, John Luther. 2012. *The Place Where You Go to Listen: In Search of an Ecology of Music*. Wesleyan University Press.

Adams, Mags D., Neil S. Bruce, William J. Davies, Rebecca Cain, Paul Jennings, Angus Carlyle, Peter Cusack et al. 2008. "Soundwalking as a Methodology for Understanding Soundscapes." In *Proceedings of the Institute of Acoustics* 30:1–8.

Adams, Mags D., Trevor Cox, Gemma Moore, Ben Croxford, Mohamed Refaee, and Steve Sharples. 2006. "Sustainable Soundscapes: Noise Policy and the Urban Experience." *Urban Studies* 43 (13). https://doi.org/10.1080/00420980600972504.

Adriasola, Ignacio, Sarah Teasley, and Jilly Traganou. 2016. "Design and Society in Modern Japan: An Introduction." *Review of Japanese Culture and Society* 28 (1): 1–50.

Agapito, Dora, Patrícia Valle, and Júlio Mendes. 2014. "The Sensory Dimension of Tourist Experiences: Capturing Meaningful Sensory-Informed Themes in Southwest Portugal." *Tourism Management* 42:224–37.

Agrawal, Arun. 1995. "Dismantling the Divide between Indigenous and Scientific Knowledge." *Development and Change* 26 (3): 413–39.

Agrawal, Rakesh, Tomasz Imielinski, and Arun Swami. 1993. "Mining Association in Large Databases." In *Proceedings of the 1993 ACM SIGMOD International Conference on Management of Data—SIGMOD '93*, 207–16.

Agrawal, Rakesh, and Ramakrishnan Srikant. "Fast Algorithms for Mining Association Rules." In *Proceedings of the 20th International Conference on Very Large Data Bases*, 487–99. San Francisco: Morgan Kaufmann.

Agrawal, Rakesh, and Ramakrishnan Srikant. 1995. "Mining Sequential Patterns." In *Proceedings of the Eleventh International Conference on Data Engineering*, 3–14.

Agresti, Alan. 2002. *Categorical Data Analysis*. 2nd ed. Wiley Series in Probability and Statistics. Wiley Interscience.

Ahern, Jack. 2013. "Urban Landscape Sustainability and Resilience: The Promise and Challenges of Integrating Ecology with Urban Planning and Design." *Landscape Ecology* 28 (6): 1203–12.

Aide, T. Mitchell, Carlos Corrada-Bravo, Marconi Campos-Cerqueira, Carlos Milan, Giovany Vega, and Rafael Alvarez. 2013. "Real-Time Bioacoustics Monitoring and Automated Species Identification." *PeerJ* 1: e103.

Aide, T. Mitchell, Andres Hernández-Serna, Marconi Campos-Cerqueira, Orlando Acevedo-Charry, and Jessica L. Deichmann. 2017. "Species Richness (of Insects) Drives the Use of Acoustic Space in the Tropics." *Remote Sensing* 9 (11): 1036.

Akiyama, Mitchell. 2010. "Transparent Listening: Soundscape Composition's Objects of Study." *RACAR: Revue d'art Canadienne/Canadian Art Review* 35 (1): 54–62.

Albanese, Davide, Roberto Visintainer, Stefano Merler, Samantha Riccadonna, Giuseppe Jurman, and Cesare Furlanello. 2012. "Mlpy: Machine Learning Python." *ArXiv* 1202.6548.

Albrecht, R. I., S. J. Goodman, W. A. Petersen, D. E. Buechler, E. C. Bruning, R. J. Blakeslee, and H. J. Christian. 2011. "The 13 Years of TRMM Lightning Imaging Sensor: From Individual Flash Characteristics to Decadal Tendencies." Paper presented at the 14th International Conference on Atmospheric Electricity, August 08–12, 2011, Rio de Janeiro, Brazil.

Aletta, Francesco, and Jian Kang. 2015. "Soundscape Approach Integrating Noise Mapping Techniques: A Case Study in Brighton, UK." *Noise Mapping* 1 (open-issue). De Gruyter.

Aletta, Francesco, and Jian Kang. 2018. "Towards an Urban Vibrancy Model: A Soundscape Approach." *International Journal of Environmental Research and Public Health* 15 (8): 1712.

Aletta, Francesco, Jian Kang, Arianna Astolfi, and Samuele Fuda. 2016a. "Differences in Soundscape Appreciation of Walking Sounds from Different Footpath Materials in Urban Parks." *Sustainable Cities and Society* 27:367–76.

Aletta, Francesco, Jian Kang, and Östen Axelsson. 2016b. "Soundscape Descriptors and a Conceptual Framework for Developing Predictive Soundscape Models." *Landscape and Urban Planning* 149:65–74.

Aletta, Francesco, Jian Kang, Samuele Fuda, and Arianna Astolfi. 2016c. "The Effect of Walking Sounds from Different Walked-on Materials on the Soundscape of Urban Parks." *Journal of Environmental Engineering and Landscape Management* 24 (3): 165–75.

Aletta, Francesco, Efstathios Margaritis, Karlo Filipan, V. Puyana Romero, Östen Axelsson, and Jian Kang. 2015. "Characterization of the Soundscape in Valley Gardens, Brighton, by a Soundwalk Prior to an Urban Design Intervention." In *Proceedings of the Euronoise Conference (2015, Maastricht, The Netherlands)*, 1547–52.

Aletta, Francesco, Tin Oberman, and Jian Kang. 2018a. "Associations between Positive Health-Related Effects and Soundscapes Perceptual Constructs: A Systematic Review." *International Journal of Environmental Research and Public Health* 15 (11): 2392.

Aletta, Francesco, Tin Oberman, and Jian Kang. 2018b. "Positive Health-Related Effects of Perceiving Urban Soundscapes: A Systematic Review." *Lancet* 392:S3.

Aletta, Francesco, Tin Oberman, Andrew Mitchell, Huan Tong, and Jian Kang. 2020. "Assessing the Changing Urban Sound Environment during the COVID-19 Lockdown Period Using Short-Term Acoustic Measurements." *Noise Mapping* 7 (1): 123–34.

Alexander, Richard D. 1968. "Life Cycle Origins, Speciation, and Related Phenomena in Crickets." *Quarterly Review of Biology* 43:1–41.

Alexander, Richard D. 1975. "The Search for a General Theory of Behavior." *Behavioral Science* 20 (2): 77–100.

Alford, Ross A., Philip M. Dixon, and Joseph H. K. Pechmann. 2001. "Global Amphibian Population Declines." *Nature* 412 (6846): 499–500.

Alías, Francesc, and Rosa Ma. Alsina-Pagès. 2019. "Review of Wireless Acoustic Sensor Networks for Environmental Noise Monitoring in Smart Cities." *Journal of Sensors.* https://doi.org/10.1155/2019/7634860.

Allen, Aaron S. 2011a. "Ecomusicology: Ecocriticism and Musicology." *Journal of the American Musicological Society* 64 (2): 391–94.

Allen, Aaron S. 2011b. "Prospects and Problems for Ecomusicology in Confronting a Crisis of Culture." *Journal of the American Musicological Society* 64 (2): 414–24.

Allen, Aaron S. 2012a. "Ecomusicology: Bridging the Sciences, Arts, and Humanities." In *Environmental Leadership: A Reference Handbook*, edited by Deborah Rigling Gallagher, 373–81. Sage.

Allen, Aaron S. 2012b. "Ecomusicology: Music, Culture, Nature . . . and Change in Environmental Studies?" *Journal of Environmental Studies and Sciences.* https://doi.org/10.1007/s13412-012-0072-1.

Allen, Aaron S. 2013. Ecomusicology. In *The Grove Dictionary of American Music*, 2nd ed., edited by Charles Hiroshi Garrett. Oxford University Press.

Allen, Aaron S. 2017. "Sustainable Futures for Music Cultures: An Ecological Perspective." *Ethnomusicology Forum* 26 (3): 400–405.

Allen, Aaron S. 2018. "Introduction. One Ecology and Many Ecologies: The Problem and Opportunity of Ecology for Music and Sound Studies." *Musicultures* 45 (1): 1–13.

Allen, Aaron S., and Kevin Dawe, eds. 2016. *Current Directions in Ecomusicology: Music, Culture, Nature.* Routledge.

Allen, Joel A. 1877. "The Influence of Physical Conditions in the Genesis of Species." *Radical Review* 1:108–40.

Allen, Timothy F. H., and Thomas B. Starr. 2017. *Hierarchy: Perspectives for Ecological Complexity.* University of Chicago Press.

Amorim, M. Clara P. 2006. "Diversity of Sound Production in Fish." *Communication in Fishes* 1:71–104.

Amoser, Sonja, and Friedrich Ladich. 2010. "Year-Round Variability of Ambient Noise in Temperate Freshwater Habitats and Its Implications for Fishes." *Aquatic Sciences* 72 (3): 371–78.

Ananiadou, Sophia, Douglas B. Kell, and Jun-ichi Tsujii. 2006. "Text Mining and Its Potential Applications in Systems Biology." *Trends in Biotechnology* 24 (12): 571–79.

Ananthakrishnan, Gopal, Robert Eklund, Gustav Peters, and Evans Mabiza. 2011. "An Acoustic Analysis of Lion Roars. II: Vocal Tract Characteristics." In *Fonetik 2011. Royal Institute of Technology, Stockholm, Sweden, 8–10 June 2011*, 5–8.

Anderson, Chris. 2008. "The End of Theory: The Data Deluge Makes the Scientific Method Obsolete." *Wired Magazine* 16 (7): 7–16.

Anderson, Dorothy H., and Perry J. Brown. 1984. "The Displacement Process in Recreation." *Journal of Leisure Research* 16 (1): 61–73.

Anderson, James R., Ernest E. Hardy, John T. Roach, and Richard E. Witmer. 1976. "A Land Use and Land Cover Classification System for Use with Remote Sensor Data." Professional Paper 964. USGS.

Anderson, Lisa M., B. Edward Mulligan, Lee S. Goodman, and H. Z. Regen. 1983. "Effects of Sounds on Preferences for Outdoor Settings." *Environment and Behavior* 15 (5): 539–66.

Andreotti, Bruno. 2004. "The Song of Dunes as a Wave-Particle Mode Locking." *Physical Review Letters* 93 (23): 238001.

Angeler, David G., Miguel Alvarez-Cobelas, and Salvador Sánchez-Carrillo. 2018. "Sonifying Social-Ecological Change: A Wetland Laments Agricultural Transformation." *Ecology and Society* 23 (2): 20.

Anthrop, Donald F. 1970. "The Noise Crisis." *University of Toronto Law Journal* 20 (1): 1–17.

Antrop, Marc. 2006. "Sustainable Landscapes: Contradiction, Fiction or Utopia?" *Landscape and Urban Planning* 75 (3–4): 187–97.

Arch, Victoria S., T. Ulmar Grafe, Marcos Gridi-Papp, and Peter M. Narins. 2009. "Pure Ultrasonic Communication in an Endemic Bornean Frog." *PLoS One* 4 (4): e5413.

Arch, Victoria S., T. Ulmar Grafe, and Peter M. Narins. 2008. "Ultrasonic Signalling by a Bornean Frog." *Biology Letters* 4 (1): 19–22.

Au, Whitlow W. L., and Kiara Banks. 1998. "The Acoustics of the Snapping Shrimp *Synalpheus parneomeris* in Kaneohe Bay." *Journal of the Acoustical Society of America* 103 (1): 41–47.

Audacity Team 2020. Audacity(R): Free Audio Editor and Recorder [Computer application]. Version 2.4.2. https://audacityteam.org.

Axelsson, Östen, Mats E. Nilsson, and Birgitta Berglund. 2010. "A Principal Components Model of Soundscape Perception." *Journal of the Acoustical Society of America* 128 (5): 2836–46.

Axelsson, Ãsten, Mats E. Nilsson, and Birgitta Berglund. 2012. "The Swedish Soundscape-Quality Protocol." *Journal of the Acoustical Society of America* 131 (4): 3476.

Ba, Meihui, and Jian Kang. 2019. "A Laboratory Study of the Sound-Odour Interaction in Urban Environments." *Building and Environment* 147:314–26.

Baboin, Maggie, and Damian O. Elias. 2019. "Anthropogenic Noise and the Bioacoustics of Terrestrial Invertebrates." *Journal of Experimental Biology* 222 (12): jeb178749.

Bailey, Robert G. 1995. "Description of the Ecoregions of the United States. USDA Forest Service. https://www.fs.usda.gov/land/ecosysmgmt/index.html.

Bailey, Winston J., and Glenn K. Morris. 1986. "Confusion of Phonotaxis by Masking Sounds in the Bushcricket *Conocephalus brevipennis* (Tettigoniidae: Conocephalinae)." *Ethology* 73 (1): 19–28.

Balantic, Cathleen, and Therese Donovan. 2020. "AMMonitor: Remote Monitoring of Biodiversity in an Adaptive Framework with R." *Methods in Ecology and Evolution* 11 (7): 869–77.

Balph, Donna Mae, and D. F. Balph. 1966. "Sound Communication of Uinta Ground Squirrels." *Journal of Mammalogy* 47 (3): 440–50.

Barclay, Leah. 2013. "Sonic Ecologies: Exploring the Agency of Soundscapes in Ecological Crisis." *Soundscape: The Journal of Acoustic Ecology* 12 (1): 29–32.

Barclay, Leah. 2014. "Biosphere Soundscapes." *Leonardo* 47 (5): 496–97.

Barclay, Leah. 2017. "Listening to Communities and Environments." *Contemporary Music Review* 36 (3): 143–58.

Barclay, Leah. 2019. "Acoustic Ecology and Ecological Sound Art: Listening to Changing Ecosystems." In *Sound, Media, Ecology*, edited by Milena Droumeva and Randolph Jordan, 153–77. Springer International.

Barclay, Leah, Toby Gifford, and Simon Linke. 2014. "River Listening: Creative Approaches to Aquatic Bioacoustics in Australian River Systems." In *Invisible Places, Sounding Cities: Sound, Urbanism and Sense of Place*, proceedings, 311–20.

Barclay, Leah, Toby Gifford, and Simon Linke. 2018. "River Listening: Acoustic Ecology and Aquatic Bioacoustics in Global River Systems." *Leonardo* 51 (03): 298–99.

Barclay, Leah, Toby Gifford, and Simon Linke. 2020. "Interdisciplinary Approaches to Freshwater Ecoacoustics." *Freshwater Science* 39 (2): 356–61.

Barklow, William E. 2004. "Amphibious Communication with Sound in Hippos, *Hippopotamus amphibius*." *Animal Behaviour* 68 (5): 1125–32.

Barth, Friedrich G. 1985. "Neuroethology of the Spider Vibration Sense." In *Neurobiology of Arachnids*, edited by Friedrich G. Barth, 203–29. Springer.

Bartholome, Etienne, and Allan S. Belward. 2005. "GLC2000: A New Approach to Global Land Cover Mapping from Earth Observation Data." *International Journal of Remote Sensing* 26 (9): 1959–77.

Baru, Chaitan, Eric H. Fegraus, Sandy J. Andelman, Sandeep Chandra, Kate Kaya, Kai Lin, and Choonhan Youn. 2012. "Cyberinfrastructure for Observatory and Monitoring Networks: A Case Study from the TEAM Network." *BioScience* 62 (7): 667–75.

Basanta, Adam. 2010. "Syntax as Sign: The Use of Ecological Models within a Semiotic Approach to Electroacoustic Composition." *Organised Sound* 15 (2): 125–32.

Basso, Keith H. 1996. *Wisdom Sits in Places: Landscape and Language among the Western Apache*. University of New Mexico Press.

Bastian, Anna, and David S. Jacobs. 2015. "Listening Carefully: Increased Perceptual Acuity for Species Discrimination in Multispecies Signalling Assemblages." *Animal Behaviour* 101:141–54.

Beason, Richard D., Rüdiger Riesch, and Julia Koricheva. 2019. "AURITA: An Affordable, Autonomous Recording Device for Acoustic Monitoring of Audible and Ultrasonic Frequencies." *Bioacoustics* 28 (4): 381–96.

Beason, Robert C. 2004. "What Can Birds Hear?" In *Proceedings of the 21st Vertebrate Pest Conference*, edited by R. M. Timm and W. P. Gorenzel, 92–96.

Beck, Hylke E., Niklaus E. Zimmermann, Tim R. McVicar, Noemi Vergopolan, Alexis Berg, and Eric F. Wood. 2018. "Present and Future Köppen-Geiger Climate Classification Maps at 1-km Resolution." *Scientific Data* 5:180–214.

Becker, Judith. 2004. *Deep Listeners: Music, Emotion, and Trancing*, vol. 1. Indiana University Press.

Bedoya, Carol L., Richard W. Hofstetter, Ximena J. Nelson, Michael Hayes, Daniel R. Miller, and Eckehard G. Brockerhoff. 2021. "Sound Production in Bark and Ambroaia Beetles." *Bioacoustics* 30 (1): 58–73.

Bedoya, Carol L., Claudia Isaza, Juan M. Daza, and José D. López. 2017. "Automatic Identification of Rainfall in Acoustic Recordings." *Ecological Indicators* 75:95–100.

Beever, Jonathan. 2020. "Sonic Liminality: Soundscapes, Semiotics, and Ecologies of Meaning." *Biosemiotics* 13 (5): 1–12.

Bejarano, Sonia, Peter J. Mumby, and Ian Sotheran. 2011. "Predicting Structural Complexity of Reefs and Fish Abundance Using Acoustic Remote Sensing." *Marine Biology* 158 (3): 489–504.

Bellisario, Kristen M. 2018. "Using Computational Musicological Approaches and Informatics to Characterize Soundscapes in Diverse Natural and Human-Dominated Ecosystems." PhD diss., Purdue University.

Bellisario, Kristen M., Taylor Broadhead, David Savage, Zhao Zhao, Hichem Omrani, Saihua Zhang, John Springer, and Bryan C. Pijanowski. 2019. "Contributions of MIR to Soundscape Ecology. Part 3: Tagging and Classifying Audio Features Using a Multi-Labeling k-Nearest Neighbor Approach." *Ecological Informatics* 51:103–11.

Bellisario, Kristen M., and Bryan C. Pijanowski. 2019. "Contributions of MIR to Soundscape Ecology. Part 1: Potential Methodological Synergies." *Ecological Informatics* 51:96–102.

Bellisario, Kristen M., Jack VanSchaik, Zhao Zhao, Amandine Gasc, Hichem Omrani, and Bryan C. Pijanowski. 2019. "Contributions of MIR to Soundscape Ecology. Part 2: Spectral Timbral Analysis for Discriminating Soundscape Components." *Ecological Informatics* 51:1–14.

Bellman, R. 1957. *Dynamic Programming*. Princeton University Press,

Bello-Orgaz, Gema, Jason J. Jung, and David Camacho. 2016. "Social Big Data: Recent Achievements and New Challenges." *Information Fusion* 28:45–59.

Bennet-Clark, H. C. 1998. "Size and Scale Effects as Constraints in Insect Sound Communication." *Philosophical Transactions of the Royal Society of London. Series B: Biological Sciences* 353 (1367): 407–19.

Beranek, Leo L. 1966. "Noise." *Scientific American* 215 (6): 66–79.

Berglund, Birgitta, Peter Hassmen, and R. F. Soames Job. 1996. "Sources and Effects of Low-Frequency Noise." *Journal of the Acoustical Society of America* 99 (5): 2985–3002.

Berglund, Birgitta, and Mats E. Nilsson. 2006. "On a Tool for Measuring Soundscape Quality in Urban Residential Areas." *Acta Acustica United with Acustica* 92 (6): 938–44.

Berkes, Fikret. 1993. "Traditional Ecological Knowledge in Perspective." In *Traditional Ecological Knowledge: Concepts and Cases*, edited by Julian T. Inglis. IDRC.

Berkes, Fikret. 1999. "Role and Significance of 'Tradition' in Indigenous Knowledge: Focus on Traditional Ecological Knowledge." *Indigenous Knowledge and Development Monitor* (Netherlands) 7 (1): 19.

Berkes, Fikret. 2009. "Evolution of Co-Management: Role of Knowledge Generation, Bridging Organizations and Social Learning." *Journal of Environmental Management* 90 (5): 1692–1702.

Berkes, Fikret. 2017. *Sacred Ecology*. Routledge.

Berkes, Fikret, Johan Colding, and Carl Folke. 2000. "Rediscovery of Traditional Ecological Knowledge as Adaptive Management." *Ecological Applications* 10 (5): 1251–62.

Berkes, Fikret, Carl Folke, and Johan Colding, eds. 2000. *Linking Social and Ecological Systems: Management Practices and Social Mechanisms for Building Resilience*. Cambridge University Press.

Berkes, Fikret, Alun Hughes, Peter J. George, Richard J. Preston, Bryan D. Cummins, and John Turner. 1995. "The Persistence of Aboriginal Land Use: Fish and Wildlife Harvest Areas in the Hudson and James Bay Lowland, Ontario." *Arctic* 48 (1): 81–93.

Bernal, Ximena E., A. Stanley Rand, and Michael J. Ryan. 2006. "Acoustic Preferences and Localization Performance of Blood-Sucking Flies (*Corethrella coquillett*) to Túngara Frog Calls." *Behavioral Ecology* 17 (5): 709–15.

Bertucci, Frédéric, K. Maratrat, Cécile Berthe, Marc Besson, A. S. Guerra, X. Raick, and Eric Parmentier. 2020. "Local Sonic Activity Reveals Potential Partitioning in a Coral Reef Fish Community." *Oecologia* 193:125–34.

Bertucci, Frédéric, Eric Parmentier, Cécile Berthe, Marc Besson, Anthony D. Hawkins, Thierry Aubin, and David Lecchini. 2017. "Snapshot Recordings Provide a First Description of the Acoustic Signatures of Deeper Habitats Adjacent to Coral Reefs of Moorea." *PeerJ* 5:e4019.

Betancourt, Ileana, and Colleen M. McLinn. 2012. "Teaching with the Macaulay Library: An Online Archive of Animal Behavior Recordings." *Journal of Microbiology & Biology Education: JMBE* 13 (1): 86.

Bignotte-Giró, Irelis, and Germán M. López-Iborra. 2019. "Acoustic Niche Partitioning in Five Cuban Frogs of the Genus *Eleutherodactylus*." *Amphibia-Reptilia* 40 (1): 1–11.

Bivand, Roger S. 2003. "Approaches to Classes for Spatial Data in R." In *Proceedings of the Third International Workshop on Distributed Statistical Computing (DSC 2003), March*, edited by Kurt Hornik, Friedrich Leisch, and Achim Zeileis, 20–22.

Bivand, Roger, Tim Keitt, Barry Rowlingson, and E. D. Z. E. R. Pebesma. 2016. "rgdal: Bindings for the Geospatial Data Abstraction Library." *R package version* 1 (10).

Bivand, Roger S., Edzer J. Pebesma, and Virgilio Gomez-Rubio. 2013. *Applied Spatial Data Analysis with R*. Vol. 2. Springer.

Blackman, Allen, Richard Morgenstern, and Libardo Montealegre Murcia. 2005. "Water Issues." In *Nanotechnology: Environmental Implications and Solutions*, edited by Louis Theodore and Robert G. Kunz, 157–200. John Wiley.

Blackman, Allen, Richard Morgenstern, Libardo Montealegre Murcia, and Juan Carlos García de Brigard. 2006. "Review of the Efficiency and Effectiveness of Colombia's Environmental Policies." Resources for the Future. http://www.mamacoca.org/docs _de_base/Fumigas/rff-rpt-coloepefficiency.pdf.

Blair, W. Frank. 1958. "Call Difference as an Isolation Mechanism in Florida Species of Hylid Frogs." *Quarterly Journal of the Florida Academy of Sciences* 21 (1): 32–48.

Blair, W. Frank. 1964. "Isolating Mechanisms and Interspecies Interactions in Anuran Amphibians." *Quarterly Review of Biology* 39 (4): 334–44.

Blaustein, Andrew R., and Joseph M. Kiesecker. 2002. "Complexity in Conservation: Lessons from the Global Decline of Amphibian Populations." *Ecology Letters* 5 (4): 597–608.

Blest, A. D., Thomas Stephen Collett, and J. D. Pye. 1963. "The Generation of Ultrasonic Signals by a New World Arctiid Moth." In *Proceedings of the Royal Society of London. Series B: Biological Sciences* 158 (971): 196–207.

Blumstein, Daniel T., Janice C. Daniel, Andrea S. Griffin, and Christopher S. Evans. 2000. "Insular Tammar Wallabies (*Macropus eugenii*) Respond to Visual but Not Acoustic Cues from Predators." *Behavioral Ecology* 11 (5): 528–35.

Boelman, Natalie T., Gregory P. Asner, Patrick J. Hart, and Roberta E. Martin. 2007. "Multi-trophic Invasion Resistance in Hawaii: Bioacoustics, Field Surveys, and Airborne Remote Sensing." *Ecological Applications* 17 (8): 2137–44.

Bohnenstiehl, D., R. Lyon, O. Caretti, S. Ricci, and David Eggleston. 2018. "Investigating the Utility of Ecoacoustic Metrics in Marine Soundscapes." *Journal of Ecoacoustics* 2: R1156L.

Bolin, Karl, and Mats Åbom. 2010. "Air-Borne Sound Generated by Sea Waves." *Journal of the Acoustical Society of America* 127 (5): 2771–79.

Bolstad, Paul. 2002. *GIS Fundamentals: A First Text on GIS*. XanEdu.

Boncoraglio, Giuseppe, and Nicola Saino. 2007. "Habitat Structure and the Evolution of Bird Song: A Meta-analysis of the Evidence for the Acoustic Adaptation Hypothesis." *Functional Ecology* 21:134–42.

Bonvallet, G. L. 1951. "Levels and Spectra of Traffic, Industrial, and Residential Area Noise." *Journal of the Acoustical Society of America* 23 (4): 435–39.

Boonman, Arjan, and Hellen Kurniati. 2011. "Evolution of High-Frequency Communication in Frogs." *Evolutionary Ecology Research* 13 (2): 197–207.

Borker, Abraham L., Rachel T. Buxton, Ian L. Jones, Heather L. Major, Jeffrey C. Williams, Bernie R. Tershy, and Donald A. Croll. 2020. "Do Soundscape Indices Predict Landscape-Scale Restoration Outcomes? A Comparative Study of Restored Seabird Island Soundscapes." *Restoration Ecology* 28 (1): 252–60.

Bormpoudakis, Dimitrios, Jérôme Sueur, and John D. Pantis. 2013. "Spatial Heterogeneity of Ambient Sound at the Habitat Type Level: Ecological Implications and Applications." *Landscape Ecology* 28:495–506.

Boström, Christoffer, Simon J. Pittman, Charles Simenstad, and Ronald T. Kneib. 2011. "Seascape Ecology of Coastal Biogenic Habitats: Advances, Gaps, and Challenges." *Marine Ecology Progress Series* 427:191–217.

Botteldooren, Dick, Bert De Coensel, and Tom De Muer. 2006. "The Temporal Structure of Urban Soundscapes." *Journal of Sound and Vibration* 292 (1–2): 105–23.

Botteldooren, Dick, Catherine Lavandier, Anna Preis, Daniele Dubois, Itziar Aspuru, Catherine Guastavino, Lex Brown et al. 2011. "Understanding Urban and Natural Soundscapes." In *Forum Acusticum 2011*, 2047–52.

Bradbury, Jack W., and Sandra L. Vehrencamp. 1998. *Principles of Animal Communication*. Sinauer Associates.

Bradfer-Lawrence, Tom, Nils Bunnefeld, Nick Gardner, Stephen G. Willis, and Daisy H. Dent. 2020. "Rapid Assessment of Avian Species Richness and Abundance Using Acoustic Indices." *Ecological Indicators* 115:106400.

Bradfer-Lawrence, Tom, Nick Gardner, Lynsey Bunnefeld, Nils Bunnefeld, Stephen G. Willis, and Daisy H. Dent. 2019. "Guidelines for the Use of Acoustic Indices in Environmental Research." *Methods in Ecology and Evolution* 10 (10): 1796–807.

Bradley, John, and Elizabeth Mackinlay. 2018. "Singing the Land, Singing the Family: Song, Place and Spirituality amongst the Yanyuwa." In *The Soundscapes of Australia: Music, Place and Spirituality*, edited by Fiona Richards, 75–91. Routledge.

Brainard, Julii S., Andrew P. Jones, Ian J. Bateman, and Andrew A. Lovett. 2004. "Exposure to Environmental Urban Noise Pollution in Birmingham, UK." *Urban Studies* 41 (13): 2581–600.

Brambilla, Giovanni, and Luigi Maffei. 2006. "Responses to Noise in Urban Parks and in Rural Quiet Areas." *Acta Acustica United with Acustica* 92 (6): 881–86.

Brandes, T. S. 2005. "Acoustic Monitoring Protocol." Tropical Ecology Assessment and Monitoring (TEAM) Initiative Biodiversity Monitoring Protocols. Conservation International. http://www.teamnetwork.org.

Brandes, T. Scott. 2008. "Automated Sound Recording and Analysis Techniques for Bird Surveys and Conservation." *Bird Conservation International* 18 (S1): S163–S173.

Brandes, T. Scott, Piotr Naskrecki, and Harold K. Figueroa. 2006. "Using Image Processing to Detect and Classify Narrow-Band Cricket and Frog Calls." *Journal of the Acoustical Society of America* 120 (5): 2950–57.

Breda, N. J. J. 2003. "Ground-Based Measurements of Leaf Area Index : A Review of

Methods, Instruments and Current Controversies." *Journal of Experimental Botany* 54 (392): 2403–17.

Bre Breiman, Leo. 2017. *Classification and Regression Trees.* Routledge.

iman, Leo, Jerome Friedman, Charles J. Stone, and Richard A. Olshen. 1984. *Classification and Regression Trees.* CRC Press.

Brillet, Charlese, and Madeline Pailette. 1991. "Acoustic Signals of the Nocturnal Lizard *Gekko gecko*: Analysis of the 'Long Complex Sequence.'" *Bioacoustics* 3 (1): 33–44.

Britzke, Eric R., E. H. Gillam, and K. L. Murray. 2013. "Current State of Understanding of Ultrasonic Detectors for the Study of Bat Ecology." *Acta Theriologica* 58 (2): 109–17.

Brookes, Kate L., Helen Bailey, and Paul M. Thompson. 2013. "Predictions from Harbor Porpoise Habitat Association Models Are Confirmed by Long-Term Passive Acoustic Monitoring." *Journal of the Acoustical Society of America* 134 (3): 2523–33.

Brooks, Bennett M. 2016. "The Soundscape Standard." In *Inter-Noise and Noise-Con Congress and Conference Proceedings*, 253:2188–92.

Brooks, Bennett M., Brigitte Schulte-Fortkamp, Kay S. Voigt, and Alex U. Case. 2014. "Exploring Our Sonic Environment through Soundscape Research & Theory." *Acoustics Today* 10 (1): 30–40.

Brown, A. L. 2010. "Soundscapes and Environmental Noise Management." *Noise Control Engineering Journal* 58 (5): 493–500.

Brown, A. Lex. 2011. "Advancing the Concepts of Soundscapes and Soundscape Planning." In *Proceedings of the Conference of the Australian Acoustical Society (Acoustics 2011).*

Brown, A. Lex. 2012. "A Review of Progress in Soundscapes and an Approach to Soundscape Planning." *International Journal of Acoustics and Vibrations* 17 (2): 73–81.

Brown, A. Lex, Truls Gjestland, and Danièle Dubois. 2016. "Acoustic Environments and Soundscapes." In *Soundscape and the Built Environment*, edited by Jian Kang and Brigitte Schulte-Fortkamp, 1–16. CRC Press.

Brown, A. Lex., Jian Kang, and Truls Gjestland. 2009. "Towards Some Standardization in Assessing Soundscape Preference." *Internoise* 2009.

Brown, A. Lex., Jian Kang, and Truls Gjestland. 2011. "Towards Standardization in Soundscape Preference Assessment." *Applied Acoustics* 72 (6): 387–92.

Brown, A. Lex., and K. C. Lam. 1987. "Urban Noise Surveys." *Applied Acoustics* 20 (1): 23–39.

Brown, Ann M. 1984. "Ultrasound in Gecko Distress Calls (Reptilia: Gekkonidae)." *Israel Journal of Ecology and Evolution* 33 (3): 95–101.

Brown, Charles H., Rafael Gomez, and Peter M. Waser. 1995. "Old World Monkey Vocalizations: Adaptation to the Local Habitat?" *Animal Behaviour* 50 (4): 945–61.

Brown, Daniel G., Kenneth M. Johnson, Thomas R. Loveland, and David M. Theobald. 2005. "Rural Land-Use Trends in the Conterminous United States, 1950–2000." *Ecological Applications* 15 (6): 1851–63.

Brown, James H. 2014. "Why Are There So Many Species in the Tropics?" *Journal of Biogeography* 41 (1): 8–22.

Brown, James H., Andrew P. Allen, and James F. Gillooly. 2007. "The Metabolic Theory of Ecology and the Role of Body Size in Marine and Freshwater Ecosystems." In *Body Size: The Structure and Function of Aquatic Ecosystems*, edited by Alan G. Hildrew, David G. Raffaelli, and Ronni Edmonds-Brown, 1–15. Cambridge University Press.

Brown, Mark T. et al. 2020. "The Quantitative Relation between Ambient Soundscapes and Landscape Development Intensity in North Central Florida." *Landscape Ecology* 35 (1): 113–27.

Brown, T. J., and P. Handford. 1996. "Acoustic Signal Amplitude Patterns: A Computer Simulation Investigation of the Acoustic Adaptation Hypothesis." *Condor* 98 (3): 608–23.

Brumm, Henrik. 2006. "Signalling through Acoustic Windows: Nightingales Avoid Interspecific Competition by Short-Term Adjustment of Song Timing." *Journal of Comparative Physiology A* 192:1279–85.

Brussard, Peter F., and John C. Tull. 2007. "Conservation Biology and Four Types of Advocacy." *Conservation Biology* 21 (1): 21–24.

Buckland, S. T., and A. Johnston. 2017. "Monitoring the Biodiversity of Regions: Key Principles and Possible Pitfalls." *Biological Conservation* 214:23–34.

Buckley, Emma M. Brinley, Andrew J. Caven, Benjamin L. Gottesman, Mary J. Harner, Bryan C. Pijanowski, and Michael L. Forsberg. 2018. "Assessing Biological and Environmental Effects of a Total Solar Eclipse with Passive Multimodal Technologies." *Ecological Indicators* 95:353–69.

Bullard, Robert Doyle, ed. 2005. *The Quest for Environmental Justice: Human Rights and the Politics of Pollution*. Sierra Club Books.

Bunkley, Jessie P., Christopher J. W. McClure, Nathan J. Kleist, Clinton D. Francis, and Jesse R. Barber. 2015. "Anthropogenic Noise Alters Bat Activity Levels and Echolocation Calls." *Global Ecology and Conservation* 3:62–71.

Burivalova, Z., F. Hua, L. P. Koh, C. Garcia, and F. Putz. 2017. "A Critical Comparison of Conventional, Certified, and Community Management of Tropical Forests for Timber in Terms of Environmental, Economic, and Social Variables." *Conservation Letters* 10 (1): 4–14.

Burivalova, Zuzana, Samantha Orndorff, Anthony Truskinger, Paul Roe, Edward T. Game et al. 2021. "The Sound of Logging: Tropical Forest Soundscape Before, During, and After Selective Timber Extraction." *Biological Conservation* 254:108812.

Burivalova, Zuzana, Michael Towsey, Tim Boucher, Anthony Truskinger, Cosmas Apelis, Paul Roe, and Edward T. Game. 2018. "Using Soundscapes to Detect Variable Degrees of Human Influence on Tropical Forests in Papua New Guinea." *Conservation Biology* 32 (1): 205–15.

Burivalova, Zuzana, Bambang Wahyudi, Timothy M. Boucher, Peter Ellis, Anthony Truskinger, Michael Towsey, Paul Roe et al. 2019. "Using Soundscapes to Investigate Homogenization of Tropical Forest Diversity in Selectively Logged Forests." *Journal of Applied Ecology* 56 (11): 2493–504.

Burtner, Matthew. 2005. "Ecoacoustic and Shamanic Technologies for Multimedia Composition and Performance." *Organised Sound* 10 (1): 3–19.

Burtner, Matthew. 2011. "EcoSono: Adventures in Interactive Ecoacoustics in the World." *Organised Sound* 16 (3): 234–44.

Buscombe, D., Paul E. Grams, and Matthew A. Kaplinski. 2014. "Characterizing Riverbed Sediment Using High-Frequency Acoustics: 1. Spectral Properties of Scattering." *Journal of Geophysical Research: Earth Surface* 119 (12): 2674–91.

Butchart, Stuart H. M., Matt Walpole, Ben Collen, Arco Van Strien, Jörn P. W. Scharlemann, Rosamunde E. A. Almond, Jonathan E. M. Baillie et al. 2010. "Global Biodiversity: Indicators of Recent Declines." *Science* 328 (5982): 1164–68.

Butler, Jack, Ed Parnell, and Ana Širović. 2017. "Who's Making All That Racket?

Seasonal Variability in Kelp Forest Soundscapes." *Journal of the Acoustical Society of America* 141 (5): 3864.

Butler, Jack, Jenni A. Stanley, and Mark J. Butler IV. 2016. "Underwater Soundscapes in Near-Shore Tropical Habitats and the Effects of Environmental Degradation and Habitat Restoration." *Journal of Experimental Marine Biology and Ecology* 479:89–96.

Buxton, Rachel T., Samira Agnihotri, V. V. Robin, Anurag Goel, Rohini Balakrishnan, et al. 2018a. "Acoustic Indices as Rapid Indicators of Avian Diversity in Different Land-Use Types in an Indian Biodiversity Hotspot." *Journal of Ecoacoustics* 2 (1): 1–17.

Buxton, Rachel T., Megan F. McKenna, Mary Clapp, Erik Meyer, Erik Stabenau, Lisa M. Angeloni, Kevin Crooks, and George Wittemyer. 2018b. "Efficacy of Extracting Indices from Large-Scale Acoustic Recordings to Monitor Biodiversity." *Conservation Biology* 32 (5): 1174–84.

Buxton, Rachel T., Megan F. McKenna, Daniel Mennitt, Kurt Fristrup, Kevin Crooks, Lisa Angeloni, and George Wittemyer. 2017. "Noise Pollution Is Pervasive in US Protected Areas." *Science* 356 (6337): 531–33.

Buxton, Rachel T., Amber L. Pearson, Claudia Allou, Kurt Fristrup, and George Wittemyer. 2021. "A Synthesis of Health Benefits of Natural Sounds and Their Distribution in National Parks." *Proceedings of the National Academy of Sciences* 118 (14): e2013097118.

Cai, Ximing, Claudia Ringler, and Jiing-Yun You. 2008. "Substitution between Water and Other Agricultural Inputs: Implications for Water Conservation in a River Basin Context." *Ecological Economics* 66 (1): 38–50.

Caldwell, Michael S. 2014. "Interactions between Airborne Sound and Substrate Vibration in Animal Communication." In *Studying Vibrational Communication*, edited by Reginald B. Cocroft, Matija Gogala, Peggy S. M. Hill, and Andreas Wessel, 65–92. Springer.

Callicott, J. Baird. 1990. "Whither Conservation Ethics?" *Conservation Biology* 4 (1): 15–20.

Campos-Cerqueira, Marconi, and T. Mitchell Aide. 2017. "Changes in the Acoustic Structure and Composition along a Tropical Elevational Gradient." *Journal of Ecoacoustics* 1:1–13.

Campos-Cerqueira, Marconi, Jose Luis Mena, Vania Tejeda-Gómez, Naikoa Aguilar-Amuchastegui, Nelson Gutierrez, and T. Mitchell Aide. 2020. "How Does FSC Forest Certification Affect the Acoustically Active Fauna in Madre de Dios, Peru?" *Remote Sensing in Ecology and Conservation* 6 (3): 274–85.

Can, Arnaud, Alain L'Hostis, Pierre Aumond, Dick Botteldooren, Margarida C. Coelho, Claudio Guarnaccia, and Jian Kang. 2020. "The Future of Urban Sound Environments: Impacting Mobility Trends and Insights for Noise Assessment and Mitigation." *Applied Acoustics* 170:107518.

Cantor, Nancy, Julie Norem, Christopher Langston, Sabrina Zirkel, William Fleeson, and Carol Cook-Flannagan. 1991. "Life Tasks and Daily Life Experience." *Journal of Personality* 59 (3): 425–51.

Cao, Xinhao, Qi Meng, and Jian Kang. 2020. "Red Soundscape Index (RSI): An Index with the Potential to Assess Soundscape Quality." In *Inter-Noise and Noise-Con Congress and Conference Proceedings*, 261:3527–39.

Cardinale, Bradley J., J. Emmett Duffy, Andrew Gonzalez, David U. Hooper, Charles Perrings, Patrick Venail, Anita Narwani et al. 2012. "Biodiversity Loss and Its Impact on Humanity." *Nature* 486 (7401): 59–67.

Carpenter, Stephen R., Ruth DeFries, Thomas Dietz, Harold A. Mooney, Stephen Po-
lasky, Walter V. Reid, and Robert J. Scholes. 2006. "Millennium Ecosystem Assessment: Research Needs." *Science* 314 (5797): 257–58.

Carpenter, Stephen R., Harold A. Mooney, John Agard, Doris Capistrano, Ruth S.
DeFries, Sandra Díaz, Thomas Dietz et al. 2009. "Science for Managing Ecosystem
Services: Beyond the Millennium Ecosystem Assessment." *Proceedings of the National
Academy of Sciences* 106 (5): 1305–12.

Carriço, Rita, Mónica Silva, Manuel Vieira, Pedro Afonso, Gui Menezes, Paulo Fonseca,
and Maria Amorim. 2020. "The Use of Soundscapes to Monitor Fish Communities:
Meaningful Graphical Representations Differ with Acoustic Environment." *Acoustics*
2 (2): 382–98.

Casacci, Luca P., Jeremy A. Thomas, Marco Sala, David Treanor, Simona Bonelli, Emilio
Balletto, and Karsten Schönrogge. 2013. "Ant Pupae Employ Acoustics to Communicate Social Status in Their Colony's Hierarchy." *Current Biology* 23 (4): 323–27.

Casey, Joan A., Rachel Morello-Frosch, Daniel J. Mennitt, Kurt Fristrup, Elizabeth L.
Ogburn, and Peter James. 2017. "Race/Ethnicity, Socioeconomic Status, Residen-
tial Segregation, and Spatial Variation in Noise Exposure in the Contiguous United
States." *Environmental Health Perspectives* 125 (7): 077017.

Cash, David W. 2001. "'In Order to Aid in Diffusing Useful and Practical Information':
Agricultural Extension and Boundary Organizations." *Science, Technology, & Human
Values* 26 (4): 431–53.

Cash, David W., William C. Clark, Frank Alcock, Nancy M. Dickson, Noelle Eckley, David
H. Guston, Jill Jäger, and Ronald B. Mitchell. 2003. "Knowledge Systems for Sustainable Development." *Proceedings of the National Academy of Sciences* 100 (14): 8086–91.

Catchpole, Clive K., and Peter J. B. Slater. 2003. *Bird Song: Biological Themes and Variations*. Cambridge University Press.

Cator, Lauren J., Ben J. Arthur, Laura C. Harrington, and Ronald R. Hoy. 2009. "Harmonic Convergence in the Love Songs of the Dengue Vector Mosquito." *Science* 323
(5917): 1077–79.

Celis-Murillo, Antonio, Jill L. Deppe, and Michael F. Allen. 2009. "Using Soundscape Recordings to Estimate Bird Species Abundance, Richness, and Composition." *Journal
of Field Ornithology* 80 (1): 64–78.

Cerdà, A. 1997. "Rainfall Drop Size Distribution in the Western Mediterranean Basin,
València, Spain." *Catena* 30 (2–3): 169–82.

Cerwén, Gunnar. 2016. "Urban Soundscapes: A Quasi-Experiment in Landscape Architecture." *Landscape Research* 41 (5): 481–94.

Cessford, Gordon R. 1999. "Recreational Noise Issues and Examples for Protected Areas." *Noise Control Engineering Journal* 47:3.

Cessford, Gordon R. 2003. "Perception and Reality of Conflict: Walkers and Mountain
Bikes on the Queen Charlotte Track in New Zealand." *Journal for Nature Conservation*
11 (4): 310–16.

Changnon, Stanley A., Jr. 1970. "Hailstreaks." *Journal of Atmospheric Sciences* 27 (1):
109–25.

Changnon, Stanley A., Jr. 1977. "The Scales of Hail." *Journal of Applied Meteorology and
Climatology* 16 (6): 626–48.

Chapin, F. Stuart, Osvaldo E. Sala, Ingrid C. Burke, J. Phillip Grime, David U. Hooper,
William K. Lauenroth, Amanda Lombard et al. 1998. "Ecosystem Consequences of
Changing Biodiversity." *BioScience* 48 (1): 45–52.

Chary, Kandlakunta Laxminarsimha, G. B. Sreekanth, M. K. Deshmukh, and Nitin Sharma. 2020. "Marine Soundscape and Fish Chorus in an Archipelago Ecosystem Comprising Bio-Diverse Tropical Islands off Goa Coast, India." *Aquatic Ecology* 54 (2): 475–93.

Chaverri, Gloriana, Leonardo Ancillotto, and Danilo Russo. 2018. "Social Communication in Bats." *Biological Reviews* 93 (4): 1938–54.

Chek, Andrew A., James P. Bogart, and Stephen C. Lougheed. 2003. "Mating Signal Partitioning in Multi-species Assemblages: A Null Model Test Using Frogs." *Ecology Letters* 6 (3): 235–47.

Chen, Jin, Per Jönsson, Masayuki Tamura, Zhihui Gu, Bunkei Matsushita, and Lars Eklundh. 2004. "A Simple Method for Reconstructing a High-Quality NDVI Time-Series Data Set Based on the Savitzky-Golay Filter." *Remote Sensing of Environment* 91 (3–4): 332–44.

Chen, Maichi, Peng Yu, Yingying Zhang, Ke Wu, and Yixin Yang. 2020. "Acoustic Environment Management in the Countryside: A Case Study of Tourist Sentiment for Rural Soundscapes in China." *Journal of Environmental Planning and Management* 64 (12): 1–23.

Chitnis, Shivam S., Samyuktha Rajan, and Anand Krishnan. 2020. "Sympatric Wren-Warblers Partition Acoustic Signal Space and Song Perch Height." *Behavioral Ecology* 31 (2): 559–67.

Christie, Alec P., Tatsuya Amano, Philip A. Martin, Silviu O. Petrovan, Gorm E. Shackelford, Benno I. Simmons, Rebecca K. Smith, David R. Williams, Claire F. R. Wordley, and William J. Sutherland. 2021. "The Challenge of Biased Evidence in Conservation." *Conservation Biology* 35 (1): 249–62.

Chu, Dezhang, and Peter H. Wiebe. 2005. "Measurements of Sound-Speed and Density Contrasts of Zooplankton in Antarctic Waters." *ICES Journal of Marine Science* 62 (4): 818–31.

Clark, Roger N., and George H. Stankey. 1979. *The Recreation Opportunity Spectrum: A Framework for Planning, Management, and Research.* Vol. 98. USDA Forest Service, Pacific Northwest Research Station.

Clarke, E., and N. Cook, eds. 2004. *Empirical Musicology: Aims, Methods, Prospects.* Oxford University Press.

Cleland, Elsa E., Isabelle Chuine, Annette Menzel, Harold A. Mooney, and Mark D. Schwartz. 2007. "Shifting Plant Phenology in Response to Global Change." *Trends in Ecology & Evolution* 22 (7): 357–65.

Cleveland, William S. 1979. "Robust Locally Weighted Regression and Smoothing Scatterplots." *Journal of the American Statistical Association* 74 (368): 829–36.

Cleveland, William S., and Susan J. Devlin. 1988. "Locally Weighted Regression: An Approach to Regression Analysis by Local Fitting." *Journal of the American Statistical Association* 83 (403): 596–610.

Cocroft, Reginald B., and Michael J. Ryan. 1995. "Patterns of Advertisement Call Evolution in Toads and Chorus Frogs." *Animal Behaviour* 49 (2): 283–303.

Cocroft, Reginald B., and Rafael L. Rodríguez. 2005. "The Behavioral Ecology of Insect Vibrational Communication." *BioScience* 55 (4): 323–34.

Cody, Martin L., and James H. Brown. 1969. "Song Asynchrony in Neighbouring Bird Species." *Nature* 222 (5195): 778–80.

Cohen-Waeber, J., R. Bürgmann, E. Chaussard, C. Giannico, and A. Ferretti. 2018. "Spatiotemporal Patterns of Precipitation-Modulated Landslide Deformation from

Independent Component Analysis of InSAR Time Series." *Geophysical Research Letters* 45 (4): 1878–87.

Coquereau, Laura, Jacques Grall, Laurent Chauvaud, Cédric Gervaise, Jacques Clavier, Aurélie Jolivet, and Lucia Di Iorio. 2016. "Sound Production and Associated Behaviours of Benthic Invertebrates from a Coastal Habitat in the North-East Atlantic." *Marine Biology* 163:1–13.

Coquereau, Laura, Julie Lossent, Jacques Grall, and Laurent Chauvaud. 2017. "Marine Soundscape Shaped by Fishing Activity." *Royal Society Open Science* 4 (1). https://doi.org/10.1098/rsos.160606.

Coss, Derek A., Kimberly L. Hunter, and Ryan C. Taylor. 2021. "Silence Is Sexy: Soundscape Complexity Alters Mate Choice in Túngara Frogs." *Behavioral Ecology* 32 (1): 49–59.

Creswell, John W. 1999. "Mixed-Method Research: Introduction and Application." In *Handbook of Educational Policy*, edited by Gregory J. Cizek, 455–72. Academic Press.

Creswell, John W., and Vicki L. Plano Clark. 2017. *Designing and Conducting Mixed Methods Research*. Sage.

Crutzen, P. J. 2006. "The 'Anthropocene.'" In *Earth System Science in the Anthropocene*, edited by Eckart Ehlers and Thomas Krafft, 13–18. Springer.

Cummings, Molly E., and John A. Endler. 2018. "25 Years of Sensory Drive: The Evidence and Its Watery Bias." *Current Zoology* 64 (4): 471–84.

Dagois-Bohy, Simon, Sandrine Ngo, Sylvain Courrech du Pont, and Stéphane Douady. 2010. "Laboratory Singing Sand Avalanches." *Ultrasonics* 50 (2): 127–32.

Dallas, Cameron, and Cristina Tollefsen. 2015. "Physical Mechanisms Underlying the Acoustic Signature of Breaking Waves." *Canadian Acoustics* 43 (3).

Darras, Kevin, Péter Batáry, Brett J. Furnas, Ingo Grass, Yeni A. Mulyani, and Teja Tscharntke. 2019. "Autonomous Sound Recording Outperforms Human Observation for Sampling Birds: A Systematic Map and User Guide." *Ecological Applications* 29 (6). https://doi.org/10.1002/eap.1954.

Darwin, Charles, and William F. Bynum. 2009. *The origin of species by means of natural selection: or, the preservation of favored races in the struggle for life*. Penguin.

Das, Tushar Kant, and P. Mohan Kumar. 2013. "Big Data Analytics: A Framework for Unstructured Data Analysis." *International Journal of Engineering Science & Technology* 5 (1): 153.

Da Silva, Fernando Rodrigues. 2010. "Evaluation of Survey Methods for Sampling Anuran Species Richness in the Neotropics." *South American Journal of Herpetology* 5 (3): 212–20.

Davies, Nicholas B., John R. Krebs, and Stuart A. West. 2012. *An Introduction to Behavioural Ecology*. John Wiley & Sons.

Davies, William J., Mags D. Adams, Neil S. Bruce, Rebecca Cain, Angus Carlyle, Peter Cusack, Deborah A. Hall et al. 2013. "Perception of Soundscapes: An Interdisciplinary Approach." *Applied Acoustics* 74 (2): 224–31.

Davis, Jesse, and Mark Goadrich. 2006. "The Relationship between Precision-Recall and ROC Curves." In *Proceedings of the 23rd International Conference on Machine Learning*, 233–40. https://doi.org/10.1145/1143844.1143874.

Deane, Grant B., and M. Dale Stokes. 2002. "Scale Dependence of Bubble Creation Mechanisms in Breaking Waves." *Nature* 418 (6900): 839–44.

De Beurs, K. M., and G. M. Henebry. 2005. "A Statistical Framework for the Analysis of Long Image Time Series." *International Journal of Remote Sensing* 26 (8): 1551–73.

de Camargo, Ulisses, Tomas Roslin, and Otso Ovaskainen. 2019. "Spatio-Temporal Scaling of Biodiversity in Acoustic Tropical Bird Communities." *Ecography* 42 (11): 1936–47.

Decker, Emilia, Brett Parker, Simon Linke, Samantha Capon, and Fran Sheldon. 2020. "Singing Streams: Describing Freshwater Soundscapes with the Help of Acoustic Indices." *Ecology and Evolution* 10 (11): 4979–89.

De Coensel, Bert, Annelies Bockstael, Luc Dekoninck, Dick Botteldooren, Brigitte Schulte-Fortkamp, Jian Kang, Mats E. Nilsson. 2010. "The Soundscape Approach for Early State Urban Planning: A Case Study." In *Noise Control Engineering, 39th International Congress, Proceedings*, 1–10.

De Coensel, Bert, Tom De Muer, Isaak Yperman, and Dick Botteldooren. 2005. "The Influence of Traffic Flow Dynamics on Urban Soundscapes." *Applied Acoustics* 66 (2): 175–94.

De Coensel, Bert, Sofie Vanwetswinkel, and Dick Botteldooren. 2011. "Effects of Natural Sounds on the Perception of Road Traffic Noise." *Journal of the Acoustical Society of America* 129 (4): EL148–EL153.

Deichmann, Jessica L., Orlando Acevedo-Charry, Leah Barclay, Zuzana Burivalova, Marconi Campos-Cerqueira, Fernando d'Horta, Edward T. Game et al. 2018. "It's Time to Listen: There Is Much to Be Learned from the Sounds of Tropical Ecosystems." *Biotropica* 50 (5): 713–18.

Deichmann, Jessica L., Andres Hernandez-Serna, Marconi Campos-Cerqueira, T. Mitchell Aide et al. 2017. "Soundscape Analysis and Acoustic Monitoring Document Impacts of Natural Gas Exploration on Biodiversity in a Tropical Forest." *Ecological Indicators* 74:39–48.

Delamont, Sara. 2004. "Ethnography and Participant Observation." *Qualitative Research Practice* 217:206–7.

De Lorenzo, Robert A., and Mark A. Eilers. 1991. "Lights and Siren: A Review of Emergency Vehicle Warning Systems." *Annals of Emergency Medicine* 20 (12): 1331–35.

Demsetz, Harold. 2000. *Toward a Theory of Property Rights*. Palgrave Macmillan.

Denscombe, Martyn. 2008. "Communities of Practice. A Research Paradigm for the Mixed Methods Approach." *Journal of Mixed Methods Research* 2 (3): 270–83.

de Oliveira, Allan G., Thiago M. Ventura, Todor D. Ganchev, Lucas N. S. Silva, Marinêz I. Marques, and Karl-L. Schuchmann. 2020. "Speeding Up Training of Automated Bird Recognizers by Data Reduction of Audio Features." *PeerJ* 8: e8407.

De Oliveira, M. C. Ferreira, and Haim Levkowitz. 2003. "From Visual Data Exploration to Visual Data Mining: A Survey." *IEEE Transactions on Visualization and Computer Graphics* 9 (3): 378–94.

Depraetere, Marion, Sandrine Pavoine, Fréderic Jiguet, Amandine Gasc, Stéphanie Duvail, and Jérôme Sueur. 2012. "Monitoring Animal Diversity Using Acoustic Indices: Implementation in a Temperate Woodland." *Ecological Indicators* 13 (1): 46–54.

Desjonquères, Camille, Toby Gifford, and Simon Linke. 2020. "Passive Acoustic Monitoring as a Potential Tool to Survey Animal and Ecosystem Processes in Freshwater Environments." *Freshwater Biology* 65 (1): 7–19.

Desjonquères, Camille, Fanny Rybak, Emmanuel Castella, Diego Llusia, and Jérôme Sueur. 2018. "Acoustic Communities Reflects Lateral Hydrological Connectivity in Riverine Floodplain Similarly to Macroinvertebrate Communities." *Scientific Reports* 8 (1): 1–11.

Desjonquères, Camille, Fanny Rybak, Marion Depraetere, Amandine Gasc, Isabelle Le

Viol, Sandrine Pavoine, and Jérôme Sueur. 2015. "First Description of Underwater Acoustic Diversity in Three Temperate Ponds." *PeerJ* 3: e1393.

Dethier, Vincent Gaston. 1971. *The Physiology of Insect Senses*. Chapman and Hall.

De Winne, Jorg, Karlo Filipan, Bart Moens, Paul Devos, Marc Leman, Dick Bottel-dooren, and Bert De Coensel. 2020. "The Soundscape Hackathon as a Methodology to Accelerate Co-creation of the Urban Public Space." *Applied Sciences (Switzerland)* 10 (6). https://doi.org/10.3390/app10061932.

Dhar, Vasant. 2013. "Data Science and Prediction." *Communications of the ACM* 56 (12): 64–73.

Dias, Fábio Felix, Helio Pedrini, and Rosane Minghim. 2021. "Soundscape Segregation Based on Visual Analysis and Discriminating Features." *Ecological Informatics* 61:101184.

Díaz, Sandra, Josef Settele, Eduardo Brondízio, H. Ngo, Maximilien Guèze, John Agard, Almut Arneth et al. 2019. *Summary for Policymakers of the Global Assessment Report on Biodiversity and Ecosystem Services*. Intergovernmental Science-Policy Platform on Biodiversity and Ecosystem Services.

Diener, E. D., Robert A. Emmons, Randy J. Larsen, and Sharon Griffin. 1985. "The Satisfaction with Life Scale." *Journal of Personality Assessment* 49 (1): 71–75.

Dietze, Michael C., Andrew Fox, Lindsay M. Beck-Johnson, Julio L. Betancourt, Mevin B. Hooten, Catherine S. Jarnevich, Timothy H. Keitt et al. 2018. "Iterative Near-Term Ecological Forecasting: Needs, Opportunities, and Challenges." *Proceedings of the National Academy of Sciences* 115 (7): 1424–32.

Diez Gaspon, Itxasne, Ibon Saratxaga, and Karmele de Ipiña. 2019. "Deep Learning for Natural Sound Classification." In *Inter-Noise and Noise-Con Congress and Conference Proceedings*, 259:5683–92.

Di Fabio, Annamaria, and Maureen E. Kenny. 2018. "Connectedness to Nature, Personality Traits and Empathy from a Sustainability Perspective." *Current Psychology* 40 (2): 1–12.

Dillman, Don A. 2000. "Procedures for Conducting Government-Sponsored Establishment Surveys: Comparisons of the Total Design Method (TDM), a Traditional Cost-Compensation Model, and Tailored Design." In *Proceedings of American Statistical Association, Second International Conference on Establishment Surveys*, 343–52.

Dirzo, Rodolfo, and Peter H. Raven. 2003. "Global State of Biodiversity and Loss." *Annual Review of Environment and Resources* 28 (1):137–67.

Dodgin, Sarah R., Carrie L. Hall, and Daniel R. Howard. 2020. "Eavesdropping on the Dead: Heterogeneity of Tallgrass Prairie Soundscapes and Their Relationship with Invertebrate Necrophilous Communities." *Biodiversity* 21 (1): 28–40.

Dominoni, Davide M., Wouter Halfwerk, Emily Baird, Rachel T. Buxton, Esteban Fernández-Juricic, Kurt M. Fristrup, Megan F. McKenna et al. 2020. "Why Conservation Biology Can Benefit from Sensory Ecology." *Nature Ecology & Evolution* 4 (4): 502–11.

Do Nascimento, Leandro A., Marconi Campos-Cerqueira, and Karen H. Beard. 2020. "Acoustic Metrics Predict Habitat Type and Vegetation Structure in the Amazon." *Ecological Indicators* 117 (June): 106679. https://doi.org/10.1016/j.ecolind.2020.106679.

Dong, Xueyan, Michael Towsey, Jinglan Zhang, Jasmine Banks, and Paul Roe. 2013. "A Novel Representation of Bioacoustic Events for Content-Based Search in Field

Audio Data." In *Proceedings of the International Conference on Digital Image Computing: Techniques and Applications (DICTA)*. IEEE.

Dooley, Jenet M., and Mark T. Brown. 2020. "The Quantitative Relation between Ambient Soundscapes and Landscape Development Intensity in North Central Florida." *Landscape Ecology* 35 (1): 113–27.

Dooling, Robert J. 1992. "Hearing in Birds." In *The Evolutionary Biology of Hearing*, edited by Douglas B. Webster, Arthur N. Popper, and Richard R. Fay, 545–59. Springer.

Dooling, Robert J., Bernard Lohr, and Micheal L. Dent. 2000. "Hearing in Birds and Reptiles." In *Comparative Hearing: Birds and Reptiles*, edited by Robert J. Dooling, Richard R. Fay, and Arthur N. Popper, 308–59. Springer.

Dooling, Robert J., and Nora H. Prior. 2017. "Do We Hear What Birds Hear in Birdsong?" *Animal Behaviour* 124:283–89.

Douady, Stephane, A. Manning, P. Hersen, H. Elbelrhiti, S. Protiere, A. Daerr, and B. Kabbachi. 2006. "Song of the Dunes as a Self-Synchronized Instrument." *Physical Review Letters* 97 (1): 018002.

Downing, J. Micah, and Eric Stusnick. 2000. "Measurement of the Natural Soundscape in National Parks." *Journal of the Acoustical Society of America* 108 (5): 2497.

Drever, John L. 2002. "Soundscape Composition: The Convergence of Ethnography and Acousmatic Music." *Organised Sound* 7 (1): 21–27.

Drewry, George E., and A. Stanley Rand. 1983. "Characteristics of an Acoustic Community: Puerto Rican Frogs of the Genus *Eleutherodactylus*." *Copeia* 1983 (4): 941–53.

Droumeva, Milena. 2021. "The Sound of the Future: Listening as Data and the Politics of Soundscape Assessment." *Sound Studies* 7 (2): 225–41.

Druck, Gregory, Burr Settles, and Andrew McCallum. 2009. "Active Learning by Labeling Features." In *Proceedings of the 2009 Conference on Empirical Methods in Natural Language Processing*, 81–90.

Duarte, Carlos M., Lucille Chapuis, Shaun P. Collin, Daniel P. Costa, Reny P. Devassy, Victor M. Eguiluz, Christine Erbe et al. 2021. "The Soundscape of the Anthropocene Ocean." *Science* 371 (6529).

Duarte, M. H. L., R. S. S. Sousa-Lima, R. J. Young, M. F. Vasconcelos, E. Bittencourt, M. D. A. Scarpelli, A. Farina, and N. Pieretti. 2021. "Changes on Soundscapes Reveal Impacts of Wildfires in the Fauna of a Brazilian Savanna." *Science of the Total Environment* 769 (June 15): 144988.

Dubois, Danièle. 2000. "Categories as Acts of Meaning: The Case of Categories in Olfaction and Audition." *Cognitive Science Quarterly* 1 (1): 35–68.

Dubois, Danièle, Catherine Guastavino, and Manon Raimbault. 2006. "A Cognitive Approach to Urban Soundscapes: Using Verbal Data to Access Everyday Life Auditory Categories." *Acta Acustica United with Acustica* 92 (6): 865–74.

Dubos, Nicolas, Christian Kerbiriou, Jean François Julien, Luc Barbaro, Kevin Barré, Fabien Claireau, Jérémy Froidevaux et al. 2021. "Going Beyond Species Richness and Abundance: Robustness of Community Specialisation Measures in Short Acoustic Surveys." *Biodiversity and Conservation* 30 (2): 343–63.

Dudgeon, David, Angela H. Arthington, Mark O. Gessner, Zen Ichiro Kawabata, Duncan J. Knowler, Christian Lévêque, Robert J. Naiman et al. 2006. "Freshwater Biodiversity: Importance, Threats, Status and Conservation Challenges." *Biological Reviews of the Cambridge Philosophical Society* 81 (2): 163–82.

Dudley, Nigel, Craig Groves, Kent H. Redford, and Sue Stolton. 2014. "Where Now for Protected Areas? Setting the Stage for the 2014 World Parks Congress." *Oryx* 48 (4): 496–503.

Duellman, William E. 1967. "Social Organization in the Mating Calls of Some Neotropical Anurans." *American Midland Naturalist* 77 (1): 156–63.

Duellman, William E., and Rebecca A. Pyles. 1983. "Acoustic Resource Partitioning in Anuran Communities." *Copeia* 1983 (3): 639–49.

Duhautpas, Frédérick, and Makis Solomos. 2014. "Hildegard Westerkamp and the Ecology of Sound as Experience. Notes on 'Beneath the Forest Floor.'" *Soundscape: The Journal of Acoustic Ecology* 13 (1): 6–10.

Dumyahn, Sarah L. 2013. "Theory and Application of Soundscape Conservation and Management: Lessons Learned from the US National Park Service." PhD diss., Purdue University.

Dumyahn, Sarah L., and Bryan C. Pijanowski. 2011a. "Beyond Noise Mitigation: Managing Soundscapes as Common-Pool Resources." In "Soundscape Ecology," ed. Bryan C. Pijanowski and Almo Farina, special issue, *Landscape Ecology* 26 (9): 1311.

Dumyahn, Sarah L., and Bryan C. Pijanowski. 2011b. "Soundscape Conservation." In "Soundscape Ecology," ed. Bryan C. Pijanowski and Almo Farina, special issue, *Landscape Ecology* 26 (9): 1327.

Duxbury, Nancy, William Francis Garrett-Petts, and David MacLennan. 2015. *Cultural Mapping as Cultural Inquiry*. Routledge.

East, M. L., and H. Hofer. 1991. "Loud-Calling in a Female-Dominated Mammalian Society: I. Structure and Composition of Whooping Bouts of Spotted Hyaenas, *Crocuta crocuta*." *Animal Behaviour* 42:637–49.

Ednie, Andrea, Trace Gale, Karen Beeftink, and Andrés Adiego. 2022. "Connecting Protected Area Visitor Experiences, Wellness Motivations, and Soundscape Perceptions in Chilean Patagonia." *Journal of Leisure Research* 53 (3): 377–403.

Eklund, Robert, Gustav Peters, Gopal Ananthakrishnan, and Evans Mabiza. 2011. "An Acoustic Analysis of Lion Roars. 1. Data Collection and Spectrogram and Waveform Analyses." In *Quarterly Progress and Status Report TMH-QPSR, Volume Fonetik 2011. Royal Institute of Technology, Stockholm, Sweden, 8–10 June 2011*, 1–4. Universitetsservice.

Ekman, Paul. 1992. "An Argument for Basic Emotions." *Cognition & Emotion* 6 (3–4): 169–200.

Eldridge, Alice, Patrice Guyot, Paola Moscoso, Alison Johnston, Ying Eyre-Walker, and Mika Peck. 2018. "Sounding Out Ecoacoustic Metrics: Avian Species Richness Is Predicted by Acoustic Indices in Temperate but Not Tropical Habitats." *Ecological Indicators* 95:939–52.

Elise, Simon, Arthur Bailly, Isabel Urbina-Barreto, Gérard Mou-Tham, Frédéric Chiroleu, Laurent Vigliola, William D. Robbins, and J. Henrich Bruggemann. 2019a. "An Optimised Passive Acoustic Sampling Scheme to Discriminate among Coral Reefs' Ecological States." *Ecological Indicators* 107:105627.

Elise, Simon, Isabel Urbina-Barreto, Romain Pinel, Vincent Mahamadaly, Sophie Bureau, Lucie Penin, Mehdi Adjeroud, Michel Kulbicki, and J. Henrich Bruggemann. 2019b. "Assessing Key Ecosystem Functions through Soundscapes: A New Perspective from Coral Reefs." *Ecological Indicators* 107:105623.

Ellis, Richard J., and Fred Thompson. 1997. "Culture and the Environment in the Pacific Northwest." *American Political Science Review* 91 (4): 885–97.

Emerson, Robert M., Rachel I. Fretz, and Linda L. Shaw. 2011. *Writing Ethnographic Fieldnotes*. University of Chicago Press.

Endler, John A. 1992. "Signals, Signal Conditions, and the Direction of Evolution." *American Naturalist* 139: S125–53.

Endler, John A. 1993. "Some General Comments on the Evolution and Design of Animal Communication Systems." *Philosophical Transactions of the Royal Society of London. Series B: Biological Sciences* 340 (1292): 215–25.

Erbe, Christine, and Micheal L. Dent. 2017. "Animal Bioacoustics." *Acoustics Today* 13 (2): 65–67.

Erbe, Christine, Arti Verma, Robert McCauley, Alexander Gavrilov, and Iain Parnum. 2015. "The Marine Soundscape of the Perth Canyon." *Progress in Oceanography* 137:38–51.

European Union. 2002. *Directive 2002/49/EC of the European Parliament and of the Council*. EN-02002L0049. www.journal.uta45jakarta.ac.id.

Everest, F. Alton, and Ken C. Pohlmann. 2022. *Master Handbook of Acoustics*. McGraw-Hill Education.

Ewing, Arthur W. 1989. *Arthropod Bioacoustics: Neurobiology and Behaviour*. Comstock.

Ey, Elodie, and Julia Fischer. 2009. "The 'Acoustic Adaptation Hypothesis'—A Review of the Evidence from Birds, Anurans and Mammals." *Bioacoustics* 19 (1–2): 21–48.

Eyring, Carl F. 1946. "Jungle Acoustics." *Journal of the Acoustical Society of America* 18 (2): 257–70.

Fahrig, Lenore. 2013. "Rethinking Patch Size and Isolation Effects: The Habitat Amount Hypothesis." *Journal of Biogeography* 40 (9): 1649–63.

Fahrig, Lenore, Jacques Baudry, Lluís Brotons, Françoise G. Burel, Thomas O. Crist, Robert J. Fuller, Clelia Sirami, Gavin M. Siriwardena, and Jean-Louis Martin. 2011. "Functional Landscape Heterogeneity and Animal Biodiversity in Agricultural Landscapes." *Ecology Letters* 14 (2): 101–12.

Fairbrass, Alison J., Michael Firman, Carol Williams, Gabriel J. Brostow, Helena Titheridge, and Kate E. Jones. 2019. "CityNet—Deep Learning Tools for Urban Ecoacoustic Assessment." *Methods in Ecology and Evolution* 10 (2): 186–97.

Fairbrass, Alison J., Peter Rennert, Carol Williams, Helena Titheridge, and Kate E. Jones. 2017. "Biases of Acoustic Indices Measuring Biodiversity in Urban Areas." *Ecological Indicators* 83:169–77.

Falk, Benjamin, Joseph Kasnadi, and Cynthia F. Moss. 2015. "Tight Coordination of Aerial Flight Maneuvers and Sonar Call Production in Insectivorous Bats." *Journal of Experimental Biology* 218 (22): 3678–88.

Farina, Almo, and Stuart H. Gage. 2017. *Ecoacoustics: The Ecological Role of Sounds*. John Wiley & Sons.

Farina, Almo, Stuart H. Gage, and Paolo Salutari. 2018. "Testing the Ecoacoustics Event Detection and Identification (EEDI) Approach on Mediterranean Soundscapes." *Ecological Indicators* 85:698–715.

Farina, Almo, Philip James, C. Bobryk, Nadia Pieretti, Emanuele Lattanzi, and J. McWilliam. 2014. "Low Cost (Audio) Recording (LCR) for Advancing Soundscape Ecology towards the Conservation of Sonic Complexity and Biodiversity in Natural and Urban Landscapes." *Urban Ecosystems* 17 (4): 923–44.

Farina, Almo, and Nadia Pieretti. 2012. "The Soundscape Ecology: A New Frontier of Landscape Research and Its Application to Islands and Coastal Systems." *Journal of Marine and Island Cultures* 1 (1): 21–26.

Farina, Almo, and Nadia Pieretti. 2014. "Sonic Environment and Vegetation Structure: A Methodological Approach for a Soundscape Analysis of a Mediterranean Maqui." *Ecological Informatics* 21:120–32.

Farina, Almo, Nadia Pieretti, P. Salutari, E. Tognari, and A. Lombardi. 2016. "The Application of the Acoustic Complexity Indices (ACI) to Ecoacoustic Event Detection and Identification (EEDI) Modeling." *Biosemiotics* 9 (2): 227–46.

Farina, Almo, R. Righini, S. Fuller, P. Li, and G. Pavan. 2021. "Acoustic Complexity Indices Reveal the Acoustic Communities of the Old-Growth Mediterranean Forest of Sasso Fratino Integral Natural Reserve (Central Italy)." *Ecological Indicators* 120:106927.

Fay, Richard. 2009. "Soundscapes and the Sense of Hearing of Fishes." *Integrative Zoology* 4 (1): 26–32.

Fayyad, Usama, Gregory Piatetsky-Shapiro, and Padhraic Smyth. 1996. "From Data Mining to Knowledge Discovery in Databases." *AI Magazine* 17 (3): 37.

Feinerer, Ingo. 2008. "An Introduction to Text Mining in R." *R News* 8 (2): 19–22.

Feisst, Sabine. 2016. "Music and Ecology." *Contemporary Music Review* 35 (3): 293–95.

Feld, Steven. 1982. *Sound and Sentiment: Birds, Weeping, Poetics, and Song in Kaluli Expression*. University of Pennsylvania Press.

Feld, Steven. 1984. "Sound Structure as Social Structure." *Ethnomusicology* 28 (3): 383–409.

Feld, Steven. 1986. Review of *The Study of Ethnomusicology*, by Bruno Nettl. *Latin American Music Review / Revista de Música Latinoamericana* 7 (2): 375–78.

Feld, Steven. 1986. *Sound and Sentiment: Birds, Weeping, Poetics, and Song in Kaluli Expression*. University of Pennsylvania Press.

Feld, Steven. 1994. "From Ethnomusicology to Echo-Muse-Ecology: Reading R. Murray Schafer in the Papua New Guinea Rainforest." *The Soundscape Newsletter* 8 (6): 9–13.

Feld, Steven. 2001. "Thoughts on Recording Soundscapes: Interview with Carlos Palombini." In *The Sound World of Bosavi*. https://aeinews.org/aeiarchive/edu/educurrbosavi.html#Anchor-Thoughts-11481.

Feld, Steven. 2004. "Places Sensed, Senses Placed: Toward a Sensuous Epistemology of Environments Routledge." In *Empire of the Senses*, edited by David Howes, 179–91..

Feld, Steven. 2012. *Sound and Sentiment: Birds, Weeping, Poetics, and Song in Kaluli Expression*. 3rd ed., with a new introduction by the author. Duke University Press.

Feld, Steven, and Keith H. Basso, eds. 1996. *Senses of Place*. School of American Research Press.

Feld, Steven, and Aaron A. Fox. 1994. "Music and Language." *Annual Review of Anthropology* 23:25–53.

Feldman, Ronen, and Ido Dagan. 1995. "Knowledge Discovery in Textual Databases (KDT)." In *KDD-95 Proceedings*, 112–17.

Feldman, Ronen, James Sanger et al. 2007. *The Text Mining Handbook: Advanced Approaches in Analyzing Unstructured Data*. Cambridge University Press.

Feng, Albert S., Jim C. Hall, and David M. Gooler. 1990. "Neural Basis of Sound Pattern Recognition in Anurans." *Progress in Neurobiology* 34 (4): 313–29.

Feng, Albert S., Peter M. Narins, Chun-He Xu, Wen-Yu Lin, Zu-Lin Yu, Qiang Qiu, Zhi-Min Xu, and Jun-Xian Shen. 2006. "Ultrasonic Communication in Frogs." *Nature* 440 (7082): 333–36.

Fenton, M. Brock. 2013. "Questions, Ideas and Tools: Lessons from Bat Echolocation." *Animal Behaviour* 85 (5): 869–79.

Fenton, M. Brock, Alan D. Grinnell, Arthur N. Popper, and Richard R. Fay, eds. 2016. *Bat Bioacoustics*. Springer Handbook of Auditory Research 54. Springer.

Fernández-Juricic, Esteban, Rachael Poston, Karin De Collibus, Timothy Morgan, Bret Bastain, Cyndi Martin, Kacy Jones, and Ronald Treminio. 2005. "Microhabitat Selection and Singing Behavior Patterns of Male House Finches (*Carpodacus mexicanus*) in Urban Parks in a Heavily Urbanized Landscape in the Western US." *Urban Habitats* 3 (1): 49–69.

Ferrara, Camila R., Richard C. Vogt, Jacqueline C. Giles, and Gerald Kuchling. 2014. "Chelonian Vocal Communication." In *Biocommunication of Animals*, edited by Guenther Witzany, 261–74. Springer.

Ferraro, Danielle M., Zachary D. Miller, Lauren A. Ferguson, B. Derrick Taff, Jesse R. Barber, Peter Newman, and Clinton D. Francis. 2020. "The Phantom Chorus: Birdsong Boosts Human Well-Being in Protected Areas." *Proceedings of the Royal Society of London. Series B: Biological Sciences* 287 (1941): 20201811.

Few, Arthur A. 1969. "Power Spectrum of Thunder." *Journal of Geophysical Research* 74 (28): 6926–34.

Few, A. A. 1974. "Thunder Signatures." *Eos, Transactions American Geophysical Union* 55 (5): 508–14.

Few, Arthur A. 1975. "Thunder." *Scientific American* 233 (1): 80–91.

Few, A. A., A. J. Dessler, Don J. Latham, and M. Brook. 1967. "A Dominant 200-Hertz Peak in the Acoustic Spectrum of Thunder." *Journal of Geophysical Research* 72 (24): 6149–54.

Ficken, Robert W., Millicent S. Ficken, and Jack P. Hailman. 1974. "Temporal Pattern Shifts to Avoid Acoustic Interference in Singing Birds." *Science* 183 (4126): 762–63.

Fielding, Alan H., and John F. Bell. 1997. "A Review of Methods for the Assessment of Prediction Errors in Conservation Presence/Absence Models." *Environmental Conservation* 24 (1): 38–49.

Fijn, Natasha. 2011. *Living with Herds: Human-Animal Coexistence in Mongolia*. Cambridge University Press.

Firth, Raymond. 1985. "Degrees of Intelligibility." In *Reason and Morality*, edited by Joanna Overing, 29–46. Taylor & Francis.

Fischer, Julia, Philip Wadewitz, and Kurt Hammerschmidt. 2017. "Structural Variability and Communicative Complexity in Acoustic Communication." *Animal Behaviour* 134:229–37.

Fisher, Jessica Claris, Katherine Nesbitt Irvine, Jake Emmerson Bicknell, William Michael Hayes, Damian Fernandes, Jayalaxshmi Mistry, and Zoe Georgina Davies. 2021. "Perceived Biodiversity, Sound, Naturalness and Safety Enhance the Restorative Quality and Wellbeing Benefits of Green and Blue Space in a Neotropical City." *Science of the Total Environment* 755:143095.

Fitch, W. Tecumseh, and Gregory Kramer. 1994. "Sonifying the Body Electric: Superiority of an Auditory over a Visual Display in a Complex, Multivariate System." *Santa Fe Institute Studies in the Sciences of Complexity* 18:307.

Fleming, A. J., A. A. Lindeman, A. L. Carroll, and J. E. Yack. 2013. "Acoustics of the Mountain Pine Beetle (*Dendroctonus ponderosae*) (Curculionidae, Scolytinae): Sonic, Ultrasonic, and Vibration Characteristics." *Canadian Journal of Zoology* 91 (4): 235–44.

Fletcher, Harvey. 1940. "Auditory Patterns." *Reviews of Modern Physics* 12 (1): 47.

Floyd, Donald A., and Jay E. Anderson. 1987. "A Comparison of Three Methods for Estimating Plant Cover." *Journal of Ecology* 75:221–28.

Foley, Jonathan A., Ruth DeFries, Gregory P. Asner, Carol Barford, Gordon Bonan, Stephen R. Carpenter, F. Stuart Chapin et al. 2005. "Global Consequences of Land Use." *Science* 309 (5734): 570–74.

Forman, Richard T. T. 1983. "An Ecology of the Landscape." *BioScience* 33 (9): 535–35.

Forman, Richard T. T. 1995. "Some General Principles of Landscape and Regional Ecology." *Landscape Ecology* 10 (3): 133–42.

Forman, Richard T. T. 2014a. "Land Mosaics: The Ecology of Landscapes and Regions (1995)." In *The Ecological Design and Planning Reader*, ed. Forster Ndubisi, 217–34. Island Press.

Forman, Richard T. T. 2014b. *Urban Ecology: Science of Cities*. Cambridge University Press.

Forman, Richard T. T., and Michel Godron. 1981. "Patches and Structural Components for a Landscape Ecology." *BioScience* 31 (10): 733–40.

Forman, Richard T. T., and Michel Godron. 1984. "Landscape Ecology Principles and Landscape Function." In *Methodology in Landscape Ecological Research and Planning: Proceedings, 1st Seminar, International Association of Landscape Ecology, Roskilde, Denmark, Oct 15–19, 1984*, edited by J. Brandt, P. Agger, 4–15.

Forrest, T. G. 1982. "Acoustic Communication and Baffling Behaviors of Crickets." *Florida Entomologist* 65 (1): 33–44.

Forrester Jay, W. 1961. *Industrial Dynamics*. MIT Press.

Forster, Johann Reinhold. 1778. *Observations Made during a Voyage round the World* [. . .].

Foster, Ian. 2006. "Globus Toolkit Version 4: Software for Service-Oriented Systems." *Journal of Computer Science and Technology* 21 (4): 513–20.

Fouks, B., Patrizia d'Ettorre, and Volker Nehring. 2011. "Brood Adoption in the Leaf-Cutting Ant *Acromyrmex echinatior*: Adaptation or Recognition Noise?" *Insectes sociaux* 58 (4): 479.

Fradkin, Dmitriy, and Fabian Mörchen. 2015. "Mining Sequential Patterns for Classification." *Knowledge and Information Systems* 45 (3): 731–49.

Francis, Clinton D., and Jesse R. Barber. 2013. "A Framework for Understanding Noise Impacts on Wildlife: An Urgent Conservation Priority." *Frontiers in Ecology and the Environment* 11 (6): 305–13.

Francis, Clinton D., Nathan J. Kleist, Catherine P. Ortega, and Alexander Cruz. 2012. "Noise Pollution Alters Ecological Services: Enhanced Pollination and Disrupted Seed Dispersal." *Proceedings of the Royal Society of London. Series B: Biological Sciences* 279 (1739): 2727–35.

Francis, Clinton D., Peter Newman, B. Derrick Taff, Crow White, Christopher A. Monz, Mitchell Levenhagen, Alissa R. Petrelli et al. 2017. "Acoustic Environments Matter: Synergistic Benefits to Humans and Ecological Communities." *Journal of Environmental Management* 203:245–54.

Francis, Clinton D., Catherine P. Ortega, and Alexander Cruz. 2009. "Noise Pollution Changes Avian Communities and Species Interactions." *Current Biology* 19 (16): 1415–19.

Francis, Clinton D., Catherine P. Ortega, and Alexander Cruz. 2011. "Noise Pollution Filters Bird Communities Based on Vocal Frequency." *PLoS One* 6 (11): e27052.

Franco, Lara S., Danielle F. Shanahan, and Richard A. Fuller. 2017. "A Review of the

Benefits of Nature Experiences: More Than Meets the Eye." *International Journal of Environmental Research and Public Health* 14 (8): 864.

Francomano, Dante, Benjamin L. Gottesman, and Bryan C. Pijanowski. 2020. "Biogeographical and Analytical Implications of Temporal Variability in Geographically Diverse Soundscapes." *Ecological Indicators* 112:105845.

Francomano, Dante, Mayra I. Rodríguez González, Alejandro E. J. Valenzuela, Zhao Ma, Andrea N. Raya Rey, Christopher B. Anderson, and Bryan C. Pijanowski. 2022. "Human-Nature Connection and Soundscape Perception: Insights from Tierra del Fuego, Argentina." *Journal for Nature Conservation* 65:126110.

Frawley, William J., Gregory Piatetsky-Shapiro, and Christopher J. Matheus. 1992. "Knowledge Discovery in Databases: An Overview." *AI Magazine* 13 (3): 57.

Fredrickson, Barbara L. 2001. "The Role of Positive Emotions in Positive Psychology: The Broaden-and-Build Theory of Positive Emotions." *American Psychologist* 56 (3): 218.

Freeman, Lauren A., and Simon E. Freeman. 2016. "Rapidly Obtained Ecosystem Indicators from Coral Reef Soundscapes." *Marine Ecology Progress Series* 561:69–82.

Frick, Winifred F. 2013. "Acoustic Monitoring of Bats, Considerations of Options for Long-Term Monitoring." *Therya* 4 (1): 69–70.

Frings, Hubert, and Mable Frings. 1958. "Uses of Sounds by Insects." *Annual Review of Entomology* 3:87–106.

Frisby, Emily M., and H. W. Sansom. 1967. "Hail Incidence in the Tropics." *Journal of Applied Meteorology and Climatology* 6 (2): 339–54.

Fromm, E. 1964. *The Heart of Man*. Harper & Row.

Frost, David M., Sara I. McClelland, Jennifer B. Clark, and Elizabeth A. Boylan. 2014. "Phenomenological Research Methods in the Psychological Study of Sexuality." In *APA Handbook of Sexuality and Psychology*, vol. 1, *Person-Based Approaches*, edited by D. L. Tolman, L. M. Diamond, J. A. Bauermeister, W. H. George, J. G. Pfaus, and L. M. Ward, 121–41. American Psychological Association.

Frumkin, Howard, Gregory N. Bratman, Sara Jo Breslow, Bobby Cochran, Peter H. Kahn Jr., Joshua J. Lawler, Phillip S. Levin et al. 2017. "Nature Contact and Human Health: A Research Agenda." *Environmental Health Perspectives* 125 (7): 075001.

Fu, Zhouyu, Guojun Lu, Kai Ming Ting, and Dengsheng Zhang. 2011. "A Survey of Audio-Based Music Classification and Annotation." *IEEE Transactions on Multimedia* 13 (2): 303–19.

Fuller, Rebecca C., and John A. Endler. 2018. "A Perspective on Sensory Drive." *Current Zoology* 64 (4): 465–70.

Fuller, Richard A., Philip H. Warren, and Kevin J. Gaston. 2007. "Daytime Noise Predicts Nocturnal Singing in Urban Robins." *Biology Letters* 3 (4): 368–70.

Fuller, Susan, Anne C. Axel, David Tucker, and Stuart H. Gage. 2015. "Connecting Soundscape to Landscape: Which Acoustic Index Best Describes Landscape Configuration?" *Ecological Indicators* 58:207–15.

Furumo, Paul R., and T. Mitchell Aide. 2019. "Using Soundscapes to Assess Biodiversity in Neotropical Oil Palm Landscapes." *Landscape Ecology* 34 (4): 911–23.

Gage, Stuart H., and Anne C. Axel. 2014. "Visualization of Temporal Change in Soundscape Power of a Michigan Lake Habitat over a 4-Year Period." *Ecological Informatics* 21:100–109.

Gage, Stuart H., Jason Wimmer, Tom Tarrant, and Peter R. Grace. 2017. "Acoustic Patterns at the Samford Ecological Research Facility in South East Queensland,

Australia: The Peri-Urban SuperSite of the Terrestrial Ecosystem Research Network." *Ecological Informatics* 38:62–75.

Galloway, Kate. 2014. "Ecotopian Spaces: Soundscapes of Environmental Advocacy and Awareness." *Social Alternatives* 33 (3): 71–79.

Galvin, C. J. 1968. Breaker Type Classification on Three Laboratory Beaches. *Journal of Geophysical Research* 73:3651–59.

Ganchev, Todor. 2017. *Computational Bioacoustics: Biodiversity Monitoring and Assessment*. Vol. 4. Walter de Gruyter.

Gandomi, Amir, and Murtaza Haider. 2015. "Beyond the Hype: Big Data Concepts, Methods, and Analytics." *International Journal of Information Management* 35 (2): 137–44.

Gardner, Robert H., and Dean L. Urban. 2003. "Model Validation and Testing: Past Lessons, Present Concerns, Future Prospects." In *Models in Ecosystem Science*, edited by Charles D. Canham, Jonathan J. Cole, and William K. Lauenroth, 184–203. Princeton University Press.

Garreta, Raul, and Guillermo Moncecchi. 2013. *Learning scikit-learn: Machine Learning in Python*. Packt Publishing.

Garrioch, David. 2003. "Sounds of the City: The Soundscape of Early Modern European Towns." *Urban History* 30 (1): 5–25.

Gasc, Amandine, J. Anso, Jérôme Sueur, Herve Jourdan, and L. Desutter-Grandcolas. 2018. "Cricket Calling Communities as an Indicator of the Invasive Ant *Wasmannia auropunctata* in an Insular Biodiversity Hotspot." *Biological Invasions* 20 (5): 1099–111.

Gasc, Amandine, Dante Francomano, John B. Dunning, and Bryan C. Pijanowski. 2017. "Future Directions for Soundscape Ecology: The Importance of Ornithological Contributions." *Auk* 134 (1): 215–28.

Gasc, Amandine, Benjamin L. Gottesman, Dante Francomano, Jinha Jung, Mark Durham, Jason Mateljak, and Bryan C. Pijanowski. 2019. "Soundscapes Reveal Disturbance Impacts: Biophonic Response to Wildfire in the Sonoran Desert Sky Islands." *Landscape Ecology* 33 (8): 1399–415.

Gasc, Amandine, S. Pavoine, L. Lellouch, P. Grandcolas, and Jérôme Sueur. 2015. "Acoustic Indices for Biodiversity Assessments: Analyses of Bias Based on Simulated Bird Assemblages and Recommendations for Field Surveys." *Biological Cons ervation* 191:306–12.

Gasc, Amandine, Jérôme Sueur, Sandrine Pavoine, Roseli Pellens, and Philippe Grandcolas. 2013. "Biodiversity Sampling Using a Global Acoustic Approach: Contrasting Sites with Microendemics in New Caledonia." *PLoS One* 8 (5): e65311.

Gaston, Kevin J., Sarah F. Jackson, Lisette Cantú-Salazar, and Gabriela Cruz-Piñón. 2008. "The Ecological Performance of Protected Areas." *Annual Review of Ecology, Evolution, and Systematics* 39:93–113.

Gause, Georgii Frantsevitch. 1934. "Experimental Analysis of Vito Volterra's Mathematical Theory of the Struggle for Existence." *Science* 79 (2036): 16–17.

Gaver, William W. 1993. "What in the World Do We Hear?: An Ecological Approach to Auditory Event Perception." *Ecological Psychology* 5 (1): 1–29.

Ge, Jian, and Kazunori Hokao. 2004. "Research on the Sound Environment of Urban Open Space from the Viewpoint of Soundscape—A Case Study of Saga Forest Park, Japan." *ACTA Acustica United with Acustica* 90 (3): 555–63.

Geissmann, Thomas. 2000. "Gibbon Songs and Human Music from an Evolutionary

Perspective." In *The Origins of Music*, edited by N. L. Wallin, B. Merker, and S. Brown, 103–23. MIT Press.

Genuit, Klaus, André Fiebig, and Brigitte Schulte-Fortkamp. 2012. "Relationship between Environmental Noise, Sound Quality, Soundscape." *Journal of the Acoustical Society of America* 132 (3): 1924.

Ghadiri Khanaposhtani, Maryam. 2018. "Soundscape Ecology: Rich Contexts for Investigating Conservation Biology and the Effect of Informal Environmental Experiences on Youth's Conceptual Understanding, Interest and Identity Development." PhD diss., Purdue University.

Ghadiri Khanaposhtani, Maryam, ChangChia James Liu, Benjamin L. Gottesman, Daniel Shepardson, and Bryan Pijanowski. 2018. "Evidence That an Informal Environmental Summer Camp Can Contribute to the Construction of the Conceptual Understanding and Situational Interest of STEM in Middle-School Youth." *International Journal of Science Education, Part B* 8 (3): 227–49.

Ghadiri Khanaposhtani, Maryam, Amandine Gasc, Dante Francomano, Luis J. Villanueva-Rivera, Jinha Jung, Michael J. Mossman, and Bryan C. Pijanowski. 2019. "Effects of Highways on Bird Distribution and Soundscape Diversity around Aldo Leopold's Shack in Baraboo, Wisconsin, USA." *Landscape and Urban Planning* 192:103666.

Gibb, Rory, Ella Browning, Paul Glover-Kapfer, and Kate E. Jones. 2019. "Emerging Opportunities and Challenges for Passive Acoustics in Ecological Assessment and Monitoring." *Methods in Ecology and Evolution* 10 (2): 169–85.

Gibbons, Michael, Camille Limoges, Helga Nowotny, Simon Schwartzman, Peter Scott, and Martin Trow. 1994. *The New Production of Knowledge: The Dynamics of Science and Research in Contemporary Societies*. Sage.

Gibson, James J. 1950. *The Perception of the Visual World*. Houghton Mifflin.

Gibson, James J. 1954. "The Visual Perception of Objective Motion and Subjective Movement." *Psychological Review* 61 (5): 304.

Gibson, James J. 1960. "The Concept of the Stimulus in Psychology." *American Psychologist* 15 (11): 694.

Gibson, James J. 1966. *The Senses Considered as Perceptual Systems*. Houghton Mifflin.

Gibson, James J. 1968. "What Gives Rise to the Perception of Motion?" *Psychological Review* 75 (4): 335.

Gibson, James J. 1979. *The Ecological Approach to Visual Perception*. Houghton Mifflin.

Gibson, James J. 2014. *The Ecological Approach to Visual Perception*. Classic edition. Psychology Press.

Gifford, Toby, Simon Linke, and Leah Barclay. 2014. "Listening to the Thames." *Electronic Visualisation and the Arts (EVA 2014)*, 190–91.

Giles, Jacqueline C., Jenny A. Davis, Robert D. McCauley, and Gerald Kuchling. 2009. "Voice of the Turtle: The Underwater Acoustic Repertoire of the Long-Necked Freshwater Turtle, *Chelodina oblonga*." *Journal of the Acoustical Society of America* 126 (1): 434–43.

Gill, Sharon A., Erin E. Grabarczyk, Kathleen M. Baker, Koorosh Naghshineh, and Maarten J. Vonhof. 2017. "Decomposing an Urban Soundscape to Reveal Patterns and Drivers of Variation in Anthropogenic Noise." *Science of the Total Environment* 599–600:1191–201.

Gillam, Erin, and M. Brock Fenton. 2016. "Roles of Acoustic Social Communication in the Lives of Bats." In *Bat Bioacoustics*, edited by M. Brock, Fenton, Alan D. Grinnell, Arthur N. Popper, and Richard R. Fay. 117–39. Springer Handbook of Auditory Research 54. Springer.

Glaser, Barney G., and Anselm Strauss. 1967. *The Discovery of Grounded Theory: Strategies for Qualitative Research*. Aldine.

Glaser, B. G. 1992. *Basics of Grounded Theory Analysis: Emergence vs. Forcing*. Sociology Press.

Gobster, Paul H., Joan I. Nassauer, Terry C. Daniel, and Gary Fry. 2007. "The Shared Landscape: What Does Aesthetics Have to Do with Ecology?" *Landscape Ecology* 22 (7): 959–72.

Goeau, Herve, Stefan Kahl, Hervé Glotin, Robert Planqué, Willem-Pier Vellinga, and Alexis Joly. 2018. "Overview of BirdCLEF 2018: Monospecies vs. Soundscape Bird Identification." CLEF 2018—Conference and Labs of the Evaluation Forum, September 2018, Avignon, France.

Gomes, Dylan G. E., Clinton D. Francis, and Jesse R. Barber. 2021. "Using the Past to Understand the Present: Coping with Natural and Anthropogenic Noise." *BioScience* 71 (3): 223–34.

Goodchild, Michael F. 1992. "Geographical Information Science." *International Journal of Geographical Information Systems* 6 (1): 31–45.

Goodchild, Michael F. 2017. "Big Geodata." In *GIS Applications for Socio-Economics and Humanity*, 19–25. Elsevier. https://doi.org/10.1016/B978-0-12-409548-9.09595-6.

Goodchild, Michael F., and Robert P. Haining. 2004. "GIS and Spatial Data Analysis: Converging Perspectives." *Papers in Regional Science* 83 (1): 363–85.

Goodchild, Michael F., May Yuan, and Thomas J. Cova. 2007. "Towards a General Theory of Geographic Representation in GIS." *International Journal of Geographical Information Science* 21 (3): 239–60.

Göpfert, Martin C., and Daniel Robert. 2001a. "Active Auditory Mechanics in Mosquitoes." *Proceedings of the Royal Society of London. Series B: Biological Sciences* 268 (1465): 333–39.

Göpfert, Martin C., and Daniel Robert. 2001b. "Turning the Key on *Drosophila* Audition." *Nature* 411 (6840): 908.

Gottesman, Benjamin L., Dante Francomano, Zhao Zhao, Kristen Bellisario, Maryam Ghadiri, Taylor Broadhead, Amandine Gasc, and Bryan C. Pijanowski. 2020a. "Acoustic Monitoring Reveals Diversity and Surprising Dynamics in Tropical Freshwater Soundscapes." *Freshwater Biology* 65 (1): 117–32.

Gottesman, Benjamin L., Joshua Sprague, David J. Kushner, Kristen Bellisario, David Savage, Megan F. McKenna, David L. Conlin et al. 2020b. "Soundscapes Indicate Kelp Forest Condition." *Marine Ecology Progress Series* 654:35–52.

Goutte, Sandra, Alain Dubois, Sam David Howard, Rafael Márquez, J. J. L. Rowley, J. M. Dehling, P. Grandcolas, R. C. Xiong, and Frédéric Legendre. 2018. "How the Environment Shapes Animal Signals: A Test of the Acoustic Adaptation Hypothesis in Frogs." *Journal of Evolutionary Biology* 31 (1): 148–58.

Goutte, Sandra, Alain Dubois, Samuel D. Howard, Rafael Marquez, Jodi J. L. Rowley, J. Maximilian Dehling, Philippe Grandcolas, Xiong Rongchuan, and Frédéric Legendre. 2016. "Environmental Constraints and Call Evolution in Torrent-Dwelling Frogs." *Evolution* 70 (4): 811–26.

Gozalo, G. Rey, J. Trujillo Carmona, J. M. Barrigón Morillas, Rosendo Vilchez-Gómez, and V. Gómez Escobar. 2015. "Relationship between Objective Acoustic Indices and Subjective Assessments for the Quality of Soundscapes." *Applied Acoustics* 97:1–10.

Gramann, James. 1999. "Research Review: The Effect of Mechanical Noise and Natural Sound on Visitor Experiences in Units of the National Park System." *Social Science Research Review* 1:1–16.

Granö, J. G. 1929. *Reine Geographie. Acta Geographica* 2. Houghton Mifflin.

Granö, Johannes Gabriel. 1997. *Pure Geography*. Johns Hopkins University Press.

Gray, Claudia L., Samantha L. L. Hill, Tim Newbold, Lawrence N. Hudson, Luca Börger, Sara Contu, Andrew J. Hoskins, Simon Ferrier, Andy Purvis, and Jörn P. W. Scharlemann. 2016. "Local Biodiversity Is Higher inside Than outside Terrestrial Protected Areas Worldwide." *Nature Communications* 7 (1): 12306.

Green, Marc, and Damian Murphy. 2020. "Environmental Sound Monitoring Using Machine Learning on Mobile Devices." *Applied Acoustics* 159:107041.

Greene, Jennifer C. 2007. *Mixed Methods in Social Inquiry*. Research Methods for the Social Sciences, vol. 9. John Wiley & Sons.

Greenfield, M. D. 2002. *Signalers and Receivers: Mechanisms and Evolution of Arthropod Communication*. Oxford University Press.

Greenhalgh, Jack A., Martin J. Genner, Gareth Jones, and Camille Desjonquères. 2020. "The Role of Freshwater Bioacoustics in Ecological Research." *Wiley Interdisciplinary Reviews: Water* 7 (3): e1416.

Grenier, L. 1998. *Working with Indigenous Knowledge: A Guide for Researchers*. IDRC.

Grinnell, Joseph. 1917. "The Niche-Relationships of the California Thrasher." *Auk* 34 (4): 427–33.

Groves, Craig R., Deborah B. Jensen, Laura L. Valutis, Kent H. Redford, Mark L. Shaffer, J. Michael Scott, Jeffrey V. Baumgartner et al. 2002. "Planning for Biodiversity Conservation: Putting Conservation Science into Practice." *BioScience* 52 (6): 499–512.

Guastavino, Catherine. 2006. "The Ideal Urban Soundscape: Investigating the Sound Quality of French Cities." *Acta Acustica United with Acustica* 92 (6): 945–51.

Guerra, Vinicius, Nathane de Queiroz Costa, Diego Llusia, Rafael Marquez, and Rogério P. Bastos. 2020. "Nightly Patterns of Calling Activity in Anuran Assemblages of the Cerrado, Brazil." *Community Ecology* 21(1): 1–10.

Gustafson, Eric J. 1998. "Quantifying Landscape Spatial Pattern: What Is the State of the Art?" *Ecosystems* 1 (2): 143–56.

Gustafson, Eric J., and George R. Parker. 1992. "Relationships between Landcover Proportion and Indices of Landscape Spatial Pattern." *Landscape Ecology* 7 (2): 101–10.

Guston, David H. 2001. "Boundary Organizations in Environmental Policy and Science: An Introduction." *Science, Technology, & Human Values* 26 (4): 399–408.

Guston, David H. 2007. *Between Politics and Science: Assuring the Integrity and Productivity of Research*. Cambridge University Press.

Guy, Nancy. 2009. "Flowing down Taiwan's Tamsui River: Towards an Ecomusicology of the Environmental Imagination." *Ethnomusicology* 53 (2): 218–48.

Guyette, Margaret Q., and Jennifer C. Post. 2016. "Ecomusicology, Ethnomusicology, and Soundscape Ecology: Scientific and Musical Responses to Sound Study." In *Current Directions in Ecomusicology*, edited by Aaron S. Allen and Kevin Dawe, 48–64. Routledge.

Haimoff, Elliott H., and S. Paul Gittins. 1985. "Individuality in the Songs of Wild Agile Gibbons (*Hylobates agilis*) of Peninsular Malaysia." *American Journal of Primatology* 8 (3): 239–47.

Haines-Young, Roy. 2009. "Land Use and Biodiversity Relationships." *Land Use Policy* 26: S178–S186.

Hallmann, Caspar A., Martin Sorg, Eelke Jongejans, Henk Siepel, Nick Hofland, Heinz Schwan, Werner Stenmans et al. 2017. "More than 75 Percent Decline over 27 Years in Total Flying Insect Biomass in Protected Areas." *PLoS One* 12 (10): e0185809.

Hamet, J. F., and P. Klein. 2000. "Road Texture and Tire Noise." In *Proceedings Inter-Noise*, 178–83.

Hammersley, Martyn. 2018. "What Is Ethnography? Can It Survive? Should It?" *Ethnography and Education* 13 (1): 1–17.

Han, Jiawei, Jian Pei, and Yiwen Yin. 2000. "Mining Frequent Patterns without Candidate Generation." In *Proceedings of the 2000 ACM SIGMOD International Conference on Management of Data— SIGMOD'2000*, 1–12.

Haney, Matthew M., Alexa R. Van Eaton, John J. Lyons, Rebecca L. Kramer, David Fee, Alexandra M. Iezzi, Robert P. Dziak et al. 2020. "Characteristics of Thunder and Electromagnetic Pulses from Volcanic Lightning at Bogoslof Volcano, Alaska." *Bulletin of Volcanology* 82 (2): 15.

Haney, Matthew M., Alexa R. Van Eaton, John J. Lyons, Rebecca L. Kramer, David Fee, and Alexandra M. Iezzi. 2018. "Volcanic Thunder from Explosive Eruptions at Bogoslof Volcano, Alaska." *Geophysical Research Letters* 45 (8): 3429–35.

Hanke, Michael, Yaroslav O. Halchenko, Per B. Sederberg, Emanuele Olivetti, Ingo Fründ, Jochem W. Rieger, Christoph S. Herrmann, James V. Haxby, Stephen J. Hanson, and Stefan Pollmann. 2009. "Pymvpa: A Unifying Approach to the Analysis of Neuroscientific Data." *Frontiers in Neuroinformatics* (2009): 3.

Hanslowe, Thomas. 2015. "Composing the Soundscape Ecological Listening in the Music of John Luther Adams." PhD diss., Tufts University.

Hanson, Jeffrey O., Jonathan R. Rhodes, Stuart H. M. Butchart, Graeme M. Buchanan, Carlo Rondinini, Gentile F. Ficetola, and Richard A. Fuller. 2020. "Global Conservation of Species' Niches." *Nature* 580 (7802): 232–34.

Hao, Zezhou, Cheng Wang, Zhenkai Sun, Cecil Konijnendijk van den Bosch, Dexian Zhao, Baoqiang Sun, Xinhui Xu et al. 2021. "Soundscape Mapping for Spatial-Temporal Estimate on Bird Activities in Urban Forests." *Urban Forestry & Urban Greening* 57:126822.

Harbrow, Michael A., G. R. Cessford, B. J. Kazmierow et al. 2011. *The Impact of Noise on Recreationists and Wildlife in New Zealand's Natural Areas: A Literature Review.* Science for Conservation, no. 314. New Zealand Department of Conservation.

Hardin, Garrett. 1960. "The Competitive Exclusion Principle." *Science* 131 (3409): 1292–97.

Hardin, Garrett. 1968. "The Tragedy of the Commons: The Population Problem Has No Technical Solution; It Requires a Fundamental Extension in Morality." *Science* 162 (3859): 1243–48.

Harrington, Fred H., and Cheryl S. Asa, 2003. "Wolf Communication." Chapter 3 in *Wolves: Behavior, Ecology, and Conservation*, edited by L. David Mech and Luigi Boitani, 66–103. University of Chicago Press.

Harris, Sydney A., Nick T. Shears, and Craig A. Radford. 2016. "Ecoacoustic Indices as Proxies for Biodiversity on Temperate Reefs." *Methods in Ecology and Evolution* 7 (6): 713–24.

Harrison, Klisala. 2020. "Indigenous Music Sustainability during Climate Change." *Current Opinion in Environmental Sustainability* 43:28–34.

Hartig, Terry, Kalevi Korpela, Gary W. Evans, and Tommy Gärling. 1997. "A Measure of Restorative Quality in Environments." *Scandinavian Housing and Planning Research* 14 (4): 175–94.

Hartmann, William M. 1998. *Signals, Sound, and Sensation*. Springer Science & Business Media.

Harvey, Ben P., Koetsu Kon, and Sylvain Agostini. 2020. "Survey Techniques in Marine Ecology." In *Japanese Marine Life: A Practical Training Guide in Marine Biology*, edited by Kazuo Inaba and Jason M. Hall-Spencer, 263–71. Springer.

Hastings, Philip A., and Ana Širović. 2015. "Soundscapes Offer Unique Opportunities for Studies of Fish Communities." *Proceedings of the National Academy of Sciences* 112 (19): 5866–67.

Hausmann, Anna, R. O. B. Slotow, Jonathan K. Burns, and Enrico Di Minin. 2016. "The Ecosystem Service of Sense of Place: Benefits for Human Well-Being and Biodiversity Conservation." *Environmental Conservation* 43 (2): 117–27.

Haver, Samara M., Michelle E. Fournet, Robert P. Dziak, Christine Gabriele, Jason Gedamke, Leila T. Hatch, Joseph Haxel et al. 2019. "Comparing the Underwater Soundscapes of Four U.S. National Parks and Marine Sanctuaries." *Frontiers in Marine Science* 6 (500): 1–14.

Hawkins, Charles P., John R. Olson, and Ryan A. Hill. 2010. "The Reference Condition: Predicting Benchmarks for Ecological and Water-Quality Assessments." *Journal of the North American Benthological Society* 29 (1): 312–43.

Hayashi, Kiyotada, Vita Dhian Lelyana, Kohji Yamamura et al. 2020. "Acoustic Dissimilarities between an Oil Palm Plantation and Surrounding Forests: Analysis of Index Time Series for Beta-Diversity in South Sumatra, Indonesia." *Ecological Indicators* 112:106086.

Hayes, Lauren, and Julian Stein. 2018. 2019. "Desert and Sonic Ecosystems: Incorporating Environmental Factors within Site-Responsive Sonic Art." *Applied Sciences* 8 (1): 111.

He, Fengzhi, Christiane Zarfl, Vanessa Bremerich, Jonathan N. W. David, Zeb Hogan, Gregor Kalinkat, Klement Tockner, and Sonja C. Jähnig. 2019. "The Global Decline of Freshwater Megafauna." *Global Change Biology* 25 (11): 3883–92.

Hearst, Marti A. 1999. "Untangling Text Data Mining." In *Proceedings of the 37th Annual Meeting of the Association for Computational Linguistics*, 3–10.

Hearst, Marti A. 2003. "What Is Text Mining?" Summer Institute for the Mathematical Sciences, University of California, Berkeley, 5. https://people.ischool.berkeley.edu/~hearst/text-mining.html.

Heenehan, Heather, Joy E. Stanistreet, Peter J. Corkeron, Laurent Bouveret, Julien Chalifour, Genevieve E. Davis, Angiolina Henriquez et al. 2019. "Caribbean Sea Soundscapes: Monitoring Humpback Whales, Biological Sounds, Geological Events, and Anthropogenic Impacts of Vessel Noise." *Frontiers in Marine Science* 6:347.

Heinz, Kevin M., Michael P. Parrella, and Julie P. Newman. 1992. "Time-Efficient Use of Yellow Sticky Traps in Monitoring Insect Populations." *Journal of Economic Entomology* 85 (6): 2263–69.

Henry, Kenneth S., and Jeffrey R. Lucas. 2010. "Auditory Sensitivity and the Frequency Selectivity of Auditory Filters in the Carolina Chickadee, *Poecile carolinensis*." *Animal Behaviour* 80 (3): 497–507.

Hermann, Thomas. 2002. "Sonification for Exploratory Data Analysis." PhD diss., Bielefeld University.

Hermann, Thomas. 2008. "Taxonomy and Definitions for Sonification and Auditory Display." In *Proceedings of the 14th International Conference on Auditory Display*, 1–8. ICAD.

Hermann, Thomas, and Helge Ritter. 1999. "Listen to Your Data: Model-Based Sonification for Data Analysis." In *Advances in Intelligent Computing and Multimedia Systems*, edited by G. E. Lasker and M. R. Syed, 189–94. International Institute for Advanced Studies in System Research and Cybernetics.

Herrera-Montes, Maria Isabel. 2018. "Protected Area Zoning as a Strategy to Preserve Natural Soundscapes, Reduce Anthropogenic Noise Intrusion, and Conserve Biodiversity." *Tropical Conservation Science* 11 (1). https://doi.org/10.1177/1940082918804344.

Herzog, Sebastian K., Michael Kessler, and Thomas M. Cahill. 2002. "Estimating Species Richness of Tropical Bird Communities from Rapid Assessment Data." *Auk* 119 (3): 749–69.

Hessels, Laurens K., and Harro Van Lente. 2008. "Re-thinking New Knowledge Production: A Literature Review and a Research Agenda." *Research Policy* 37 (4): 740–60.

Hey, Anthony J. G., Stewart Tansley, Kristin Michele Tolle et al., eds. 2009. *The Fourth Paradigm: Data-Intensive Scientific Discovery*. Vol. 1. Microsoft Research.

Hey, T., and Anne Trefethen. 2003. "The Data Deluge: An e-Science Perspective." In *Grid Computing: Making the Global Infrastructure a Reality*, edited by Fran Berman, Geoffrey Fox, and Tony Hey, 809–24. Wiley Online Library.

Hey, T., and Anne Trefethen. 2020. "The Fourth Paradigm 10 Years On." *Informatik Spektrum* 42 (6): 441–47.

Hickling, Robert, and Richard L. Brown. 2000. "Analysis of Acoustic Communication by Ants." *Journal of the Acoustical Society of America* 108 (4): 1920–29.

Hijmans, Robert J., Jacob Van Etten, Joe Cheng, Matteo Mattiuzzi, Michael Sumner, Jonathan A. Greenberg, Oscar Perpinan Lamigueiro et al. 2015. "Package 'Raster.'" *R Package* 734:473.

Hilal, AlYahmady Hamed, and Saleh Said Alabri. 2013. "Using NVivo for Data Analysis in Qualitative Research." *International Interdisciplinary Journal of Education* 2 (2): 181–86.

Hildebrand, John A. 2009. "Anthropogenic and Natural Sources of Ambient Noise in the Ocean." *Marine Ecology Progress Series* 395:5–20.

Hingston, Andrew B., Timothy J. Wardlaw, Susan C. Baker, and Gregory J. Jordan. 2018. "Data Obtained from Acoustic Recording Units and from Field Observer Point Counts of Tasmanian Forest Birds Are Similar but Not the Same." *Australian Field Ornithology* 35:30–39.

Hinton, Geoffrey E., and Sam Roweis. 2002. "Stochastic Neighbor Embedding." In *Advances in Neural Information Processing Systems* 15: Proceedings of the 2002 Conference, edited by Suzanna Baker, Sebastian Thrun, and Klaus Obermayer . MIT.

Hobbs, Richard. 1997. "Future Landscapes and the Future of Landscape Ecology." *Landscape and Urban Planning* 37 (1–2): 1–9.

Hobson, Keith A., Robert S. Rempel, Hamilton Greenwood, Brian Turnbull, and Steven L. Van Wilgenburg. 2002. "Acoustic Surveys of Birds Using Electronic Recordings: New Potential from an Omnidirectional Microphone System." *Wildlife Society Bulletin* 30 (3): 709–20.

Hödl, Walter. 1977. "Call Differences and Calling Site Segregation in Anuran Species from Central Amazonian Floating Meadows." *Oecologia* 28: 351–63.

Holekamp, Kay, Sarah Benson-Amram, Kevin Theis, and Keron Greene. 2007. "Sources of Variation in the Long-Distance Vocalizations of Spotted Hyenas." *Behaviour* 144 (5): 557–84.

Holling, Crawford S. 1992. "Cross-Scale Morphology, Geometry, and Dynamics of Eco-systems." *Ecological Monographs* 62 (4): 447–502.

Holt, Daniel E., and Carol E. Johnston. 2015. "Traffic Noise Masks Acoustic Signals of Freshwater Stream Fish." *Biological Conservation* 187:27–33.

Hong, Joo Young, and Jin Yong Jeon. 2013. "Designing Sound and Visual Components for Enhancement of Urban Soundscapes." *Journal of the Acoustical Society of America* 134 (3): 2026–36.

Hong, Joo Young, and Jin Yong Jeon. 2015. "Influence of Urban Contexts on Sound-scape Perceptions: A Structural Equation Modeling Approach." *Landscape and Urban Planning* 141:78–87.

Hong, Joo Young, Bhan Lam, Zhen Ting Ong, Kenneth Ooi, Woon Seng Gan, Jian Kang, Samuel Yeong, Irene Lee, and Sze Tiong Tan. 2021. "A Mixed-Reality Approach to Soundscape Assessment of Outdoor Urban Environments Augmented with Natural Sounds." *Building and Environment* 194 (May 2021): 107688.

Hong, Joo Young, Zhen-Ting Ong, Bhan Lam, Kenneth Ooi, Woon-Seng Gan, Jian Kang, Jing Feng, and Sze-Tiong Tan. 2020. "Effects of Adding Natural Sounds to Urban Noises on the Perceived Loudness of Noise and Soundscape Quality." *Science of the Total Environment* 711:134571.

Hong, Xin-Chen, Jiang Liu, Guangyu Wang, Yu Jiang, Shuting Wu, and Si-Ren Lan. 2019a. "Factors Influencing the Harmonious Degree of Soundscapes in Urban For-ests: A Comparison of Broad-Leaved and Coniferous Forests." *Urban Forestry & Urban Greening* 39:18–25.

Hong, Xin-Chen, Guang-Yu Wang, Jiang Liu, Lei Song, and Ernest T. Y. Wu. 2021. "Mod-eling the Impact of Soundscape Drivers on Perceived Birdsongs in Urban Forests." *Journal of Cleaner Production* 292:125315.

Hong, Xin-Chen, Zhi-Peng Zhu, Jiang Liu, De-Hui Geng, Guang-Yu Wang, and Si-Ren Lan. 2019b. "Perceived Occurrences of Soundscape Influencing Pleasantness in Ur-ban Forests: A Comparison of Broad-Leaved and Coniferous Forests." *Sustainability* 11 (17): 4789.

Hopson, Adrienne, and Ferenc de Szalay. 2021. "Alteration of Above and Below-Water Soundscapes by Roads." *Wetlands* 41 (1). https://doi.org/10.1007/s13157-021-01407-8.

Hotelling, H. 1933. "Analysis of a Complex of Statistical Variables into Principal Compo-nents." *Journal of Educational Psychology* 24 (6): 417–41.

Hotho, Andreas, Andreas Nürnberger, and Gerhard Paaß. 2005. "A Brief Survey of Text Mining." *LDV Forum* 20:19–62.

Houlahan, Jeff E., C. Scott Findlay, Benedikt R. Schmidt, Andrea H. Meyer, and Sergius L. Kuzmin. 2000. "Quantitative Evidence for Global Amphibian Population De-clines." *Nature* 404 (6779): 752–55.

Howard, David M., and Jamie Angus. 2017. *Acoustics and Psychoacoustics*. 5th ed. Routledge.

Howe, Bruce M., Jennifer Miksis-Olds, Eric Rehm, Hanne Sagen, Peter F. Worcester, and Georgios Haralabus. 2019. "Observing the Oceans Acoustically." *Frontiers in Marine Science* 6:426.

Howell, Andrew J., Raelyne L. Dopko, Holli-Anne Passmore, and Karen Buro. 2011. "Nature Connectedness: Associations with Well-Being and Mindfulness." *Personality and Individual Differences* 51 (2): 166–71.

Howes, David, ed. 1991. *The Varieties of Sensory Experience: A Sourcebook in the Anthropol-ogy of the Senses*. University of Toronto Press,.

Howes, David. 2014. "Introduction to Sensory Museology." *Senses and Society* 9 (3): 259–67.

Howes, David, and Constance Classen. 2013. *Ways of Sensing: Understanding the Senses in Society*. Routledge. https://nsidc.org/cryosphere/glaciers/questions/what.html.

Hu, Marian Y., Hong Young Yan, Wen-Sung Chung, Jen-Chieh Shiao, and Pung-Pung Hwang. 2009. "Acoustically Evoked Potentials in Two Cephalopods Inferred Using the Auditory Brainstem Response (ABR) Approach." *Comparative Biochemistry and Physiology Part A: Molecular & Integrative Physiology* 153 (3): 278–83.

Hu, Wen, Nirupama Bulusu, Chun Tung Chou, Sanjay Jha, Andrew Taylor, and Van Nghia Tran. 2005. "A Hybrid Sensor Network for Cane-Toad Monitoring." In *Proceedings of the 3rd International Conference on Embedded Networked Sensor Systems*, 305.

Hughes, Robert M., David P. Larsen, and James M. Omernik. 1986. "Regional Reference Sites: A Method for Assessing Stream Potentials." *Environmental Management* 10 (5): 629–35.

Humboldt, Alexander von. 1856. *Cosmos: A Sketch or a Physical Description of the Universe*. Translated by E. C. Otté. Harper & Brothers.

Humboldt, Alexandro de. 1808. *Conspectus Longitudinum et Latitudinum Geographicarum per Decursum Annorum 1799 ad 1804 in Plaga Aequinoctiali*. Schoell,

Huntington, Henry P. 2000. "Using Traditional Ecological Knowledge in Science: Methods and Applications." *Ecological Applications* 10 (5): 1270–74.

Hutchins, K. G. 2019. "Like a Lullaby: Song as Herding Tool in Rural Mongolia." *Journal of Ethnobiology* 39 (3): 445. https://doi.org/10.2993/0278-0771-39.3.445.

Hutchinson, G. Evelyn. 1957. *A Treatise on Limnology*. Wiley.

Hutchinson, G. Evelyn. 1959. "Homage to Santa Rosalia, or Why Are There So Many Kinds of Animals?" *American Naturalist* 93 (870): 145–59.

Inglis, Julian, ed. 1993. *Traditional Ecological Knowledge: Concepts and Cases*. IDRC.

Ingold, Tim. 2000. *The Perception of the Environment: Essays on Livelihood, Dwelling and Skill*. Psychology Press.

Ingold, Tim. 2002. "Culture and the Perception of the Environment." In *Bush Base, Forest Farm*, ed. Elisabeth Croll and David Parkin, 38–56. Routledge.

Ingold, Tim. 2003. "Globes and Spheres: The Topology of Environmentalism." In *Environmentalism*, edited by Kay Milton, 39–50. Routledge.

Ingram, David. 2010. "The Jukebox in the Garden: Ecocriticism and American Popular Music since 1960." *Nature, Culture and Literature* 7:4.

International Union for Conservation of Nature and Natural Resources. 2009. *Wildlife in a Changing World: An Analysis of the 2008 IUCN Red List of Threatened Species*. IUCN.

International Union for Conservation of Nature and Natural Resources. 2012. *Protected Planet Report 2012*. IUCN.

Ishay, Jacob S., and Dror Sadeh. 1982. "The Sounds of Honey Bees and Social Wasps Are Always Composed of a Uniform Frequency." *Journal of the Acoustical Society of America* 72 (3): 671–75.

Ishay, Jacob. 1976. "Comb Building by the Oriental Hornet (*Vespa orientalis*)." *Animal Behaviour* 24 (1): 72–83.

Ismail, Mostafa Refat. 2014. "Sound Preferences of the Dense Urban Environment: Soundscape of Cairo." *Frontiers of Architectural Research* 3 (1): 55–68.

Jaeger, Jochen A. G. 2000. "Landscape Division, Splitting Index, and Effective Mesh Size: New Measures of Landscape Fragmentation." *Landscape Ecology* 15 (2): 115–30.

Jain, Anil K. 2010. "Data Clustering: 50 Years beyond K-Means." *Pattern Recognition Letters* 31 (8): 651–66.

Jain, Manjari, and Rohini Balakrishnan. 2011. "Microhabitat Selection in an Assemblage of Crickets (Orthoptera: Ensifera) of a Tropical Evergreen Forest in Southern India." *Insect Conservation and Diversity* 4 (2): 152–58.

Jain, Manjari, Swati Diwakar, Jimmy Bahuleyan, Rittik Deb, and Rohini Balakrishnan. 2014. "A Rain Forest Dusk Chorus: Cacophony or Sounds of Silence?" *Evolutionary Ecology* 28 (1): 1–22.

Janik, Vincent M., and Peter J. B. Slater. 1997. "Vocal Learning in Mammals." *Advances in the Study of Behaviour* 26:59–100.

Janzen, Daniel H. 1967. "Why Mountain Passes Are Higher in the Tropics." *American Naturalist* 101 (919): 233–49.

Janzen, Daniel H. 1973. "Sweep Samples of Tropical Foliage Insects: Description of Study Sites, with Data on Species Abundances and Size Distributions." *Ecology* 54 (3): 659–86.

Janzen, Daniel H. 1973. "Sweep Samples of Tropical Foliage Insects: Effects of Seasons, Vegetation Types, Elevation, Time of Day, and Insularity." *Ecology* 54 (3): 687–708.

Järviluoma, Helmi. 2016. "The Art and Science of Sensory Memory Walking." In *The Routledge Companion to Sounding Art*, edited by Marcel Cobussen, Vincent Meelberg, and Barry Truax, 211–24. Routledge.

Jaszczak, Agnieszka, Natalia Małkowska, Katarina Kristianova, Sebastian Bernat, and Ewelina Pochodyła. 2021. "Evaluation of Soundscapes in Urban Parks in Olsztyn (Poland) for Improvement of Landscape Design and Management." *Land* 10 (1): 66.

Jetz, Walter, Chris Carbone, Jenny Fulford, and James H. Brown. 2004. "The Scaling of Animal Space Use." *Science* 306 (5694): 266–68.

Jeon, Jin Yong, and Joo Young Hong. 2015. "Classification of Urban Park Soundscapes through Perceptions of the Acoustical Environments." *Landscape and Urban Planning* 141:100–111.

Jeon, Jin Yong, Pyoung Jik Lee, Jin You, and Jian Kang. 2012. "Acoustical Characteristics of Water Sounds for Soundscape Enhancement in Urban Open Spaces." *Journal of the Acoustical Society of America* 131 (3): 2101–9.

Jia, Yihong, Hui Ma, and Jian Kang. 2020. "Characteristics and Evaluation of Urban Soundscapes Worthy of Preservation." *Journal of Environmental Management* 253:109722.

Jin, Xiaolong, Benjamin W. Wah, Xueqi Cheng, and Yuanzhuo Wang. 2015. "Significance and Challenges of Big Data Research." *Big Data Research* 2 (2): 59–64.

Jiskoot, Hester. 2011. "Dynamics of Glaciers." *Physical Research* 92 (B9): 9083–100.

Job, R. F. S., and J. Hatfield. 2001. "The Impact of Soundscape, Enviroscape, and Psychscape on Reaction to Noise: Implications for Evaluation and Regulation of Noise Effects." *Noise Control Engineering Journal* 49 (3): 120–24.

Johannes, Robert Earle, ed. 1989. *Traditional Ecological Knowledge: A Collection of Essays*. IUCN.

Johnson, Matthew F., and Stephen P. Rice. 2014. "Animal Perception in Gravel-Bed Rivers: Scales of Sensing and Environmental Controls on Sensory Information." *Canadian Journal of Fisheries and Aquatic Sciences* 71 (6): 945–57.

Jones, G. P., and N. L. Andrew. 1992. "Temperate Reefs and the Scope of Seascape Ecology." In *Proceedings of the Second International Temperate Reef Symposium*, vol. 7, 63–76. National Institute of Water and Atmospheric Research (NIWA).

Jones, Gareth, and Emma C. Teeling. 2006. "The Evolution of Echolocation in Bats." *Trends in Ecology & Evolution* 21 (3): 149–56.

Jones, Kendall R., Oscar Venter, Richard A. Fuller, James R. Allan, Sean L. Maxwell, Pablo Jose Negret, and James E. M. Watson. 2018. "One-Third of Global Protected Land Is under Intense Human Pressure." *Science* 360 (6390): 788–91.

Jönsson, Per, and Lars Eklundh. 2004. "TIMESAT—A Program for Analyzing Time-Series of Satellite Sensor Data." *Computers and Geosciences* 30 (8): 833–45.

Joo, Wooyeong, Stuart H. Gage, and Eric P. Kasten. 2011. "Analysis and Interpretation of Variability in Soundscapes along an Urban–Rural Gradient." *Landscape and Urban Planning* 103 (3–4): 259–76.

Jordan, Michael I., and Tom M. Mitchell. 2015. "Machine Learning: Trends, Perspectives, and Prospects." *Science* 349 (6245): 255–60.

Jorgensen, Bradley S., and Richard C. Stedman. 2001. "Sense of Place as an Attitude: Lakeshore Owners Attitudes toward Their Properties." *Journal of Environmental Psychology* 21 (3): 233–48.

Juillet, J. A. 1963. "A Comparison of Four Types of Traps Used for Capturing Flying Insects." *Canadian Journal of Zoology* 41 (2): 219–23.

Jung, Jinha, and Bryan C. Pijanowski. 2015. "LiDARHub: A Free and Open Source Software Platform for Web-Based Management, Visualization and Analysis of LiDAR Data." *Geosciences Journal* 19 (4): 741–49.

Kacem, Zaccaria, Marco A. Rodríguez, Irene T. Roca, and Raphaël Proulx. 2020. "The Riverscape Meets the Soundscape: Acoustic Cues and Habitat Use by Brook Trout in a Small Stream." *Canadian Journal of Fisheries and Aquatic Sciences* 77 (6): 991–99.

Kanciruk, P. 1980. "Ecology of Juvenile and Adult Palinuridae (Spiny Lobsters)." *Biology and Management of Lobsters* 2:59–96.

Kanders, Karlis, Tom Lorimer, Florian Gomez, and Ruedi Stoop. 2017. "Frequency Sensitivity in Mammalian Hearing from a Fundamental Nonlinear Physics Model of the Inner Ear." *Scientific Reports* 7 (1): 1–8.

Kang, Jian. 2002. "Numerical Modelling of the Sound Fields in Urban Streets with Diffusely Reflecting Boundaries." *Journal of Sound and Vibration* 258 (5): 793–813.

Kang, Jian. 2005. "Numerical Modeling of the Sound Fields in Urban Squares." *Journal of the Acoustical Society of America* 117 (6): 3695–706.

Kang, Jian. 2006. *Urban Sound Environment*. CRC Press.

Kang, Jian. 2010. "From Understanding to Designing Soundscapes." *Frontiers of Architecture and Civil Engineering in China* 4 (4): 403–17.

Kang, Jian, Francesco Aletta, Truls T. Gjestland, Lex A. Brown, Dick Botteldooren, Brigitte Schulte-Fortkamp, Peter Lercher et al. 2016. "Ten Questions on the Soundscapes of the Built Environment." *Building and Environment* 108:284–94.

Kang, Jian, Francesco Aletta, Tin Oberman, Mercede Erfanian, Magdalena Kachlicka, Matteo Lionello, Andrew Mitchell et al. 2019. *Towards Soundscape Indices*. Universitätsbibliothek der RWTH Aachen.

Kang, Jian, and Brigitte Schulte-Fortkamp, eds. 2018. *Soundscape and the Built Environment*. CRC Press.

Kang, Jian, and Mei Zhang. 2010. "Semantic Differential Analysis of the Soundscape in Urban Open Public Spaces." *Building and Environment* 45 (1): 150–57.

Kantardzic, Mehmed. 2011. *Data Mining: Concepts, Models, Methods, and Algorithms*. John Wiley & Sons.

Kaplan, Maxwell B., and T. Aran Mooney. 2015. "Ambient Noise and Temporal Patterns of Boat Activity in the US Virgin Islands National Park." *Marine Pollution Bulletin* 98 (1–2): 221–28.

Kaplan, Maxwell B., and T. Aran Mooney. 2016. "Coral Reef Soundscapes May Not Be Detectable Far From The Reef." *Scientific Reports* 6 (1): 1–10.

Kaplan, Maxwell B., T. Aran Mooney, Jim Partan, and Andrew R. Solow. 2015. "Coral Reef Species Assemblages Are Associated with Ambient Soundscapes." *Marine Ecology Progress Series* 533:93–107.

Kaplan, Rachel, and Stephen Kaplan. 1989. *The Experience of Nature: A Psychological Perspective*. Cambridge University Press.

Kaplan, Stephen. 1995. "The Restorative Benefits of Nature: Toward an Integrative Framework." *Journal of Environmental Psychology* 15 (3): 169–82.

Kaplan, Stephen, and Rachel Kaplan. 1982. *Cognition and Environment: Functioning in an Uncertain World*. Greenwood.

Kareiva, Peter, and Michelle Marvier. 2012. "What Is Conservation Science?" *BioScience* 62 (11): 962–69.

Kariel, H. G. 1977. "Evaluation of Campground Sounds in Canadian Rocky Mountain National Parks." Revised version of a paper presented at the Annual Meeting of the Association of American Geographers, Salt Lake City, UT. April 14–17.

Kariel, Herbert G. 1980. "Mountaineers and the General Public: A Comparison of Their Evaluation of Sounds in a Recreational Environment." *Leisure Sciences* 3 (2): 155–67.

Kariel, Herbert G. 1990. "Factors Affecting Response to Noise in Outdoor Recreational Environments." *Canadian Geographer / Le Géographe canadien* 34 (2): 142–49.

Kasten, Eric P., Stuart H. Gage, Jordan Fox, and Wooyeong Joo. 2012. "The Remote Environmental Assessment Laboratory's Acoustic Library: An Archive for Studying Soundscape Ecology." *Ecological Informatics* 12:50–67.

Kates, Robert W. 1994. "Sustaining Life on the Earth." *Scientific American* 271 (4): 114–22.

Keitt, Timothy H., Dean L. Urban, and Bruce T. Milne. 1997. "Detecting Critical Scales in Fragmented Landscapes." *Conservation Ecology* 1 (1): 4.

Keller, Damián. 2000. "Compositional Processes from an Ecological Perspective." *Leonardo Music Journal* 10 (1): 55–60.

Kellert, Stephen R., and Edward O. Wilson, eds. 1995. *The Biophilia Hypothesis*. Island Press.

Kelley, Kate, Belinda Clark, Vivienne Brown, and John Sitzia. 2003. "Good Practice in the Conduct and Reporting of Survey Research." *International Journal for Quality in Health Care* 15 (3): 261–66.

Kelman, Ari Y. 2010. "Rethinking the Soundscape: A Critical Genealogy of a Key Term in Sound Studies." *Senses and Society* 5 (2): 212–34.

Kenyon, Todd N., Friedrich Ladich, and Hong Y. Yan. 1998. "A Comparative Study of Hearing Ability in Fishes: The Auditory Brainstem Response Approach." *Journal of Comparative Physiology A* 182 (3): 307–18.

Keogh, Brent. 2013. "On the Limitations of Music Ecology." *Journal of Music Research Online* 4.

Kerrick, Jean S., David C. Nagel, and Ricarda L. Bennett. 1969. "Multiple Ratings of Sound Stimuli." *Journal of the Acoustical Society of America* 45 (4): 1014–17.

Khait, I., U. Obolski, Y. Yovel, and L. Hadany. 2019. "Sound Perception in Plants." In *Seminars in Cell & Developmental Biology* 92:134–38.

Kight, Caitlin R., and John P. Swaddle. 2011. "How and Why Environmental Noise Impacts Animals: An Integrative, Mechanistic Review." *Ecology Letters* 14 (10): 1052–61.

Kiley, Marthe. 1972. "The Vocalizations of Ungulates, Their Causation and Function." *Zeitschrift für Tierpsychologie* 31 (2): 171–222.

King, Ronald P., and Jeffrey R. Davis. 2003. "Community Noise: Health Effects and Management." *International Journal of Hygiene and Environmental Health* 206 (2): 123–31.

Kinnear, Tyler. 2012. "Voicing Nature in John Luther Adams's 'The Place Where You Go to Listen.'" *Organised Sound* 17 (3): 230–39.

Kitchin, Rob. 2014. *The Data Revolution: Big Data, Open Data, Data Infrastructures and Their Consequences.* Sage.

Kitting, Christopher L. 1979. "The Use of Feeding Noises to Determine the Algal Foods Being Consumed by Individual Intertidal Molluscs." *Oecologia* 40:1–17.

Kloser, R. J., J. D. Penrose, and A. J. Butler. 2010. "Multi-Beam Backscatter Measurements Used to Infer Seabed Habitats." *Continental Shelf Research* 30 (16): 1772–82.

Knudsen, Vern O., R. S. Alford, and J. W. Emling. 1948. "Underwater Ambient Noise." *Journal of Marine Research* 7 (3): 410–29.

Koda, H., T. Nishimura, I. T. Tokuda, C. Oyakawa, T. Nihonmatsu, and N. Masataka. 2012. "Soprano Singing in Gibbons." *American Journal of Physical Anthropology* 149 (3): 347–55.

Kogan, Pablo, Jorge P. Arenas, Fernando Bermejo, María Hinalaf, and Bruno Turra. 2018. "A Green Soundscape Index (GSI): The Potential of Assessing the Perceived Balance between Natural Sound and Traffic Noise." *Science of the Total Environment* 642:463–72.

Kogan, Pablo, Bruno Turra, Jorge P. Arenas, and María Hinalaf. 2017. "A Comprehensive Methodology for the Multidimensional and Synchronic Data Collecting in Soundscape." *Science of the Total Environment* 580:1068–77.

Körner, Christian, and David Basler. 2010. "Phenology under Global Warming." *Science* 327 (5972): 1461–62.

Krause, Bernard L. 1987. "Bioacoustics, Habitat Ambience in Ecological Balance." *Whole Earth Review* 57 (Winter): 14–18.

Krause, Bernard L. 1993. "The Niche Hypothesis: A Virtual Symphony of Animal Sounds, the Origins of Musical Expression and the Health of Habitats." *Soundscape Newsletter* 6:6–10.

Krause, Bernard L. 2016. *Wild Soundscapes: Discovering the Voice of the Natural World.* Yale University Press.

Krause, Bernard L., Stuart H. Gage, and Wooyeong Joo. 2011. "Measuring and Interpreting the Temporal Variability in the Soundscape at Four Places in Sequoia National Park." In "Soundscape Ecology," ed. Bryan C. Pijanowski and Almo Farina, special issue, *Landscape Ecology* 26 (9): 1247–56.

Krishnan, Anand, and Krishnapriya Tamma. 2016. "Divergent Morphological and Acoustic Traits in Sympatric Communities of Asian Barbets." *Royal Society Open Science* 3 (8): 160117.

Kryter, Karl D. 1972. "Non-Auditory Effects of Environmental Noise." *American Journal of Public Health* 62 (3): 389–98.

Kryter, Karl D. 2013. *The Effects of Noise on Man.* Elsevier.

Kubisch, Christina. 2019. "Electrical Walks." In *Acoustic Turn*, edited by Petra Maria Meyer, 682–86. Wilhelm Fink.

Kuehne, Lauren M., Britta L. Padgham, and Julian D. Olden. 2013. "The Soundscapes of Lakes across an Urbanization gradient." *PLoS One* 8 (2): e55661.

Kullback, S., and R. Leibler. 1951. "On Information and Sufficiency." *Annals of Mathematical Statistics* 22 (1): 79–86.

Kuperman, Roux. 2007. "Underwater Acoustics." In *Springer Handbook of Acoustics*, edited by Thomas Rossing, 149–204. Springer Science & Business Media.

Lakes-Harlan, Reinhard, and Gerlind U. C. Lehmann. 2015. "Parasitoid Flies Exploiting Acoustic Communication of Insects—Comparative Aspects of Independent Functional Adaptations." *Journal of Comparative Physiology A* 201 (1): 123–32.

Lambin, Eric F. 1999. *Land-Use and Land-Cover Change (LUCC)—Implementation Strategy. A Core Project of the International Geosphere-Biosphere Programme and the International Human Dimensions Programme on Global Environmental Change*. International Geosphere-Biosphere Programme Secretariat.

Lambin, Eric F., Billie L. Turner, Helmut J. Geist, Samuel B. Agbola, Arild Angelsen, John W. Bruce, Oliver T. Coomes et al. 2001. "The Causes of Land-Use and Land-Cover Change: Moving beyond the Myths." *Global Environmental Change* 11 (4): 261–69.

Lamel, Lori, Lawrence Rabiner, Aaron Rosenberg, and J. Wilpon. 1981. "An Improved Endpoint Detector for Isolated Word Recognition." *IEEE Transactions on Acoustics, Speech, and Signal Processing* 29 (4): 777–85.

Landis, J. Richard, and Gary G. Koch. 1977. "The Measurement of Observer Agreement for Categorical Data." *Biometrics* 33 (1): 159.

Laney, D. 2001. "3-D Data Management: Controlling Data Volume, Velocity and Variety," META Group Research Note, February 6. http://goo.gl/B03GS.

Langbauer, W. R. Jr. 2000. "Elephant Communication." *Zoo Biology* 19 (5): 425–45.

Langbauer, William R., Katharine B. Payne, Russell A. Charif, Lisa Rapaport, and Ferrel Osborn. 1991. "African Elephants Respond to Distant Playbacks of Low-Frequency Conspecific Calls." *Journal of Experimental Biology* 157 (1): 35–46.

Larned, Scott T., and Marc Schallenberg. 2019. "Stressor-Response Relationships and the Prospective Management of Aquatic Ecosystems." *New Zealand Journal of Marine and Freshwater Research* 53 (4): 489–512.

Larsen, Ole Næsbye, Georg Gleffe, and Jan Tengö. 1986. "Vibration and Sound Communication in Solitary Bees and Wasps." *Physiological Entomology* 11 (3): 287–96.

Laub, Curtis A., Roger Ray Youngman, Kenner Love, Timothy Mize et al. 2009. *Using Pitfall Traps to Monitor Insect Activity*. Virginia Cooperative Extension.

Laundré, John W. 1981. "Temporal Variation in Coyote Vocalization Rates." *Journal of Wildlife Management* 45 (3): 767–69.

Lavandier, Catherine, and Boris Defréville. 2006. "The Contribution of Sound Source Characteristics in the Assessment of Urban Soundscapes." *Acta Acustica United with Acustica* 92 (6): 912–21.

Lavia, Lisa, Harry J. Witchel, Francesco Aletta, Jochen Steffens, André Fiebig, Jian Kang, Christine Howes, and Patrick G. T. Healey. 2018. "Non-participant Observation Methods for Soundscape Design and Urban Planning." In *Handbook of Research on Perception-Driven Approaches to Urban Assessment and Design* edited by Francesco Aletta and Jieling Xiao, 73–99. IGI Global.

LeCun, Yann, Yoshua Bengio, and Geoffrey Hinton. 2015. "Deep Learning." *Nature* 521 (7553): 436–44.

Lee, John A., and Michel Verleysen. 2007. *Nonlinear Dimensionality Reduction*. Vol. 1. Springer.

Lee, Pyoung Jik, Joo Young Hong, and Jin Yong Jeon. 2014. "Assessment of Rural Soundscapes with High-Speed Train Noise." *Science of the Total Environment* 482–83 (1): 432–39.

Leek, Jeffrey T., and Roger D. Peng. 2015. "Statistics: P Values Are Just the Tip of the Iceberg." *Nature News* 520 (7549): 612.

Lehner, Philip N. 1978. "Coyote Vocalizations: A Lexicon and Comparisons with Other Canids." *Animal Behaviour* 26:712–22.

Lemieux, Christopher J., Sean T. Doherty, Paul F. J. Eagles, Mark W. Groulx, Glen T. Hvenegaard, Joyce Gould, Elizabeth Nisbet, and Francesc Romagosa. 2016. "Policy and Management Recommendations Informed by the Health Benefits of Visitor Experiences in Alberta's Protected Areas." *Journal of Park and Recreation Administration* 34 (1). https://doi.org/10.18666/JPRA-2016-V34-I1-6800.

Leopold, Aldo. 1949. "Odyssey." In *A Sand County Almanac, and Sketches Here and There*. Oxford University Press.

Lerch, Alexander. 2012. *An Introduction to Audio Content Analysis: Applications in Signal Processing and Music Informatics*. Wiley-IEEE.

Lercher, Peter. 1996. "Environmental Noise and Health: An Integrated Research Perspective." *Environment International* 22 (1): 117–29.

Levin, Simon A. 1992. "The Problem of Pattern and Scale in Ecology: The Robert H. MacArthur Award Lecture." *Ecology* 73 (6): 1943–67.

Levin, Theodore C. 1994. *Where Rivers and Mountains Sing: Sound, Music, and Nomadism in Tuva and Beyond*. Indiana University Press.

Levin, Theodore C. 1997. *The Hundred Thousand Fools of God: Musical Travels in Central Asia (and Queens, New York)*. Indiana University Press,

Levin, Theodore C., and Michael E. Edgerton. 1999. "The Throat Singers of Tuva." *Scientific American* 281 (3): 80–87.

Levins, Richard, and Robert MacArthur. 1969. "An Hypothesis to Explain the Incidence of Monophagy." *Ecology* 50 (5): 910–11.

Lewicka, Maria. 2011. "Place Attachment: How Far Have We Come in the Last 40 Years?" *Journal of Environmental Psychology* 31 (3): 207–30.

Li, Renjie, Saurabh Garg, and Alexander Brown. 2019. "Identifying Patterns of Human and Bird Activities Using Bioacoustic Data." *Forests* 10 (10): 917.

Liang, Jingjing, Dave E. Calkin, Krista M. Gebert, Tyron J. Venn, and Robin P. Silverstein. 2008. "Factors Influencing Large Wildland Fire Suppression Expenditures." *International Journal of Wildland Fire* 17 (5): 650–59.

Liang, Jingjing, Thomas W. Crowther, Nicolas Picard, Susan Wiser, Mo Zhou, Giorgio Alberti, Ernst-Detlef Schulze et al. 2016. "Positive Biodiversity-Productivity Relationship Predominant in Global Forests." *Science* 354 (6309). https://doi.org/10.1126/science.aaf8957.

Lillis, Ashlee, DelWayne R. Bohnenstiehl, and David B. Eggleston. 2015. "Soundscape Manipulation Enhances Larval Recruitment of a Reef-Building Mollusk." *PeerJ* 3: e999.

Lillis, Ashlee, David B. Eggleston, and DelWayne R. Bohnenstiehl. 2014. "Estuarine Soundscapes: Distinct Acoustic Characteristics of Oyster Reefs Compared to Soft-Bottom Habitats." *Marine Ecology Progress Series* 505:1–17.

Lima, M. S. C. S., J. Pederassi, R. B. Pineschi, and D. B. S. Barbosa. 2019. "Acoustic Niche Partitioning in an Anuran Community from the Municipality of Floriano, Piauí, Brazil." *Brazilian Journal of Biology* 79 (4): 566–76.

Lima, Steven L., and Patrick A. Zollner. 1996. "Towards a Behavioral Ecology of Ecological Landscapes." *Trends in Ecology & Evolution* 11 (3): 131–35.

Lin, Tzu-Hao, Tomonari Akamatsu, Frederic Sinniger, and Saki Harii. 2021a. "Exploring Coral Reef Biodiversity via Underwater Soundscapes." *Biological Conservation* 253:108901.

Lin, Tzu-Hao, Tomonari Akamatsu, and Yu Tsao. 2021b. "Sensing Ecosystem Dynamics via Audio Source Separation: A Case Study of Marine Soundscapes off Northeastern Taiwan." *PLoS Computational Biology* 17 (2): 1–23.

Lin, Tzu-Hao, and Yu Tsao. 2020. "Source Separation in Ecoacoustics: A Roadmap towards Versatile Soundscape Information Retrieval." *Remote Sensing in Ecology and Conservation* 6 (3): 236–47.

Lindseth, Adelaide V., and Phillip S. Lobel. 2018. "Underwater Soundscape Monitoring and Fish Bioacoustics: A Review." *Fishes* 3 (3): 36.

Ling, Arthur. 1947. "Town Planning and Health—Part II." *Health Education Journal* 5 (2): 87–92.

Linke, Simon, Emilia Decker, Toby Gifford, and Camille Desjonquères. 2020a. "Diurnal Variation in Freshwater Ecoacoustics: Implications for Site-Level Sampling Design." *Freshwater Biology* 65 (1): 86–95.

Linke, Simon, and Jo-Anne Deretic. 2020. "Ecoacoustics Can Detect Ecosystem Responses to Environmental Water Allocations." *Freshwater Biology* 65 (1): 133–41.

Linke, S., T. Gifford, and M. Kennard, M. 2013. Listening to the Fish—Acoustic Monitoring of Freshwater Biodiversity through Audio Signature Recognition. Presented at Listening in the Wild, Queen Mary University of London, June 25.

Linke, Simon, Toby Gifford, and Camille Desjonquères. 2020b. "Six Steps towards Operationalising Freshwater Ecoacoustic Monitoring." *Freshwater Biology* 65 (1): 1–6.

Linke, Simon, Toby Gifford, Camille Desjonquères, Diego Tonolla, Thierry Aubin, Leah Barclay, Chris Karaconstantis, Mark J. Kennard, Fanny Rybak, and Jérôme Sueur. 2018. "Freshwater Ecoacoustics as a Tool for Continuous Ecosystem Monitoring." *Frontiers in Ecology and the Environment* 16 (4): 231–38.

Lionello, Matteo, Francesco Aletta, and Jian Kang. 2020. "A Systematic Review of Prediction Models for the Experience of Urban Soundscapes." *Applied Acoustics* 170:107479.

Littlejohn, M. J. 1965. "Vocal Communication in Frogs." *Australian Natural History* 15 (2): 52–55.

Littlejohn, M. J., and A. A. Martin. 1969. "Acoustic Interaction between Two Species of Leptodactylid Frogs." *Animal Behaviour* 17 (4): 785–91.

Liu, Fangfang, and Jian Kang. 2016. "A Grounded Theory Approach to the Subjective Understanding of Urban Soundscape in Sheffield." *Cities* 50:28–39.

Liu, J., T. Dietz, S. R. Carpenter, M. Alberti, C. Folke, E. Moran, A. N. Pell et al. 2007. Complexity of Coupled Human and Natural Systems. *Science* 317 (5844): 1513–16.

Liu, Jiang, Jian Kang, and Holger Behm. 2014a. "Birdsong as an Element of the Urban Sound Environment: A Case Study Concerning the Area of Warnemünde in Germany." *Acta Acustica United with Acustica* 100 (3): 458–66.

Liu, Jiang, Jian Kang, Holger Behm, and Tao Luo. 2014b. "Effects of Landscape on Soundscape Perception: Soundwalks in City Parks." *Landscape and Urban Planning* 123:30–40.

Liu, Jiang, Jian Kang, Tao Luo, Holger Behm, and Timothy Coppack. 2013. "Spatiotemporal Variability of Soundscapes in a Multiple Functional Urban Area." *Landscape and Urban Planning* 115:1–9.

Loewen, M. R., and W. K. Melville. 1991. "A Model of the Sound Generated by Breaking Waves." *Journal of the Acoustical Society of America* 90 (4): 2075–80.

Lohr, Bernard, Timothy F. Wright, and Robert J. Dooling. 2003. "Detection and Discrimination of Natural Calls in Masking Noise by Birds: Estimating the Active Space of a Signal." *Animal Behaviour* 65 (4): 763–77.

Lomolino, Mark V. 2016. "The Unifying, Fundamental Principles of Biogeography: Understanding Island Life." *Frontiers of Biogeography* 8 (2). https://doi.org/10.21425/F5FBG29920.

Lomolino, Mark V., James H. Brown, and Dov F. Sax. 2010. "Island Biogeography Theory: Reticulations and Reintegration of "a Biogeography of the Species." In *The Theory of Island Biogeography Revisited*, edited by Jonathan B. Losos and Robert E. Ricklefs, 13–51. Princeton University Press.

Lomolino, Mark V., and Bryan C. Pijanowski. 2020. "Sonoric Geography: Addressing the Silence of Biogeography." *Frontiers of Biogeography* 13 (1). https://doi.org/10.21425/F5FBG49529.

Lomolino, Mark V., Bryan C. Pijanowski, and Amandine Gasc. 2015. "The Silence of Biogeography." *Journal of Biogeography* 42 (7): 1187–96.

Lomolino, Mark V., Brett R. Riddle, and Robert J. Whittaker. 2017. *Biogeography*. 5th ed. Sinauer Associates.

Longley, Alys, and Nancy Duxbury. 2016. "Introduction: Mapping Cultural Intangibles." *City, Culture and Society* 7 (1): 1–7.

Longley, Paul A., Michael F. Goodchild, David J. Maguire, and David W. Rhind. 2005. *Geographic Information Systems and Science*. John Wiley & Sons.

Louv, Richard. 2005. "Nature Deficit." *Orion* 70:71.

Louv, Richard. 2008. *Last Child in the Woods: Saving Our Children from Nature-Deficit Disorder*. Algonquin Books.

Loveland, Thomas R., Bradley C. Reed, Jesslyn F. Brown, Donald O. Ohlen, Zhiliang Zhu, L. Yang, and James W. Merchant. 2000. "Development of a Global Land Cover Characteristics Database and IGBP DISCover from 1 Km AVHRR Data." *International Journal of Remote Sensing* 21 (6–7): 1303–30.

Lucas, Jeffrey R., Alejandro Vélez, and Kenneth S. Henry. 2015. "Habitat-Related Differences in Auditory Processing of Complex Tones and Vocal Signal Properties in Four Songbirds." *Journal of Comparative Physiology A* 201 (4): 395–410.

Luczkovich, Joseph J., David A. Mann, and Rodney A. Rountree. 2008. "Passive Acoustics as a Tool in Fisheries Science." *Transactions of the American Fisheries Society* 137 (2): 533–41.

Lugli, Marco, and Michael L. Fine. 2003. "Acoustic Communication in Two Freshwater Gobies: Ambient Noise and Short-Range Propagation in Shallow Streams." *Journal of the Acoustical Society of America* 114 (1): 512–21.

Lumsdon, A. E., I. Artamonov, M. C. Bruno, M. Righetti, K. Tockner, Diego Tonolla, and C. Zarfl. 2018. "Soundpeaking—Hydropeaking Induced Changes in River Soundscapes." *River Research and Applications* 34 (1): 3–12.

Luque, Amalia, Alejandro Carrasco, Alejandro Martín, and Ana de las Heras. 2019. "The Impact of Class Imbalance in Classification Performance Metrics Based on the Binary Confusion Matrix." *Pattern Recognition* 91:216–31.

Luther, David A. 2008. "Signaller: Receiver Coordination and the Timing of Communication in Amazonian Birds." *Biology Letters* 4 (6): 651–54.

Luther, David. 2009. "The Influence of the Acoustic Community on Songs of Birds in a Neotropical Rain Forest." *Behavioral Ecology* 20 (4): 864–71.

Lynam, Christopher Philip. 2006. "Ecological and Acoustic Investigations of Jellyfish (Scyphozoa and Hydrozoa)." PhD diss., University of St Andrews.

MacArthur, Robert, and Richard Levins. 1967. "The Limiting Similarity, Convergence, and Divergence of Coexisting Species." *American Naturalist* 101 (921): 377–85.

MacArthur, Robert H., and Edward O. Wilson. 1963. "An Equilibrium Theory of Insular Zoogeography." *Evolution* 17 (4): 373–87.

MacArthur, Robert H., and Edward O. Wilson. 1967. *The Theory of Island Biogeography.* Princeton University Press.

Machado, Ricardo B., Ludmilla Aguiar, and Gareth Jones. 2017. "Do Acoustic Indices Reflect the Characteristics of Bird Communities in the Savannas of Central Brazil?" *Landscape and Urban Planning* 162:36–43.

Maculuve, Rufus. 2020. "Soundscapes of Omuhipiti." *Eastern African Literary and Cultural Studies* 6 (4): 258–68.

Maeder, Marcus, Martin M. Gossner, Armin Keller, and Martin Neukom. 2019. "Sounding Soil: An Acoustic, Ecological & Artistic Investigation of Soil Life." *Soundscape: The Journal of Acoustic Ecology* 18 (1): 5–14.

Mahanty, Madan Mohan, Latha Ganeshan, and Raguraman Govindan. 2018. "Soundscapes in Shallow Water of the Eastern Arabian Sea." *Progress in Oceanography* 165 (June): 158–67.

Maletic, J. I, and A. Marcus. 2000. "Data Cleansing: Beyond Integrity Analysis." In *IQ2000: Fifth Conference on Information Quality*, 1–10. http://citeseerx.ist.psu.edu/viewdoc/download?doi=10.1.1.37.5212&rep=rep1&type=pdf.

Malinowski, Bronislaw. 1922. "Ethnology and the Study of Society." *Economica* (6): 208–19.

Mammides, C., Goodale, E., Dayananda, S. K., Kang, L., and Chen, J. 2017. "Do Acoustic Indices Correlate with Bird Diversity? Insights from Two Biodiverse Regions in Yunnan Province, South China. *Ecological Indicators* 82: 470–77.

Mann, David A. 2012. "Remote Sensing of Fish Using Passive Acoustic Monitoring." *Acoustics Today* 8 (3): 8–15.

Margaritis, Efstathios, and Jian Kang. 2017. "Relationship between Green Space-Related Morphology and Noise Pollution." *Ecological Indicators* 72:921–33.

Margaritis, Efstathios, Jian Kang, Karlo Filipan, and Dick Botteldooren. 2018. "The Influence of Vegetation and Surrounding Traffic Noise Parameters on the Sound Environment of Urban Parks." *Applied Geography* 94 (April): 199–212.

Margules, Christopher Robert, and Robert L. Pressey. 2000. "Systematic Conservation Planning." *Nature* 405 (6783): 243–53.

Markl, Hubert. 1983. "Vibrational Communication." In *Neuroethology and Behavioral Physiology: Roots and Growing Points*, edited by Franz Huber and Hubert Markl, 332–53. Springer.

Marler, Peter, and Linda Hobbett. 1975. "Individuality in a Long-Range Vocalization of Wild Chimpanzees." *Zeitschrift für Tierpsychologie* 38 (1): 97–109.

Marler, Peter, and John C. Mitani. 1988. "Vocal Communication in Primates and Birds: Parallels and Contrasts." In *Primate Vocal Communication*, edited by D. Todt, P. Goedeking, and D. Symmes, 3–14. Springer.

Marler, Peter, and John C. Mitani. 1989. "A Phonological Analysis of Male Gibbon Singing Behavior." *Behaviour* 109 (1–2): 20–45.

Marques, Paulo A. M., and Carlos B. De Araújo. 2014. "The Need to Document and Preserve Natural Soundscape Recordings as Acoustic Memories." In *Invisible Places, Sounding Cities: Sound, Urbanism and Sense of Place*, proceedings, 321–28.

Marques, Tiago A., Len Thomas, Stephen W. Martin, David K. Mellinger, Jessica A. Ward, David J. Moretti, Danielle Harris, and Peter L. Tyack. 2013. "Estimating Animal Population Density Using Passive Acoustics." *Biological Reviews* 88 (2): 287–309.

Marquet, Pablo A. 2002. "Of Predators, Prey, and Power Laws." *Science* 295 (5563): 2229–30.

Marra, Peter P., Charles M. Francis, Robert S. Mulvihill, and Frank R. Moore. 2005. "The Influence of Climate on the Timing and Rate of Spring Bird Migration." *Oecologia* 142 (2): 307–15.

Marten, Ken, and Peter Marler. 1977. "Sound Transmission and Its Significance for Animal Vocalization: I. Temperate Habitats." *Behavioral Ecology and Sociobiology* (2): 271–90.

Martin, Brona. 2018. "Soundscape Composition: Enhancing Our Understanding of Changing Soundscapes." *Organised Sound* 23 (1): 20.

Martyn, Patricia, and Eric Brymer. 2016. "The Relationship between Nature Relatedness and Anxiety." *Journal of Health Psychology* 21 (7): 1436–45.

Maurer, Brian A., James H. Brown, and Renee D. Rusler. 1992. "The Micro and Macro in Body Size Evolution." *Evolution* 46 (4): 939–53.

May-Collado, L. J., T. Mitchell Aide, C. Corrada, and R. Alvarez. 2010. "A Comparison of Acoustic Soundscapes Within and Among Three Tropical Habitats: Can Soundscape Heterogeneity Be Used as an Index of Alpha Diversity?" *Journal of the Acoustical Society of America* 128 (4): 2412.

Mayer, F. Stephan, and Cynthia McPherson Frantz. 2004. "The Connectedness to Nature Scale: A Measure of Individuals' Feeling in Community with Nature." *Journal of Environmental Psychology* 24 (4): 503–15.

Mayer, F. Stephan, Cynthia McPherson Frantz, Emma Bruehlman-Senecal, and Kyffin Dolliver. 2009. "Why Is Nature Beneficial? The Role of Connectedness to Nature." *Environment and Behavior* 41 (5): 607–43.

Mayring, Philipp. 2007. "Mixing Qualitative and Quantitative Methods." In *Mixed Methodology in Psychological Research*, edited by Philipp Mayring, Günter L. Huber, Leo Gürtler, and Mechthild Kiegelmann, 27–36. Brill Sense.

McCoy, Earl D. 1990. "The Distribution of Insects along Elevational Gradients." *Oikos* 58 (3): 313–22.

McDonald, Geoffrey T., and Lex Brown. 1995. "Going Beyond Environmental Impact Assessment: Environmental Input to Planning and Design." *Environmental Impact Assessment Review* 15 (6): 483–95.

McDonnell, Mark J., and Stewart T. A. Pickett. 1990. "Ecosystem Structure and Function along Urban-Rural Gradients: An Unexploited Opportunity for Ecology." *Ecology* 71 (4): 1232–37.

McDougall, Craig W., Richard S. Quilliam, Nick Hanley, and David M. Oliver. 2020. "Freshwater Blue Space and Population Health: An Emerging Research Agenda." *Science of the Total Environment* 737:140196.

McGarigal, Kevin. 1995. *FRAGSTATS: Spatial Pattern Analysis Program for Quantifying*

Landscape Structure. Vol. 351. USDA Forest Service, Pacific Northwest Research Station.

McGarigal, Kevin. 2015. "FRAGSTATS Help." FRAGSTATS Version 4 Documentation, April 21, 182.

McGregor, Iain, Alison Crerar, David Benyon, and Catriona Macaulay. 2002. "Soundfields and Soundscapes: Reifying Auditory Communities." In *Proceedings of the 2002 International Conference on Auditory Display*, 290–94. ICAD.

McGregor, Iain, Grégory Leplâtre, Alison Crerar, and David Benyon. 2006. "Sound and Soundscape Classification: Establishing Key Auditory Dimensions and Their Relative Importance." In *Proceedings of the 12th International Conference on Auditory Display*, 105–12. ICAD.

McPherson, Craig, Bruce Martin, Jeff MacDonnell, and Christopher Whitt. 2016. "Examining the Value of the Acoustic Variability Index in the Characterisation of Australian Marine Soundscapes." In *Proceedings of 2nd Australasian Acoustical Societies Conference, (Acoustics 2016)*, 9–11 November, Brisbane, Australia, 284–96.

McWilliam, Jamie N., and Anthony D. Hawkins. 2013. "A Comparison of Inshore Marine Soundscapes." *Journal of Experimental Marine Biology and Ecology* 446:166–76.

Mellinger, David K., and Christopher W. Clark. 2003. "Blue Whale (*Balaenoptera musculus*) Sounds from the North Atlantic." *Journal of the Acoustical Society of America* 114 (2): 1108–19.

Meng, Qi, Xuejun Hu, Jian Kang, and Yue Wu. 2020. "On the Effectiveness of Facial Expression Recognition for Evaluation of Urban Sound Perception." *Science of the Total Environment* 710:135484.

Meng, Qi, and Jian Kang. 2016. "Effect of Sound-Related Activities on Human Behaviours and Acoustic Comfort in Urban Open Spaces." *Science of the Total Environment* 573:481–93.

Merriam, Alan P., and Valerie Merriam. 1964. *The Anthropology of Music*. Northwestern University Press.

Merriam, Clinton Hart. 1898. *Life Zones and Crop Zones of the United States*. Bulletin no. 10. US Department of Agriculture, Division of Biological Survey.

Metcalf, Oliver C., Jos Barlow, Christian Devenish, Stuart Marsden, Erika Berenguer, and Alexander C. Lees. 2021. "Acoustic Indices Perform Better when Applied at Ecologically Meaningful Time and Frequency Scales." *Methods in Ecology and Evolution* 12 (3): 421–31.

Metcalf, Oliver C., Alexander C. Lees, Jos Barlow, Stuart J. Marsden, and Christian Devenish. 2020. "HardRain: An R Package for Quick, Automated Rainfall Detection in Ecoacoustic Datasets Using a Threshold-Based Approach." *Ecological Indicators* 109:105793.

Meyer, David, Kurt Hornik, and Ingo Feinerer. 2008. "Text Mining Infrastructure in R." *Journal of Statistical Software* 25 (5): 1–54.

Michelsen, Axel, and Kristin Rohrseitz. 1995. "Directional Sound Processing and Interaural Sound Transmission in a Small and a Large Grasshopper." *Journal of Experimental Biology* 198 (9): 1817–27.

Middlebrooks, John C. 1997. "Spectral Shape Cues for Sound Localization." In *Binaural and Spatial Hearing in Real and Virtual Environments*, edited by Robert H. Gilkey and Timothy R. Anderson, 77–97. Lawrence Erlbaum Associates.

Miksis-Olds, Jennifer L., Bruce Martin, and Peter L. Tyack. 2018. "Exploring the Ocean through Soundscapes." *Acoustics Today* 14:26–34.

Miller, B. W. 2001. "A Method for Determining Relative Activity of Free Flying Bats Using a New Activity for Acoustic Monitoring." *Acta Chiropterologica* 3 (1): 93–105.

Miller, Nicholas P. 1999. "The Effects of Aircraft Overflights on Visitors to US National Parks." *Noise Control Engineering Journal* 47 (3): 112–17.

Miller, Nicholas P. 2003. "Transportation Noise and Recreational Lands." *Noise News International* 11 (1): 9–21.

Miller, Richard S. 1967. "Pattern and Process in Competition." In *Advances in Ecological Research*, vol. 4, 1–74. Academic Press.

Mishra, Ratnesh Chandra, Ritesh Ghosh, and Hanhong Bae. 2016. "Plant Acoustics: In the Search of a Sound Mechanism for Sound Signaling in Plants." *Journal of Experimental Botany* 67 (15): 4483–94.

Mitchell, Harvey B. 2007. *Multi-sensor Data Fusion: An Introduction*. Springer Science & Business Media.

Mitchell, Simon L., Jake E. Bicknell, David P. Edwards, Nicolas J. Deere, Henry Bernard, Zoe G. Davies, and Matthew J. Struebig. 2020. "Spatial Replication and Habitat Context Matters for Assessments of Tropical Biodiversity Using Acoustic Indices." *Ecological Indicators* 119:106717.

Mitra, Sanjit K., and James F. Kaiser. 1993. *Handbook for Digital Signal Processing*. John Wiley & Sons.

Mittelbach, Gary G., and Brian J. McGill. 2019. *Community Ecology*. Oxford University Press.

Møller, Henrik, and Christian Sejer Pedersen. 2011. "Low-Frequency Noise from Large Wind Turbines." *Journal of the Acoustical Society of America* 129 (6): 3727–44.

Monacchi, David. 2013. "Fragments of Extinction: Acoustic Biodiversity of Primary Rainforest Ecosystems." *Leonardo Music Journal* 23:23–25.

Monczak, Agnieszka, Bradshaw McKinney, Claire Mueller, and Eric W. Montie. 2020. "What's All That Racket! Soundscapes, Phenology, and Biodiversity in Estuaries." *PLoS One* 15 (9): e0236874.

Montealegre-Z, Fernando, Glenn K. Morris, and Andrew C. Mason. 2006. "Generation of Extreme Ultrasonics in Rainforest Katydids." *Journal of Experimental Biology* 209 (24): 4923–37.

Montgomery, John C., and Craig A. Radford. 2017. "Marine Bioacoustics." *Current Biology* 27 (11): R502–R507.

Mooney, T. Aran, Lucia Di Iorio, Marc Lammers, Tzu Hao Lin, Sophie L. Nedelec, Miles Parsons, Craig Radford, Ed Urban, and Jenni Stanley. 2020. "Listening Forward: Approaching Marine Biodiversity Assessments Using Acoustic Methods: Acoustic Diversity and Biodiversity." *Royal Society Open Science* 7 (8). https://doi.org/10.1098/rsos.201287.

Moran, Emilio F. 2010. *Environmental Social Science: Human-Environment Interactions and Sustainability*. John Wiley & Sons.

Morehen, John, and Ian Bent. 1979. "Computer Applications in Musicology." *Musical Times* 120 (1637): 563–66.

Moreira, Miguel, Daniel S. Hayes, Isabel Boavida, Martin Schletterer, Stefan Schmutz, and António Pinheiro. 2019. "Ecologically-Based Criteria for Hydropeaking Mitigation: A Review." *Science of the Total Environment* 657:1508–22.

Moreno-Gómez, Felipe N., José Bartheld, Andrés A. Silva-Escobar, Raúl Briones, Rafael Márquez, and Mario Penna. 2019. "Evaluating Acoustic Indices in the Valdivian Rainforest, a Biodiversity Hotspot in South America." *Ecological Indicators* 103:1–8.

Moritz, Craig. 1994. "Defining 'Evolutionarily Significant Units' for Conservation." *Trends in Ecology & Evolution* 9 (10): 373–75.

Morton, Eugene S. 1975. "Ecological Sources of Selection on Avian Sounds." *American Naturalist* 109 (965): 17–34.

Moulton, James M. 1957. "Sound Production in the Spiny Lobster *Panulirus argus* (Latreille)." *Biological Bulletin* 113 (2): 286–95.

Moustakas, Clark. 1994. *Phenomenological Research Methods*. Sage.

Muirhead-Thompson, R. C. 2012. *Trap Responses of Flying Insects: The Influence of Trap Design on Capture Efficiency*. Academic Press.

Munezero, Myriam, Calkin Suero Montero, Erkki Sutinen, and John Pajunen. 2014. "Are They Different? Affect, Feeling, Emotion, Sentiment, and Opinion Detection in Text." *IEEE Transactions on Affective Computing* 5 (2): 101–11.

Munro, Jasmine, Ian Williamson, and Susan Fuller. 2018. "Traffic Noise Impacts on Urban Forest Soundscapes in South-Eastern Australia." *Austral Ecology* 43 (2): 180–90.

Murray, Daniel G. 2013. *Tableau Your Data! Fast and Easy Visual Analysis with Tableau Software*. John Wiley & Sons.

Musacchio, Laura R. 2013. "Cultivating Deep Care: Integrating Landscape Ecological Research into the Cultural Dimension of Ecosystem Services." *Landscape Ecology* 28 (6): 1025–38.

Myers, David, Håkan Berg, and Giorgos Maneas. 2019. "Comparing the Soundscapes of Organic and Conventional Olive Groves: A Potential Method for Bird Diversity Monitoring." *Ecological Indicators* 103:642–49.

Myers, Norman, Russell A. Mittermeier, Cristina G. Mittermeier, Gustavo A. B. Da Fonseca, and Jennifer Kent. 2000. "Biodiversity Hotspots for Conservation Priorities." *Nature* 403 (6772): 853–58.

Myers, Norman. 1988. "Threatened Biotas: 'Hot Spots' in Tropical Forests." *Environmentalist* 8 (3): 187–208.

Mysore, Gautham J. 2010. *A Non-negative Framework for Joint Modeling of Spectral Structure and Temporal Dynamics in Sound Mixtures*. PhD thesis, Stanford University.

Nakano, Ryo, Takuma Takanashi, and Annemarie Surlykke. 2015. "Moth Hearing and Sound Communication." *Journal of Comparative Physiology A* 201 (1): 111–21.

Narins, Peter M. 1990. "Seismic Communication in Anuran Amphibians." *BioScience* 40 (4): 268–74.

Nassauer, Joan Iverson. 1995a. "Culture and Changing Landscape Structure." *Landscape Ecology* 10 (4): 229–37.

Nassauer, Joan Iverson. 1995b. "Messy Ecosystems, Orderly Frames." *Landscape Journal* 14 (2): 161–70.

National Park Service. 2022. "Quick History of the National Park Service." Last updated August 24, 2022. https://www.nps.gov/articles/quick-nps-history.htm.

Naughton-Treves, Lisa, Margaret Buck Holland, and Katrina Brandon. 2005. "The Role of Protected Areas in Conserving Biodiversity and Sustaining Local Livelihoods." *Annual Review of Environment and Resources* (30): 219–52.

Naveh, Zev. 1984. "Towards a Transdisciplinary Conceptual Framework of Landscape Ecology." In *Methodology in Landscape Ecological Research and Planning: Proceedings, 1st seminar, International Association of Landscape Ecology, Roskilde, Denmark, Oct 15–19, 1984*, edited by J. Brandt and P. Agger. Roskilde University Centre.

Naveh, Zev, and Arthur S. Lieberman. 2013. *Landscape Ecology: Theory and Application*. Springer Science & Business Media.

Nedelec, Sophie L., Stephen D. Simpson, Marc Holderied, Andrew N. Radford, Gael Lecellier, Craig Radford, and David Lecchini. 2015. "Soundscapes and Living Communities in Coral Reefs: Temporal and Spatial Variation." *Marine Ecology Progress Series* 524:125–35.

Nelson, Douglas A., and Peter Marler. 1990. *The Perception of Birdsong and an Ecological Concept of Signal Space*. John Wiley & Sons.

Nemeth, Erwin, and Henrik Brumm. 2010. "Birds and Anthropogenic Noise: Are Urban Songs Adaptive?" *American Naturalist* 176 (4): 465–75.

Nettl, Bruno. 1983. *The Study of Ethnomusicology: Twenty-Nine Issues and Concepts*. University of Illinois Press, 1983.

Nettl, Bruno. 2013. *Becoming an Ethnomusicologist: A Miscellany of Influences*. Scarecrow.

Ng, May-Le, Nathan Butler, and Nina Woods. 2018. "Soundscapes as a Surrogate Measure of Vegetation Condition for Biodiversity Values: A Pilot Study." *Ecological Indicators* 93:1070–80.

Niessen, Maria, Caroline Cance, and Danièle Dubois. 2010. "Categories for Soundscape: Toward a Hybrid Classification." In *Inter-Noise and Noise-Con Congress and Conference Proceedings*, 2010:5816–29.

Nisbet, Elizabeth K. L. 2005. "The Human-Nature Connection: Increasing Nature Relatedness, Environmental Concern, and Well-Being through Education." PhD diss., Carleton University, 2005.

Nisbet, Elizabeth K., and John M. Zelenski. 2013. "The NR-6: A New Brief Measure of Nature Relatedness." *Frontiers in Psychology* 4:813.

Nisbet, Elizabeth K., John M. Zelenski, and Steven A. Murphy. 2009. "The Nature Relatedness Scale: Linking Individuals' Connection with Nature to Environmental Concern and Behavior." *Environment and Behavior* 41 (5): 715–40.

Nisbet, Elizabeth K., John M. Zelenski, and Steven A. Murphy. 2011. "Happiness Is in Our Nature: Exploring Nature Relatedness as a Contributor to Subjective Well-Being." *Journal of Happiness Studies* 12 (2): 303–22.

Norman, Katharine. 2012. "Listening Together, Making Place." *Organised Sound* 17 (3): 257.

Norström, Albert V., Christopher Cvitanovic, Marie F. Löf, Simon West, Carina Wyborn, Patricia Balvanera, Angela T. Bednarek et al. 2020. "Principles for Knowledge Co-Production in Sustainability Research." *Nature Sustainability* 3 (3): 182–90.

Noss, Reed F. 1990. "Indicators for Monitoring Biodiversity: A Hierarchical Approach." *Conservation Biology* 4 (4): 355–64.

Noss, Reed, Roderick Nash, Paul Paquet, and Michael Soulé. 2013. "Humanity's Domination of Nature Is Part of the Problem: A Response to Kareiva and Marvier." *BioScience* 63 (4): 241–42.

Nottebohm, Fernando. 1980. "Brain Pathways for Vocal Learning in Birds." *Progress in Psychobiology and Physiological Psychology* 9:86–125.

Nowotny, Helga, Peter B. Scott, and Michael T. Gibbons. 2013. *Re-Thinking Science: Knowledge and the Public in an Age of Uncertainty*. John Wiley & Sons.

Nunez-Mir, Gabriela C., Basil V. Iannone III, Bryan C. Pijanowski, Ningning Kong, and Songlin Fei. 2016. "Automated Content Analysis: Addressing the Big Literature Challenge in Ecology and Evolution." *Methods in Ecology and Evolution* 7 (11): 1262–72.

Nystuen, Jeffrey A. 2001. "Listening to Raindrops from Underwater: An Acoustic Disdrometer." *Journal of Atmospheric and Oceanic Technology* 18 (10): 1640–57.

Nystuen, Jeffrey A., Charles C. McGlothin, and Michael S. Cook. 1993. "The Underwater

Sound Generated by Heavy Rainfall." *Journal of the Acoustical Society of America* 93 (6): 3169–77.

Obrist, Martin K., Gianni Pavan, Jérôme Sueur, Klaus Riede, Diego Llusia, and Rafael Márquez. 2010. "Bioacoustics Approaches in Biodiversity Inventories." In *Manual on Field Recording Techniques and Protocols for All Taxa Biodiversity Inventories*, edited by J. Eymann, J. Degreef, C. Häuser, J. C. Monje, Y. Samyn, and D. VandenSpiegel, 68–99. ABC Taxa, no. 8. ISSUU.

O'Callaghan, James. 2013. "Orchestration of Ecology, as Ecology." In *Proceedings of the Music and Ecologies of Sound Conference*, vol. 201.

O'Connell, Allan F., James D. Nichols, and K. Ullas Karanth. 2010. *Camera Traps in Animal Ecology: Methods and Analyses*. Springer Science & Business Media.

Odum, Eugene P., 1971. *Fundamentals of Ecology*. 3rd ed. Saunders.

Odum, Eugene P., and H. T. Odum. 1953. *Fundamentals of Ecology*. B. Saunders (1963).

Odum, Howard T. 1983. *Systems Ecology: An Introduction*. U.S. Department of Energy Office of Scientific and Technical Information.

Officer, Charles B. 1958. *Introduction to the Theory of Sound Transmission: With Application to the Ocean*. McGraw-Hill.

Ohlson, Birger. 1976. "Sound Fields and Sonic Landscapes in Rural Environments." *Fennia: International Journal of Geography* 148 (1). https://fennia.journal.fi/article/view/9210.

Oliphant, Travis E. 2006. *A Guide to NumPy*. Vol. 1. Trelgol Publishing.

Oliphant, Travis E. 2007. "Python for Scientific Computing." *Computing in Science & Engineering* 9 (3): 10–20.

Olson, Christopher R., Marcela Fernández-Vargas, Christine V. Portfors, and Claudio V. Mello. 2018. "Black Jacobin Hummingbirds Vocalize Above the Known Hearing Range of Birds." *Current Biology* 28 (5): R204–R205.

Olson, David M., and Eric Dinerstein. 2002. "The Global 200: Priority Ecoregions for Global Conservation." *Annals of the Missouri Botanical Garden* 89 (2): 199–224.

Olson, David M., Eric Dinerstein, Eric D. Wikramanayake, Neil D. Burgess, George V. N. Powell, Emma C. Underwood, Jennifer A. D'amico et al. 2001. "Terrestrial Ecoregions of the World: A New Map of Life on Earth: A New Global Map of Terrestrial Ecoregions Provides an Innovative Tool for Conserving Biodiversity." *BioScience* 51 (11): 933–38.

O'Neill, Robert V., Donald Lee DeAngelis, Jack B. Waide, Timothy F. H. Allen, and Garland E. Allen. 1986. *A Hierarchical Concept of Ecosystems*. Monographs in Population Biology, 23. Princeton University Press.

O'Neill, Robert. V., J. R. Krummel, R. H. Gardner, G. Sugihara, B. Jackson, Donald Lee DeAngelis, B. T. Milne et al. 1988. "Indices of Landscape Pattern." *Landscape Ecology* 1 (3): 153–62.

Opaev, Alexey, Svetlana Gogoleva, Igor Palko, Van Thinh Nguyen, and Viatcheslav Rozhnov. 2021. "Annual Acoustic Dynamics Are Associated with Seasonality in a Monsoon Tropical Forest in South Vietnam." *Ecological Indicators* 122 (March 2020): 107269.

Opdam, Paul, Eveliene Steingröver, and Sabine Van Rooij. 2006. "Ecological Networks: A Spatial Concept for Multi-Actor Planning of Sustainable Landscapes." *Landscape and Urban Planning* 75 (3–4): 322–32.

Organization for Tropical Studies. n.d. https://tropicalstudies.org/. Accessed November 29, 2020.

Osborne, Patrick E., Giles M. Foody, and Susana Suárez-Seoane. 2007. "Non-Stationarity and Local Approaches to Modelling the Distributions of Wildlife." *Diversity and Distributions* 13 (3): 313–23.

O'Shea, Thomas J., and Lynn B. Poché Jr. 2006. "Aspects of Underwater Sound Communication in Florida Manatees (*Trichechus manatus latirostris*)." *Journal of Mammalogy* 87 (6): 1061–71.

Ostrom, Elinor. 1990. *Governing the Commons: The Evolution of Institutions for Collective Action*. Cambridge University Press.

Ostrom, Elinor. 2009. "A General Framework for Analyzing Sustainability of Social-Ecological Systems." *Science* 325 (5939): 419–22.

Otondo, Felipe. 2018. "Listening to Wetland Soundscapes." *Leonardo Music Journal* 28:50–52.

Otte, Daniel. 1970. *A Comparative Study of Communicative Behavior in Grasshoppers*. Miscellaneous Publications, no. 141. University of Michigan Museum of Zoology.

Palacios, Vicente, Enrique Font, and Rafael Márquez. 2007. "Iberian Wolf Howls: Acoustic Structure, Individual Variation, and a Comparison with North American Populations." *Journal of Mammalogy* 88 (3): 606–13.

Park, Tae Hong. 2017. "Mapping Urban Soundscapes via Citygram." In *Seeing Cities through Big Data: Research, Methods and Applications in Urban Informatics*, edited by Piyushimita (Vonu) Thakuriah, Nebiyou Tilahun, and Moira Zellner, 491–513. Springer.

Park, Tae Hong, Johnathan Turner, Michael Musick, Jun Hee Lee, Christopher Jacoby, Charlie Mydlarz, and Justin Salamon. 2014. "Sensing Urban Soundscapes." In *EDBT/ICDT Workshops* 1113:375–82.

Parks, Susan E., Jennifer L. Miksis-Olds, and Samuel L. Denes. 2014. "Assessing Marine Ecosystem Acoustic Diversity across Ocean Basins." *Ecological Informatics* 21:81–88.

Passilongo, D., A. Buccianti, F. Dessi-Fulgheri, A. Gazzola, M. Zaccaroni, and M. Apollonio. 2010. "The Acoustic Structure of Wolf Howls in Some Eastern Tuscany (Central Italy) Free Ranging Packs." *Bioacoustics* 19 (3): 159–75.

Passmore, N. I. 1977. "Mating Calls and Other Vocalizations of Five Species of Ptychadena (Anura: Ranidae)." *South African Journal of Science* 73 (7): 212.

Patek, Sheila N. 2001. "Spiny Lobsters Stick and Slip to Make Sound." *Nature* 411 (6834): 153–54.

Pavan, Gianni. 2017. "Fundamentals of Soundscape Conservation." In *Ecoacoustics: The Ecological Role of Sounds:*, edited by Almo Farina, Stuart H. Gage, 235–58. John Wiley & Sons.

Payne, Sarah R. 2008. "Are Perceived Soundscapes within Urban Parks Restorative." *Journal of the Acoustical Society of America* 123 (5): 3809.

Payne, Sarah R. 2013. "The Production of a Perceived Restorativeness Soundscape Scale." *Applied Acoustics* 74 (2): 255–63.

Payne, Sarah R., and Catherine Guastavino. 2018. "Exploring the Validity of the Perceived Restorativeness Soundscape Scale: A Psycholinguistic Approach." *Frontiers in Psychology* 9:2224.

Pedregosa, Fabian, Gaël Varoquaux, Alexandre Gramfort, Vincent Michel, Bertrand Thirion, Olivier Grisel, Mathieu Blondel et al. 2011. "Scikit-learn: Machine Learning in Python." *Journal of Machine Learning Research* 12:2825–30.

Peeters, G., & Rodet, X. 2002. September. "Automatically Selecting Signal Descriptors for Sound Classification." In *Proceedings of the 2002 International Computer Music Conference, Sweden*. ICMC.

Peeters, Geoffroy. 2003. "Automatic Classification of Large Musical Instrument Databases Using Hierarchical Classifiers with Inertia Ratio Maximization." Paper presented at the 115th Audio Engineering Society Convention, October 1.

Pekin, Burak K., Jinha Jung, Luis J. Villanueva-Rivera, Bryan C. Pijanowski, and Jorge A. Ahumada. 2012. "Modeling Acoustic Diversity Using Soundscape Recordings and LIDAR-Derived Metrics of Vertical Forest Structure in a Neotropical Rainforest." *Landscape Ecology* 27 (10): 1513–22.

Pekin, Burak K., and Bryan C. Pijanowski. 2012. "Global Land Use Intensity and the Endangerment Status of Mammal Species." *Diversity and Distributions* 18 (9): 909–18.

Pelletier, Catherine, Robert B. Weladji, Louis Lazure, and Patrick Paré. 2020. "Zoo Soundscape: Daily Variation of Low-to-High-Frequency Sounds." *Zoo Biology* 39 (6): 374–81.

Pérez-Granados, C., D. la Rosa, J. Gómez-Catasús, A. Barrero, I. Abril-Colón, and J. Traba. 2018. "Autonomous Recording Units as Effective Tool for Monitoring of the Rare and Patchily Distributed Dupont's Lark *Chersophilus duponti*." *Ardea* 106 (2): 139–46.

Peters, Winfried S., and Dieter Stefan Peters. 2010. "Sexual Size Dimorphism Is the Most Consistent Explanation for the Body Size Spectrum of *Confuciusornis sanctus*." *Biology Letters* 6 (4): 531.

Petter, J. J., and P. Charles-Dominique. 1979. "Vocal Communication in Prosimians." In *The Study of Prosimian Behavior*, edited by G. S. Doyle, 247–305. Elsevier.

Phillips, Yvonne F., and Michael W. Towsey. 2017. "The Clustering of Acoustic Indices Derived from Long-Duration Recordings of the Environment." QUT ePrints.

Phillips, Yvonne F., Michael Towsey, and Paul Roe. 2018. "Revealing the Ecological Content of Long-Duration Audio-Recordings of the Environment through Clustering and Visualisation." *PLoS One* 13 (3): e0193345.

Piatetsky-Shapiro, Gregory. 2000. "Knowledge Discovery in Databases: 10 Years After." *ACM SIGKDD Explorations Newsletter* 1 (2): 59–61.

Pickett, Steward T. A., Mary L. Cadenasso, J. Morgan Grove, Peter M. Groffman, Lawrence E. Band, Christopher G. Boone, William R. Burch et al. 2008. "Beyond Urban Legends: An Emerging Framework of Urban Ecology, as Illustrated by the Baltimore Ecosystem Study." *BioScience* 58 (2): 139–50.

Pickett, Steward T. A., and Mark J. McDonnell. 1993. "Human as Components of Ecosystems: A Synthesis." In *Humans as Components of Ecosystems: The Ecology of Subtle Human Effects and Populated Areas*, edited by Mark J. McDonnell and Steward T.A. Pickett, 310–16. Springer.

Pickett, Steward T. A., V. Thomas Parker, and Peggy L. Fiedler. 1992. "The New Paradigm in Ecology: Implications for Conservation Biology Above the Species Level." In *Conservation Biology*, edited by P. L. Fiedler and S. K. Jain, 65–88. Springer.

Pieretti, Nadia, and Roberto Danovaro. 2020. "Acoustic Indexes for Marine Biodiversity Trends and Ecosystem Health." *Philosophical Transactions of the Royal Society. Series B: Biological Sciences* 375 (1814): 20190447.

Pieretti, N., M. H. L. Duarte, R. S. Sousa-Lima, M. Rodrigues, R. J. Young, and Almo Farina. 2015. "Determining Temporal Sampling Schemes for Passive Acoustic Studies in Different Tropical Ecosystems." *Tropical Conservation Science* 8 (1): 215–34.

Pieretti, Nadia, Almo Farina, and Davide Morri. 2011. "A New Methodology to Infer the Singing Activity of an Avian Community: The Acoustic Complexity Index (ACI)." *Ecological Indicators* 11 (3): 868–73.

Pieretti, Nadia, M. Lo Martire, A. Farina, and R. Danovaro. 2017. "Marine Soundscape as an Additional Biodiversity Monitoring Tool: A Case Study from the Adriatic Sea (Mediterranean Sea)." *Ecological Indicators* 83:13–20.

Pijanowski, B. C. 1991. "The Adaptive Significance of Brood Reduction and Clutch Size in the Tree Swallow." PhD diss., Michigan State University.

Pijanowski, Bryan C., Daniel G. Brown, Bradley A. Shellito, and Gaurav A. Manik. 2002. "Using Neural Networks and GIS to Forecast Land Use Changes: A Land Transformation Model." *Computers, Environment and Urban Systems* 26 (6): 553–75.

Pijanowski, Bryan C., and Almo Farina. 2011. "Introduction to the Special Issue on Soundscape Ecology." In "Soundscape Ecology," ed. Bryan C. Pijanowski and Almo Farina, special issue, *Landscape Ecology*, 26 (9): 1209–11.

Pijanowski, Bryan C., Almo Farina, Stuart H. Gage, Sarah L. Dumyahn, and Bernard L. Krause. 2011b. "What Is Soundscape Ecology? An Introduction and Overview of an Emerging New Science." In "Soundscape Ecology," ed. Bryan C. Pijanowski and Almo Farina, special issue, *Landscape Ecology* 26 (9): 1213–32.

Pijanowski, B. C., A. Gasc, K. Bellisario, and D. Francomano. 2018. Hands-on Training for the Use of R to Analyze Soundscape Recordings: Global Sustainable Soundscape Network." Center for Global Soundscapes, Purdue University.

Pijanowski, Bryan C., Amin Tayyebi, Jarrod Doucette, Burak K. Pekin, David Braun, and James Plourde. 2014. "A Big Data Urban Growth Simulation at a National Scale: Configuring the GIS and Neural Network Based Land Transformation Model to Run in a High Performance Computing (HPC) Environment." *Environmental Modelling and Software* 51:250–68.

Pijanowski, Bryan C., Luis J. Villanueva-Rivera, Sarah L. Dumyahn, Almo Farina, Bernard L. Krause, Brian M. Napoletano, Stuart H. Gage, and Nadia Pieretti. 2011a. "Soundscape Ecology: The Science of Sound in the Landscape." *BioScience* 61 (3): 203–16.

Pilcher, Ericka J., Peter Newman, and Robert E. Manning. 2009. "Understanding and Managing Experiential Aspects of Soundscapes at Muir Woods National Monument." *Environmental Management* 43 (3): 425.

Pine, M. K., C. A. Radford, and A. G. Jeffs. 2015. "Eavesdropping on the Kaipara Harbour: Characterising Underwater Soundscapes within a Seagrass Bed and a Subtidal Mudflat." *New Zealand Journal of Marine and Freshwater Research* 49 (2): 247–58.

Pink, Sarah. 2009. *Doing Sensory Ethnography*. Sage.

Pink, Sarah. 2013. "Engaging the Senses in Ethnographic Practice: Implications and Advances." *Senses and Society* 8 (3): 261–67.

Pink, Sarah. 2015. *Doing Sensory Ethnography*. 2nd ed. Sage.

Pittman, Simon J., ed. 2017. *Seascape Ecology*. John Wiley & Sons.

Pittman, Simon J., Ronald T. Kneib, and Charles A. Simenstad. 2011. "Practicing Coastal Seascape Ecology." *Marine Ecology Progress Series* 427:187–90.

Planque, Robert, and Hans Slabbekoorn. 2008. "Spectral Overlap in Songs and Temporal Avoidance in a Peruvian Bird Assemblage." *Ethology* 114. (3): 262–71.

Plutchik, Robert. 1982. "A Psychoevolutionary Theory of Emotions." *Social Science Information* 21 (4/5): 529–53.

Plutchik, Robert. 2001. "The Nature of Emotions: Human Emotions Have Deep Evolutionary Roots, A Fact That May Explain Their Complexity and Provide Tools for Clinical Practice." *American Scientist* 89 (4): 344–50.

Pocock, Douglas. 1989. "Sound and the Geographer." *Geography* 74 (3): 193–200.

Polli, Andrea. 2005. "Atmospherics/Weather Works: A Spatialized Meteorological Data Sonification Project." *Leonardo* 38 (1): 31–36.

Polli, Andrea. 2012. "Soundscape, Sonification, and Sound Activism." *AI & Society* 27 (2): 257–68.

Pollock, Kenneth H., James D. Nichols, Theodore R. Simons, George L. Farnsworth, Larissa L. Bailey, and John R. Sauer. 2002. "Large Scale Wildlife Monitoring Studies: Statistical Methods for Design and Analysis." *Environmetrics* 13 (2): 105–19.

Poole, J. H., Peter L. Tyack, A. S. Stoeger-Horwath, and S. Watwood. 2005. "Elephants Are Capable of Vocal Learning." *Nature* 434 (7032): 455–56.

Popp, James W., Robert W. Ficken, and James A. Reinartz. 1985. "Short-Term Temporal Avoidance of Interspecific Acoustic Interference among Forest Birds." Auk 102 (4): 744–48.

Popper, Arthur N., and Robert J. Dooling. 2002. "History of Animal Bioacoustics." *Journal of the Acoustical Society of America* 112 (5): 2368.

Popper, Arthur N., and Richard R. Fay. 1973. "Sound Detection and Processing by Teleost Fishes: A Critical Review." *Journal of the Acoustical Society of America* 53 (6): 1515–29.

Popper, Arthur N., Richard R. Fay, Christopher Platt, and Olav Sand. 2003. "Sound Detection Mechanisms and Capabilities of Teleost Fishes." In *Sensory Processing in Aquatic Environments*, edited by Shaun P. Collin and N. Justin Marshall, 3–38. Springer.

Popper, Arthur N., and M. C. Hastings. 2009. "The Effects of Anthropogenic Sources of Sound on Fishes." *Journal of Fish Biology* 75 (3): 455–89.

Popper, Arthur N., and Anthony D. Hawkins. 2019. "An Overview of Fish Bioacoustics and the Impacts of Anthropogenic Sounds on Fishes." *Journal of Fish Biology* 94 (5): 692–713.

Porteous, J. Douglas, and Jane F. Mastin. 1985. "Soundscape." *Journal of Architectural and Planning Research* 2 (3): 169–86.

Post, Jennifer C. 2013. *Ethnomusicology: A Contemporary Reader*. Routledge.

Post, Jennifer C. 2019. "Songs, Settings, Sociality: Human and Ecological Well-Being in Western Mongolia." *Journal of Ethnobiology* 39 (3): 371.

Post, Jennifer C. 2020. "21st-Century Trading Routes in Mongolia: Changing Pastoral Soundscapes and Lifeways." In *Silk Roads: From Local Realities to Global Narratives*, edited by Yaohua Shi and Jeffrey D. Lerner, 177. Oxbow.

Post, Jennifer C., and Bryan C. Pijanowski. 2018. "Coupling Scientific and Humanistic Approaches to Address Wicked Environmental Problems of the Twenty-first Century: Collaborating in an Acoustic Community Nexus." *MUSICultures* 45 (1–2).

Prabhakaran, S. 2019. *Top 50 ggplot2 Visualizations—The Master List*. r-statistics.co. http://r-statistics.co/Top50-Ggplot2-Visualizations-MasterList-R-Code.html.

Press, Gil. 2012. "A Short History of Big Data." What's the Big Data?: The Evolving IT Landscape, posted June 6, 2012. https://whatsthebigdata.com/2012/06/06/a-very -short-history-of-big-data/.

Pressey, Robert L., Mar Cabeza, Matthew E. Watts, Richard M. Cowling, and Kerrie A. Wilson. 2007. "Conservation Planning in a Changing World." *Trends in Ecology & Evolution* 22 (11): 583–92.

Price, Benjamin, and Ed Baker. 2016. "Nightlife: A Cheap, Robust, LED Based Light Trap for Collecting Aquatic Insects in Remote Areas." *Biodiversity Data Journal* 4:e7648.

Primeau, Kristy E., and David E. Witt. 2018. "Soundscapes in the Past: Investigating Sound at the Landscape Level." *Journal of Archaeological Science: Reports* 19 (February 2017): 875–85.

Priyadarshani, Nirosha, Stephen Marsland, and Isabel Castro. 2018. "Automated Bird-song Recognition in Complex Acoustic Environments: A Review." *Journal of Avian Biology* 49 (5): 1–27.

Proppe, Darren S., Christopher B. Sturdy, and Colleen Cassady St. Clair. 2013. "Anthropogenic Noise Decreases Urban Songbird Diversity and May Contribute to Homogenization." *Global Change Biology* 19 (4): 1075–84.

Proshansky, Harold M. 1978. "The City and Self-Identity." *Environment and Behavior* 10 (2): 147–69.

Proulx, Raphael, Jessica Waldinger, and Nicola Koper. 2019. "Anthropogenic Landscape Changes and Their Impacts on Terrestrial and Freshwater Soundscapes." *Current Landscape Ecology Reports* 4 (3): 41–50.

Provost, Foster, and Tom Fawcett. 2013. "Data Science and Its Relationship to Big Data and Data-Driven Decision Making." *Big Data* 1 (1): 51–59.

Pumphrey, R. J. 1950. "Hearing." In *Symposia of the Society for Experimental Biology, IV: Physiological Mechanism in Animal Behavior*, 1–18. Academic Press.

Punge, H. J., and M. Kunz. 2016. "Hail Observations and Hailstorm Characteristics in Europe: A Review." *Atmospheric Research* 176:159–84.

Purkis, Hubert John. 1964. "Transport Noise and Town Planning." *Journal of Sound and Vibration* 1 (3): 323–34.

Putland, R. L., Rochelle Constantine, and C. A. Radford. 2017. "Exploring Spatial and Temporal Trends in the Soundscape of an Ecologically Significant Embayment." *Scientific Reports* 7 (1): 5713.

Putland, R. L., and A. F. Mensinger. 2020. "Exploring the Soundscape of Small Freshwater Lakes." *Ecological Informatics* 55 (April 2019). https://doi.org/10.1016/j.ecoinf.2019.101018.

Pytte, C. L., M. S. Ficken, and A. Moiseff. 2004. "Ultrasonic Singing by the Blue-Throated Hummingbird: A Comparison between Production and Perception." *Journal of Comparative Physiology A* 190 (8): 665–73.

Qi, Jiaguo, S. H. Gage, Wooyeong Joo, Brian Napoletano, and S. Biswas. 2008. "Soundscape Characteristics of an Environment: A New Ecological Indicator of Ecosystem Health." In *Wetland and Water Resource Modeling and Assessment*, edited by Wei Ji, 201–11. CRC Press.

Qiu, MengYuan, Jie Zhang, HongLei Zhang, Li Li, Hui Zhang et al. 2017. "The Driving Mechanism of Tourists' Pro-environment Behavior Based on Cognition of Tourism Soundscapes: A Case of Kulangsu." *Tourism Tribune* 32 (11): 105–15.

Quinlan, J. Ross. 1986. "Induction of Decision Trees." *Machine Learning* 1:81–106.

Quinlan, J. Ross. 2014. *C4. 5: Programs for Machine Learning*. Elsevier.

Rabiner, Lawrence R., and Bernard Gold. 1975. *Theory and Application of Digital Signal Processing*. Prentice-Hall.

Radicchi, Antonella, Dietrich Henckel, and Martin Memmel. 2018. "Citizens as Smart, Active Sensors for a Quiet and Just City: The Case of the 'Open Source Soundscapes' Approach to Identify, Assess and Plan 'Everyday Quiet Areas' in Cities." *Noise Mapping* 5 (1): 1–20.

Raimbault, Manon, and Danièle Dubois. 2005. "Urban Soundscapes: Experiences and Knowledge." *Cities* 22 (5): 339–50.

Rajan, Sajeev C., K. Athira, R. Jaishanker, N. P. Sooraj, and V. Sarojkumar. 2019. "Rapid Assessment of Biodiversity Using Acoustic Indices." *Biodiversity and Conservation* 28 (8): 2371–83.

Ramsier, Marissa A., Andrew J. Cunningham, Gillian L. Moritz, James J. Finneran, Cathy V. Williams, Perry S. Ong, Sharon L. Gursky-Doyen, and Nathaniel J. Dominy. 2012. "Primate Communication in the Pure Ultrasound." *Biology Letters* 8 (4): 508–11.

Ranft, R. 2004. "Natural Sound Archives: Past, Present, and Future." *Anais da Academia Brasileira de Ciências* 76 (2): 455–65.

Ranspot, Tamara. 2019. "The Relational Nature of Song in Musical Human-Animal Interactions in Tr'ondëk Hwëch'in Traditional Territory, Yukon." *Journal of Ethnobiology* 39 (3): 478–91.

Rappaport, Danielle I., J. Andrew Royle, and Douglas C. Morton. 2020. "Acoustic Space Occupancy: Combining Ecoacoustics and Lidar to Model Biodiversity Variation and Detection Bias across Heterogeneous Landscapes." *Ecological Indicators* 113:106172.

Ratcliffe, Eleanor. 2021. "Sound and Soundscape in Restorative Natural Environments: A Narrative Literature Review." *Frontiers in Psychology* 12:963.

Ratcliffe, Eleanor, Birgitta Gatersleben, and Paul T. Sowden. 2013. "Bird Sounds and Their Contributions to Perceived Attention Restoration and Stress Recovery." *Journal of Environmental Psychology* 36:221–28.

Ratcliffe, Eleanor, Birgitta Gatersleben, and Paul T. Sowden. 2016. "Associations with Bird Sounds: How Do They Relate to Perceived Restorative Potential?" *Journal of Environmental Psychology* 47:136–44.

Ratcliffe, Eleanor, Birgitta Gatersleben, and Paul T. Sowden. 2020. "Predicting the Perceived Restorative Potential of Bird Sounds through Acoustics and Aesthetics." *Environment and Behavior* 52 (4): 371–400.

Raynor, Edward J., Cara E. Whalen, Mary Bomberger Brown, and Larkin A. Powell. 2017. "Grassland Bird Community and Acoustic Complexity Appear Unaffected by Proximity to a Wind Energy Facility in the Nebraska Sandhills." *Condor* 119 (3): 484–96.

Rehding, Alexander. 2002. "Eco-musicology." *Journal of the Royal Musical Association* 127 (2): 305–20.

Reich, Megan A. 2016. "Soundscape Composition as Environmental Activism and Awareness: An Ecomusicological Approach." *Summer Research*, 282. http://soundideas.pugetsound.edu/summer_research/282.

Reid, Walter V. 1998. "Biodiversity Hotspots." *Trends in Ecology & Evolution* 13 (7): 275–80.

Relph, Edward. 1976. *Place and Placelessness*. Pion.

Remillard, Wilfred J. 1961. "The History of Thunder Research." *Weather* 16 (8): 245–53.

Rempel, Robert S., Keith A. Hobson, George Holborn, Steve L. Van Wilgenburg, and Julie Elliott. 2005. "Bioacoustic Monitoring of Forest Songbirds: Interpreter Variability and Effects of Configuration and Digital Processing Methods in the Laboratory." *Journal of Field Ornithology* 76 (1): 1–11.

Reyes-García, Victoria, and Álvaro Fernández-Llamazares. 2019. "Sing to Learn: The Role of Songs in the Transmission of Indigenous Knowledge among the Tsimane' of Bolivian Amazonia." *Journal of Ethnobiology* 39 (3): 460–77.

Ribeiro, José Wagner, Larissa Sayuri Moreira Sugai, and Marconi Campos-Cerqueira. 2017. "Passive Acoustic Monitoring as a Complementary Strategy to Assess Biodiversity in the Brazilian Amazonia." *Biodiversity and Conservation* 26 (12): 2999–3002.

Rice, Aaron N., Melissa S. Soldevilla, and John A. Quinlan. 2017. "Nocturnal Patterns in Fish Chorusing off the Coasts of Georgia and Eastern Florida." *Bulletin of Marine Science* 93 (2): 455–74.

Rice, Timothy. 1987. "Toward the Remodeling of Ethnomusicology." *Ethnomusicology* 31 (3): 469–88.

Rice, Tom. 2018. "Acoustemology." In *International Encyclopedia of Anthropology*, edited by Hilary Callan, 1–7. John Wiley & Sons.

Rice, William L., Peter Newman, Zachary D. Miller, and B. Derrick Taff. 2020. "Protected Areas and Noise Abatement: A Spatial Approach." *Landscape and Urban Planning* 194:103701.

Richards, Douglas G., and R. Haven Wiley. 1980. "Reverberations and Amplitude Fluctuations in the Propagation of Sound in a Forest: Implications for Animal Communication." *American Naturalist* 115 (3): 381–99.

Richards, Fiona. 2018. *The Soundscapes of Australia: Music, Place and Spirituality.* Routledge.

Riede, Klaus. 1993. "Monitoring Biodiversity: Analysis of Amazonian Rainforest Sounds." *Ambio* 22 (8): 546–48.

Riede, Klaus. 1998. "Acoustic Monitoring of Orthoptera and Its Potential for Conservation." *Journal of Insect Conservation* 2 (3): 217–23.

Riede, Klaus. 2018. "Acoustic Profiling of Orthoptera." *Journal of Orthoptera Research* 27 (2): 203–15.

Riitters, Kurt H., R. V. O'Neill, C. T. Hunsaker, James D. Wickham, D. H. Yankee, S. P. Timmins, K. B. Jones, and B. L. Jackson. 1995. "A Factor Analysis of Landscape Pattern and Structure Metrics." *Landscape Ecology* 10 (1): 23–39.

Risser, P. G., J. R. Karr, and R. T. T. Forman. 1984. *Landscape Ecology: Directions and Approaches.* Special Publication no. 2. Illinois Natural History Survey.

Robert, Alois, Thierry Lengagne, Martim Melo, Vanessa Gardette, Sacha Julien, Rita Covas, Doris Gomez, and Claire Doutrelant. 2019. "The Theory of Island Biogeography and Soundscapes: Species Diversity and the Organization of Acoustic Communities." *Journal of Biogeography* 46 (9): 1901–11.

Roca, Irene T., Pierre Magnan, and Raphaël Proulx. 2020. "Use of Acoustic Refuges by Freshwater Fish: Theoretical Framework and Empirical Data in a Three-Species Trophic System." *Freshwater Biology* 65 (1): 45–54.

Roca, Irene T., and Ilse Van Opzeeland. 2020. "Using Acoustic Metrics to Characterize Underwater Acoustic Biodiversity in the Southern Ocean." *Remote Sensing in Ecology and Conservation* 6 (3): 262–73.

Rockström, Johan, Will Steffen, Kevin Noone, Åsa Persson, F. Stuart Chapin, Eric F. Lambin, Timothy M. Lenton et al. 2009. "A Safe Operating Space for Humanity." *Nature* 461 (7263): 472–75.

Rodaway, Paul. 1964. *Sensuous Geographies: Body, Sense and Place.* Classic edition. Psychology Press.

Rodaway, Paul. 1994. *Sensuous Geographies: Body, Sense and Place.* Routledge.

Roddy, Stephen, and Brian Bridges. 2018. "Sound, Ecological Affordances and Embodied Mappings in Auditory Display." In *New Directions in Third Wave Human-Computer In-*

teraction, vol. 2, *Methodologies*, edited by Michael Filimowicz and Veronika Tzankova, 231–58. Springer.

Rodrigues, Ana S. L., Sandy J. Andelman, Mohamed I. Bakarr, Luigi Boitani, Thomas M. Brooks, Richard M. Cowling, Lincoln DC Fishpool et al. 2004. "Effectiveness of the Global Protected Area Network in Representing Species Diversity." *Nature* 428 (6983): 640–43.

Rosales. 1990. "Auditory Characteristics of the Cicada Stridulation." *Journal of Entomology* 12 (3): 67–72.

Rosen, George. 1974. "A Backward Glance at Noise Pollution." *American Journal of Public Health* 64 (5): 514–17.

Rosenberg, K. V., A. M. Dokter, P. J. Blancher, J. R. Sauer, A. C. Smith, P. A. Smith, J. C. Stanton et al. 2019. "Decline of the North American Avifauna." *Science* 366 (6461): 120–24.

Rosenzweig, Mark R., Donald A. Riley, and David Krech. 1955. "Evidence for Echolocation in the Rat." *Science* 121 (3147): 600.

Rossi, Tullio, Sean D. Connell, and Ivan Nagelkerken. 2016. "Silent Oceans: Ocean Acidification Impoverishes Natural Soundscapes by Altering Sound Production of the World's Noisiest Marine Invertebrate." *Proceedings of the Royal Society of London. Series B: Biological Sciences* 283 (1826): 20153046.

Rossi, Tullio, Sean D. Connell, and Ivan Nagelkerken. 2017. "The Sounds of Silence: Regime Shifts Impoverish Marine Soundscapes." *Landscape Ecology* 32 (2): 239–48.

Rountree, Rodney A., Marta Bolgan, and Francis Juanes. 2019. "How Can We Understand Freshwater Soundscapes Without Fish Sound Descriptions?" *Fisheries* 44 (3): 137–43.

Rountree, Rodney A., R. Grant Gilmore, Clifford A. Goudey, Anthony D. Hawkins, Joseph J. Luczkovich, and David A. Mann. 2006a. "Listening to Fish: Applications of Passive Acoustics to Fisheries Science." *Fisheries* 31 (9): 433–46.

Rountree, Rodney A., Francis Juanes, and Marta Bolgan. 2017. "Fish Sound Production in Freshwater Habitats of New England: Widespread Occurrence of Air Movements Sounds." *Journal of the Acoustical Society of America* 141 (5): 4003.

Rountree, Rodney A., Francis Juanes, and Marta Bolgan. 2018. "Air Movement Sound Production by Alewife, White Sucker, and Four Salmonid Fishes Suggests the Phenomenon Is Widespread among Freshwater Fishes." *PLoS One* 13. https://doi.org/10.1371/journal.pone.0204247.

Rountree, Rodney A., Francis Juanes, and Marta Bolgan. 2020. "Temperate Freshwater Soundscapes: A Cacophony of Undescribed Biological Sounds Now Threatened by Anthropogenic Noise." *PLoS One* 15 (3): 1–26.

Rountree, Rodney, Francis Juanes, and Cliff Goudey. 2006b. "Listening to Fish: Applications of Passive Acoustics to Fisheries." *Journal of the Acoustical Society of America* 119 (5): 3277.

Rountree, Rodney A., Paul J. Perkins, Robert D. Kenney, and Kenneth R. Hinga. 2002. "Sounds of Western North Atlantic Fishes—Data Rescue." *Bioacoustics* 12 (2–3): 242–44.

Rousseeuw, Peter J. 1987. "Silhouettes: A Graphical Aid to the Interpretation and Validation of Cluster Analysis." *Journal of Computational and Applied Mathematics* 20 (C): 53–65.

Rowe, Benjamin, Jinglan Zhang, Michael Towsey, Paul Roe, and Margot Brereton. 2018. "Ecosound-Explorer: A Method for Large Scale Interactive Visual Navigation of Environmental Acoustic Data." In *Proceedings of the 30th Australian Conference on Computer-Human Interaction*, 539–43.

Rowley, Jennifer. 2012. "Conducting Research Interviews." *Management Research Review* 35 (3/4): 260–71.

Roy, David P., Valeriy Kovalskyy, Hankui Zhang, Lin Yan, and Indrani Kommareddy. 2015. "The Utility of Landsat Data for Global Long Term Terrestrial Monitoring." In *Remote Sensing Time Series: Revealing Land Surface Dynamics*, edited by Claudia Kuenzer, Stefan Dech, and Wolfgang Wagner, 289–305. Springer.

Royle, J. Andrew, and James D. Nichols. 2003. "Estimating Abundance from Repeated Presence-Absence Data or Point Counts." *Ecology* 84 (3): 777–90.

Ruddle, Kenneth. 1993. "The Transmission of Traditional Ecological Knowledge." In *Traditional Ecological Knowledge: Concepts and Cases*, edited by Julian T. Inglis, 17–31. Canadian Museum of Nature and International Development Research Centre.

Rudinsky, J. A., and R. R. Michael. 1972. "Sound Production in Scolytidae: Chemostimulus of Sonic Signal by the Douglas-Fir Beetle." *Science* 175 (4028): 1386–90.

Ruppé, Laëtitia, Gaël Clément, Anthony Herrel, Laurent Ballesta, Thierry Décamps, Loïc Kéver, and Eric Parmentier. 2015. "Environmental Constraints Drive the Partitioning of the Soundscape in Fishes." *Proceedings of the National Academy of Sciences* 112 (19): 6092–97.

Russell, James A. 1980. "A Circumplex Model of Affect." *Journal of Personality and Social Psychology* 39 (6): 1161.

Rychtáriková, Monika, and Gerrit Vermeir. 2013. "Soundscape Categorization on the Basis of Objective Acoustical Parameters." *Applied Acoustics* 74 (2): 240–47.

Ryff, Carol D. 1989. "Happiness Is Everything, or Is It? Explorations on the Meaning of Psychological Well-Being." *Journal of Personality and Social Psychology* 57 (6): 1069.

Sachs, Jeffrey D., Jonathan E. M. Baillie, William J. Sutherland, Paul R. Armsworth, Neville Ash, John Beddington, Tim M. Blackburn et al. 2009. "Biodiversity Conservation and the Millennium Development Goals." *Science* 325 (5947): 1502–3.

Sah, Hanyrol H. Ahmad, and T. Ulmar Grafe. 2019. "Amphibian Species Diversity in the Proposed Extension of the Bukit Teraja Protection Forest, Brunei Darussalam." *Scientia Bruneiana* 18 (1): 11–23.

Sakakibara, Ken-Ichi, Hiroshi Imagawa, Tomoko Konishi, Kazumasa Kondo, Emi Zuiki Murano, Masanobu Kumada, and Seiji Niimi. 2001. "Vocal Fold and False Vocal Fold Vibrations in Throat Singing and Synthesis of Khöömei." In *Proceedings of the International Computer Music Conference, Havana, Cuba*, 135–38.

Salafsky, Nick. 2011. "Integrating Development with Conservation: A Means to a Conservation End, or a Mean End to Conservation?" *Biological Conservation* 144 (3): 973–78.

Salamon, Justin, Christopher Jacoby, and Juan Pablo Bello. 2014. "A Dataset and Taxonomy for Urban Sound Research." In *Proceedings of the 22nd ACM International Conference on Multimedia*, 1041–44.

Sales, Gillian, and David Pye. 1974. "Ultrasound in Rodents." In *Ultrasonic Communication by Animals*, 149–201. Springer.

Sánchez-Bayo, Francisco, and Kris A. G. Wyckhuys. 2019. "Worldwide Decline of the Entomofauna: A Review of Its Drivers." *Biological Conservation* 232:8–27.

Sánchez-Giraldo, Camilo, Carol L. Bedoya, Raúl A. Morán-Vásquez, Claudia V. Isaza, and Juan M. Daza. 2020. "Ecoacoustics in the Rain: Understanding Acoustic Indices under the Most Common Geophonic Source in Tropical Rainforests." *Remote Sensing in Ecology and Conservation* 6 (3): 248–61.

Sandberg, Ulf. 2003. "The Multi-Coincidence Peak around 1000 Hz in Tyre/Road Noise Spectra." *Euronoise Naples* 89 (1–8): 2019.

Sandberg, Ulf, and Guy Descornet. 1980. "Road Surface Influence on Tire/Road Noise." Paper presented at the International Conference on Noise Control Engineering, Miami, December 8–10.

Sarkar, Sahotra, and Patricia Illoldi-Rangel. 2010. "Systematic Conservation Planning: An Updated Protocol." *Natureza & Conservação* 8 (1): 19–26.

Sarma, Hiren Kumar Deva, and Swapnil Mishra. 2016. "Mining Time Series Data with Apriori Tid Algorithm." In *International Conference on Information Technology (ICIT)*, 160–64. IEEE.

Saunders, Manu E., Jasmine K. Janes, and James C. O'Hanlon. 2020. "Moving On from the Insect Apocalypse Narrative: Engaging with Evidence-Based Insect Conservation." *BioScience* 70 (1): 80–89.

Scannell, Leila, and Robert Gifford. 2010. "Defining Place Attachment: A Tripartite Organizing Framework." *Journal of Environmental Psychology* 30 (1): 1–10.

Scarpelli, Marina D. A., Milton Cezar Ribeiro, and Camila P. Teixeira. 2021. "What Does Atlantic Forest Soundscapes Can Tell Us about Landscape?" *Ecological Indicators* 121:107050.

Schafer, R. Murray. 1969. *The New Soundscape*. BMI Canada Limited.

Schafer, R. Murray. 1977. *The Tuning of the World: Toward a Theory of Soundscape Design*. University of Pennsylvania Press.

Schafer, R. Murray. 1985. "Acoustic Space." In *Dwelling, Place and Environment: Towards a Phenomenology of Person and World*, edited by David Seamon and Robert Mugerauer, 87–98. Springer.

Schafer, R. Murray. 1994. *The Soundscape: Our Sonic Environment and the Tuning of the World*. Rochester: Destiny Books.

Schafer, R. Murray. 2012. *My Life on Earth and Elsewhere*. Porcupine's Quill.

Schaul, Tom, Justin Bayer, Daan Wierstra, Yi Sun, Martin Felder, Frank Sehnke, Thomas Rückstieß, and Jürgen Schmidhuber. 2010. "PyBrain." *Journal of Machine Learning Research* 11: 743–46.

Schevill, William Edward. 1964. *Underwater Sounds of Cetaceans*. Pergamon.

Schmidt, Arne K. D., and Rohini Balakrishnan. 2015. "Ecology of Acoustic Signalling and the Problem of Masking Interference in Insects." *Journal of Comparative Physiology A* 201:133–42.

Schmidt, Arne K. D., Klaus Riede, and Heiner Römer. 2011. "High Background Noise Shapes Selective Auditory Filters in a Tropical Cricket." *Journal of Experimental Biology* 214(10): 1754–62.

Schmidt, Arne K. D., Heiner Römer, and Klaus Riede. 2013. "Spectral Niche Segregation and Community Organization in a Tropical Cricket Assemblage." *Behavioral Ecology* 24 (2): 470–80.

Schnitzler, Hans-Ulrich, and Elisabeth K. V. Kalko. 2001. "Echolocation by Insect-Eating Bats: We Define Four Distinct Functional Groups of Bats and Find Differences in Signal Structure That Correlate with the Typical Echolocation Tasks Faced by Each Group." *Bioscience* 51 (7): 557–69.

Schoener, Thomas W. 1974. "Resource Partitioning in Ecological Communities: Research on How Similar Species Divide Resources Helps Reveal the Natural Regulation of Species Diversity." *Science* 185 (4145): 27–39.

Scholz, Christopher H. 2019. *The Mechanics of Earthquakes and Faulting*. Cambridge University Press.

Schulte-Fortkamp, Brigitte. 1999. "Noise from Combined Sources: How Attitudes towards Environment and Sources Influence the Assessment." In *Proceedings, Inter-Noise 1999*, 1383–86.

Schulte-Fortkamp, Brigitte. 2018. "Soundscape, Standardization, and Application." In Conference *Proceedings, Euronoise 2018*, 2445–49.

Schulte-Fortkamp, Brigitte, Bennett M. Brooks, and Wade R. Bray. 2007. "Soundscape: An Approach to Rely on Human Perception and Expertise in the Post-Modern Community Noise Era." *Acoustics Today* 3 (1): 7–15.

Schulte-Fortkamp, Brigitte, and Jian Kang. 2013. "Introduction to the Special Issue on Soundscapes." *Journal of the Acoustical Society of America* 134 (1): 765–66.

Schulze, Katharina, Kathryn Knights, Lauren Coad, Jonas Geldmann, Fiona Leverington, April Eassom, Melitta Marr, Stuart H. M. Butchart, Marc Hockings, and Neil D. Burgess. 2018. "An Assessment of Threats to Terrestrial Protected Areas." *Conservation Letters* 11 (3): e12435.

Schultz, P. Wesley. 2002. "Inclusion with Nature: The Psychology of Human-Nature Relations." In *Psychology of Sustainable Development*, edited by Peter Schmuck and Wesley P. Schultz, 61–78. Kluwer Academic Publishers.

Schwartz, Mark D. 1998. "Green-Wave Phenology." *Nature* 394 (6696): 839–40.

Schwartz, Mark D., ed. 2003. *Phenology: An Integrative Environmental Science*. Kluwer Academic Publishers.

Schwartz, Mark D., Rein Ahas, and Anto Aasa. 2006. "Onset of Spring Starting Earlier across the Northern Hemisphere." *Global Change Biology* 12 (2): 343–51.

Schwarz, Diemo. 1998. "Spectral Envelopes in Sound Analysis and Synthesis." PhD diss., Universität Stuttgart, Fakultät Informatik.

Sebastián-González, Esther, Richard J. Camp, Ann M. Tanimoto, Priscilla M. de Oliveira, Bruna B. Lima, Tiago A. Marques, and Patrick J. Hart. 2018."Density Estimation of Sound-Producing Terrestrial Animals Using Single Automatic Acoustic Recorders and Distance Sampling." *Avian Conservation and Ecology* 13 (2), 7.

Sedláček, Ondřej, Jana Vokurková, Michal Ferenc, Eric Nana Djomo, Tomáš Albrecht, and David Hořák. 2015. "A Comparison of Point Counts with a New Acoustic Sampling Method: A Case Study of a Bird Community from the Montane Forests of Mount Cameroon." *Ostrich* 86 (3): 213–20.

Seeger, Anthony. 1986. "The Role of Sound Archives in Ethnomusicology Today." *Ethnomusicology* 30 (2): 261–76.

Seeger, Anthony. 1987. "Do We Need to Remodel Ethnomusicology?" *Ethnomusicology* 31 (3): 491–95.

Seidman, Irving. 2019. *Interviewing as Qualitative Research: A Guide for Researchers in Education and the Social Sciences*. Teachers College Press.

Semidor, Catherine. 2006. "Listening to a City with the Soundwalk Method." *Acta Acustica United with Acustica* 92 (6): 959–64.

Serres, Michel. 2008. *The Five Senses: A Philosophy of Mingled Bodies*. Bloomsbury Publishing.

Servick, Kelly. 2014. "Eavesdropping on Ecosystems." *Science* 343 (6173): 834–37.

Sethi, Sarab S., Robert M. Ewers, Nick S. Jones, Jani Sleutel, Adi Shabrani, Nursyamin

Zulkifli, and Lorenzo Picinali. 2022. "Soundscapes Predict Species Occurrence in Tropical Forests." *Oikos* 2022 (3). https://doi.org/10.1111/oik.08525.

Sethi, Sarab S., Nick S. Jones, Ben D. Fulcher, Lorenzo Picinali, Dena Jane Clink, Holger Klinck, C. David L. Orme, Peter H. Wrege, and Robert M. Ewers. 2020. "Characterizing Soundscapes across Diverse Ecosystems Using a Universal Acoustic Feature Set." *Proceedings of the National Academy of Sciences* 117 (29): 17049–55.

Seto, Karen C., Michail Fragkias, Burak Güneralp, and Michael K. Reilly. 2011. "A Meta-analysis of Global Urban Land Expansion." *PLoS One* 6 (8): e23777.

Settles, Burr. 2009. *Active Learning Literature Survey*. CS Technical Report 1648. University of Wisconsin–Madison, Department of Computer Sciences. http://digital.library.wisc.edu/1793/60660.

Shamon, Hila, Zoe Paraskevopoulou, Justin Kitzes, Emily Card, Jessica L. Deichmann, Andy J. Boyce, and William J. McShea. 2021. "Using Ecoacoustics Metrices to Track Grassland Bird Richness across Landscape Gradients." *Ecological Indicators* 120:06928.

Shannon, Graeme, Megan F. McKenna, Lisa M. Angeloni, Kevin R. Crooks, Kurt M. Fristrup, Emma Brown, Katy A. Warner et al. 2016. "A Synthesis of Two Decades of Research Documenting the Effects of Noise on Wildlife." *Biological Reviews* 91 (4): 982–1005.

Shen, Jun-Xian, Zhi-Min Xu, Albert S. Feng, and Peter M. Narins. 2011. "Large Odorous Frogs (*Odorrana graminea*) Produce Ultrasonic Calls." *Journal of Comparative Physiology A* 197 (10): 1027.

Shvachko, Konstantin V. 2010. "HDFS Scalability: The Limits to Growth." *Login: The Magazine of USENIX & SAGE* 35 (2): 6–16.

Shvachko, K., Kuang, H., Radia, S., & Chansler, R. 2010. The Hadoop Distributed File System. In *Proceedings of the 2010 IEEE 26th Symposium on Mass Storage Systems and Technologies (MSST)*, edited by Mohammed G. Khatib, Xubin He, and Michael Factor, 1–10. IEEE.

Siemann, Evan, and James H. Brown. 1999. "Gaps in Mammalian Body Size Distributions Reexamined." *Ecology* 80 (8): 2788–92.

Siemers, Björn M., and Hans-Ulrich Schnitzler. 2004. "Echolocation Signals Reflect Niche Differentiation in Five Sympatric Congeneric Bat Species." *Nature* 429 (6992): 657–61.

Silver, Christina, and Ann Lewins. 2010. *Computer Assisted Qualitative Data Analysis*. Elsevier.

Simberloff, Daniel. 1998. "Flagships, Umbrellas, and Keystones: Is Single-Species Management Passé in the Landscape Era?" *Biological Conservation* 83 (3): 247–57.

Simmel, Georg. 1903. *Soziologie des Raumes*. Duncker & Humblot.

Simmel, Georg. 2012. "The Metropolis and Mental Life." In *The Urban Sociology Reader*, edited by Jan Lin and Christopher Mele, 37–45. Routledge.

Simmons, Kayelyn R., David B. Eggleston, and Del Wayne R. Bohnenstiehl. 2021. "Hurricane Impacts on a Coral Reef Soundscape." *PLoS One* 16 (2): 1–27.

Sinsch, Ulrich, Katrin Lümkemann, Katharina Rosar, Christiane Schwarz, and Maximilian Dehling. 2012. "Acoustic Niche Partitioning in an Anuran Community Inhabiting an Afromontane Wetland (Butare, Rwanda)." *African Zoology* 47 (1): 60–73.

Schnitzler, Hans, and Niels Bouton. 2008. "Soundscape Orientation: A New Field in Need of Sound Investigation." *Animal Behaviour* 4 (76): e5–e8.

Smith, Christopher C. 1978. "Structure and Function of the Vocalizations of Tree Squirrels (*Tamiasciurus*)." *Journal of Mammalogy* 59 (4): 793–808.

Smith, Craig A., and Richard S. Lazarus. 1990. "Emotion and Adaptation." In *Handbook of Personality: Theory and Research*, edited by L. A. Pervin, 609–37. Guilford Press.

Smith, Jansen A., John C. Handley, and Gregory P. Dietl. 2018. "Effects of Dams on Downstream Molluscan Predator–Prey Interactions in the Colorado River Estuary." *Proceedings of the Royal Society B: Biological Sciences* 285 (1879): 20180724.

Smith, Jordan W., and Bryan C. Pijanowski. 2014. "Human and Policy Dimensions of Soundscape Ecology." *Global Environmental Change* 28:63–74.

Smith, Susan J. 1994. "Soundscape." *Area* 16:232–40.

Smith, W. Brad. 2002. "Forest Inventory and Analysis: A National Inventory and Monitoring Program." *Environmental Pollution* 116: S233–S242.

Snelson, Chareen L. 2016. "Qualitative and Mixed Methods Social Media Research: A Review of the Literature." *International Journal of Qualitative Methods* 15 (1). https://doi.org/10.1177/1609406915624574.

Sodhi, Navjot S., David Bickford, Arvin C. Diesmos, Tien Ming Lee, Lian Pin Koh, Barry W. Brook, Cagan H. Sekercioglu, and Corey J. A. Bradshaw. 2008. "Measuring the Meltdown: Drivers of Global Amphibian Extinction and Decline." *PLoS One* 3 (2): e1636.

Sofaer, Helen R., Jennifer A. Hoeting, and Catherine S. Jarnevich. 2019. "The Area under the Precision-Recall Curve as a Performance Metric for Rare Binary Events." *Methods in Ecology and Evolution* 10 (4): 565–77.

Sokal, Robert R., and F. James Rohlf. 1987. *Introduction to Biostatistics*. 2nd ed. W. H. Freeman.

Sonnenburg, Sören, Gunnar Rätsch, Sebastian Henschel, Christian Widmer, Jonas Behr, Alexander Zien, Fabio de Bona et al. 2010. "The SHOGUN Machine Learning Toolbox." *Journal of Machine Learning Research* 11:1799–802.

Soulé, Michael E. 1985. "What Is Conservation Biology?." *BioScience* 35 (11): 727–34.

Sourdril, Anne, Luc Barbaro, Marc Deconchat, Eric Garine, and Christine Raimond. 2018. "Listening to Birds: How Local Populations Understand Environmental Changes through Everyday Sounds and Soundscapes?" Annual Meeting SFAA 2018. https://hal.inrae.fr/hal-02785312.

Sousa-Lima, Renata S., Deborah P. Fernandes, Thomas F. Norris, and Julie N. Oswald. 2013. "A Review and Inventory of Fixed Autonomous Recorders for Passive Acoustic Monitoring of Marine Mammals: 2013 State-of-the-Industry." 2013 IEEE/OES Acoustics in Underwater Geosciences Symposium, RIO Acoustics 2013. IEEE. https://doi.org/10.1109/RIOAcoustics.2013.6683984.

Southworth, Michael Frank. 1967. "The Sonic Environment of Cities." PhD diss., Massachusetts Institute of Technology.

Srikant, Ramakrishnan, and Rakesh Agrawal. 1993. *Mining Generalized Association Rules: Teaching Statistics*, vol. 15. https://doi.org/10.1111/j.1467-9639.1993.tb00256.x.

Srikant, Ramakrishnan, and Rakesh Agrawal. 1996. "Mining Quantitative Association Rules in Large Relational Tables." In *Proceedings of the 1996 ACM SIGMOD International Conference on Management of Data*, edited by Jennifer Widom, 1–12..

Staaterman, Erica, Matthew B. Ogburn, Andrew H. Altieri, Simon J. Brandl, Ross Whippo, Janina Seemann, Michael Goodison, and J. Emmett Duffy. 2017. "Bioacoustic Measurements Complement Visual Biodiversity Surveys: Preliminary Evidence from Four Shallow Marine Habitats." *Marine Ecology Progress Series* 575:207–15.

Staaterman, Erica, Claire B. Paris, Harry A. DeFerrari, David A. Mann, Aaron N. Rice, and Evan K. D'Alessandro. 2014. "Celestial Patterns in Marine Soundscapes." *Marine Ecology Progress Series* 508:17–32.

Staaterman, Erica, A. N. Rice, D. A. Mann, and Claire B. Paris. 2013. "Soundscapes from a Tropical Eastern Pacific Reef and a Caribbean Sea Reef." *Coral Reefs* 32 (2): 553–57.

Stanley, Jenni A., Craig A. Radford, and Andrew G. Jeffs. 2010. "Induction of Settlement in Crab Megalopae by Ambient Underwater Reef Sound." *Behavioral Ecology* 21 (1): 113–20.

Steel, Robert George Douglas, and James Hiram Torrie. 1980. *Principles and Procedures of Statistics, a Biometrical Approach.* McGraw-Hill.

Steffen, Will, Katherine Richardson, Johan Rockström, Sarah E. Cornell, Ingo Fetzer, Elena M. Bennett, Reinette Biggs et al. 2015. "Planetary Boundaries: Guiding Human Development on a Changing Planet." *Science* 347 (6223): 1259855.

Stein, Peter J., and Patrick Edson. 2016. "Active Acoustic Monitoring of Aquatic Life." In *The Effects of Noise on Aquatic Life II*, edited by Arthur N. Popper and Anthony Hawkins, 1113–21. Springer.

Steinbach, Michael, George Karypis, and Vipin Kumar. 2000. *A Comparison of Document Clustering Techniques.* Technical Report #00-034. Department of Computer Science and Engineering, University of Minnesota.

Stelmachowicz, Patricia G., Kathryn A. Beauchaine, Ann Kalberer, and Walt Jesteadt. 1989. "Normative Thresholds in the 8- to 20-kHz Range as a Function of Age." *Journal of the Acoustical Society of America* 86 (4): 1384–91.

Stern, Paul C., Thomas Dietz, and Gregory A. Guagnano. 1995. "The New Ecological Paradigm in Social-Psychological Context." *Environment and Behavior* 27 (6): 723–43.

Stoichita, Victor A., and Bernd Brabec de Mori. 2017. "Postures of Listening." *Terrain* (2007): 1–25.

Stollery, Pete. 2013. "Capture, Manipulate, Project, Preserve: A Compositional Journey." *Journal of Music, Technology & Education* 6 (3): 285–98.

Strauss, Anselm L. 1987. *Qualitative Analysis for Social Scientists.* Cambridge University Press.

Strayer, David L., and David Dudgeon. 2010. "Freshwater Biodiversity Conservation: Recent Progress and Future Challenges." *Journal of the North American Benthological Society* 29 (1): 344–58.

Stuart, Simon N., Janice S. Chanson, Neil A. Cox, Bruce E. Young, Ana S. L. Rodrigues, Debra L. Fischman, and Robert W. Waller. 2004. "Status and Trends of Amphibian Declines and Extinctions Worldwide." *Science* 306 (5702): 1783–86.

Sudarsono, Anugrah Sabdono, Yiu W. Lam, and William J. Davies. 2016. "The Effect of Sound Level on Perception of Reproduced Soundscapes." *Applied Acoustics* 110:53–60.

Suddaby, Roy. 2006. "From the Editors: What Grounded Theory Is Not." *Academy of Management Journal* 49 (4). https://doi.org/10.5465/amj.2006.22083020.

Sueur, Jérôme. 2002. "Cicada Acoustic Communication: Potential Sound Partitioning in a Multispecies Community from Mexico (Hemiptera: Cicadomorpha: Cicadidae)." *Biological Journal of the Linnean Society* 75 (3): 379–94.

Sueur, Jérôme, and Almo Farina. 2015. "Ecoacoustics: The Ecological Investigation and Interpretation of Environmental Sound." *Biosemiotics* 8 (3): 493–502.

Sueur, Jérôme, Almo Farina, Amandine Gasc, Nadia Pieretti, and Sandrine Pavoine. 2014. "Acoustic Indices for Biodiversity Assessment and Landscape Investigation." *Acta Acustica United with Acustica* 100 (4): 772–81.

Sueur, Jérôme, Sandrine Pavoine, Olivier Hamerlynck, and Stéphanie Duvail. 2008. "Rapid Acoustic Survey for Biodiversity Appraisal." *PLoS One* 3 (12): e4065.

Sugai, Larissa Sayuri Moreira, Camille Desjonquères, Thiago Sanna Freire Silva, and Diego Llusia. 2020. "A Roadmap for Survey Designs in Terrestrial Acoustic Monitoring." *Remote Sensing in Ecology and Conservation* 6 (3): 220–35.

Sugai, Larissa Sayuri Moreira, and Diego Llusia. 2019. "Bioacoustic Time Capsules: Using Acoustic Monitoring to Document Biodiversity." *Ecological Indicators* 99 (April 2019): 149–52. https://doi.org/10.1016/j.ecolind.2018.12.021.

Sugai, Larissa Sayuri Moreira, Thiago Sanna Freire Silva, José Wagner Ribeiro, and Diego Llusia. 2019. "Terrestrial Passive Acoustic Monitoring: Review and Perspectives." *BioScience* 69 (1): 5–11.

Sun, Jennifer W. C., and Peter M. Narins. 2005. "Anthropogenic Sounds Differentially Affect Amphibian Call Rate." *Biological Conservation* 121 (3): 419–27.

Supreeth, H. V., Sumukh Rao, K. S. Chethan, and U. Purushotham. 2020. "Identification of Ambulance Siren Sound and Analysis of the Signal Using Statistical Method." In *2020 International Conference on Intelligent Engineering and Management (ICIEM)*, 198–202. IEEE.

Sutherland, Louis C. 1999. "Natural Quiet: An Endangered Environment: How to Measure, Evaluate, and Preserve It." *Noise Control Engineering Journal* 47 (3): 82–86.

Sutton, Steve. 1998. *Visitor Perceptions of Aircraft Activity and Crowding at Franz Josef and Fox Glaciers*. New Zealand Department of Conservation.

Swanson, Don R. 1986. "Fish Oil, Raynaud's Syndrome, and Undiscovered Public Knowledge." *Perspectives in Biology and Medicine* 30 (1): 7–18.

Taheri, Mahsa, Néhémy Lim, and Johannes Lederer. 2006. "Balancing Statistical and Computational Precision and Applications to Penalized Linear Regression with Group Sparsity." In *Proceedings of the 23rd International Conference on Machine Learning*, 233–40. http://arxiv.org/abs/1609.07195.

Tan, Ah-Hwee, et al. 1999. "Text Mining: The State of the Art and the Challenges." In *Proceedings of the PAKDD 1999 Workshop on Knowledge Discovery from Advanced Databases*, 8:65–70.

Tan, Lizhe, and Jean Jiang. 2018. *Digital Signal Processing: Fundamentals and Applications*. Academic Press.

Tan, Pang-Ning, Michael Steinbach, and Vipin Kumar. 2013. "Data Mining Cluster Analysis: Basic Concepts and Algorithms." In *Introduction to Data Mining*, 487–533. Pearson Education India.

Tan, Pang-Ning, Michael Steinbach, and Vipin Kumar. 2016. *Introduction to Data Mining*. Pearson Education India.

Tashakkori, Abbas, and Charles B. Teddlie. 1998. *Mixed Methodology: Combining Qualitative and Quantitative Approaches*. Applied Social Research Methods, vol. 46. Sage.

Tauber, Eran, and Daniel F. Eberl. 2003. "Acoustic Communication in *Drosophila*." *Behavioural Processes* 64 (2): 197–210.

Tavolga, William N. 1971. "Sound Production and Detection." In *Fish Physiology*, vol. 5, edited by W. S. Hoar and D. J. Randall, 135–205. Academic Press.

Taylor, L. R. 1962. "The Efficiency of Cylindrical Sticky Insect Traps and Suspended Nets." *Annals of Applied Biology* 50 (4): 681–85.

Teer, Thomas L., and A. A. Few. 1974. "Horizontal Lightning." *Journal of Geophysical Research* 79 (24): 3436–41.

Tenaza, Richard R. 1976. "Songs, Choruses and Countersinging of Kloss' Gibbons

(*Hylobates klossii*) in Siberut Island, Indonesia." *Zeitschrift für Tierpsychologie* 40 (1): 37–52.

ter Hofstede, Hannah M., Elisabeth K. V. Kalko, and James H. Fullard. 2010. "Auditory-Based Defence against Gleaning Bats in Neotropical Katydids (Orthoptera: Tettigoniidae)." *Journal of Comparative Physiology A* 196 (5): 349–58.

ter Hofstede, Hannah M., and John M. Ratcliffe. 2016. "Evolutionary Escalation: The Bat–Moth Arms Race." *Journal of Experimental Biology* 219 (11): 1589–602.

Thomas, Chris D., T. Hefin Jones, and Sue E. Hartley. 2019. "'Insectageddon'": A Call for More Robust Data and Rigorous Analyses." *Global Change Biology* 25 (6): 1891–92.

Thoret, Etienne, Léo Varnet, Yves Boubenec, Régis Férriere, François-Michel Le Tourneau, Bernard L. Krause, and Christian Lorenzi. 2020. "Characterizing Amplitude and Frequency Modulation Cues in Natural Soundscapes: A Pilot Study on Four Habitats of a Biosphere Reserve." *Journal of the Acoustical Society of America* 147 (5): 3260–74.

Tilman, David. 2004. "Niche Tradeoffs, Neutrality, and Community Structure: A Stochastic Theory of Resource Competition, Invasion, and Community Assembly." *Proceedings of the National Academy of Sciences* 101 (30): 10854–61.

Titon, Jeff Todd. 2009. "Music and Sustainability: An Ecological Viewpoint." *World of Music* 51 (1): 119–37.

Titon, Jeff Todd. 2012. "A Sound Commons for All Living Creatures." *Smithsonian Folkways Magazine*, Fall/Winter.

Titon, Jeff Todd. 2013. "The Nature of Ecomusicology." *Música e Cultura: revista da ABET* 8 (1): 8–18.

Titon, Jeff Todd. 2015. "Exhibiting Music in a Sound Community." *Ethnologies* 37 (1): 23–41.

Titon, Jeff Todd. 2018. "Ecomusicology and the Problems in Ecology." *MUSICultures* 45 (1/2): 255–64.

Tobias, Joseph A., Robert Planqué, Dominic L. Cram, and Nathalie Seddon. 2014. "Species Interactions and the Structure of Complex Communication Networks." *Proceedings of the National Academy of Sciences* 111 (3): 1020–25.

Toliver, Brooks. 2004. "Eco-ing in the Canyon: Ferde Grofé's Grand Canyon Suite and the Transformation of Wilderness." *Journal of American Musicological Society* 57 (2): 325–68.

Tonolla, Diego, Vicenç Acuña, Mark S. Lorang, Kurt Heutschi, and Klement Tockner. 2010. "A Field-Based Investigation to Examine Underwater Soundscapes of Five Common River Habitats." *Hydrological Processes* 24 (22): 3146–56.

Tonolla, Diego, Mark S. Lorang, Kurt Heutschi, Chris C. Gotschalk, and Klement Tockner. 2011. "Characterization of Spatial Heterogeneity in Underwater Soundscapes at the River Segment Scale." *Limnology and Oceanography* 56 (6): 2319–33.

Tonolla, Diego, Mark S. Lorang, Kurt Heutschi, and Klement Tockner. 2009. "A Flume Experiment to Examine Underwater Sound Generation by Flowing Water." *Aquatic Sciences* 71 (4): 449–62.

Torigoe, Keiko. 1982. "A Study of the World Soundscape Project." MFA thesis, York University.

Tosi, Patrizia, Paola Sbarra, and Valerio De Rubeis. 2012. "Earthquake Sound Perception." *Geophysical Research Letters* 39 (24).

Towsey, Michael, Stuart Parsons, and Jérôme Sueur. 2014a. "Ecology and Acoustics at a Large Scale." *Ecological Informatics* 21:1–3.

Towsey, Michael, Jason Wimmer, Ian Williamson, and Paul Roe. 2014b. "The Use of Acoustic Indices to Determine Avian Species Richness in Audio-Recordings of the Environment." *Ecological Informatics* 21 (100): 110–19.

Towsey, Michael, Liang Zhang, Mark Cottman-Fields, Jason Wimmer, Jinglan Zhang, and Paul Roe. 2014c. "Visualization of Long-Duration Acoustic Recordings of the Environment." *Procedia Computer Science* 29:703–12.

Truax, Barry. 1974. "Soundscape Studies: An Introduction to the World Soundscape Project." *Numus West* 5:36–39.

Truax, Barry. 1988. "Real-Time Granular Synthesis with a Digital Signal Processor." *Computer Music Journal* 12 (2): 14–26.

Truax, Barry. 1992. "Composing with Time-Shifted Environmental Sound." *Leonardo Music Journal* 2 (1): 37–40.

Truax, Barry. 1996. "Soundscape, Acoustic Communication and Environmental Sound Composition." *Contemporary Music Review* 15 (1–2): 49–65.

Truax, Barry. 2001. *Acoustic Communication*. Greenwood Publishing Group.

Truax, Barry. 2002. "Genres and Techniques of Soundscape Composition as Developed at Simon Fraser University." *Organised Sound* 7 (1): 5–14.

Truax, Barry. 2008. "Soundscape Composition as Global Music: Electroacoustic Music as Soundscape." *Organised Sound* 13 (2): 103.

Truax, Barry. 2012a. "From Soundscape Documentation to Soundscape Composition." In *Acoustics 2012*.

Truax, Barry. 2012b. "Music, Soundscape and Acoustic Sustainability." *Moebius Journal* 1 (1).

Truax, Barry. 2012c. "Sound, Listening and Place: The Aesthetic Dilemma." *Organised Sound* 17 (3): 193–201.

Truax, Barry. 2013. "The World Soundscape Project." https://www.sfu.ca/~truax/wsp .html.

Truax, Barry. 2019. "Acoustic Ecology and the World Soundscape Project." In *Sound, Media, Ecology*, edited by Milena Droumeva and Randolph Jordan, 21–44. Palgrave Macmillan.

Truax, Barry, and Gary W. Barrett. 2011. "Soundscape in a Context of Acoustic and Landscape Ecology." In "Soundscape Ecology," ed. Bryan C. Pijanowski and Almo Farina, special issue, *Landscape Ecology* 26 (9): 1201–7.

Truskinger, Anthony, Michael Towsey, and Paul Roe. 2015. "Decision Support for the Efficient Annotation of Bioacoustic Events." *Ecological Informatics* 25:14–21.

Tsaligopoulos, Aggelos, Stella Kyvelou, Nefta Eleftheria Votsi, Aimilia Karapostoli, Chris Economou, and Yiannis G. Matsinos. 2021. "Revisiting the Concept of Quietness in the Urban Environment—Towards Ecosystems' Health and Human Well-Being." *International Journal of Environmental Research and Public Health* 18 (6): 1–19.

Tscharntke, Teja, Jason M. Tylianakis, Tatyana A. Rand, Raphael K. Didham, Lenore Fahrig, Péter Batáry, Janne Bengtsson et al. 2012. "Landscape Moderation of Biodiversity Patterns and Processes—Eight Hypotheses." *Biological Reviews* 87 (3): 661–85.

Tuan, Yi-Fu. 1975. "Place: An Experiential Perspective." *Geographical Review* 65 (2): 151–65.

Tuan, Yi-Fu. 1977. *Space and Place: The Perspective of Experience*. University of Minnesota Press.

Tuan, Yi-Fu. 1990. *Topophilia: A Study of Environmental Perceptions, Attitudes, and Values*. Columbia University Press.

Tucker, David, Stuart H. Gage, Ian Williamson, and Susan Fuller. 2014. "Linking Ecological Condition and the Soundscape in Fragmented Australian Forests." *Landscape Ecology* 29 (4): 745–58.

Tuneu-Corral, Carme, Xavier Puig-Montserrat, Carles Flaquer, Maria Mas, Ivana Budinski, and Adrià López-Baucells. 2020. "Ecological Indices in Long-Term Acoustic Bat Surveys for Assessing and Monitoring Bats' Responses to Climatic and Land-Cover Changes." *Ecological Indicators* 110:105849.

Turner, Billie, W. C. Clark, R. W. Kates, J. F. Richards, J. T. Mathews, and W. B. Meyer. 1990. *The Earth as Transformed by Human Action: Global and Regional Changes in the Biosphere over the Past 300 Years*. Cambridge University Press, with Clark University.

Turner, Monica G. 1989. "Landscape Ecology: The Effect of Pattern on Process." *Annual Review of Ecology and Systematics* 20:171–97.

Turner, Monica G. 1990. "Spatial and Temporal Analysis of Landscape Patterns." *Landscape Ecology* 4 (1): 21–30.

Turner, Monica G., Robert H. Gardner, and Robert V. O'Neill. 2001. *Landscape Ecology in Theory and Practice: Pattern and Process*. Springer.

Ulloa, Juan Sebastian, Angélica Hernández-Palma, Orlando Acevedo-Charry, Bibiana Gómez-Valencia, Cristian Cruz-Rodríguez, Yenifer Herrera-Varón, Margarita Roa et al. 2021. "Listening to Cities during the COVID-19 Lockdown: How Do Human Activities and Urbanization Impact Soundscapes in Colombia?" *Biological Conservation* 255 (March 2021). https://doi.org/10.1016/j.biocon.2021.108996.

Ulrich, Roger S. 1983. "Aesthetic and Affective Response to Natural Environment." In *Behavior and the Natural Environment*, edited by Irwin Altman and Joachim F. Wohwill, 85–125. Springer.

Uman, Martin A. 2001. *The Lightning Discharge*. Dover.

Uman, Martin A. 2012. *Lightning*. Courier Corporation.

United Nations. 2007. *World Urbanization Prospects: The 2007 Revision*. United Nations.

Urban, Dean L., Sarah Goslee, Ken Pierce, and Todd Lookingbill. 2002. "Extending Community Ecology to Landscapes." *Ecoscience* 9 (2): 200–212.

Urban, Dean L., Robert V. O'Neill, and Herman H. Shugart Jr. 1987. "A Hierarchical Perspective Can Help Scientists Understand Spatial Patterns." *BioScience* 37 (2): 119–27.

van der Lee, Gea H., Camille Desjonquères, Jérôme Sueur, Michiel H. S. Kraak, and Piet F. M. Verdonschot. 2020. "Freshwater Ecoacoustics: Listening to the Ecological Status of Multi-Stressed Lowland Waters." *Ecological Indicators* 113:106252.

Van der Maaten, Laurens, and Geoffrey Hinton. 2008. "Visualizing Data Using t-SNE." *Journal of Machine Learning Research* 9 (11): 2579–605.

Van Kempen, Elise E. M. M., Hanneke Kruize, Hendriek C. Boshuizen, Caroline B. Ameling, Brigit A. M. Staatsen, and Augustinus E. M. de Hollander. 2002. "The Association between Noise Exposure and Blood Pressure and Ischemic Heart Disease: A Meta-analysis." *Environmental Health Perspectives* 110 (3): 307–17.

Van Kranenburg, Peter, Jörg Garbers, Anja Volk, Frans Wiering, Louis P. Grijp, and Remco C. Veltkamp. 2010. "Collaboration Perspectives for Folk Song Research and Music Information Retrieval: The Indispensable Role of Computational Musicology." *jims* 2009:030.

Van Manen, Max. 2016. *Phenomenology of Practice: Meaning-Giving Methods in Phenomenological Research and Writing*. Routledge.

Van Renterghem, Timothy, and Dick Botteldooren. 2010. "The Importance of Roof

Shape for Road Traffic Noise Shielding in the Urban Environment." *Journal of Sound and Vibration* 329 (9): 1422–34.

Vavrek, R. James, R. Kithil, R. L. Holle, J. Allsopp, and Mary Ann Cooper. 2006. "The Science of Thunder." *Earth Science* 22 (3): 5–9.

Veits, Marine, Itzhak Khait, Uri Obolski, Eyal Zinger, Arjan Boonman, Aya Goldshtein, Kfir Saban et al. 2019. "Flowers Respond to Pollinator Sound within Minutes by Increasing Nectar Sugar Concentration." *Ecology Letters* 22 (9): 1483–92.

Vellend, Mark, Lander Baeten, Isla H. Myers-Smith, Sarah C. Elmendorf, Robin Beauséjour, Carissa D. Brown, Pieter De Frenne, Kris Verheyen, and Sonja Wipf. 2013. "Global Meta-analysis Reveals No Net Change in Local-Scale Plant Biodiversity over Time." *Proceedings of the National Academy of Sciences* 110 (48): 19456–59.

Venables, William N., David M. Smith, and R Development Core Team. 2021. "An Introduction to R."

Vermeij, Mark JA, Kristen L. Marhaver, Chantal M. Huijbers, Ivan Nagelkerken, and Stephen D. Simpson. 2010. "Coral Larvae Move toward Reef Sounds." *PloS one* 5 (5): e10660.

Versluis, Michel, Barbara Schmitz, Anna von der Heydt, and Detlef Lohse. 2000. "How Snapping Shrimp Snap: Through Cavitating Bubbles." *Science* 289 (5487): 2114–17.

Vick, K. W., R. W. Mankin, R. R. Cogburn, M. Mullen, J. E. Throne, V. F. Wright, and L. D. Cline. 1990. "Review of Pheromone-Baited Sticky Traps for Detection of Stored-Product Insects." *Journal of the Kansas Entomological Society* 63 (4) 526–32.

Villanueva-Rivera, Luis J., and Bryan C. Pijanowski. 2012. "Pumilio: A Web-Based Management System for Ecological Recordings." *Bulletin of the Ecological Society of America* 93 (1): 71–81.

Villanueva-Rivera, Luis J. 2014. "Eleutherodactylus Frogs Show Frequency but No Temporal Partitioning: Implications for the Acoustic Niche Hypothesis." *PeerJ* 2:e496.

Villanueva-Rivera, Luis J., and Bryan C. Pijanowski. 2018. "Package 'Soundecology.'" R package version 1.3.3.

Villanueva-Rivera, Luis J., Bryan C. Pijanowski, Jarrod Doucette, and Burak Pekin. 2011. "A Primer of Acoustic Analysis for Landscape Ecologists." In "Soundscape Ecology," ed. Bryan C. Pijanowski and Almo Farina, special issue, *Landscape Ecology* 26 (9): 1233–46.

Vinyeta, K., and K. Lynn. 2013. *Exploring the Role of Traditional Ecological Knowledge in Climate Change Initiatives*. USDA Forest Service, Pacific Northwest Research Station.

Vitousek, Peter M., and Pamela A. Matson. 1991. "Gradient Analysis of Ecosystems." In *Comparative Analyses of Ecosystems: Patterns, Mechanisms, and Theories*, edited by Jonathan Cole, Gary Lovett, and Stuart Findlay, 287–98. Springer.

Vitousek, Peter M., Harold A. Mooney, Jane Lubchenco, and Jerry M. Melillo. 1997. "Human Domination of Earth's Ecosystems." *Science* 277 (5325): 494–99.

Vokes, Richard. 2021. "(Re) constructing the Field through Sound: Actor-networks, Ethnographic Representation and 'Radio Elicitation' in South-western Uganda." In *Creativity and Cultural Improvisation*, edited by Elizabeth Hallam and Tim Ingold, 285–303. Routledge.

Volk, Anja, Frans Wiering, and Peter van Kranenburg. 2011. "Unfolding the Potential of Computational Musicology." In *Proceedings of the 13th International Conference on Informatics and Semiotics in Organisations*.

Von Bertalanffy, Ludwig. 1950. "The Theory of Open Systems in Physics and Biology." *Science* 111 (2872): 23–29.

Wagner, David L. 2019. "Insect Declines in the Anthropocene." *Annual Review of Entomology* 65:467–80.

Wald, Lucien. 1999. "Some Terms of Reference in Data Fusion." *IEEE Transactions on Geoscience and Remote Sensing* 37 (3): 1190–93.

Wallace, Alfred R. 1880. "The Origin of Species and Genera." *The Nineteenth Century and After: A Monthly Review* 7 (35): 93–106.

Walters, Carl J. 1986. *Adaptive Management of Renewable Resources*. Macmillan.

Walters, Carl J., and Ray Hilborn. 1978. "Ecological Optimization and Adaptive Management." *Annual Review of Ecology and Systematics* 9 (1): 157–88.

Wang, Bo, and Jian Kang. 2011. "Effects of Urban Morphology on the Traffic Noise Distribution through Noise Mapping: A Comparative Study between UK and China." *Applied Acoustics* 72 (8): 556–68.

Wang, Qi, Yue Ma, Kun Zhao, and Yingjie Tian. 2020. "A comprehensive survey of loss functions in machine learning." *Annals of Data Science* 9:187–212.

Wang, Yun, and Zhiyong Deng. 2011. "Soundscape: In the View of Music." In *Proceedings, Inter-Noise 2011*, 4–7.

Ward, J. V. 1998. "Riverine Landscapes: Biodiversity Patterns, Disturbance Regimes, and Aquatic Conservation." *Biological Conservation* 83 (3): 269–78.

Ward, Michelle, Santiago Saura, Brooke Williams, Juan Pablo Ramírez-Delgado, Nur Arafeh-Dalmau, James R. Allan, Oscar Venter, Grégoire Dubois, and James E. M. Watson. 2020. "Just Ten Percent of the Global Terrestrial Protected Area Network Is Structurally Connected via Intact Land." *Nature Communications* 11 (1): 4563.

Ware, Heidi E., Christopher J. W. McClure, Jay D. Carlisle, and Jesse R. Barber. 2015. "A Phantom Road Experiment Reveals Traffic Noise Is an Invisible Source of Habitat Degradation." *Proceedings of the National Academy of Sciences* 112 (39): 12105–9.

Warren, Joe D., A. R. Jennings, and Timothy D. Griffiths. 2005. "Analysis of the Spectral Envelope of Sounds by the Human Brain." *Neuroimage* 24 (4): 1052–57.

Warren, Paige S., Madhusudan Katti, Michael Ermann, and Anthony Brazel. 2006. "Urban Bioacoustics: It's Not Just Noise." *Animal Behaviour* 71 (3): 491–502.

Warren, William H., and Robert R. Verbrugge. 1984. "Auditory Perception of Breaking and Bouncing Events: A Case Study in Ecological Acoustics." *Journal of Experimental Psychology: Human Perception and Performance* 10 (5): 704.

Watanabe. W. 1985. *Pattern Recognition: Human and Mechanical*. Wiley.

Watson, David, Lee Anna Clark, and Auke Tellegen. 1988. "Development and Validation of Brief Measures of Positive and Negative Affect: The PANAS Scales." *Journal of Personality and Social Psychology* 54 (6): 1063.

Watson, Dianne L., Euan S. Harvey, Marti J. Anderson, and Gary A. Kendrick. 2005. "A Comparison of Temperate Reef Fish Assemblages Recorded by Three Underwater Stereo-Video Techniques." *Marine Biology* 148 (2): 415–25.

Watson, James E. M., Nigel Dudley, Daniel B. Segan, and Marc Hockings. 2014. "The Performance and Potential of Protected Areas." *Nature* 515 (7525): 67–73.

Watts, Greg, Abdul Miah, and Rob Pheasant. 2013. "Tranquility and Soundscapes in Urban Green Spaces—Predicted and Actual Assessments from a Questionnaire Survey." *Environment and Planning B: Planning and Design* 40 (1): 170–81.

Wavey, Robert. 1993. "International Workshop on Indigenous Knowledge and Community Based Resource Management: Keynote Address." In *Traditional Ecological Knowledge: Concepts and Cases*, edited by Julian T. Inglis, 11–16. IDRC.

Wedding, Lisa M., Christopher A. Lepczyk, Simon J. Pittman, Alan M. Friedlander, and

Stacy Jorgensen. 2011. "Quantifying Seascape Structure: Extending Terrestrial Spatial Pattern Metrics to the Marine Realm." *Marine Ecology Progress Series* 427:219–32.

Welch, Peter. 1967. "The Use of Fast Fourier Transform for the Estimation of Power Spectra: A Method Based on Time Averaging over Short, Modified Periodograms." *IEEE Transactions on Audio and Electroacoustics* 15 (2): 70–73.

Wenz, Gordon M. 1962. "Acoustic Ambient Noise in the Ocean: Spectra and Sources." *Journal of the Acoustical Society of America* 34 (12): 1936–56.

West, Geoffrey B., James H. Brown, and Brian J. Enquist. 1997. "A General Model for the Origin of Allometric Scaling Laws in Biology." *Science* 276 (5309): 122–26.

Westerkamp, Hildegard. 1974. "Soundwalking." *Sound Heritage* 3 (4): 18–27.

Westerkamp, Hildegard. 1988. "Listening and Soundmaking: A Study of Music-as-Environment." PhD diss., Simon Fraser University.

Westerkamp, Hildegard. 1991. "The World Soundscape Project." *Soundscape Newsletter*, no. 1., August 1991. https://www.sfu.ca/sonic-studio-webdav/WSP_Doc/Newsletters/Number1.pdf

Westerkamp, Hildegard. 1999. "Soundscape Composition: Linking Inner and Outer Worlds." https://www.hildegardwesterkamp.ca/writings/writingsby/?post_id=19&title=%E2%80%8Bsoundscape-composition:-linking-inner-and-outer-worlds-.

Westerkamp, Hildegard. 2002. "Linking Soundscape Composition and Acoustic Ecology." *Organised Sound* 7 (1): 51–56.

Westerkamp, Hildegard. 2006. "Soundwalking as Ecological Practice." *The West Meets the East in Acoustic Ecology*, edited by Tadahiko Imada, Kozo Hiramatsu and Keiko Torigoe, 2–4. Japanese Association for Sound Ecology (JASE) / Hirosaki University International Music Centre (HIMC).

Westerkamp, Hildegard. 2010. "What's in a Soundwalk?" Unpublished paper prepared for Sonic Acts XIII Conference "The Poetics of Space," Amsterdam.

Westman, Jack C., and James R. Walters. 1981. "Noise and Stress: A Comprehensive Approach." *Environmental Health Perspectives* 41:291–309.

Wheeland, Laura J., and George A. Rose. 2015. "Quantifying Fish Avoidance of Small Acoustic Survey Vessels in Boreal Lakes and Reservoirs." *Ecology of Freshwater Fish* 24 (1): 67–76.

Wheeland, Laura J., and George A. Rose. 2016. "Acoustic Measures of Lake Community Size Spectra." *Canadian Journal of Fisheries and Aquatic Sciences* 73 (4): 557–64.

White, Tom. 2012. *Hadoop: The Definitive Guide*. O'Reilly Media, Inc.

Whittaker, Robert J. 2014. "Developments in Biogeography." *Journal of Biogeography* 41 (1): 1–5.

Whittaker, Robert J., Miguel B. Araújo, Paul Jepson, Richard J. Ladle, James E. M. Watson, and Katherine J. Willis. 2005. "Conservation Biogeography: Assessment and Prospect." *Diversity and Distributions* 11 (1): 3–23.

Whittaker, Robert J., Joaquín Hortal, Dov F. Sax, David J. Currie, David M. Richardson, Alycia L. Stigall, and Michael N. Dawson. 2018. "*Frontiers of Biogeography*: Taking Its Place as a Journal of Choice for the Publication of High Quality Biogeographical Research Articles." *Frontiers of Biogeography* 10 (1–2): e40499.

Wickham, Hadley. 2011. "Ggplot2." *Wiley Interdisciplinary Reviews: Computational Statistics* 3 (2): 180–85.

Wickham, Hadley, and Garrett Grolemund. 2016. *R for Data Science: Import, Tidy, Transform, Visualize, and Model Data*. O'Reilly Media, Inc.

Wickham, Lionel. 2008. "Schism and Reconciliation in a Sixth-Century Trinitarian

Dispute: Damian of Alexandria and Peter of Callinicus on 'Properties, Roles and Relations.'" *International Journal for the Study of the Christian Church* 8 (1): 3–15.

Wiens, John A. 1989. "Spatial Scaling in Ecology." *Functional Ecology* 3 (4): 385–97.

Wiens, John A., and Bruce T. Milne. 1989. "Scaling of 'Landscapes' in Landscape Ecology, or, Landscape Ecology from a Beetle's Perspective." *Landscape Ecology* 3 (2): 87–96.

Wilcock, William S. D., Kathleen M. Stafford, Rex K. Andrew, and Robert I. Odom. 2014. "Sounds in the Ocean at 1–100 Hz." *Annual Review of Marine Science* 6:117–40.

Wilczynski, Walter, Harold H. Zakon, and Eliot A. Brenowitz. 1984. "Acoustic Communication in Spring Peepers." *Journal of Comparative Physiology A* 155 (5): 577–84.

Wiley, R. Haven, and Douglas G. Richards. 1978. "Physical Constraints on Acoustic Communication in the Atmosphere: Implications for the Evolution of Animal Vocalizations." *Behavioral Ecology and Sociobiology* 3:69–94.

Wilkinson, Leland. 2012. "The Grammar of Graphics." In *Handbook of Computational Statistics: Concepts and Methods*, edited by James E. Gentle, Wolfgang Härdle, and Yuichi Mori, 375–414. Springer.

Willis, Trevor J., and Russell C. Babcock. 2000. "A Baited Underwater Video System for the Determination of Relative Density of Carnivorous Reef Fish." *Marine and Freshwater Research* 51 (8): 755–63.

Wilkins, Matthew R., Nathalie Seddon, and Rebecca J. Safran. 2013. "Evolutionary Divergence in Acoustic Signals: Causes and Consequences." *Trends in Ecology & Evolution* 28 (3): 156–66.

Wilson, Edward O. 1986. *Biophilia*. Harvard University Press.

Wilson, Jackson D., Nancy McGinnis, Pavlina Latkova, Patrick Tierney, and Aiko Yoshino. 2016. "Urban Park Soundscapes: Association of Noise and Danger with Perceived Restoration." *Journal of Park and Recreation Administration* 34 (3). https://doi.org/10.18666/JPRA-2016-V34-I3-6927.

Winters, P. R. 1960. "Forecasting Sales by Exponentially Weighted Moving Averages." *Management Science* 6 (3): 324–42.

Witt, Peter N., and Jerome S. Rovner. 2014. *Spider Communication: Mechanisms and Ecological Significance*. Princeton University Press.

Wong, L. P. 2008. "Data Analysis in Qualitative Research: A Brief Guide to Using NVIVO." *Malaysian Family Physician* 3 (1): 14–20.

Wood, Connor M., Holger Klinck, Michaela Gustafson, John J. Keane, Sarah C. Sawyer, R. J. Gutiérrez, and M. Zachariah Peery. 2021. "Using the Ecological Significance of Animal Vocalizations to Improve Inference in Acoustic Monitoring Programs." *Conservation Biology* 35 (1): 336–45.

Woodward, Guy, Bo Ebenman, Mark Emmerson, Jose M. Montoya, Jens M. Olesen, Alfredo Valido, and Philip H. Warren. 2005. "Body Size in Ecological Networks." *Trends in Ecology & Evolution* 20 (7): 402–9.

Woolley, S. C., and M. H. Kao. 2015. "Variability in Action: Contributions of a Songbird Cortical-Basal Ganglia Circuit to Vocal Motor Learning and Control." *Neuroscience* 296:39–47.

Woolley, Sarah M. N., and John H. Casseday. 2004. "Response Properties of Single Neurons in the Zebra Finch Auditory Midbrain: Response Patterns, Frequency Coding, Intensity Coding, and Spike Latencies." *Journal of Neurophysiology* 91 (1): 136–51.

World Health Organization. 2011. *Burden of Disease from Environmental Noise: Quantifi-*

cation of Healthy Life Years Lost in Europe. World Health Organization, Regional Office for Europe.

World Health Organization. 2021. *World Report on Hearing*. World Health Organization.

Wright, Laura. 2010. *Wilderness into Civilized Shapes: Reading the Postcolonial Environment*. University of Georgia Press

Wrightson, Kendall. 2000. "An Introduction to Acoustic Ecology." *Soundscape: The Journal of Acoustic Ecology* 1 (1): 10–13.

Wu, Jianguo. 2004. "Effects of Changing Scale on Landscape Pattern Analysis: Scaling Relations." *Landscape Ecology* 19 (2): 125–38.

Wu, Jianguo. 2010. "Landscape of Culture and Culture of Landscape: Does Landscape Ecology Need Culture?" *Landscape Ecology* 25:1147–50.

Wu, Jianguo. 2013. "Key Concepts and Research Topics in Landscape Ecology Revisited: 30 Years after the Allerton Park Workshop." *Landscape Ecology* 28:1–11.

Wu, Jianguo. 2014. "Urban Ecology and Sustainability: The State-of-the-Science and Future Directions." *Landscape and Urban Planning* 125:209–21.

Wu, Jianguo, Chunyang He, Ganlin Huang, and Deyong Yu. 2013. "Urban Landscape Ecology: Past, Present, and Future." In *Landscape Ecology for Sustainable Environment and Culture*, edited by Bojie Fu and K. Bruce Jones, 37–53. Springer.

Wu, Xindong, Vipin Kumar, Quinlan J. Ross, Joydeep Ghosh, Qiang Yang, Hiroshi Motoda, Geoffrey J. McLachlan et al. 2008. "Top 10 Algorithms in Data Mining." *Knowledge and Information Systems* 14:1–37. https://doi.org/10.1007/s10115-007-0114-2.

Wyburn, George M., Ralph W. Pickford, and Rodney Julian Hirst. 1964. *Human Senses and Perception*. University of Toronto Press.

Yager, David D., and Hayward G. Spangler. 1995. "Characterization of Auditory Afferents in the Tiger Beetle, *Cicindela marutha* Dow." *Journal of Comparative Physiology A* 176:587–99.

Yang, Ming, and Jian Kang. 2013. "Psychoacoustical Evaluation of Natural and Urban Sounds in Soundscapes." *Journal of the Acoustical Society of America* 134 (1): 840–51.

Yang, Wei, and Jian Kang. 2005a. "Acoustic Comfort Evaluation in Urban Open Public Spaces." *Applied Acoustics* 66 (2): 211–29.

Yang, Wei, and Jian Kang. 2005b. "Soundscape and Sound Preferences in Urban Squares: A Case Study in Sheffield." *Journal of Urban Design* 10 (1): 61–80.

Ye, Jiaxing, Takumi Kobayashi, and Masahiro Murakawa. 2017. "Urban Sound Event Classification Based on Local and Global Features Aggregation." *Applied Acoustics* 117:246–56.

Yeh, Yow-Tyng, Moises Rivera, and Sarah M. N. Woolley. 2023. "Auditory Sensitivity and Vocal Acoustics in Five Species of Estrildid Songbirds." *Animal Behaviour* 195:107–16.

Yong Jeon, Jin, Joo Young Hong, and Pyoung Jik Lee. 2013. "Soundwalk Approach to Identify Urban Soundscapes Individually." *Journal of the Acoustical Society of America* 134 (1): 803–12.

Yoon, Sunmin. 2018. "What's in the Song? Urtyn Duu as Sonic "Ritual" among Mongolian Herder-Singers." *MUSICultures* 45 (1–2).

Yu, Lei, and Jian Kang. 2008. "Effects of Social, Demographical and Behavioral Factors on the Sound Level Evaluation in Urban Open Spaces." *Journal of the Acoustical Society of America* 123 (2): 772–83.

Yu, Lei, and Jian Kang. 2009. "Modeling Subjective Evaluation of Soundscape Quality in Urban Open Spaces: An Artificial Neural Network Approach." *Journal of the Acoustical Society of America* 126 (3): 1163–74.

Yu, Lei, and Jian Kang. 2010. "Factors Influencing the Sound Preference in Urban Open Spaces." *Applied Acoustics* 71 (7): 622–33.

Yu, Xiaoyuan, Jiangping Wang, Roland Kays, Patrick A. Jansen, Tianjiang Wang, and Thomas Huang. 2013. "Automated Identification of Animal Species in Camera Trap Images." *EURASIP Journal on Image and Video Processing* 2013 (1): 1–10.

Zagorinsky, A. A., R. D. Zhantiev, and O. S. Korsunovskaya. 2012. "The Sound Signals of Hawkmoths (Lepidoptera, Sphingidae)." *Entomological Review* 92:601–4.

Zaharia, Matei, Mosharaf Chowdhury, Tathagata Das, Ankur Dave, Justin Ma, Murphy Mccauley, M. Franklin, Scott Shenker, and Ion Stoica. 2012. "Fast and Interactive Analytics over Hadoop Data with Spark." *Usenix Login* 37 (4): 45–51.

Zahle, Julie. 2012. "Practical Knowledge and Participant Observation." *Inquiry* 55 (1): 50–65.

Zahorian, Stephen A., and Amir Jalali Jagharghi. 1993. "Spectral-Shape Features versus Formants as Acoustic Correlates for Vowels." *Journal of the Acoustical Society of America* 94. (4): 1966–82.

Zelenski, John M., and Elizabeth K. Nisbet. 2014. "Happiness and Feeling Connected: The Distinct Role of Nature Relatedness." *Environment and Behavior* 46 (1): 3–23.

Zhang, Sai Hua, Zhao Zhao, Zhi Yong Xu, Kristen Bellisario, and Bryan C. Pijanowski. 2018. "Automatic Bird Vocalization Identification Based on Fusion of Spectral Pattern and Texture Features." In *ICASSP, IEEE International Conference on Acoustics, Speech and Signal Processing—Proceedings* 2018 (April): 271–75. IEEE.

Zhang, Tianhao, Jie Yang, Deli Zhao, and Xinliang Ge. 2007. "Linear Local Tangent Space Alignment and Application to Face Recognition." *Neurocomputing* 70 (7–9): 1547–53.

Zhang, Xu, Meihui Ba, Jian Kang, and Qi Meng. 2018. "Effect of Soundscape Dimensions on Acoustic Comfort in Urban Open Public Spaces." *Applied Acoustics* 133:73–81.

Zollinger, Sue Anne, and Henrik Brumm. 2011. "The Lombard Effect." *Current Biology* 21 (16): R614–R615.

Zwick, Rebecca. 1988. "Another Look at Interrater Agreement." *Psychological Bulletin* 103 (3): 374–78.

Zwicker, Eberhard, and Hugo Fastl. 2013. *Psychoacoustics: Facts and Models*. Springer Science & Business Media.

Index

Page numbers in italics refer to figures or tables.